ROBUST CONTROL
AND FILTERING FOR
TIME-DELAY SYSTEMS

CONTROL ENGINEERING

A Series of Reference Books and Textbooks

Editor

NEIL MUNRO, PH.D., D.SC.

Professor
Applied Control Engineering
University of Manchester Institute of Science and Technology
Manchester, United Kingdom

1. Nonlinear Control of Electric Machinery, *Darren M. Dawson, Jun Hu, and Timothy C. Burg*
2. Computational Intelligence in Control Engineering, *Robert E. King*
3. Quantitative Feedback Theory: Fundamentals and Applications, *Constantine H. Houpis and Steven J. Rasmussen*
4. Self-Learning Control of Finite Markov Chains, *A. S. Poznyak, K. Najim, and E. Gómez-Ramírez*
5. Robust Control and Filtering for Time-Delay Systems, *Magdi S. Mahmoud*
6. Classical Feedback Control: With MATLAB, *Boris J. Lurie and Paul J. Enright*

Additional Volumes in Preparation

ROBUST CONTROL AND FILTERING FOR TIME-DELAY SYSTEMS

Magdi S. Mahmoud
Kuwait University
Safat, Kuwait

CRC Press
Taylor & Francis Group
Boca Raton   London   New York

CRC Press is an imprint of the
Taylor & Francis Group, an **informa** business

CRC Press
Taylor & Francis Group
6000 Broken Sound Parkway NW, Suite 300
Boca Raton, FL 33487-2742

First issued in paperback 2019

ISBN-13: 978-0-8247-0327-1 (hbk)
ISBN-13: 978-0-367-39895-8 (pbk)

Library of Congress Cataloging-in-Publication Data

Mahmoud, Magdi S.
  Robust control and filtering for time-delay systems / Magdi S. Mahmoud.
    p. cm. -- (Control engineering : 5)
  Includes bibliographical references and index.
  ISBN: 0-8247-0327-8
    1. Robust control. 2. Time delay systems. I. Title. II. Control engineering (Marcel Dekker) ; 5.

TJ217.2 M34 2000
629.8'312--dc21

99-054346

Visit the Taylor & Francis Web site at
http://www.taylorandfrancis.com

and the CRC Press Web site at
http://www.crcpress.com

To the biggest S's of my life:
my mother SAKINA and my wife SALWA
for their unique style, devotion,
and overwhelming care

MSM

Series Introduction

Many textbooks have been written on control engineering, describing new techniques for controlling systems, or new and better ways of mathematically formulating existing methods to solve the ever-increasing complex problems faced by practicing engineers. However, few of these books fully address the applications aspects of control engineering. It is the intention of this series to redress this situation.

The series will stress applications issues, and not just the mathematics of control engineering. It will provide texts that present not only both new and well-established techniques, but also detailed examples of the application of these methods to the solution of real-world problems. The authors will be drawn from both the academic world and the relevant applications sectors.

There are already many exciting examples of the application of control techniques in the established fields of electrical, mechanical (including aerospace), and chemical engineering. We have only to look around in today's highly automated society to see the use of advanced robotics techniques in the manufacturing industries; the use of automated control and navigation systems in air and surface transport systems; the increasing use of intelligent control systems in the many artifacts available to the domestic consumer market; and the reliable supply of water, gas, and electrical power to the domestic consumer and to industry. However, there are currently many challenging problems that could benefit from wider exposure to the applicability of control methodologies, and the systematic systems-oriented basis inherent in the application of control techniques.

This series will present books that draw on expertise from both the academic world and the applications domains, and will be useful not only as academically recommended course texts but also as handbooks for practitioners in many applications domains.

Professor Mahmoud is to be congratulated for another outstanding contribution to the series.

Neil Munro

0.1 Preface

In many physical, industrial and engineering systems, **delays** occur due to the finite capabilities of information processing and data transmission among various parts of the system. Delays could arise as well from inherent physical phenomena like mass transport flow or recycling. Also, they could be by-products of computational delays or could intentionally be introduced for some design consideration. Such delays could be constant or time-varying, known or unknown, deterministic or stochastic depending on the system under consideration. In all of these cases, the time-delay factors have, by and large, counteracting effects on the system behavior and most of the time lead to poor performance. Therefore, the subject of **Time-Delay Systems (TDS)** has been investigated as functional differential equations over the past three decades. This has occupied a separate discipline in mathematical sciences falling between differential and difference equations. For example, the books by Hale [1], Kolmanovskii and Myshkis [2], Gorecki et al [3] and Hale and Lunel [4] provide modest coverage on the fundamental mathematical notions and concepts related to TDS; the book by Malek-Zavarei and Jamshidi [5] presents different topics of modeling and control related to TDS with constant delay and the book by Stepan [6] gives a good account of classical stability methods of TDS.

Due to the fact that almost all existing systems are subject to uncertainties, due to component aging, parameter variations or modeling errors, the concepts of **robustness, robust performance** and **robust design** have recently become common phrases in engineering literature and constitute integral part of control systems research. In turn, this has naturally brought into focus an important class of systems: **Uncertain Time-Delay Systems (UTDS)**. During the last decade, we have witnessed increasingly growing interest on the subject of UTDS and numerous results have appeared in conferences and/or published in technical journals. Apart from these scattered results and the volume edited very recently by Dugard and Verriest [7] however, there is no single book written exclusively on the analysis, design, filtering and control of uncertain time-delay systems. It is therefore believed that a book that aims at bridging this gap is certainly needed.

This book is about UTDS. It is directed towards providing a pool of methods and approaches that deal with uncertain time-delay systems. In so doing, it is intended to familiarize the reader with various aspects of the control and filtering of different uncertain time-delay systems. This will range from linear to some classes of nonlinear, from continuous-time to discrete-time and from time-invariant to time-varying systems. Throughout the book, I have endeavored to stress mathematical formality in a way to spring intuitive understanding and to explain how things work. I hope that this approach will attract the attention of a wide spectrum of readership.

The book consists of ten chapters and is organized as follows. Chapter 1 is an introduction to UTDS. It gives an overview of the related issues in addition to some systems examples. The remaining nine chapters are divided into two major parts. Part I deals with **robust control** and consists of Chapters 2 through 7. Part II treats **robust filtering** and is divided into Chapters 8 to 10. The book is supplemented by appendices containing some standard lemmas and mathematical results that are repeatedly used throughout the different chapters.

The material included makes it adequate for use as a text for one-year (two-semesters) courses at the graduate level in Engineering. The prerequisites are linear system theory, modern control theory and elementary matrix theory. As a textbook, it does not purport to be a compendium of all known results on the subject. Rather, it puts more emphasis on the recent robust results of control and filtering of time-delay systems.

Outstanding features of the book are:
(1) It brings together the recent ideas and methodologies of dealing with uncertain time-delay systems.
(2) It adopts a state-space approach in the system representation and analysis throughout.
(3) It provides a unification of results on control design and filtering.
(4) It presents the material systematically all the way from stability analysis, stabilization, control synthesis and filtering.
(5) It includes the treatment of continuous-time and discrete-time systems side-by-side.

Magdi S. Mahmoud

Bibliography

[1] Hale, J., "**Theory of Functional Differential Equations**," Springer-Verlag, New York, 1977.

[2] Kolomanovskii, V. and A. Myshkis, "**Applied Theory of Functional Differential Equations**," Kluwer Academic Pub., New York, 1992.

[3] Gorecki, H., S. Fuska, P. Garbowski and A. Korytowski, "**Analysis and Synthesis of Time-Delay Systems**," J. Wiley, New York, 1989.

[4] Hale, J. and S. M. V. Lunel, "**Introduction to Functional Differential Equations**," vol. 99 , Applied Math. Sciences, Springer-Verlag, New-York, 1991.

[5] Malek-Zavarei, M. and M. Jamshidi, "**Time-Delay Systems: Analysis Optimization and Applications**," North-Holland, Amsterdam, 1987.

[6] Stepan, G., "**Retarded Dynamical Systems: Stability and Characteristic Functions**," Longman Scientific & Technical, Essex, 1989.

[7] Dugard, L. and E. I. Verriest (Editors), "**Stability and Control of Time-Delay Systems**," Springer-Verlag, New York, 1997.

0.2 Acknowledgments

In writing this book on time-delay systems that aims at providing a unified view of a large number of results obtained over two decades or more, I faced the difficult problem of acknowledging the contributions of the individual researchers. After several unsuccessful attempts and barring the question of priority, I settled on the approach of referring to papers and/or books which I believed taught me a particular approach and then adding some notes at the end of each chapter to shed some light on the various papers. I apologize, in advance, in case I committed some injustices and assure the researchers that the mistake was unintentional. Although the book is an outgrowth of my academic activities for more than twelve years, most of the material has been compiled while I was on sabbatical leave from Kuwait University (KU), KUWAIT and working as a visiting professor at Nanyang Technological University (NTU), SINGAPORE. I am immensely pleased for such an opportunity which generated the proper environment for producing this volume. In particular, I am gratefully indebted to the excellent library services provided by KU and NTU.

Over the course of my career, I have enjoyed the opportunity of interacting with several colleagues who have stimulated my thinking and research in the systems engineering field. In some cases, their technical contributions are presented explicitly in this volume; in other cases, their influence has been more subtle. Among these colleagues are Professors A. A. Kamal and A. Y. Bilal (Cairo University), Professor M. I. Younis (National Technology Program, EGYPT), Professors W. G. Vogt and M. H. Mickle (University of Pittsburgh), Professor M. G. Singh (UK), Professor M. Jamshidi (University of New Mexico), Professor A. P. Sage (George Mason University), Professor H. K. Khalil (Michigan State University), Dr. M. Zribi, Dr. L. Xie, Dr. A. R. Leyman and Dr. A. Yacin (NTU) and Dr. S. Kotob (Kuwait Institute for Scientific Research, KUWAIT). I have also enjoyed the encouragement and patience of my family (Salwa, Medhat, Monda and Mohamed) who were very supportive, as time working on this book was generally time spent away from them. Finally, I owe a measure of gratitude to Cairo University (EGYPT) and Kuwait University (KUWAIT) for providing the intellectual environment that encourages me to excel further in the area of systems engineering.

Magdi S. Mahmoud

Contents

II ROBUST FILTERING

Chapter 1

Introduction

An integral part of systems science and engineering is that of modeling. By observing certain phenomena, the immediate task consists of two parts: we wish to describe it and then determine its subsequent behavior. It is well known, in many important cases, that a useful and convenient representation of the system state is by means of a finite-dimensional vector at a particular instant of time. This constitutes a state-space modeling via ordinary differential equations, which has formed a great deal of the literature on dynamical systems. On another dimension, due to increasing complexity and interconnection of many physical systems to suit growing demand, other factors have seemingly been taken into account in the process of modeling. One important factor is that the rate of change of several physical systems depends not only on their present state, but also in their past history or delayed information among system components.

Delays thus occur in many physical, industrial and engineering systems as a direct consequence of the finite capabilities of information processing and data transmission among various parts of the system [2,3]. They could arise as well from inherent physical phenomena like mass transport flow or recycling [8]. Also, delays could be intentionally introduced for some design consideration. Such delays could be constant or time-varying, known or unknown, deterministic or stochastic depending on the system under consideration. This brings about a distinct class of dynamical systems: **Time-Delay Systems (TDS)**. Indeed, proper modeling of such systems and examining their structural properties establish important prerequisites for adequate control systems design. The subject of **TDS** is interesting, as it is difficult. In addition to the fact that many practical industrial installations and de-

vices possess delays which cannot be ignored, it is interesting since it offers
many open research topics. It is, by and large, difficult because the behavior
of such systems can be complex and analysis intricate.

Broadly speaking, there are two classes of TDS [3]:
1. Systems with lumped delays
2. Systems with distributed delays.

As we show later on, examples of class 1 include: conveyor belts, rolling
steel mills and some population models. In all of these, a finite number of
parameters can be identified which encapsulate all delay phenomena; hence
the terminology "lumped delays." Mathematically, the description involves
ordinary differential equations with delays, like:

$$T\,\dot{x} \;=\; -x \;+\; u(t - \tau)$$

where $x = x(t)$ is the state, t is the time, $u = u(t)$ is the input, τ is a single,
lumped delay and T is a time constant.

Class 2 is best represented by heat exchanging systems, whose spatial
extent makes it difficult to identify a finite number of delays which would
fully describe the heat propagation phenomena. They are frequently termed
"systems with distributed delays" and are described by partial differential
equations (PDEs). As a typical example, consider a heat exchanger of the
pipe-in-pipe type which can be represented by a system of PDEs:

$$\frac{\partial T_1}{\partial t} + h_1 \frac{\partial T_1}{\partial \ell} \;=\; k_{s1}(T_s - T_1)$$

$$\frac{\partial T_2}{\partial t} + h_2 \frac{\partial T_2}{\partial \ell} \;=\; k_{s2}(T_s - T_2)$$

$$\frac{\partial T_s}{\partial t} \;=\; k_{1s}(T_1 - T_s) + k_{2s}(T_2 - T_s)$$

where T_1 is the temperature of the first medium, T_2 is the temperature of the
second medium, T_s is the temperature of the partition wall, $k_{s1}, k_{s2}, k_{1s}, k_{2s}$
are the heat exchange coefficients and h_1, h_2 are flow coefficient. In the
model, time delays are distributed and are given by partial derivatives $\partial T_1/\partial t$,
$\partial T_2/\partial t$, $\partial T_s/\partial t$, which are functions of t and thus infinite-dimensional systems. Note also that spatial delays are also distributed and given by $\partial T_1/\partial t\ell$,
$\partial T_2/\partial t\ell$, so they are infinite-dimensional, as well. In this book, we are going to focus on class 1 dynamical systems, that is, only the lumped delay
systems are considered.

By tracing the technical approaches to mathematical representation of
TDS, we identify the following:

(1) Infinite Dimensional Systems Theory
Here the approach is based on embedding the class of TDS into a larger class of dynamical systems for which the state evolution is described by appropriate operators in infinite-dimensional spaces. On one hand, this approach presents quite a general modeling approach. On the other hand, it should be further strengthened to incorporate structural concepts like detectability and stabilizability. For a detailed coverage of this approach, the reader is referred to [10-12].

(2) Algebraic Systems Theory
In this approach, the evolution of delay-differential systems is provided in terms of linear systems over rings. Here the issues of modeling and analysis are easily described [15] but the control design is still at early stages of development.

(3) Functional Differential Systems
By incorporating the influence of the hereditary effects of system dynamics on the rate of change of the system, this approach [3-7,9] provides an appropriate mathematical structure in which the system state evolves either in finite-dimensional space [5,6] or in functional space [5,9].

Strictly speaking, there has been extensive work and research results based on the foregoing approaches. This book focuses exclusively on the third approach. As it will become clear throughout the various chapters, this approach facilitates the use of the wealth theory of finite-dimensional systems. In particular, we adopt the view of treating the delay factors as *"additional parameters"* of the system under consideration and closely examining their effects on the system behavior and performance. The reader is advised to consult [1] for a lucid discussion on the foregoing approaches.

1.1 Notations and Definitions

1.1.1 Notations

The notations followed throughout the book are quite standard. Matrices are represented by capital letters while vectors and scalars are represented by lower case letters. $f(t)$ denotes a scalar-valued function of time t. The quantities $\dot{x}$, and $\ddot{x}$ are the first and second derivative of x with respect to time, respectively. $(.,.), (.,.], [.,.]$ denote, respectively, open, semiclosed, and closed intervals; that is t in the interval $(a,b] \equiv t \in (a,b] \equiv a < t \leq b$. $\Re, \Re_+$ denote the set of real and positive real numbers, respectively, $\mathcal{C}^-$ denotes the proper left half of the complex plane, $\mathcal{C}_+ := \{s : Re(s) > 0\}$ denotes the open proper left half of the complex plane, $C_+ := \{s : Re(s) > 0\}$ denotes the open right half plane with $\bar{C}_+$ being its closure, Z, Z_+ denote, respectively, the set of integers and positive integers, $\Re^n$ denotes the n-dimensional Euclidean space over the reals equipped with the norm $\|.\|$ and $\Re^{n \times m}$ denotes the set of all $n \times m$ real matrices. The Lebsegue space $\mathcal{L}_2[0, \infty)$ consists of square-integrable functions on the interval $[0, \infty)$ and equipped with the norm

$$\|p\|_2^2 := \left(\int_0^\infty p^t(\tau)p(\tau)d\tau \right)$$

Similarly, the Lebsegue space $\ell_2[0, \infty)$ consists of square-summable functions on the interval $[0, \infty)$ and equipped with the norm

$$\|q\|_2^2 := \left(\sum_{j=0}^\infty q^t(j)q(j) \right)$$

For any square matrix W, W^t, W^{-1}, $\lambda(W)$, $tr(W)$, $r(W)$, $det(W)$, $\lambda_M(W)$, $\lambda_m(W)$ and $\rho(W) := max_j|\lambda_j(W)|$ denote the transpose, the inverse, the spectrum (set of eigenvalues), the trace, the rank and the determinant, the maximum and minimum eigenvalue and the spectral radius, respectively. For any real symmetric matrix W, $W > 0$ ($W < 0$) stands for positive-(negative-) definite matrix. When a matrix $W(\theta)$, $\theta \in \Re^r$ depends affinely on parameters $(\theta_1, ..., \theta_r)$, it means that

$$W(\theta) := W_o + \theta_1 W_1 + ... + \theta_r W_r$$

where $W_o, W_1, ..., W_r$ are known fixed matrices. $\|W\|$ denotes the induced matrix norm given by $\lambda_M(WW^t)^{1/2}$, $diag(W_1, W_2, ..., W_n)$ denotes the block-diagonal matrix

$$\begin{bmatrix} W_1 & 0 & ..... & 0 \\ 0 & W_2 & ..... & 0 \\ 0 & ... & ..... & 0 \\ 0 & ... & ..... & W_n \end{bmatrix}$$

I stands for the unit matrix with appropriate dimension and $\mu(W)$ denotes the matrix measure of a square matrix W defined by:

$$\mu(W) := \lim_{\epsilon \to 0^+} \frac{\|I + \epsilon W\| - 1}{\epsilon}$$

If matrix W^* denotes the complex conjugate transpose of W, then $\mu(W)$ is given by:

$$\mu(W) = \max_{j \in [1,n]} \lambda_j(W + W^*)$$

which possesses the following properties:

$$Re\lambda_j(W) \leq \mu(W)$$
$$\mu(W) \leq \|W\|$$
$$\mu(W + Y) \leq \mu(W) + \mu(Y)$$
$$\mu(\epsilon W) \leq \epsilon\mu(W)$$

$C_{n,\tau} = C([-\tau, 0], \Re^n)$ denotes the banach space of continuous vector functions mapping the interval $[-\tau, 0]$ into $\Re^n$ with the topology of uniform convergence and designate the norm of an element ϕ in $C_{n,\tau}$ by

$$\|\phi\|_* = \sup_{\theta \in [-\tau,0]} \|\phi(\theta)\|_*$$

Sometimes, the arguments of a function will be omitted in the analysis when no confusion can arise.

1.1.2 Definitions

In what follows we collect information and mathematical definitions related to *functional differential equations* (FDEs). Unless stated otherwise, all quantities and variables under consideration are real.

It is well-known from mathematical sciences that, an *ordinary differential equation* (ODE) is an equation connecting the values of an unknown function and some of its derivatives for one and the same argument value, for example $H(t, x, \dot{x}, \ddot{x}) = 0$. Following [4,6], a *functional equation* (FE) is an equation involving an unknown function of different argument values. For example, $x(t) + 3x(4t) = 2, x(t) = sin(t)x(t+2) + cos(t+1)x^2(t-3) = 2$ are FEs. By combining the notions of differential and functional equations, we obtain the notion of a *functional differential equation* (FDE). Thus, FDE is an equation connecting the unknown function and some of its derivatives for, generally speaking, different argument values. Looked at in this light, the notion of FDE generalizes all equations of mathematical analysis for functions of a continuous argument. This assertion is greatly justified by examining models of several applications [6-13]. We take note that all fundamental properties of ODE are carried over to FDE including *order, periodicity and time-invariance*.

Next, we introduce some mathematical machinery. If $\alpha \in \Re, d \geq 0$ and $x \in C([\alpha - \tau, \alpha + d], \Re^n)$ then for any $t \in [\alpha, \alpha + d]$, we let $x_t \in C$ be defined by $x_t(\theta) := x(t + \theta)$, $-\tau \leq \theta \leq 0$. If $\mathcal{D} \subset \Re \times C$, $f : \mathcal{D} \to \Re^n$ is a given function, we say [4-6] that the relation

$$\dot{x}(t) \; = \; f(t, x_t) \tag{1.1}$$

is a retarded functional differential equation (RFDE) on $\mathcal{D}$ where $x_t(t), t \geq t_o$ denotes the restriction of $x(.)$ to the interval $[t - \tau, t]$ translated to $[-\tau, 0]$. Here, $\tau > 0$ is termed the *delay factor*. A function x is said to be a *solution* of (1.1) on $[\alpha - \tau, \alpha + d]$ if there are $\alpha \in \Re$ and $d > 0$ such that

$$x \in C([\alpha - \tau, \alpha + d], \Re^n), \quad (t, x_t) \in \mathcal{D}, \quad t \in [\alpha, \alpha + d]$$

and $x(t)$ satisfies (1.1) for $t \in [\alpha, \alpha + d]$. For a given $\alpha \in \Re$, $\phi \in C$, $x(\alpha, \phi, f)$ is said to be a *solution* of (1.1) with *initial value* ϕ at α. Alternatively, $x(\alpha, \phi, f)$ is a *solution* through (α, ϕ) if there is an $d > 0$ such that $x(\alpha, \phi, f)$ is a *solution* of (1.1) on $[\alpha - \tau, \alpha + d]$ and $x_\alpha(\alpha, \phi, f) = \phi$.

Of paramount importance is the nature of the *equilibrium solution* $x_t \equiv 0$. For this purpose, we let

$$\Upsilon_\rho := \{\phi \in C[-\tau, 0] \mid \|\phi\| < \rho\}$$

Following [4,6], the *equilibrium solution* $x_t \equiv 0$ of (1.1) is said to be *stable* if for any $\epsilon > 0$m there is a $\rho = \rho(\epsilon) > 0$, such that $|x(\alpha, \phi, f)| \leq \epsilon$ for any initial value $\phi \in \Upsilon_\rho$, and $\forall t > 0$. Otherwise it is *unstable*. The *equilibrium solution* is called *asymptotically stable* if it is *stable* and there is a $\kappa > 0$ such that for any $\nu > 0$ there is a $\tau(\nu, \kappa) > 0$ such that $|x(\alpha, \phi, f)| \leq \nu, \forall t \geq \tau(\nu, \kappa)$ and $\phi \in \Upsilon_\kappa$. The *equilibrium solution* is called *exponentially stable* if there are constants $\kappa > 0, r_1 > 0, r_2 > 0$, such that for any $\phi \in \Upsilon_\kappa$, the solution $x(\alpha, \phi, f)$ of system (1.1) satisfies the inequality

$$|x(\alpha, \phi, f)| \; \leq \; r_1 \|\phi\| \, e^{-r_2 t}, \quad 0 \leq t \leq \infty \tag{1.2}$$

Consider Γ as the class of scalar nondecreasing functions $\beta \in C([0, \infty], \Re)$ with the properties $\beta(s) > 0, s > 0$ and $\beta(0) = 0$. Let $V : \Upsilon_\kappa \to \Re$ be a continuous functional with the properties $V(0) \equiv 0$. The functional $V : \omega \to V(\omega)$ is called *positive definite* if there is a function $\beta \in \Gamma$ such that $V(\omega) \geq \beta(|\omega(0)|) \, \forall \omega \in \Upsilon_\kappa$. It then follows [4,6] when $V : \Upsilon_\kappa \to \Re$ has $\dot{V} \leq 0$

that the *equilibrium solution* of (1.1) is *stable*. On the other hand, for some $\tau > 0$ if there exists a *positive-definite continuous functional* $(\omega \rightarrow V(\omega):$ $\Upsilon_\kappa \rightarrow \Re$ such that $|V(\omega)| \leq \beta(\|\omega\|)$ $\forall \omega \in \Upsilon_\kappa$ and $\dot{V} \leq 0$ on Υ_κ, then the equilibrium solution of (1.1) is *asymptotically stable*. Finally, a necessary and sufficient condition of the exponential stability of the equilibrium solution of (1.1) is that there exists a continuous functional $V : \Upsilon_\kappa \rightarrow \Re$ such that for some positive constants k_1, k_2, k_3, k_4 , ω and $\xi \in \Upsilon_\kappa$:

$$
\begin{aligned}
k_1 \|\omega\| &\leq V(\omega) \leq k_2\|\omega\| \\
\dot{V}(\omega) &\leq -k_3 \|\omega\| \\
|V(\omega) - V(\xi)| &\leq k_4 \|\omega - \xi\|
\end{aligned}
\tag{1.3}
$$

In view of the generality of (1.1), we now extract two relevant important cases. First, if $\tau = 0$ then (1.1) reduces to the ordinary differential equation $\dot{x}(t) = F(x(t))$. Second, letting $0 \leq \eta_j(t) \leq \tau$; $j = 1, ..., s$ expresses (1.1) in the form:

$$
\dot{x}(t) = f(t, x(t), x(t - \eta_1(t)), ..., x(t - \eta_s(t)))
\tag{1.4}
$$

which is frequently called *differential-difference equation* (DDE). Other forms can also be obtained from (1.1) including integro-differential equations and integro-difference equations. We are not going to discuss these any further and the interested reader is referred to [4-7].

Next, suppose $\Omega \subseteq \Re \times C$ is open, $f : \Omega \rightarrow \Re^n$, $M : \Omega \rightarrow \Re^n$ are given continuous functions with M having continuous derivatives at origin. Then, the relation

$$
\dot{M}(t, x_t) = f(t, x_t)
\tag{1.5}
$$

is called the *neutral functional differential equation* (NFDE(M,f)) and M is the *difference operator*. In line of RFDE, a function x is said to be a *solution* of (1.5) if there are $\alpha \in \Re$ and $d > 0$ such that

$$
x \in C([\alpha - \tau, \alpha + d], \Re^n), \quad (t, x_t) \in \Omega, \quad t \in [\alpha, \alpha + d]
$$

M is continuously differentiable and satisfies (1.3) on $[\alpha, \alpha + d]$. For a given $\alpha \in \Re$, $\phi \in C$ and $(\alpha, \phi) \in \Omega$, $x(\alpha, \phi, M, f)$ is said to be a *solution* of (1.3) with *initial value* ϕ at α. Alternatively, $x(\alpha, \phi, M, f)$ is a *solution* through (α, ϕ) if there is an $d > 0$ such that $x(\alpha, \phi, M, f)$ is a *solution* on $[\alpha - \tau, \alpha + d]$ and $x_\alpha(\alpha, \phi, M, f) = \phi$.

Most of the materials contained in this volume are restricted to classes of RFDEs with some few classes of NFDEs.

1.2   Time-Delay Systems

We focus attention on the role of the *delay factor* τ. In one case when $\tau > 0$ is a scalar, we obtain a point (single) -delay type of FDE. Note that τ could be constant or variable with its value being known *a priori* or it is unknown-but-bounded with known upper bound. On another case when we have several delay factors, we get a multiple (distributed) delay type of FDE or DDE of the form (1.4). To unify the terminology, we use from now onwards the phrase *time-delay systems* to denote physical and engineering systems with mathematical models represented either by single-delay FDEs, DDEs, or multiple-delay FDEs. Thus the time-delay system

$$\dot{x}(t) \quad = \quad Ax(t) + A_d x(t - \eta) \tag{1.6}$$
$$x(t_o + \theta) \quad = \quad \phi(\theta) \quad , \theta \in [-\eta, 0] \tag{1.7}$$

represents a free (unforced) linear FDE with a single constant delay factor η. Also, the time-delay system

$$\dot{x}(t) \quad = \quad Ax(t) + A_{d1}x(t - \eta_1) + \ldots + A_{ds}x(t - \eta_s)$$
$$= \quad Ax(t) + \sum_{j=1}^{s} A_{dj}x(t - \eta_j) \tag{1.8}$$

represents a free (unforced) linear FDE with s constant and different delay factors $(\eta_1, ..., \eta_s)$ with initial condition

$$x(t_o + \theta) \quad = \quad \phi(\theta), \quad \theta \in [-\hat{\eta}, 0]$$
$$\hat{\eta} \quad = \quad \max_{j \in [1,s]} \eta_j, \quad (t_o, \phi) \in \Re_+ \times C_{n,\tau} \tag{1.9}$$

Intuitively, setting $s = 1$ in (1.8) yields (1.6). Being unforced, models (1.6) and (1.8) are quite suitable for stability studies. In (1.8), the matrices $(A_{d1}, ..., A_{ds})$ reflect the strength of the delayed states on the system dynamics. In some cases, this strength may help in boosting the system growth toward satisfactory behavior. In other cases, such strength may counteract the system behavior thereby yielding destabilizing effects. These issues justify the direction that stability analysis of time-delay systems should include information about the size of the delayed-state matrices $(A_{d1}, ..., A_{ds})$. When a particular stability condition is derived which depends on the size of the delay factors as well, the obtained result is called a *delay-dependent stability* condition. This case means that the system stability is only preserved within a prespecified range. On the other hand, when the derived condition does not depend on the delay size, we eventually get *delay-independent stability* condition. Now suppose that the latter case holds. It therefore means that it holds for all positive and finite values of the delays. In turn, this implies a sort of robustness against the delay factor as a parameter. The crucial point to observe is that one must first examine the original system without delay before inferring any subsequent result. These issues and the main differences between both delay-independent and delay-dependent stability conditions will be discussed in later chapters.

When attending to control system design, system (1.6) with the forcing term taking one of several forms:

$$\dot{x}(t) \quad = \quad Ax(t) \ + \ A_d x(t-\eta) + Bu(t) \qquad (1.10)$$
$$x(t_o + \theta) \quad = \quad \phi(\theta), \quad \theta \in [-\eta, 0] \qquad (1.11)$$

which is the 'standard' linear FDE with a single constant delay factor η,

$$\dot{x}(t) \quad = \quad Ax(t) \ + \ B_d u(t-\pi) \qquad (1.12)$$
$$u(t_o + \vartheta) \quad = \quad \varphi(\vartheta), \quad \vartheta \in [-\pi, 0] \qquad (1.13)$$

which represents an input-delayed FDE, or

$$\dot{x}(t) \quad = \quad Ax(t) \ + \ A_d x(t-\eta) + B_d u(t-\pi) \qquad (1.14)$$
$$x(t_o + \theta) \quad = \quad \phi(\theta) \ , \theta \in [-\eta, 0] \qquad (1.15)$$
$$u(t_o + \vartheta) \quad = \quad \varphi(\vartheta) \ , \vartheta \in [-\pi, 0] \qquad (1.16)$$

which represents a linear FDE with state- and input-delays with $\eta \neq \pi$. Discussions about control design methods for models (1.10), (1.12) and (1.14) will be the subject of Chapters 2 through 6.

1.3 Uncertain Time-Delay Systems

Due to the fact that almost all existing physical and engineering systems are subject to uncertainties, due to component aging; parameter variations or modeling errors, the concepts of **robustness, robust performance,** and **robust design** have recently become common phrases in engineering literature and constitute an integral part of control systems research. By incorporating the uncertainties in the modeling of time-delay systems, we naturally obtain uncertain time-delay system (UTDS). This is a major theme

of the book, that is to study problems of analysis and control of UTDS. We
will mainly adopt state-space modeling tools. Motivated by models of TDS,
we provide hereafter corresponding models of UTDS. For example

$$\dot{x}(t) = (A + \Delta A)x(t) + (A_d + \Delta A_d)x(t - \eta) \tag{1.17}$$

represents model (1.6) with additive uncertainty. Also, the time-delay system (1.8) with parametric uncertainty becomes

$$
\begin{aligned}
\dot{x}(t) &= (A + \Delta A)x(t) + (A_{d1} + \Delta A_{d1})x(t - \eta_1) + ... \\
&+ (A_{ds} + \Delta A_{ds})x(t - \eta_s) \\
&= (A + \Delta A)x(t) + \sum_{j=1}^{s}(A_{dj} + \Delta A_{dj})x(t - \eta_j)
\end{aligned} \tag{1.18}
$$

Robust stability of models (1.17) and (1.18) are closely examined in later
chapters and suitable stability testing methods are presented.

By considering the forced TDS, we may obtain the following systems:

$$
\begin{aligned}
\dot{x}(t) &= (A + \Delta A)x(t) + (A_d + \Delta A_d)x(t - \eta) + (B + \Delta B)u(t) \tag{1.19} \\
\dot{x}(t) &= (A + \Delta A)x(t) + (B_d + \Delta B_d)u(t - \pi) \tag{1.20} \\
\dot{x}(t) &= (A + \Delta A)x(t) + (A_d + \Delta A_d)x(t - \eta) \\
&+ (B_d + \Delta B_d)u(t - \pi) \tag{1.21}
\end{aligned}
$$

where $\Delta A, \Delta B, \Delta A_d, \Delta B_d$ are matrices of uncertain parameters. In the literature, the uncertainty may be unstructured in the sense that it is only
bounded in magnitude:

$$\|\Delta A\| \leq \gamma_A, \quad \gamma_A > 0$$

with the bound γ_A being known *a priori*. Alternatively when the uncertainty
is structured, it then may take one of several forms, each of which has its
own merits and demerits. Some of the most frequently used in the context
of time-delay systems are:

(1) Matched Uncertainties
In this case, the uncertainties are assumed to be accessible to the control
input and hence related to the input matrix B by

$$\|\Delta A\| = B E$$

where E is a known constant matrix.

matched part.

(3) Norm-Bounded Uncertainties

Here, the uncertainty matrix ΔA is assumed to be represented in factored form as:

$$\Delta A = H F L, \quad \|F\| \leq 1$$

where F is a matrix of uncertain parameters and the matrices H, L are constants with compatible dimensions.

Indeed, there are other uncertainty structures that have been considered in the literature. Thses includes *linear fractional transformation* (LFT)[25] and *integral quadratic constraint* (IQC) [26]. However, their use in UTDS has been so far quite limited.

1.4 System Examples

In this section we present models of some typical systems featuring time-delay behavior. These systems have the common property that the growth of some parts (future states) of the underlying model depend not only on the present state, but also on the delayed state (past history) and/or delayed input. Therefore we provide in the sequel some representative system models.

1.4.1 Stream Water Quality

In practice, it is important to keep water quality in streams standard. This can be measured by the concentrations of some water biochemical constituents [28]. Let $z(t)$ and $q(t)$ be the concentrations per unit volume of biological oxygen demand (BOD) and dissolved oxygen (DO), respectively, at time t. For simplicity, we consider that the stream has a constant flow rate and the water is well mixed. We further assume that there exists $\tau > 0$ such that the (BOD, OD) concentrations entering at time t are equal to the corresponding concentrations τ time units ago.

Using mass balance concentration, the growth of (BOD, OD) can be expressed as:

$$
\begin{aligned}
\dot{z}(t) &= -k_c(t)\, z(t) + v^{-1}[Q_e(m + u_1(t)) + Q_s z(t - \tau) - (Q_s + Q_e)z(t)] \\
&\quad + \xi_1(t) \qquad\qquad\qquad\qquad\qquad\qquad\qquad\qquad\qquad\qquad (1.22) \\
\dot{q}(t) &= -k_d(t)\, z(t) + k_r(t)[q_d - q(t)] + v^{-1}[Q_s q(t - \tau) - (Q_s + Q_e)q(t)] \\
&\quad + u_2(t) + \xi_2(t) \qquad\qquad\qquad\qquad\qquad\qquad\qquad\qquad (1.23)
\end{aligned}
$$

Using mass balance concentration, the growth of (BOD, OD) can be expressed as:

$$
\begin{aligned}
\dot{z}(t) &= -k_c(t)\, z(t) + v^{-1}[Q_e(m + u_1(t)) + Q_s z(t - \tau) - (Q_s + Q_e)z(t)] \\
&\quad + \xi_1(t) \tag{1.22} \\
\dot{q}(t) &= -k_d(t)\, z(t) + k_r(t)[q_d - q(t)] + v^{-1}[Q_s q(t - \tau) - (Q_s + Q_e)q(t)] \\
&\quad + u_2(t) + \xi_2(t) \tag{1.23}
\end{aligned}
$$

where $k_c(t)$ is the BOD decay rate, $k_r(t)$ is the BO re-aeration rate, $k_c(t)$ is the BOD deoxygenation rate, q_d is the DO saturation concentration, Q_s is the stream flow rate, Q_e is the effluent flow rate, v is the constant volume of water in stream, m is constant, $u_1(t), u_2(t)$ are the controls and $\xi_1(t), \xi_2(t)$ are random disturbances affecting the growth of BOD and DO.

Using state-space format, model (1.22)-(1.23) can be cast into:

$$
\dot{x}(t) = f[x(t), x(t - \tau), u(t)] \tag{1.24}
$$

which represents a nonlinear system with time-varying state-delay.

1.4.2   Vehicle Following Systems

A simple version of vehicle following models for throttle control purposes can be described by [27]:

$$
\begin{aligned}
\dot{x}(t) &= v(t) \tag{1.25} \\
\dot{v}(t) &= m^{-1}[T_n(t) - T_L] \tag{1.26}
\end{aligned}
$$

where $x(t)$ is the position of vehicle, $v(t)$ is the speed of vehicle, $T_n(t)$ is the force produced by the vehicle engine, m is the mass of the vehicle and $T_L(t)$ is the total load torque on the engine. For simplicity, consider that T_L is constant. In terms of the throttle input $u(t)$, the engine dynamics can be expressed as dynamics:

$$
\dot{T_n}(t) = -\tau^{-1}[T_n + u(t)] \tag{1.27}
$$

Here, τ represents the vehicle's engine time-constant when the vehicle is travelling with a speed v. Combining (1.26) and (1.27), we obtain:

$$
\dot{T_n}(t) = -\tau^{-1}[m\dot{v} + T_L + u(t)] \tag{1.28}
$$

To proceed further, we differentiate (1.28) and setting $\dot{T}_L \equiv 0$ to get:

$$\dot{a}(t) = -\tau^{-1}a(t) + (m\tau)^{-1}[u(t) - T_L] \qquad (1.29)$$

By incorporating the effect of actuator delay, due to fuelling delay and transport factor, we express (1.25),(1.26) and (1.29) into the form:

$$
\begin{bmatrix} \dot{x}(t) \\ \dot{v}(t) \\ \dot{a}(t) \end{bmatrix}
=
\begin{bmatrix} 0 & 1 & 0 \\ 0 & 0 & 1 \\ 0 & 0 & -\tau^{-1} \end{bmatrix}
\begin{bmatrix} x(t) \\ v(t) \\ a(t) \end{bmatrix}
$$
$$
+
\begin{bmatrix} 0 \\ 0 \\ (m\tau)^{-1} \end{bmatrix} u(t-h)
+
\begin{bmatrix} 0 \\ 0 \\ -(m\tau)^{-1} \end{bmatrix} T_L \qquad (1.30)
$$

where h is the total throttle delay. Model (1.30) represents a linear system with constant input-delay.

1.4.3 Continuous Stirred Tank Reactors with Recycling

This example is considered in [29] and it represents an industrial jacketed continuous stirred tank reactor (JCSTR) of volume V gallons with a delayed recycle stream. The reactions within the JCSTR are assumed unimolecular and irreversible (exothermic). Also, perfect mixing is assumed and the heat losses are neglected. The reactor accepts a feed of reactant which contains a substance A in initial concentration C_{Ao}. The feed enters at a rate F and at a temperature T_o. Cooling of the tank is achieved by a flow of water around the jacket and the water flow in the jacket F_J is controlled by actuating a valve. Suppose that fresh feed of pure (C_A) is to be mixed with a recycled stream of unreacted (C_A) with a recycle flow rate $(1-c)$ where $0 \le c \le 1$ is the coefficient of recirculation. The amount of transport delay in the recycled stream is d. The change of concentration arises from three terms: the amount of A that is added with feed under recycling, the amount of A that leaves with the product flow, and the amount of A that is used up in the reaction. The change in the temperature of the fluid arises from four terms: a term for the heat that enters with the feed flow under recycling, a term for the heat that leaves with the product flow, a term for the heat created by the reaction and finally a term for the heat that is transferred to the cooling jacket. There are three terms associated with the changes of the temperature of the fluid in the jacket: one term representing the heat entering the jacket with cooling fluid flow, another term accounting for the heat leaving the jacket with the outflow of cooling liquid and a third term representing the heat transferred from the fluid in the reaction tank to the fluid in the jacket. Under the conditions of constant holdup, constant densities and perfect mixing, the energy and material balances can be expressed mathematically as:

$$
\begin{aligned}
\dot{C}_A(t) &= (FV^{-1})[c\,C_{Ao} - c\,C_A(t) + (-c)C_A(t-d)] - k_1 C_A(t)\,e^{-k_2/T(t)} \\
&= f_C(C_A, T) \tag{1.31} \\
\dot{T}(t) &= (FV^{-1})[c\,T_o - c\,T(t) + (-c)T(t-d)] - k_1 k_3 C_A(t)\,e^{-k_2/T(t)} \\
&\quad - k_4[T(t) - T_J(t)] \\
&= f_T(C_A, T) \tag{1.32} \\
\dot{T}_J(t) &= (F_J V_J^{-1})[T_{Jo} - T_J(t)] - k_5[T(t) - T_J(t)] \\
&= f_J(T, T_J, F_J) \tag{1.33}
\end{aligned}
$$

By defining C_A, T, T_J as a state vector and F_J as a control input, it is easy to see that models (1.31)-(1.33) represent a nonlinear time-delay system.

1.4.4 Power Systems

A simple model of a single-area power control is given by [30]:

$$
\begin{bmatrix} \Delta \dot{f}(t) \\ \Delta \dot{P}_g(t) \\ \Delta \dot{X}_g(t) \\ \Delta \dot{E}(t) \end{bmatrix}
=
\begin{bmatrix}
-T_p^{-1} & 0 & 0 & 0 \\
0 & -T_T^{-1} & T_T^{-1} & 0 \\
-(RT_g)^{-1} & 0 & -T_G^{-1} & -T_G^{-1} \\
K_E & 0 & 0 & 0
\end{bmatrix}
\begin{bmatrix} \Delta f(t) \\ \Delta P_g(t) \\ \Delta X_g(t) \\ \Delta E(t) \end{bmatrix}
$$
$$
+
\begin{bmatrix}
0 & 0 & 0 & 0 \\
0 & K_p T_p^{-1} & 0 & 0 \\
0 & 0 & 0 & 0 \\
0 & 0 & 0 & 0
\end{bmatrix}
\begin{bmatrix} \Delta f(t-\tau) \\ \Delta P_g(t-\tau) \\ \Delta X_g(t-\tau) \\ \Delta E(t-\tau) \end{bmatrix}
$$
$$
+
\begin{bmatrix} 0 \\ 0 \\ T_G^{-1} \\ 0 \end{bmatrix} u(t)
+
\begin{bmatrix} -K_p T_p^{-1} \\ 0 \\ 0 \\ 0 \end{bmatrix} \Delta P_d \tag{1.34}
$$

where $\Delta f(t), \Delta P_g(t), \Delta X_g(t), \Delta E(t)$ are the incremental changes in frequency (Hz), generator output (pu MW), governor valve position and integral control, respectively. P_d is the load disturbance, τ is the engine dead-time, T_G is the governor time constant, T_T is the turbine time constant, T_p is the plant model time constant, K_p is the plant gain and R is the speed regulation due to governor action. Model (1.34) can be put in the state-space form:

$$\begin{aligned}
\dot{x}(t) &= Ax(t) + Bu(t) + A_d x(t - \tau) + Dw(t) \\
z(t) &= Ex(t)
\end{aligned} \qquad (1.35)$$

where $x(t) := [\Delta f(t), \Delta P_g(t), \Delta X_g(t), \Delta E(t)]^t$ is the state vector and thus it has the TDS form given by (1.12).

1.4.5 Some Biological Models

The evolution of biological systems depends basically on the whole previous history of the system and therefore provides good candidates for FDE modeling [16,22,23]. Essentially, biological systems involve the interaction among processes of birth, death and growth. We mention here some typical models. The first model concerns the evolution of a single species struggling for a common food. By considering the case of limited self-renewing food resources, the logistic model [16] describes the species populations in the form

$$\dot{x}(t) = \gamma \left[1 - K^{-1} x(t - h) \right] x(t) \qquad (1.36)$$

where $x(t)$ is the size of the species, h is the production time of food resources (average age of producers), γ is called the *Malthus coefficient* of linear growth which represents the difference between birth and death rates and the constant K is the average production number which reflects the ability of the environment to accomodate the population. The meaning of $h > 0$ is that the food resources at time t are determined by the population size at time $t - h$. Despite the simplicity of model (1.36), it has all the ingredients of FDE and is amenable to numerious extensions and applications [6,16]. Another biological system is that of a predator-prey model represented by:

$$\begin{aligned}
\dot{x}_1(t) &= a_1 x_1(t) - a_2 x_1(t)x_2(t) - a_3 x_1^2(t) \\
\dot{x}_2(t) &= -a_4 x_2(t) + a_5 x_1(t - h)x_t(t - h)
\end{aligned} \qquad (1.37)$$

where $x_1(t)$, $x_2(t)$ are the number of predators and preys, respectively; $a_j > 0$ are constants and the delay $h > 0$ stands for the average period between death of prey and the birth of a subsequent number of predators. A third model is that of competing micro-organisms surviving on a single nutrient. Allowing finite delays in birth and death processes, a suitable model is given by [205]:

$$
\begin{aligned}
\dot{x}_o(t) &= 1 - x_o(t) - x_1(t)\, f_1(x_o(t)) - x_2\, f_2(x_o(t)) \\
\dot{x}_1(t) &= [f_1(x_o(t - h_1)) - 1]\, x_1(t) \\
\dot{x}_2(t) &= [f_2(x_o(t - h_2)) - 1]\, x_2(t)
\end{aligned}
\tag{1.38}
$$

where $x_o(t)$ is the nutrient concentration, $x_1(t), x_2(t)$ are the concentrations of competing micro-organisms and h_1, h_2 are constant delays.

Despite the fact that models (1.36)-(1.38) are nonlinear FDEs, they serve the purpose of illustrating the natural existence of delays in system applicaions.

1.5   Discrete-Time Delay Systems

We have seen in section 1.1.2 that FDE results from emerging ODE and FE. Since a *difference equation* connects the value of an unknown function with its previous values at different time instants, we are therefore encouraged to combine the notions of *functional equation* and *difference equation* to yield what is called *functional difference equation* or *delay-difference equation* (DDE). Much like (2.1), the relation

$$
x(k + 1) = f(k, x_k)
\tag{1.39}
$$

describes a functional difference equation where $k \in \mathcal{Z}$ and x_k denotes the restriction of $x(k)$ to the interval $[k - \tau, k]$ translated to $[-\tau, 0]$, that is $x_k(\eta) := x(k + \eta), \forall \eta \in [-\tau, -\tau + 1, ...0]$. With appropriate modifications in the mathematical language, most of the definitions of section 1.1.2 carry over here. A class of uncertain discrete-time systems with s distributed (multiple) delays takes that form:

$$
\begin{aligned}
x(k + 1) &= (A + \Delta A)x(k) + A_{d1} + \Delta A_{d1})x(k - \tau_1) \\
&+ \quad .... + (A_{ds} + \Delta A_{ds})x(k - \tau_s)
\end{aligned}
\tag{1.40}
$$

where in (1.40) $\tau_1 > 0, ..., \tau_s > 0$ are integers that represent the amount of delay units in the state. Obviously, the case of $s = 1$ corresponds to a single-delay uncertain discrete-time system . In order to motivate the analysis of discrete-time delay systems, we present two control system applications described by discrete-time models in which the time-delay appears quite natural. This will enable us in the sequel to develop results for continuous-time and discrete-time systems side-by-side. It can be argued that by state augmentation, one can convert system (1.40) to a non-delay system with higher-order state vector. We do not follow this approach here for the following reasons:

(1) It opposes the common trend in system analysis and design that seeks reduced-order models.
(2) It suppresses the effect of delay factor which might carry valuable information.
(3) It does not yield the discrete-version of the results on continuous time-delay systems.
(4) It adds undue complications to the uncertainty structure.

1.5.1 Example 1.1

Planning constitutes a crucial part in the decision-making of manufacturing systems. It requires careful modeling of the underlying processes of sales, inventory and production. Due to the nature of manufacturing systems, there are inherent time-lags between production on one hand and sales plus inventory on the other hand. In addition, there are uncertainties in the identification of the various economic ratios and coefficients. Following [31], we consider a factory that produces two kinds of products ($j = 1, 2$) sharing common resources and raw materials like color TV and black/white TV, PC and laptop computer. During the kth period (quarter or season), we let:

s_{jk}: amount of sales of product j
a_{jk}: advertisement cost spent for product j
c_{jk}: amount of inventory of product j
p_{jk}: production of product j

Let

$$x(k) := \begin{bmatrix} s_{1j} \\ s_{2j} \\ c_{1j} \\ p_{2j} \end{bmatrix}, \quad u(k) := \begin{bmatrix} p_{1,k+1} \\ p_{2,k+1} \\ a_{1k} \\ a_{2k} \end{bmatrix} \tag{1.41}$$

The effect of advertisements on sales in the marketing process and the inter-link between inventory and production in the production process (assuming one-period gestation lag) can then be expressed dynamically by a linear model of the form:

$$x(k+1) = A_o x(k) + A_m x(k - m + 1) + \Delta A_o x(k)$$

$$+ \quad \Delta A_m x(k - m + 1) + Bu(k) + \Delta Bu(k - m + 1)$$

where $\Delta A_o x(k), \Delta A_m x(k - m + 1), \Delta Bu(k - m + 1)$ denote, respectively, the uncertain amount of sales, inventory of product and change in advertisements cost and $m \geq 2$ stands for the amount of delay between making a decision and realizing its effect on production. It is readily seen that the above model fits nicely into the discrete-time delay format.

1.5.2   Example 1.2

Consider a three-stand cold rolling mill represented by [2]:

$$x(k + 1) = A_o(r)x(k) + A_1 x(k - h) + A_2 x(k - 2h) + B(r)u(k)$$

where the delay h denotes the transit time of the strip from the outlet of one stand to the inlet of the next one and the parameter r is the winding or pay-off reel radius. Note that for an n-stand cold rolling mill, there will be $(n-1)$ time-delays. The state vector represents field currents of the different motors as well as the angular velocities of rotating reels. The above model can be cast into the framework of discrete-time delay systems. In practice, it is expected that the matrices A_o, A_1, A_2, B may contain uncertainties due to variations in system parameters.

Indeed, there are other sources of delay in discrete-time systems. These include computational delays in digital systems and delays due to finite separation among arrays in signal processing.

1.6   Outline of the Book

The objective of the book is to present robust control and filtering methods that cope with classes of time-delay systems with uncertain parameters. For ease in exposition, it is divided into two parts: Part I deals with robust control and Part II treats robust filtering.

Part I is organized into six chapters as follows. The topic of robust stability and stabilization occupies Chapter 2, which includes results on delay-independent as well as delay-dependent stability for both continuous-time and discrete-time systems. Different stabilization schemes are then discussed in Chapter 3 using state-feedback and dynamic output feedback controllers with emphasis on linear matrix inequality (LMI) formulation. In addition, the case of multiple-delays is treated. Methods based on robust $\mathcal{H}_\infty$ and guaranteed cost control are the subject of Chapters 4 and 5, respectively. Again, results on continuous-time and discrete-time systems are presented side-by-side. Chapter 6 is devoted to the study of passivity analysis and synthesis for TDS and UTDS. Control design for interconnected UTDS is provided in Chapter 7.

In Part II, the main focus on state-estimation (filtering) where robust Kalman filter is developed for uncertain linear time-delay systems (Chapter 8), robust $\mathcal{H}_\infty$ filtering are constructed for linear as well as classes of nonlinear TDS (Chapter 9) and finally robust $\mathcal{H}_\infty$ filtering for interconnected TDS is covered in Chapter 10.

For ease in exposition, we follow a five-step methodology throughout the book:

Step 1: Mathematical Modelling in which the system under consideration is represented by a mathematical model
Step 2: Assumptions or Definitions where we state the constraints on the model variables or furnish the basis for the subsequent analysis
Step 3: Analysis which signifies the core of the respective sections
Step 4: Results which are provided most of the time in the form of theorems, lemmas and corollaries
Step 5: Remarks which are given to discuss the developed results in relation to other published work

Theorems (lemmas, corollaries) are keyed to chapters and stated in *italic* font, for example, *Theorem 3.2* means Theorem 2 in Chapter 3 and so on. By this way, we believe that the material covered will attract the attention of a wide-spectrum of readership. Emphasis is placed on one major approach and reference is then made to other available approaches. For convenience, the references are subdivided into three bibliographies with partial overlapping and are located at the end of Chapter 1, Part I and Part II, respectively. The book is supplemented by appendices containing some of the fundamental mathematical results and reference to any of these results is made in the text using bold face, for example, **A.2** means the second result in Appendix A and so on. Simulation examples using **MATLAB** are provided at the end of each chapter. For purpose of completeness, a brief summary of the **LMI**-Control software is provided in Appendix E. In addition, relevant notes and research issues are offered for the purpose of stimulating the reader.

1.7   Notes and References

The basic technical background of TDS are contained in [4-7,9-14] which constitute major sources of knowledge to mathematically inclined engineers and researchers interested in control systems. Treatment of some advanced topics are included in [21,22]. A wide-spectrum of system applications are considered in [3,6,8,15-17, 22,23]. Conventional control system design methods are the main subjects of [18,19] by focusing on constant delay (lag) systems. The books [2,3] are considered integrated volumes by treating the topics of mathematical modeling, analysis, control and optimization of TDS and providing several interesting applications. Different issues and approaches related to both TDS and UTDS are thoroughly discussed in the edited volume [1] which provides vast breadth of techniques addressing various problems of time-delay systems.

Bibliography

[1] Dugard, L. and E. I. Verriest (Editors), "Stability and Control of Time-Delay Systems," Springer-Verlag, New York, 1997.

[2] Malek-Zavarei, M. and M. Jamshidi, "Time-Delay Systems: Analysis, Optimization and Applications," North-Holland, Amsterdam, 1987.

[3] Gorecki, H., S. Fuska, P. Garbowski and A. Korytowski, "Analysis and Synthesis of Time-Delay Systems," J. Wiley, New York, 1989.

[4] Hale, J., "Theory of Functional Differential Equations," Springer-Verlag, New York, 1977.

[5] Hale, J. and S. M. V. Lunel, "Introduction to Functional Differential Equations," vol. 99 , Applied Math. Sciences, Springer-Verlag, New York, 1991.

[6] Kolomanovskii, V. and A. Myshkis, "Applied Theory of Functional Differential Equations," Kluwer Academic Pub., New York, 1992.

[7] Lakshmikantham, V., and S. Leela, "Differential and Integral Inequalities Theory and Applications: Vols I and II," Academic Press, New York, 1969.

[8] Stepan, G., "Retarded Dynamical Systems: Stability and Characteristic Functions," Longman Scientific & Technical, Essex, 1989.

[9] Burton, T. A., "Stability and Periodic Solutions of Ordinary and Functional Differential Equations," Academic Press, New York, 1985.

[10] Bensoussan, A., D. Prato, M. C. Defour and S. K. Mitter, "Representation and Control of Infinite Dimensional-Systems and Control: Foundations and Applications," Vols I , II, Birkhauser, Boston, 1993.

[11] Curtain, R. F. and A. J. Pritchard, "Infinite-Dimensional Linear Systems Theory," vol. 8 of Control and Information Sciences, Springer-Verlag, Berlin, 1978.

[12] Diekmann, O., S. A. van Gils, S. M. V. Lunel and O. H. Walther, "Delay Equations: Functional, Complex and Nonlinear Analysis," vol. 110 of Appl. Math. Sciences, Springer-Verlag, New York, 1995.

[13] Bellman, R. and J. M. Danskin, "A Survey of Mathematical Theory of Time-Lag, Retarded Control and Hereditary Processes," RAND Corp. Rept. No. R-256, 1956.

[14] Kamen, E. W., "Lectures on Algebraic System Theory: Linear Systems over Rings," NASA Contractor Report 3016, USA, 1978.

[15] Marshall, J. E., H. Gorecki, K. Walton and A. Korytowski, "Time-Delay Systems: Stability and Performance Criteria with Applications," Ellis Horwood, New York, 1992.

[16] Murray, J. D., "Mathematical Biology," Springer, New York, 1989.

[17] Halanay, A. "Differential Equations: Stability, Oscillations, Time Lags," Academic Press, New York, 1966.

[18] Oguztoreli, M. N., "Time-Lag Control Systems," Academic Press, New York, 1966.

[19] Marshall, J. E., "Control of Time-Delay Systems," IEE Series in Control Engineering, vol. 10, London, 1979.

[20] Boyd, S., L. El-Ghaoui, E. Feron and V. Balakrishnan, "Linear Matrix Inequalities in System and Control Theory," vol. 15, SIAM Studies in Appl. Math., Philadelphia, 1994.

[21] Busenberg, S. and M. Martelli (Editors), "Delay Differential Equations and Dynamical Systems," Springer-Verlag, Berlin, 1991.

[22] Kuang, Y., "Delay Differential Equations with Applications in Population Dynamics," Academic Press, Boston, 1993.

[23] MacDonald, N., "Time-Lags in Biological Models," Springer-Verlag, Berlin, 1978.

[24] Gahinet, P., A. Nemirovski, A. J. Laub and M. Chilali, "LMI-Control Toolbox," The MathWorks, Mass., 1995.

[25] Packard, A. and J. C. Doyle, "Quadratic Stability with Real and Complex Perturbations," **IEEE Trans. Automatic Control,** vol. 35, 1990, pp. 198-201.

[26] Savkin, A. V., "Absolute Stability of Nonlinear Control Systems with Nonstationary Linear Part," **Automation and Remote Control,** vol. 41, 1991, pp. 362-367.

[27] Huang, S. and W. Ren, "Longitudinal Control with Time-Delay in Platooning," **Proc. IEE Control Theory Appl.,** vol. 148, 1998, pp. 211-217.

[28] Lee, C. S. and G. Leitmann, "Continuous Feedback Guaranteeing Uniform Ultimate Boundedness for Uncertain Linear Delay Systems: An Application to River Pollution Control," **Computer and Mathematical Modeling,** vol. 16, 1988, pp. 929-938.

[29] Mahmoud, M. S., "Robust Stability and Stabilization of a Class of Uncertain Nonlinear Systems with Delays," **J. Mathematical Problems in Engineering,** vol. 3, 1997, pp. 1-22.

[30] Wang, Y., R. Zhou and C. Wen, "Load-Frequency Controller Design for Power Systems," **Proc. IEE Part C,** vol. 140, 1993, pp. 11-17.

[31] Tamura, H., "Decentralized Optimization for Distributed-Lag Models of Discrete Systems," **Automatica,** vol. 11, 1975, pp. 593-602.

Part I

ROBUST CONTROL

Chapter 2

Robust Stability

Motivated by the fact that *stability* is the prime objective in control system design, we present in this chapter methods for analyzing stability behavior of classes of linear time-delay systems. We will pay attention to methods in the time-domain more than in the frequency-domain due to the availability of efficient computer software [64]. The main focus is on issues of *robust stability*. Specifically, we are interested in examining to what extent the time-delay system (TDS) or uncertain time-delay system (UTDS) under consideration remains stable in the face of unknown delay factor (constant or time-varying) and/or parameteric uncertainties. We refer the reader to the basic stability definitions in section 1.1.2 and the stability theorems (Appendix C).

Problems of stability analysis and stabilization of dynamical systems with delay factors in the state variables and/or control inputs have received considerable interests for more than three decades; see [1,2] for a modest coverage of the subject. There exists a voluminous literature dealing with stability and stabilization for TDS and UTDS; see the bibliography at the end of Part I. Although all numerical simulations are based on linear matrix inequalities formalism [3] using **LMI-Control** Toolbox [4] (see also Appendix E), the theoretical analysis is pursued for both algebraic Riccati and linear matrix inequalities and are presented side-by-side. The benefit is purely technical since some of the known results are only available in algebraic Ricatti forms.

2.1 Stability Results of Time-Delay Systems

In this section, we develop stability results of Time-Delay Systems (TDS). We start by continuous-time models and then treat discrete-time models.

Both delay-independent and delay-dependent stability conditions are established. Our approach stems mainly from the application of *Lyapunov's Second Method* and is based on the constructive use of appropriate *Lyapunov-Krasovskii functionals* in the parameter space. For the sake of completeness, we will describe in later sections some methods based on *Lyapunov-Razumikhin functions* in the function space. The mathematical statements of both stability theorems are included in Appendix C. In effect, the delay-independent and delay-dependent stability conditions are transformed into the *existence* of a symmetric and positive-definite solution of the *parameterized* mathematical (algebraic Riccati or linear matrix) inequality, where the parameters are given by positive-definite matrices and/or positive scalars.

2.1.1 Stability Conditions of Continuous-Time Systems

The class of linear time-invariant state-delay systems under consideration is represented by:

$$(\Sigma_{dc}): \quad \dot{x}(t) = Ax(t) + A_d x(t - \tau) \tag{2.1}$$

where $x \in \Re^n$ is the state, $A \in \Re^{n \times n}$, $A_d \in \Re^{n \times n}$ are real constant matrices and τ is an unknown time-varying delay factor satisfying

$$0 \le \tau(t) \le \tau^o, \qquad 0 \le \dot{\tau}(t) \le \tau^+ \le 1 \tag{2.2}$$

where τ^o, τ^+ are known bounds. Since the stability of system (2.1) is a crucial step for control design of TDS, we first develop in the sequel two different stability criteria: one is delay-independent and the other is delay-dependent.

Delay-Independent Stability

Here, we focus on the nominal system only and consider the following:

Assumption 2.1: $\lambda(A) \in C^-$.

A delay-independent stability result concerning the system (Σ_{dc}) is summarized below:

Theorem 2.1: *Subject to Assumption 2.1, the time delay system (Σ_{dc}) is globally asymptotically stable independent of delay if one of the following two equivalent conditions holds:*

(1) *There exist matrices $0 < P = P^t \in \Re^{n \times n}$ and $0 < Q = Q^t \in \Re^{n \times n}$ with $Q_\tau = (1 - \tau^+)Q$ satisfying the algebraic Riccati inequality (ARI)*

$$PA + A^t P + P A_d \, Q_\tau^{-1} A_d^t P + Q \quad < \quad 0 \tag{2.3}$$

(2) *There exist matrices* $0 < P = P^t \in \Re^{n \times n}$ *and* $0 < Q = Q^t \in \Re^{n \times n}$
satisfying the linear matrix inequality (LMI)

$$\begin{bmatrix} PA + A^tP + Q & PA_d \\ A_d^tP & -Q_\tau \end{bmatrix} \quad < \quad 0 \tag{2.4}$$

Proof: (1) Introduce a Lyapunov-Krasovskii functional $V_1(x_t)$ of the form
[6]:

$$V_1(x_t) = x^t(t)Px(t) + \int_{t-\tau}^{t} x^t(\theta)Qx(\theta) \; d\theta \tag{2.5}$$

where $0 < P = P^t \in \Re^{n \times n}$ and $0 < Q = Q^t \in \Re^{n \times n}$ are weighting matrices.
By differentiating $V_1(x_t)$ along the solutions of (2.1) and arranging terms,
we get:

$$\begin{aligned} \dot{V}_1(x_t) &= x^t(t)\left[PA + A^tP + Q\right]x(t) + x^t(t)PA_dx(t - \tau) \\ &+ x^t(t - \tau)A_d^tPx(t) - (1 - \dot{\tau})x^t(t - \tau)A_d^tPA_dx(t - \tau) \end{aligned} \tag{2.6}$$

By a standard *completion of the squares* argument and using (2.2) into (2.6),
it reduces to

$$\dot{V}_1(x_t) < x^t(t)\left[PA + A^tP + PA_dQ_\tau^{-1}A_d^tP + Q\right]x(t) \tag{2.7}$$

If $\dot{V}_1(x_t) < 0$, when $x \neq 0$ then $x(t) \to 0$ as $t \to \infty$ and the time-delay
system (Σ_{dc}) is globally asymptotically stable independent of delay. From
(2.7), this stability condition is guaranteed if inequality (2.4) holds.

(2) By **A.1** , LMI (2.4) is equivalent to ARI (2.3).

Remark 2.1: Note that, apart from the knowledge of τ^+ satisfying
(2.2), **Theorem 2.1** provides a sufficient delay-independent stability condition. This condition is expressed as a feasibility of an LMI for which there
exist computationally efficient solution methods [3]. Looked at in this light,
it establishes a *robust* result since it implies that no matter what is $\tau(t)$,
system (2.1) satisfying **Assumption 2.1** is asymptotically stable so long as
$\tau(t)$ satisfies (2.2). Admittedly, it is a conservative result since it does not
carry enough information about τ. This will be shortly discussed.

Corollary 2.1: *Subject to* **Assumption 2.1**, *the time delay system*

$$(\Sigma_f): \quad \dot{x}(t) = Ax(t) + A_dx(t - d)$$

where $d > 0$ is an unknown constant delay is globally asymptotically stable independent of delay if one of the following two equivalent conditions holds:

(1) *There exist matrices $0 < P = P^t \in \Re^{n \times n}$ and $0 < Q = Q^t \in \Re^{n \times n}$ satisfying the ARI*

$$PA + A^tP + P A_d Q^{-1} A_d^t P + Q \quad < \quad 0 \qquad (2.8)$$

(2) *There exist matrices $0 < P = P^t \in \Re^{n \times n}$ and $0 < Q = Q^t \in \Re^{n \times n}$ satisfying the LMI*

$$\begin{bmatrix} PA + A^tP + Q & PA_d \\ A_d^t P & -Q \end{bmatrix} \quad < \quad 0 \qquad (2.9)$$

Proof: Follows from **Theorem 2.1** by simply setting $\tau^+ = 0$.

Remark 2.2: By comparing the ARIs (2.3) and (2.8) using the same system data, we find that $A_d Q_\tau^{-1} A_d^t > A_d Q^{-1} A_d^t$ which means that a P satisfying (2.3) is larger than the one satisfying (2.8).

Remark 2.3: By considering system (Σ_{dc}) in case the delay factor is constant and the state-delay matrix $A_d \in \Re^{n \times n}$ can be factored into $A_d = D_d F_d$ where $D_d \in \Re^{n \times q}$ and $F_d \in \Re^{q \times n}$ such that $r(A_d) = q \leq n$. Instead of (2.5), we choose a Lyapunov-Krasovskii functional of the form

$$V_{1f}(x_t) = x^t(t)Px(t) + \int_{t-\tau}^t x^t(\theta) F_d^t R F_d x(\theta) \; d\theta$$

for $0 < P = P^t, 0 < R = R^t$ as weighting matrices. Following the procedure of **Theorem 2.1**, it is readily seen that a sufficient condition for asymptotic stability independent-of-delay amounts to the existence of matrices P and R satisfying the LMI:

$$\Omega_{ll} = \begin{bmatrix} PA + A^tP + F_d^t R F_d & PD_d \\ D_d^t P & -R \end{bmatrix} \quad < \quad 0$$

The apparent benefit is that Ω_{ll} has dimension $(n+q) \times (n+q)$ whereas the corresponding matrix in (2.9) has dimension $2n \times 2n$ which might provide saving in computer simulation. This reduction in size should be contrasted with the restriction imposed by factorization. On the other hand, the above factorization would be useful in feedback control design of TDS by taking

$A_d = B F_d$ where $B \in \Re^{n \times m}$ as the input matrix. In this regard, we say that the delayed state matrix lies in the range space of the input matrix and hence it is accessible to the control input. Similar results are derived in [12,165].

2.1.2 Example 2.1

Consider the following time-delay system

$$\dot{x}(t) = \begin{bmatrix} -3 & -2 \\ 1 & 0 \end{bmatrix} x(t) + \begin{bmatrix} 0 & 0.3 \\ -0.3 & -0.2 \end{bmatrix} x(t - \tau(t))$$

such that $\dot{\tau} \leq \tau^+$ and consider $\tau^+ \in [0.1, 0.3, 0.5]$. First observe that the system without delay is internally stable since $\lambda(A) = \{-1, -2\}$. Using the LMI-Toolbox, the solution of inequality (2.4) with $Q = diag[1 \ 1]$ is given by

$$P = \begin{bmatrix} 0.4912 & 0.4489 \\ 0.4489 & 2.0849 \end{bmatrix}, \quad \tau^+ = 0.1$$

$$P = \begin{bmatrix} 0.4574 & 0.4180 \\ 0.4180 & 1.9415 \end{bmatrix}, \quad \tau^+ = 0.3$$

$$P = \begin{bmatrix} 0.5032 & 0.5920 \\ 0.5920 & 2.2628 \end{bmatrix}, \quad \tau^+ = 0.5$$

In case $\tau^+ = 0.9$, there was no feasible solution with the given data. However, with $Q = diag[0.2 \ 0.2]$, a feasible solution is obtained as:

$$P = \begin{bmatrix} 0.1541 & 0.1653 \\ 0.1653 & 0.5590 \end{bmatrix}$$

Now suppose that

$$A_d = \begin{bmatrix} -0.45 & -0.5 \\ -0.15 & -0.1 \end{bmatrix}$$

and consider $\tau^+ \in [0.1, 0.3, 0.5, 0.9]$ as before. The solution of inequality (2.4) with $Q = diag[1 \ 1]$ is given by

$$P = \begin{bmatrix} 0.4826 & 0.4244 \\ 0.4244 & 2.1197 \end{bmatrix}, \quad \tau^+ = 0.1$$

$$P = \begin{bmatrix} 0.5636 & 0.5628 \\ 0.5628 & 2.2182 \end{bmatrix}, \quad \tau^+ = 0.3$$

$$P = \begin{bmatrix} 0.5149 & 0.5128 \\ 0.5128 & 1.9912 \end{bmatrix}, \quad \tau^+ = 0.5$$

With $\tau^+ = 0.9$ and $Q = diag[0.2 \ 0.2]$, a feasible solution is obtained as:

$$P = \begin{bmatrix} 0.1511 & 0.1293 \\ 0.1293 & 0.4721 \end{bmatrix}$$

A comparison between the results of the two cases when $\tau^+ = 0.5$ clarifies the role of the delay matrix on the stability of the time-delay matrix.

Delay-Dependent Stability

In order to reduce conservatism in the stability analysis of TDS, we now focus on a delay-dependent stability measure. This requires the following assumption:

Assumption 2.2: $\lambda(A + A_d) \in C^-$.

Note that **Assumption 2.2** corresponds to the stability condition when $\tau = 0$. Hence it is necessary for the system (2.1) to be stable for any $\tau \geq 0$.

Theorem 2.2: *Consider the time-delay system* (Σ_{dc}) *satisfying* **Assumption 2.2**. *Then given a scalar* $\tau^* > 0$, *the system* (Σ_{dc}) *is globally asymptotically stable for any constant time-delay* τ *satisfying* $0 \leq \tau \leq \tau^*$ *if one of the following two equivalent conditions holds:*

(1) There exist matrix $0 < X = X^t \in \Re^{n \times n}$ *and scalars* $\varepsilon > 0$ *and* $\alpha > 0$ *satisfying the ARI*

$$(A + A_d)X + X(A + A_d)^t + \tau^* \varepsilon^{-1} X A^t A X + \tau^* \alpha^{-1} X A_d A_d^t X$$
$$+ \tau^* (\varepsilon + \alpha) A_d A_d^t < 0 \tag{2.10}$$

(2) There exist matrix $0 < X = X^t \in \Re^{n \times n}$ *and scalars* $\varepsilon > 0$ *and* $\alpha > 0$ *satisfying the LMI:*

$$\begin{bmatrix} (A + A_d)X + X(A + A_d)^t + \tau^*(\varepsilon + \alpha) A_d A_d^t & \tau^* X A^t & \tau^* X A_d^t \\ \tau^* A X & -(\tau^* \varepsilon)I & 0 \\ \tau^* A_d X & 0 & -(\tau^* \alpha)I \end{bmatrix} < 0 \tag{2.11}$$

Proof: (1) Introduce a Lyapunov-Krasovskii functional $V_2(x_t)$ of the form:

$$V_2(x_t) = x^t(t)Px(t) + \int_{t-\tau}^{t}\int_{t+\theta}^{t} r_1[x^t(s)A^tAx(s)]dsd\theta$$

$$+ \int_{t-\tau}^{t}\int_{t-\tau+\theta}^{t} r_2[x^t(s)A_d^tA_dx(s)]dsd\theta \qquad (2.12)$$

where $0 < P = P^t \in \Re^{n\times n}$ and $r_1 > 0$, $r_2 > 0$ are weighting factors. First from (2.1) we have

$$x(t-\tau) = x(t) - \int_{-\tau}^{0} \dot{x}(t+\theta)d\theta$$

$$= x(t) - \int_{-\tau}^{0} Ax(t+\theta)d\theta - \int_{-\tau}^{0} A_dx(t-\tau+\theta)d\theta \quad (2.13)$$

Substituting (2.13) back into (2.1) we get:

$$\dot{x}(t) = (A+A_d)x(t) - A_d\left\{\int_{-\tau}^{0} Ax(t+\theta)d\theta + \int_{-\tau}^{0} A_dx(t-\tau+\theta)d\theta\right\} \quad (2.14)$$

Now by differentiating $V_2(x_t)$ along the solutions of (2.14) and arranging terms, we obtain:

$$\dot{V}_2(x_t) = x^t[P(A+A_d) + (A+A_d)^tP]x - 2x^tPA_d\int_{-\tau}^{0} Ax(t+\theta)d\theta$$

$$- 2x^tPA_d\int_{-\tau}^{0} A_dx(t-\tau+\theta)d\theta - \int_{-\tau}^{0} r_1[x^t(t+\theta)A^tAx(t+\theta)]d\theta$$

$$+ \tau r_1 x^t A^tAx + \tau r_2 x^t A_d^tA_dx(t)$$

$$- \int_{-\tau}^{0} r_2[x^t(t-\tau+\theta)A_d^tA_dx(t-\tau+\theta)]d\theta \qquad (2.15)$$

By applying B.1.1, we have

$$-2x^t(t)PA_d\int_{-\tau}^{0} Ax(t+\theta)d\theta$$

$$\leq r_1^{-1}\int_{-\tau}^{0}[x^t(t)PA_dA_d^tPx(t)]d\theta$$

$$+r_1\int_{-\tau}^{0}[x^t(t+\theta)A^tAx(t+\theta)]d\theta$$

$$= \tau r_1^{-1}x^t(t)PA_dA_d^tPx(t)$$

$$+r_1\int_{-\tau}^{0}[x^t(t+\theta)A^tAx(t+\theta)]d\theta \qquad (2.16)$$

Similarly

$$-2x^t(t)PA_d \int_{-\tau}^0 A_d x(t-\tau+\theta)d\theta \ \leq \ \tau r_2^{-1} x^t(t)PA_d A_d^t Px(t)$$
$$+r_2 \int_{-\tau}^0 [x^t(t-\tau+\theta)A_d^t A_d x(t-\tau+\theta)]d\theta \tag{2.17}$$

Hence, it follows from (2.15) and (2.16)-(2.17) that

$$\dot{V}_2(x_t) \ = \ x^t[P(A+A_d)+(A+A_d)^t P+\tau r_1 A^t A+\tau r_2 A_d^t A_d$$
$$+ \ \tau r_1^{-1} PA_d A_d^t P + \tau r_2^{-1} PA_d^t A_d P]x \tag{2.18}$$

If $\dot{V}_2(x_t) < 0$ when $x \neq 0$, then $x(t) \rightarrow 0$ as $t \rightarrow \infty$ and the time-delay system (Σ_{dc}) is globally asymptotically stable. By defining $r_1 = \varepsilon^{-1}$ and $r_2 = \alpha^{-1}$, then it follows from (2.18) for any $\tau \in [0, \tau^*]$ that the stability condition is satisfied if

$$P(A+A_d)+(A+A_d)^t P+\tau^*(\varepsilon+\alpha)PA_d^t A_d P+\tau^*\varepsilon^{-1}A^t A+\tau^*\alpha^{-1}A_d^t A_d < 0$$
$$\tag{2.19}$$

Premultiplying (2.19) by P^{-1}, postmultiplying the result by P^{-1} and letting $X = P^{-1}$, we obtain the ARI (2.10) as desired.

(2) By **A.1**, LMI (2.11) is equivalent to ARI (2.10).

Remark 2.4: The motivation behind expression (2.12) is purely technical in order to take care of some terms appearing in the Lyapunov derivative later on. Note that while system (2.1) has initial value over $[-\tau, 0]$, system (2.14) requires initial date on $[-2\tau, 0]$. In [5,13,17], alternative results are derived using different approaches.

2.1.3 Example 2.2

A time-delay system of the type (2.1) has the following matrices

$$A = \begin{bmatrix} -3 & -2 \\ 1 & 0 \end{bmatrix} x(t) \ , \ A_d = \begin{bmatrix} 0 & 0.3 \\ -0.3 & -0.2 \end{bmatrix}$$

We wish to examine its delay-dependent stability. For this purpose, we first note that $\lambda(A + A_d) = \{-0.7225, -2.4775\}$ and hence **Assumption 2.2** is satisfied. Then, we proceed to solve inequality (2.11) using the **LMI-Toolbox**. The result for $\epsilon = 0.2, \alpha = 0.1$ is given by

$$X = \left[\begin{array}{cc} 0.3793 & -0.3954 \\ -0.3954 & 0.8844 \end{array} \right] \quad , \quad \tau^* = 0.2105$$

which means that the time-delay system is asymptotically stable for any constant time-delay τ satisfying $0 \leq \tau 0.2105$.

2.1.4 Stability Conditions of Discrete-Time Systems

Keeping with our objective, this section is dedicated to stability results of discrete-time systems with state-delay. Essentially, the results are parallel to those of section 2.1.1 but the mathematical machinery is quite different and has its own flavor.

The class of linear discrete-time state-delay systems under consideration is represented by:

$$(\Sigma_{dd}): \quad x(k+1) = Ax(k) + A_d x(k - \tau) \tag{2.20}$$

where $x \in \Re^n$ is the state, $A \in \Re^{n \times n}$, $A_d \in \Re^{n \times n}$ are real constant matrices and τ is an unknown integer representing the amount of delay units in the state. In the sequel, we derive stability conditions of system (Σ_{dd}).

Delay-Independent Stability

For system (Σ_{dd}), we require the following assumption:
Assumption 2.3: $|\lambda(A)| < 1$

This implies that the free nominal matrix without delay has to be a Schur-matrix, which is the discrete analog of **Assumption 2.1**. The following delay-independent stability result is then established.

Theorem 2.3: *Subject to* **Assumption 2.3**, *the time delay system* (Σ_{dd}) *is globally asymptotically stable independent-of-delay if one of the following two equivalent conditions holds:*

(1) *There exist matrices* $0 < P = P^t \in \Re^{n \times n}$ *and* $0 < Q = Q^t \in \Re^{n \times n}$ *satisfying the LMIs*

$$\left[\begin{array}{ccc} A^t P A - P + Q & A^t P A_d & 0 \\ A_d^t P A & -Q & A_d^t \\ 0 & A_d & -P^{-1} \end{array} \right] \quad < \quad 0$$

$$A_d^t P A_d \; - \; Q \quad < \quad 0 \qquad (2.21)$$

(2) *There exist matrices* $0 < P = P^t \in \Re^{n \times n}$ *and* $0 < Q = Q^t \in \Re^{n \times n}$
such that $Q - A_d^t P A_d > 0$ *satisfying the ARI*

$$A^t P A \; - \; P \; + A^t P A_d (Q - A_d^t P A_d)^{-1} A_d^t P A \; + \; Q \quad < \quad 0 \qquad (2.22)$$

Proof: (1) Introduce a discrete-type Lyapunov-Krasovskii functional $V_3(x_k)$
of the form:

$$V_3(x_k) = x^t(k) P x(k) + \sum_{\theta=k-\tau}^{k-1} x^t(\theta) Q x(\theta) \qquad (2.23)$$

where x_k has a meaning similar to x_t with respect to the discrete-time,
$0 < P = P^t \in \Re^{n \times n}$ and $0 < Q = Q^t \in \Re^{n \times n}$ are weighting matrices. By
evaluating the first-forward difference $\Delta V_3(x_k) = V_3(x_{k+1}) - V_3(x_k)$ along
the solutions of (2.20) and arranging terms, we get:

$$\Delta V_3(x_k) = \begin{bmatrix} x^t(k) & x^t(k-\tau) \end{bmatrix} \; \Omega_f \; \begin{bmatrix} x(k) \\ x(k-\tau) \end{bmatrix} \qquad (2.24)$$

where

$$\Omega_f \; = \; \begin{bmatrix} A^t P A - P + Q & A^t P A_d \\ A_d^t P A & -(Q - A_d^t P A_d) \end{bmatrix} \qquad (2.25)$$

If $\Delta V_3(x_k) < 0$, when $x \neq 0$ then $x(k) \to 0$ as $k \to \infty$ and the time-delay
system (Σ_{dd}) is globally asymptotically stable independent-of-delay. From
(2.24)-(2.25), this stability condition is guaranteed if $\Omega_f < 0$. By simple
rearrangement of (2.25) we obtain the LMI (2.21).

(2) In the same way, by **A.1**, ARI (2.22) is equivalent to LMI (2.21).

Remark 2.5: Sometimes it is argued that (2.21) is not in the standard
LMI format since it contains P and P^{-1}. To overcome this delicate issue,
we apply the Schur complements again to express (2.21) as

$$\begin{bmatrix} -P + Q & 0 & A^t \\ 0 & -Q & A_d^t \\ A & A_d & -P^{-1} \end{bmatrix} < 0$$

which can then be transformed via the matrix $T_1 = diag[I, I, P]$ to:

$$\begin{bmatrix} -P + Q & 0 & A^t P \\ 0 & -Q & A_d^t P \\ P A & P A_d & -P \end{bmatrix} < 0 \qquad (2.26)$$

The inequality is now a standard LMI in matrices P and Q.

2.1.5 Example 2.3

With reference to model (2.20), the following matrices are considered for simulation

$$A = \begin{bmatrix} 0.4 & 0.3 \\ -0.1 & 0.7 \end{bmatrix}, \quad A_d = \begin{bmatrix} 0.2 & -0.2 \\ 0.2 & -0.1 \end{bmatrix}$$

Observe that Assumption 2.3 is satisfied since $\lambda(A) = 0.55 + j0.0866, 0.55 - j0.0866 \in [0, 1)$. Selecting $Q = 0.3\, I$, we solve the LMI (2.26) to yield

$$P = \begin{bmatrix} 13.2129 & -11.7458 \\ -11.7458 & 25.6601 \end{bmatrix}$$

Since $P = P^t > 0$, then the system is asymptotically stable independent-of-delay.

Delay-Dependent Stability

Now we focus on a delay-dependent stability which requires the following assumption:

Assumption 2.4: $|\lambda(A + A_d)| < 1$.

Note that **Assumption 2.4** corresponds to the stability condition when $\tau = 0$. Hence it is necessary for the system (1) to be stable for any $\tau \geq 0$. It is the discrete version of **Assumption 2.2**.

Theorem 2.4: *Consider the time-delay system* (Σ_{dd}) *satisfying* **Assumption 2.4**. *Then given a scalar* $\tau^* > 0$, *the system* (Σ_{dd}) *is globally asymptotically stable for any constant time-delay* τ *satisfying* $0 \leq \tau \leq \tau^*$ *if one of the following two equivalent conditions holds:*

(1) *There exist a matrix* $0 < P = P^t \in \Re^{n \times n}$ *and* $\varepsilon_1 > 0$, $\varepsilon_2 > 0$, $\varepsilon_3 > 0$ *and* $\varepsilon_4 > 0$, *satisfying the ARI*

$$\begin{aligned} &(A + A_d)^t P(A + A_d) - P + \tau^*(\varepsilon_1 + \varepsilon_2)(A + A_d)^t P(A + A_d) + \\ &\tau^* \varepsilon_3\, A^t A_d^t P A_d A + \tau^* \varepsilon_4\, A_d^t A_d^t P A_d A_d < 0 \end{aligned} \qquad (2.27)$$

(2) *There exist a matrix $0 < P = P^t \in \Re^{n \times n}$ and scalars $\varepsilon_1 > 0$, $\varepsilon_2 > 0$, $\varepsilon_3 > 0$ and $\varepsilon_4 > 0$ satisfying the LMI:*

$$\begin{bmatrix} \Upsilon_1 & \Upsilon_2 \\ \Upsilon_2^t & \Upsilon_3 \end{bmatrix} < 0 \tag{2.28}$$

where

$$\Upsilon_1 = \begin{bmatrix} (A + A_d)^t P(A + A_d) - P & \tau^*(A + A_d)^t & \tau^*(A + A_d)^t \\ \tau^*(A + A_d) & -\tau^*(\varepsilon_1 P)^{-1} & 0 \\ \tau^*(A + A_d) & 0 & -\tau^*(\varepsilon_2 P)^{-1} \end{bmatrix}$$

$$\Upsilon_2 = \begin{bmatrix} \tau^*(A^t A_d^t) & \tau^*(A^t A_d^t) \\ 0 & 0 \\ 0 & 0 \end{bmatrix}$$

$$\Upsilon_3 = \begin{bmatrix} -\tau^*(\varepsilon_3 P)^{-1} & 0 \\ 0 & -\tau^*(\varepsilon_4 P)^{-1} \end{bmatrix} \tag{2.29}$$

Proof: (1) Introduce a discrete-type Lyapunov-Krasovskii functional $V_4(x_k)$ of the form:

$$\begin{aligned} V_4(x_k) &= x^t(k)Px(k) + \sum_{\theta=-\tau}^{0} \{\rho_1 \sum_{j=k+\theta-1}^{k-2} \Delta x^t(j)A^t A_d^t P A_d A \Delta x(j)\} \\ &\quad + \{\sum_{\theta=-\tau}^{0} \rho_2 \sum_{j=k+\theta-\tau-1}^{k-2} \Delta x^t(j)A_d^t A_d^t P A_d A_d \Delta x(j)\} \end{aligned} \tag{2.30}$$

where $0 < P = P^t \in \Re^{n \times n}$ and $\rho_1 > 0$, $\rho_2 > 0$ are weighting factors to be selected. First, from (2.20) we have

$$\begin{aligned} x(k - \tau) &= x(k) - \sum_{\theta=-\tau}^{0} \Delta x(k + \theta) \\ &= x(k) - \sum_{\theta=-\tau}^{0} A\Delta x(k + \theta - 1) \\ &\quad - \sum_{\theta=-\tau}^{0} A_d \Delta x(k + \theta - \tau - 1) \end{aligned} \tag{2.31}$$

Substituting (2.31) back into (2.20) we get:

$$x(k + 1) = (A + A_d)x(k) - A_d \xi_1(k) - A_d \xi_2(k) \tag{2.32}$$

where

$$\xi_1(k) = A_d \sum_{\theta=-\tau}^{0} A \triangle x(k + \theta - 1)$$

$$\xi_2(k) = A_d \sum_{\theta=-\tau}^{0} A_d \triangle x(k + \theta - \tau - 1) \qquad (2.33)$$

Now by evaluating the first-forward difference $\triangle V_4(x_k)$ along the solutions of (2.32)-(2.33) and arranging terms, we obtain:

$$\triangle V_4(x_k) = -\rho_1 \sum_{\theta=-\tau}^{0} \triangle x^t(k + \theta - 1) A^t A_d^t P A_d A \triangle x(k + \theta - 1)$$

$$- \rho_2 \sum_{\theta=-\tau}^{0} \triangle x^t(k + \theta - \tau - 1) A_d^t A_d^t P A_d A_d \triangle x(k + \theta - \tau - 1)$$

$$+ x^t(k)[(A + A_d)^t P(A + A_d) - P]x(k)$$

$$- 2x^t(k)(A + A_d)^t P A_d \xi_2(k) - 2\xi_1^t(k) A_d^t P A_d \xi_2(k)$$

$$+ \xi_1^t(k) A_d^t P A_d \xi_1(k) + \tau\rho_2 \triangle x^t(k - 1) A_d^t A_d^t P A_d A_d \triangle x(k - 1)$$

$$+ \tau\rho_1 \triangle x^t(k - 1) A^t A_d^t P A_d A \triangle x(k - 1)$$

$$- 2x^t(k)(A + A_d)^t P A_d \xi_1(k) + \xi_2^t(k) A_d^t P A_d \xi_2(k) \qquad (2.34)$$

Note that application of **B.1.1** yields

$$-2x^t(k)(A + A_d)^t P A_d \xi_1(k) :=$$

$$-2x^t(k)(A + A_d)^t P A_d \sum_{\theta=-\tau}^{0} A \triangle x(k + \theta - 1)$$

$$\leq r_1^{-1} \sum_{\theta=-\tau}^{0} [x^t(k)(A + A_d)^t P(A + A_d)x(k)]$$

$$+ r_1 \sum_{\theta=-\tau}^{0} \triangle x^t(k + \theta - 1) A^t A_d^t P A_d A \triangle x(k + \theta - 1)$$

$$= \tau r_1^{-1} x^t(k)(A + A_d)^t P(A + A_d)x(k)$$

$$+ r_1 \sum_{\theta=-\tau}^{0} [\triangle x^t(k + \theta - 1) A^t A_d^t P E_d A \triangle x(t + \theta - 1)] \qquad (2.35)$$

By the same way

$$-2x^t(k)(A + A_d)^t P A_d \xi_2(k) :=$$

$$-2x^t(k)(A + A_d)^t P A_d \sum_{\theta=-\tau}^{0} A_d \triangle x(k + \theta - \tau - 1)$$

$$\leq \tau r_2^{-1} x^t(k)(A + A_d)^t P(A + A_d) x(k)$$

$$+ r_2 \sum_{\theta=-\tau}^{0} \triangle x^t(k + \theta - \tau - 1) A_d^t A_d^t P A_d A_d \triangle x(t + \theta - \tau - 1)$$

$$(2.36)$$

and

$$-2\xi_1^t(k) A_d^t P A_d \xi_2(k) =$$

$$-2 \sum_{\theta=-\tau}^{0} A \triangle x^t(k + \theta - 1) A_d^t P A_d \sum_{\theta=-\tau}^{0} A \triangle x(k + \theta - \tau - 1)$$

$$\leq r_3^{-1} \sum_{\theta=-\tau}^{0} \triangle x^t(k + \theta - 1) A^t A_d^t P A_d A \triangle x(k + \theta - 1)$$

$$+ r_3 \sum_{\theta=-\tau}^{0} \triangle x^t(k + \theta - \tau - 1) A_d^t A_d^t P A_d A_d \triangle x(t + \theta - \tau - 1)$$

$$(2.37)$$

Hence, it follows by substituting (2.35)-(2.37) into (2.34) while letting $\rho_1 = 1 + r_1 + r_3^{-1}$ and $\rho_2 = 1 + r_2 + r_3$ that

$$
\begin{aligned}
\triangle V_4(x_k) &= x^t(k)[(A + A_d)^t P(A + A_d) - P] x(k) \\
&+ \tau r_1^{-1} x^t(k)(A + A_d)^t P(A + A_d) x(k) \\
&+ \tau r_2^{-1} x^t(k)(A + A_d)^t P(A + A_d) x(k) \\
&+ \tau \rho_1 \triangle x^t(k - 1) A^t A_d^t P A_d A \triangle x(k - 1) \\
&+ \tau \rho_2 \triangle x^t(k - 1) A_d^t A_d^t P A_d A_d \triangle x(k - 1) \\
&= \xi^t(k) \, \Pi(P) \, \xi(k)
\end{aligned}
$$

$$(2.38)$$

where

$$\Pi(P) = \begin{bmatrix} R_1 & -R_2 \\ -R_1 & R_2 \end{bmatrix} \tag{2.39}$$

$$\xi(k) = [x^t(k) \quad x^t(k - 1)]^t \tag{2.40}$$

and

$$
\begin{aligned}
R_1 &= (A + A_d)^t P(A + A_d) - P + \tau r_1^{-1}(A + A_d)^t P(A + A_d) \\
&+ \tau r_2^{-1}(A + A_d)^t P(A + A_d) + \tau \rho_1 A^t A_d^t P A_d A \\
&+ \tau \rho_2 A_d^t A_d^t P A_d A_d && (2.41) \\
R_2 &= \tau \rho_1 A^t A_d^t P A_d A + \tau \rho_2 A_d^t A_d^t P A_d A_d && (2.42)
\end{aligned}
$$

If $\Delta V_4(k) < 0$ when $x \neq 0$, then $x(k) \to 0$ as $k \to \infty$ and the time-delay system (Σ_{dd}) is globally asymptotically stable. In view of (2.39)-(2.42), this is guaranteed if $\Pi(P) < 0$. Observing that since $R_2 > 0$, then necessary and sufficient conditions for $\Pi(P) < 0$ are:

$$
R_1 < 0 \quad , \quad R_1 - R_2 < 0 \tag{2.43}
$$

By defining $\varepsilon_1 = r_1^{-1}$, $\varepsilon_2 = r_2^{-1}$, $\varepsilon_3 = \rho_1$ and $\varepsilon_4 = \rho_2$ then it follows from (2.39) using (2.41)-(2.43) for any $\tau \in [0, \tau^*]$ that the stability condition (2.43) corresponds to (2.27).

(2) A simple rearrangement of (2.27) yields (2.28)-(2.29).

Remark 2.6: In computer implementation, it would appear that the LMI (2.28) is not jointly linear in $\varepsilon_1, ..., \varepsilon_4$ and P. To overcome this difficulty, we can employ a simple gridding procedure. First, we rescale ε_j; $j = 1, .., 4$ to get $\sigma_j = \varepsilon_j(1 + \varepsilon_j)^{-1}$ and observe that $\varepsilon_j > 0$ if and only if $\sigma_j \in (0, 1)$. Second, we assign a uniform grid on each σ_j. For each grid point of $(\sigma_1, ..., \sigma_4)$, we can search for a solution of (2.28) which is now a standard LMI.

Remark 2.7: Little attention has been paid to the stability of discrete-time systems with state-delay. We hope that the foregoing material fills up this gap.

2.1.6   Example 2.4

With reference to model (2.20), the following matrices are considered for simulation

$$
A = \begin{bmatrix} 0.4 & 0.3 \\ -0.1 & 0.7 \end{bmatrix} \quad , \quad A_d = \begin{bmatrix} 0.2 & -0.2 \\ 0.25 & -0.1 \end{bmatrix}
$$

Observe that **Assumption 2.4** is satisfied since $\lambda(A+A_d) = \{0.7225, 0.4775\} \in$ $[0,1)$. It has been found, using **LMI** toolbox, that the solution of (2.28)-(2.29) is given by

$$P = \begin{bmatrix} 63.3765 & 14.8446 \\ 14.8446 & 60.8363 \end{bmatrix}, \quad \tau^* = 4.3138$$

which means that the system under consideration is globally asymptotically stable for any constant delay $\tau \leq 4.3138$.

2.2   Robust Stability of UTDS

Here, we develop stability results for uncertain time-delay systems (UTDS) using linear models. Again, we start by continuous-time systems and then move to discrete-time systems. In view of the interlink between TDS and UTDS, the material of this section draws heavily on those of the previous sections. For analytical tractibility, we will limit ourselves to the case of norm-bounded uncertainties. This is a valid point since the matched and mismatched methods of representing uncertainties require information about the input matrix which is not yet available.

2.2.1   Stability Conditions for Continuous-Time Systems

Consider the following linear model of UTDS:

$$(\Sigma_{\Delta c}): \quad \dot{x}(t) \;=\; (A + \Delta A(t))x(t) \,+\, (A_d + \Delta A_d(t))x(t-\tau)$$
$$=\; A_\Delta(t)x(t) + A_{d\Delta}x(t-\tau) \tag{2.44}$$

where $x \in \Re^n$ is the state, $A \in \Re^{n \times n}$, $A_d \in \Re^{n \times n}$ are real constant matrices and τ is an unknown time-varying delay factor satisfying (2.2). The uncertain matrices $\Delta A(t) \in \Re^{n \times n}$ and $\Delta A_d(t) \in \Re^{n \times n}$ are represented by the norm-bounded structure (see section 1.3):

$$[\Delta A(t) \;\; \Delta A_d(t)] \;=\; H\Delta(t)[E \;\; E_d] \tag{2.45}$$

where $E \in \Re^{\beta \times n}$, $E_d \in \Re^{\beta \times n}$ and $H \in \Re^{n \times \alpha}$ are known constant matrices and $\Delta(t) \in \Re^{\alpha \times \beta}$ is an unknown matrix of uncertain parameters satisfying

$$\Delta^t(t)\, \Delta(t) \;\leq\; I\,, \qquad \forall t \tag{2.46}$$

Delay-Independent Stability

A delay-independent stability result concerning the system $(\Sigma_{\Delta c})$ is summarized below:

Lemma 2.1: *Subject to* **Assumption 2.1,** *the time delay system* $(\Sigma_{\Delta c})$ *is robustly asymptotically stable independent of delay if one of the following two equivalent conditions holds:*

(1) *There exist matrices* $0 < P = P^t \in \Re^{n \times n}$ *and* $0 < Q = Q^t \in \Re^{n \times n}$ *with* $Q_\tau = (1 - \tau^+)Q$ *and scalars* $\nu > 0, \epsilon > 0$ *such that* $Q_\tau - \nu^{-1} E_d^t E_d > 0$ *satisfying the ARI*

$$PA + A^t P + Q + \epsilon^{-1} E^t E +$$
$$P[(\epsilon + \nu)HH^t + A_d(Q_\tau - \nu^{-1} E_d^t E_d)^{-1} A_d^t]P < 0 \qquad (2.47)$$

(2) *There exist matrices* $0 < P = P^t \in \Re^{n \times n}$ *and* $0 < Q = Q^t \in \Re^{n \times n}$ *with* $Q_\tau = (1 - \tau^+)Q$ *and scalars* $\nu > 0, \epsilon > 0, \mu > 0, \rho > 0$ *satisfying the LMIs*

$$\begin{bmatrix} PA + A^t P + Q + \mu E^t E & PH & PH & PA_d \\ H^t P & -\mu I & 0 & 0 \\ H^t P & 0 & -\rho I & 0 \\ A_d^t P & 0 & 0 & -(Q_\tau - \rho E_d^t E_d) \end{bmatrix} < 0$$

$$\rho E_d^t E_d - Q_\tau < 0 \ , \quad \begin{bmatrix} \epsilon & 1 \\ 1 & \mu \end{bmatrix} \geq 0 \ , \quad \begin{bmatrix} \nu & 1 \\ 1 & \rho \end{bmatrix} \geq 0$$

$$(2.48)$$

Proof: (1) Consider the Lyapunov-Krasovskii functional $V_1(x_t)$ given by (2.5) where $0 < P = P^t \in \Re^{n \times n}$ and $0 < Q = Q^t \in \Re^{n \times n}$ are weighting matrices. Now by differentiating $V_1(x_t)$ along the solutions of (2.44) with some manipulations, we get:

$$\dot{V}_1(x_t) < x^t(t) \left[PA_\Delta + A_\Delta^t P + PA_{d\Delta} Q_\tau^{-1} A_{d\Delta}^t P + Q \right] x(t) \qquad (2.49)$$

To remove the uncertainty from (2.49), we use (2.45)-(2.46) and rely on applying **B.1.2** and **B.1.3.** The result for some $\nu > 0, \epsilon > 0$ is:

$$P\Delta A(t) + \Delta A^t(t)P = PH\Delta(t)E + E^t \Delta^t(t)H^t P$$
$$\leq \epsilon PHH^t P + \epsilon^{-1} E^t E \qquad (2.50)$$

$$P[A_d + \Delta A_d(t)]Q_\tau^{-1}[A_d + \Delta A_d(t)]^t P \;=\;$$
$$P[A_d + H\Delta(t)E_d]Q_\tau^{-1}[A_d + H\Delta(t)E_d]^t P \;\leq\;$$
$$\nu PHH^t P + PA_d(Q_\tau - \nu^{-1}E_d^t E_d)^{-1}A_d^t P \qquad (2.51)$$

Substituting (2.50)-(2.51) into (2.49), it becomes

$$\dot{V}_1(x_t) < x^t[PA + A^t P + Q +$$
$$P[(\epsilon + \nu)HH^t + \epsilon^{-1}E^t E + A_d(Q_\tau - \nu^{-1}E_d^t E_d)^{-1}A_d^t P]x \qquad (2.52)$$

If $\dot{V}_1(x_t) < 0$, when $x \neq 0$ then $x(t) \to 0$ as $t \to \infty$ and the time-delay system $(\Sigma_{\Delta c})$ is globally asymptotically stable independent of delay for all admissible uncertainties satisfying (2.46). From (2.52), this stability condition is guaranteed if inequality (2.47) holds.

(2) Applying **A.1**, it is easy to see that (2.47) and using the coupling constraints $\epsilon \, \mu = 1$, $\nu \, \rho = 1$ is equivalent to the LMIs (2.48).

Remark 2.8: Observe that the LMI conditions (2.48) are convex in the variables $P, \mu, \rho, \epsilon, \nu$. To check the feasibility of these conditions, one can employ a multi-dimensional search procedure for μ, ρ, ϵ, ν and solve for P. Alternatively, one can implement the following iterative scheme:

(1) Find $P, Q, \mu_o, \rho_o, \epsilon_o, \nu_o$ that satisfy the inequality constraints (2.48). If the problem is infeasible, stop. Otherwise, set the iteration counter $k = 1$.
(2) Find $\mu_k, \rho_k, \epsilon_k, \nu_k$ that solve the LMI problem

$$\min \phi_k \;:=\; \mu_{k-1}\epsilon + \epsilon_{k-1}\mu + \nu_{k-1}\rho + \rho_{k-1}\nu$$
$$subject \; to \; LMIs \; (2.48)$$

(3) If ϕ_k has reached a stationary point, stop. Otherwise, set $k \leftarrow k+1$ and go to **(2)**.

Although the above scheme is heuristic, it has been found in practice that it has a guaranteed convergence.

2.2.2 Example 2.5

Consider the following time-delay system

$$\dot{x}(t) \;=\; [\begin{bmatrix} -3 & -2 \\ 1 & 0 \end{bmatrix} + \begin{bmatrix} 0.2 \\ 0.1 \end{bmatrix} \Delta(t)[0.2 \; 0.4]]x(t)$$

$$+ \; \left[\begin{array}{cc} -0.5 & -0.6 \\ -0.2 & -0.1 \end{array} \right] + \left[\begin{array}{c} 0.2 \\ 0.1 \end{array} \right] \Delta(t)[0.1 \; 0.3]]x(t - \tau(t))$$

such that $\dot{\tau} \leq \tau^+$. First observe that the system without delay is internally stable since $\lambda(A) = \{-1, -2\}$. Using the LMI-Toolbox, the solution of inequality (2.48) with $Q = diag[1.8 \; 1.8]$, $\epsilon = 0.3$, $\nu = 0.5$ is given by

$$P = \left[\begin{array}{cc} 2.9930 & 2.3143 \\ 2.3143 & 5.2018 \end{array} \right]$$

Since $P = P^t > 0$, **Lemma 2.1** is verified.

Delay-Dependent Stability

Reference is made to system $(\Sigma_{\Delta c})$ with (2.45)-(2.46). We invoke **Assumption 2.2** for (2.44) and in line with section 2.2.2 we express the delayed state $x(t - \tau)$ in the form:

$$\begin{aligned} x(t - \tau) & = x(t) - \int_{-\tau}^{0} \dot{x}(t + \theta)d\theta \\ & = x(t) - \int_{-\tau}^{0} A_{\Delta}(t + \theta)x(t + \theta)d\theta \\ & \quad - \int_{-\tau}^{0} A_{d\Delta}(t + \theta)x(t - \tau + \theta)d\theta \end{aligned} \qquad (2.53)$$

and thus the system dynamics become:

$$\dot{x}(t) = (A_{\Delta}(t) + A_{d\Delta}(t))x(t) + A_{d\Delta}(t)\eta(x_t, t) \qquad (2.54)$$

where

$$\eta(x_t, t) = -\{ \int_{-\tau}^{0} A_{\Delta}x(t + \theta)d\theta + \int_{-\tau}^{0} A_{d\Delta}x(t - \tau + \theta)d\theta \} \qquad (2.55)$$

The following result is an extension of **Theorem 2.2**.

Lemma 2.2: *Consider the time-delay system $(\Sigma_{\Delta c})$ satisfying* **Assumption 2.2**. *Then given a scalar $\tau^* > 0$, the system (Σ_{dc}) is robustly asymptotically stable for any constant time-delay τ satisfying $0 \leq \tau \leq \tau^*$ if one of the following two equivalent conditions hold:*

(1) *There exist matrix $0 < X = X^t \in \Re^{n \times n}$ and scalars $\epsilon_1 > 0, ..., \epsilon_7 > 0$ such that $(I - \epsilon_5 H H^t) > 0$, $(I - \epsilon_6 H H^t) > 0$ and $(I - \epsilon_7 E_d^t E_d) > 0$ satisfying the ARI*

$$
\begin{aligned}
\Xi(P, \epsilon, \tau^*) \;=\; & P(A + A_d) + (A + A_d)^t P + \epsilon_1^{-1}(E + E_d)^t (E + E_d) \\
& + \tau^* \epsilon_3 (1 + \epsilon_4^{-1})[\epsilon_5^{-1} E^t E + A^t (I - \epsilon_5 H H^t)^{-1} A] + \epsilon_1 P H H^t P \\
& + \tau^* \epsilon_3 (1 + \epsilon_4)[\epsilon_6^{-1} E_d^t E_d + A_d^t (I - \epsilon_6 H H^t)^{-1} A_d] \\
& + \tau^* \epsilon_3^{-1} P[\epsilon_7^{-1} H H^t + A_d (I - \epsilon_7 E_d^t E_d)^{-1} A_d^t] P \;<\; 0 \qquad (2.56)
\end{aligned}
$$

(2) *There exist matrix $0 < X = X^t \in \Re^{n \times n}$ and scalars $\alpha_1 > 0, ..., \alpha_5 > 0$ satisfying the LMI:*

$$
\begin{bmatrix}
S(X) + H J_1 H^t & X(E + E_d)^t & \tau^* X M^t & \tau^* N \\
(E + E_d)X & -J_1 & 0 & 0 \\
\tau^* M X & 0 & -\tau^* J_2 & 0 \\
\tau^* N^t & 0 & 0 & -\tau^* J_3
\end{bmatrix} < 0 \qquad (2.57)
$$

where

$$
\begin{aligned}
S(X) &= (A + A_d)X + X(A + A_d)^t, \quad J_1 = \alpha_1 I \\
M &= [A^t \; E^t \; A_d^t \; E_d^t]^t, \quad N = [A_d \; H] \\
J_2 &= diag[\alpha_2 I - \alpha_3 H H^t, \alpha_3 I, I - \alpha_3 I - \alpha_4 H H^t, \alpha_4 I] \\
J_3 &= diag[I - \alpha_5 E_d^t E_d, \alpha_5 I] \qquad (2.58)
\end{aligned}
$$

Proof: (1) Introduce a Lyapunov-Krasovskii functional $V_5(x_t)$ of the form:

$$
V_5(x_t) \;=\; x^t(t) P x(t) + W(x_t, t) \qquad (2.59)
$$

$$
\begin{aligned}
W(x_t, t) \;=\; & \int_{-\tau}^{0} \epsilon_3 (1 + \epsilon_4^{-1}) \int_{t+\theta}^{t} \|A_\Delta(s) x(s)\|^2 ds d\theta \\
& + \int_{-\tau}^{0} \epsilon_3 (1 + \epsilon_4) \int_{t+\theta}^{t} \|A_{d\Delta}(s + \tau) x(s)\|^2 ds d\theta \qquad (2.60)
\end{aligned}
$$

where $0 < P = P^t \in \Re^{n \times n}$ and $\epsilon_3 > 0, \epsilon_4 > 0$ are scalars to be chosen. By evaluating the time-derivative of $V_5(x_t)$ along the solution of (2.54)-(2.55), we obtain:

$$
\begin{aligned}
\dot{V}_5(x_t) \;=\; & x^t(t)[P(A_\Delta(t) + A_{d\Delta}(t)) + (A_\Delta(t) + A_{d\Delta}(t))^t P] x(t) \\
& + 2x^t(t) P A_{d\Delta}^t(t) \eta(x_t, t) + \dot{W}(x_t, t) \qquad (2.61)
\end{aligned}
$$

Application of **B.1.1** yields:

$$\begin{aligned}
2x^t(t)PA_{d\Delta}^t(t)\eta(x_t,t) &\leq \tau\epsilon_3^{-1}x^t(t)PA_{d\Delta}(t)A_{d\Delta}^t(t)Px(t) \\
&+ \epsilon_3\tau^{-1}\eta^t(x_t,t)\eta(x_t,t)
\end{aligned} \quad (2.62)$$

for some scalar $\epsilon_3 > 0$. On the other hand, using **B.2.1** gives:

$$\begin{aligned}
\eta^t(x_t,t)\eta(x_t,t) &\leq (1+\epsilon_4^{-1})\|\int_{-\tau}^0 A_\Delta(t+\theta)x(t+\theta)d\theta\|^2 \\
&+ (1+\epsilon_4)\|\int_{-\tau}^0 A_{d\Delta}(t+\tau)x(t-\tau+\theta)d\theta\|^2 \\
&\leq \tau(1+\epsilon_4^{-1})\int_{-\tau}^0 \|A_\Delta(t+\theta)x(t+\theta)\|^2 d\theta \\
&+ \tau(1+\epsilon_4)\int_{-\tau}^0 \|A_{d\Delta}(t+\tau)x(t-\tau+\theta)\|^2 d\theta
\end{aligned} \quad (2.63)$$

for some scalar $\epsilon_4 > 0$. Now turning to (2.60), we have:

$$\begin{aligned}
\dot{W}(x_t,t) &= \tau\epsilon_3(1+\epsilon_4^{-1})\|A_\Delta(t)x(t)\|^2 \\
&+ \tau\epsilon_3(1+\epsilon_4)\|A_{d\Delta}(t+\tau)x(t)\|^2 \\
&- \epsilon_3(1+\epsilon_4^{-1})\int_{-\tau}^0 \|A_\Delta(t+\theta)x(t+\theta)\|^2 d\theta \\
&- \epsilon_3(1+\epsilon_4)\int_{-\tau}^0 \|A_{d\Delta}(t+\tau)x(t-\tau+\theta)\|^2 d\theta
\end{aligned} \quad (2.64)$$

By grouping (2.62)-(2.64) into (2.61) and arranging terms we arrive at:

$$\begin{aligned}
\dot{V}_5(x_t) &\leq x^t[P(A_\Delta + A_{d\Delta}) + (A_\Delta + A_{d\Delta})^t P \\
&+ \tau\epsilon_3^{-1}PA_{d\Delta}A_{d\Delta}^t P + \tau\epsilon_3(1+\epsilon_4^{-1})A_\Delta^t A_\Delta \\
&+ \tau\epsilon_3(1+\epsilon_4)A_{d\Delta}^t(t+\tau)A_{d\Delta}(t+\tau)]x
\end{aligned} \quad (2.65)$$

Next we focus on the uncertain terms of (2.65). First, by inequalities **B.1.2**-**B.1.3**, we get:

$$\begin{aligned}
P(A_\Delta + A_{d\Delta}) &+ (A_\Delta + A_{d\Delta})^t P := \\
P(A+A_d) &+ (A+A_d)^t P + PH\Delta(E+E_d) + (E+E_d)^t\Delta^t H^t P \\
&\leq P(A+A_d) + (A+A_d)^t P + \epsilon_1 PHH^t P + \epsilon_1^{-1}(E+E_d)^t(E+E_d)
\end{aligned} \quad (2.66)$$

$$(A + H\Delta E)^t(A + H\Delta E) \leq \epsilon_5^{-1}E^tE + A^t(I - \epsilon_5 HH^t)^{-1}A \qquad (2.67)$$

$$(A_d + H\Delta E_d)^t(A_d + H\Delta E_d) \leq \epsilon_6^{-1}E_d^tE_d + A_d^t(I - \epsilon_6 HH^t)^{-1}A_d \qquad (2.68)$$

$$P(A_d + H\Delta E_d)(A_d + H\Delta E_d)^tP \leq P[\epsilon_7^{-1}HH^t + A_d(I - \epsilon_7 E_d^tE_d)^{-1}A_d^t]P$$
$$(2.69)$$

for any scalars $\epsilon_1 > 0, ..., \epsilon_7 > 0$ satisfying $(I - \epsilon_5 HH^t) > 0, (I - \epsilon_6 HH^t) > 0, (I - \epsilon_7 E_d^tE_d) > 0$. Finally, we use (2.66)-(2.69) in (2.65) and manipulating to reach:

$$\dot{V}_5(x_t) \leq x^t(t)\Xi(P, \epsilon, \tau)x(t) \qquad (2.70)$$

where

$$\begin{aligned}
\Xi(P, \epsilon, \tau) &= P(A + A_d) + (A + A_d)^tP + \epsilon_1^{-1}(E + E_d)^t(E + E_d) \\
&+ \tau\epsilon_3(1 + \epsilon_4^{-1})[\epsilon_5^{-1}E^tE + A^t(I - \epsilon_5 HH^t)^{-1}A] + \epsilon_1 PHH^tP \\
&+ \tau\epsilon_3(1 + \epsilon_4)[\epsilon_6^{-1}E_d^tE_d + A_d^t(I - \epsilon_6 HH^t)^{-1}A_d] \\
&+ \tau\epsilon_3^{-1}P[\epsilon_7^{-1}HH^t + A_d(I - \epsilon_7 E_d^tE_d)^{-1}A_d^t]P \qquad (2.71)
\end{aligned}$$

where ϵ denotes $(\epsilon_1, ..., \epsilon_7)$. In view of the monotonic nondecreasing behavior of $\Xi(P, \epsilon, \tau)$ with respect to τ, the asymptotic stability is guaranteed for $0 \leq \tau \leq \tau^*$.

(2) By making the change of variables

$$\tilde{P} = \epsilon_3^{-1}P, \quad \alpha_1 = \epsilon_1\epsilon_3, \quad \alpha_2 = \epsilon_4(1 + \epsilon_4)^{-1}$$
$$\alpha_3 = \epsilon_5\epsilon_4(1 + \epsilon_4)^{-1}, \quad \alpha_4 = \epsilon_6(1 + \epsilon_4)^{-1}, \quad \alpha_5 = \epsilon_7 \qquad (2.72)$$

in (2.71), letting $X = \tilde{P}^{-1}$ and arranging, we get:

$$\begin{aligned}
\Xi(X, \epsilon, \tau) &= (A + A_d)X + X(A + A_d)^t + \alpha_1^{-1}X(E + E_d)^t(E + E_d)X \\
&+ \tau X[\alpha_3^{-1}E^tE + A^t(\alpha_2 I - \alpha_3 HH^t)^{-1}A]X + \alpha_1 HH^t \\
&+ \tau X[\alpha_4^{-1}E_d^tE_d + A_d^t(I - \alpha_3 I - \alpha_4 HH^t)^{-1}A_d] \\
&+ \tau[\alpha_5^{-1}HH^t + A_d(I - \alpha_5 E_d^tE_d)^{-1}A_d^t] \qquad (2.73)
\end{aligned}$$

In terms of (2.58) and applying **A.1**, inequality (2.73) is equivalent to LMI (2.57).

Remark 2.9: Interestingly enough, **Lemma 2.2** in the absence of uncertainties $H = 0, E = 0, E_d$ reduces to **Theorem 2.2** despite the slight difference in the analysis. It draws heavily on the results reported in [46,78].

2.2.3 Example 2.6

In order to illustrate the application of **Lemma 2.2**, we consider a time-delay system of the type (2.44) with

$$A = \begin{bmatrix} 0 & 0 \\ 0 & 1 \end{bmatrix}, A_d = \begin{bmatrix} -2 & -0.5 \\ 0 & -1 \end{bmatrix}, H = \begin{bmatrix} 0.45 & 0 \\ 0 & 0.45 \end{bmatrix}$$

$$E = \begin{bmatrix} 0.45 & 0 \\ 0 & 0.45 \end{bmatrix}, E_d = \begin{bmatrix} 0.45 & 0 \\ 0 & 0.45 \end{bmatrix}$$

Using the software **LMI** to solve (2.57), it has been found that the system is robustly asymptotically stable for any constant time-delay $\tau \le 0.3$. It is interesting to observe that A is unstable but $A + A_d$ is stable.

2.2.4 Example 2.7

We consider the time-delay system

$$\begin{aligned}
\dot{x}(t) &= \begin{bmatrix} -2 & 0 \\ 1 & -3 \end{bmatrix} x(t) + \begin{bmatrix} -1 & 0 \\ -0.8 & -1 \end{bmatrix} x(t-\tau) \\
&+ \begin{bmatrix} 0.2\, sint & 0 \\ 0 & 0.2\, sint \end{bmatrix} x(t) + \begin{bmatrix} 0.2\, cost & 0 \\ 0 & 0.2\, cost \end{bmatrix} x(t-\tau)
\end{aligned}$$

We first note that both A and $A + A_d$ are asymptotically stable. Now we identify

$$H = \begin{bmatrix} 0.2 & 0 \\ 0 & 0.2 \end{bmatrix}, E = \begin{bmatrix} 1 & 0 \\ 0 & 1 \end{bmatrix}, E_d = \begin{bmatrix} 1 & 0 \\ 0 & 1 \end{bmatrix}$$

Then we proceed to solve (2.57) using the software **LMI**. The result is that the system is robustly asymptotically stable for any constant time-delay $\tau \le 0.501$.

2.2.5 Stability Conditions for Discrete-Time Systems

The class of discrete-time systems with state-delay and parameteric uncertainties of interest is described by:

$$(\Sigma_{\Delta d}): \quad x(k+1) = A_\Delta x(k) + A_{d\Delta} x(k-\tau) \tag{2.74}$$

where $x \in \Re^n$ is the state and τ is an unknown time delay within a known interval $0 \leq \tau \leq \tau^*$. The uncertain matrices $A_\Delta \in \Re^{n \times n}$ and $A_{d\Delta} \in \Re^{p \times n}$ are given by:

$$[A_\Delta \quad A_{d\Delta}] = [A \quad A_d] + H \Delta(k) [E \quad E_d]$$
$$\Delta^t(k) \, \Delta(k) \leq I, \quad \forall k \tag{2.75}$$

where $A \in \Re^{n \times n}$, $A_d \in \Re^{n \times n}$, $H \in \Re^{n \times \alpha}$, $E \in \Re^{\beta \times n}$ and $E_d \in \Re^{\beta \times n}$ are known real constant matrices and $\Delta(k) \in \Re^{\alpha \times \beta}$ is a matrix of unknown parameters.

For ease in exposition, we restrict ourselves to the case of delay-independent stability leaving the case of delay-dependent stability for the interested reader to pursue.

Delay-Independent Stability

Lemma 2.3: *Subject to* **Assumption 2.3,** *the time delay system* $(\Sigma_{\Delta d})$ *is robustly asymptotically stable independent-of-delay if one of the following two equivalent conditions hold:*

(1) There exist matrices $0 < P = P^t \in \Re^{n \times n}$, $0 < Q = Q^t \in \Re^{n \times n}$ and a scalar $\epsilon > 0, \mu > 0$ such that $\Lambda_1 = Q - \mu E_d^t E_d > 0$ and $\Lambda_2 = P^{-1} - \epsilon H H^t > 0$ satisfying the LMIs

$$\begin{bmatrix} -P + Q + \mu E^t E & \mu E^t E_d & A^t \\ \mu E_d^t E & -\Lambda_1 & A_d^t \\ A & A_d & -\Lambda_2 \end{bmatrix} \quad < \quad 0$$

$$\mu E_d^t E_d - Q \; < \; 0 \, , \; \epsilon H H^t - P^{-1} \; < \; 0, \; \begin{bmatrix} \epsilon & 1 \\ 1 & \mu \end{bmatrix} \; \geq \; 0 \tag{2.76}$$

(2) There exist matrices $0 < P = P^t \in \Re^{n \times n}$, $0 < Q = Q^t \in \Re^{n \times n}$ and a scalar $\epsilon > 0$ such that $\Lambda_1 = Q - \epsilon^{-1} E_d^t E_d > 0$ and $\Lambda_2 = P^{-1} - \epsilon H H^t > 0$ satisfying the ARI

$$A^t(\Lambda_2 - A_d^t \Lambda_2^{-1} A_d)^{-1} A - P + Q + \epsilon^{-1} E^t E$$
$$+ \epsilon^{-1} E^t [I - \epsilon^{-1} E_d(\Lambda_2 - A_d^t \Lambda_2^{-1} A_d)^{-1} E_d^t] E$$
$$+ \epsilon^{-1} E^t E_d(\Lambda_2 - A_d^t \Lambda_2^{-1} A_d)^{-1} A_d^t \Lambda_2^{-1} A$$
$$+ \epsilon^{-1} A^t \Lambda_2^{-1} A_d(\Lambda_2 - A_d^t \Lambda_2^{-1} A_d)^{-1} E_d^t E \; < \; 0 \tag{2.77}$$

Proof: (1) Introduce a discrete-type Lyapunov-Krasovskii functional $V_6(x_k)$ of the form:

$$V_6(x_k) = x^t(k)Px(k) + \sum_{\theta=k-\tau}^{k-1} x^t(\theta)Qx(\theta) \tag{2.78}$$

where $0 < P = P^t \in \Re^{n \times n}$ and $0 < Q = Q^t \in \Re^{n \times n}$ are weighting matrices. By evaluating the first-forward difference $\Delta V_6(x_k) = V_6(x_{k+1}) - V_6(x_k)$ along the solutions of (2.74) and arranging terms, we get:

$$\Delta V_6(x_k) = \begin{bmatrix} x^t(k) & x^t(k-\tau) \end{bmatrix} \Omega_{ff} \begin{bmatrix} x(k) \\ x(k-\tau) \end{bmatrix}$$

$$\Omega_{ff} = \begin{bmatrix} A_\Delta^t PA_\Delta - P + Q & A_\Delta^t PA_{d\Delta} \\ A_{d\Delta}^t PA_\Delta & -(Q - A_{d\Delta}^t PA_{d\Delta}) \end{bmatrix} \tag{2.79}$$

If $\Delta V_6(x_k) < 0$, when $x \neq 0$ then $x(k) \to 0$ as $k \to \infty$ and the time-delay system $(\Sigma_{\Delta d})$ is globally asymptotically stable independent-of-delay. From (2.80)-(2.81), this stability condition is guaranteed if $\Omega_{ff} < 0$. By the Schur complement formula, we expand the inequality $\Omega_{ff} < 0$ into:

$$\begin{bmatrix} -P+Q & 0 & A_\Delta^t \\ 0 & -Q & A_{d\Delta}^t \\ A_\Delta & A_{d\Delta} & -P^{-1} \end{bmatrix} < 0$$

for all admissible uncertainties satisfying (2.75). The above inequality holds if and only if

$$\begin{bmatrix} -P+Q & 0 & A^t \\ 0 & -Q & A_d^t \\ A & A_d & -P^{-1} \end{bmatrix} + \begin{bmatrix} 0 \\ 0 \\ H \end{bmatrix} \Delta(k)[E \quad E_d \quad 0] +$$

$$\begin{bmatrix} E^t \\ E_d^t \\ 0 \end{bmatrix} \Delta^t(k)[0 \quad 0 \quad H^t] < 0 \tag{2.80}$$

By **B.I.4**, (2.80) holds if for some $\epsilon > 0$:

$$\begin{bmatrix} -P+Q+\epsilon^{-1}E^tE & \epsilon^{-1}E^tE_d & A^t \\ \epsilon^{-1}E_d^tE & -(Q-\epsilon^{-1}E_d^tE_d) & A_d^t \\ A & A_d & -(P^{-1}-\epsilon HH^t) \end{bmatrix} < 0 \tag{2.81}$$

Imposing the coupling constraint $\epsilon\,\mu = 1$ and letting $\Lambda_1 = Q - \mu E_d^t E_d$ and $\Lambda_2 = P^{-1} - \epsilon HH^t$ in (2.81) immediately yields the LMIs (2.76).

(2) By repeated applications of **A.1** to (2.76), we get:

$$-P + Q + \epsilon^{-1}E^t E + A^t \Lambda_2^{-1} A$$
$$(\epsilon^{-1}E^t E_d + A^t \Lambda_2^{-1} A_d)(\Lambda_1 - A_d^t \Lambda_2^{-1} A_d)^{-1}(\epsilon^{-1}E^t E_d + A^t \Lambda_2^{-1} A_d)^t \ < \ 0$$
$$(2.82)$$

Applying the matrix inversion lemma to (2.82) yields ARI (2.77).

Had we considered the system

$$(\Sigma_{\Delta D}): \quad x(k+1) = (A + H\Delta E)x(k) + A_d x(k - \tau) \qquad (2.83)$$

we would have arrived at the following result:

Corollary 2.2: *Subject to* **Assumption 2.3**, *the time delay system* $(\Sigma_{\Delta D})$ *is robustly asymptotically stable independent-of-delay if one of the following two equivalent conditions holds:*
(1) *There exist matrices* $0 < P = P^t \in \Re^{n \times n}$, $0 < Q = Q^t \in \Re^{n \times n}$ *and a scalar* $\epsilon > 0, \mu > 0$ *such that* $\Lambda = P^{-1} - \epsilon H H^t > 0$ *satisfying the LMIs*

$$\begin{bmatrix} -P + Q + \mu E^t E & 0 & A^t \\ 0 & -Q & A_d^t \\ A & A_d & -\Lambda \end{bmatrix} \ < \ 0$$
$$\epsilon H H^t - P^{-1} \ < \ 0 \qquad (2.84)$$

(2) *There exist matrices* $0 < P = P^t \in \Re^{n \times n}$, $0 < Q = Q^t \in \Re^{n \times n}$ *and a scalar* $\epsilon > 0$ *such that* $\Lambda = P^{-1} - \epsilon H H^t > 0$ *satisfying the ARI*

$$A^t \left\{ P^{-1} - \epsilon H H^t - A_d (P^{-1} - \epsilon H H^t)^{-1} A_d^t \right\} A$$
$$-P + Q + \epsilon^{-1} E^t E \quad < \ 0 \qquad (2.85)$$

Remark 2.10: It is important to observe that by suppressing the uncertainties ($H = 0, E = 0, E_d = 0$), the uncertain system (2.74) reduces to system (2.20) and consequently the ARI (2.22) can be recovered from ARI (2.77). Note that in implementing either the LMIs (2.76) or (2.84), we can rely on the iterative scheme described in **Remark 2.8**.

2.2.6 Example 2.8

We consider the uncertain time-delay system of the type (2.74) with

$$A = \begin{bmatrix} 0.1 & 0 & -0.1 \\ 0.05 & 0.3 & 0 \\ 0 & 0.2 & 0.6 \end{bmatrix}, \ A_d = \begin{bmatrix} -0.2 & 0 & 0 \\ 0 & -0.1 & 0.1 \\ 0 & 0 & -0.2 \end{bmatrix}$$

$$H = \begin{bmatrix} 0.1 \\ 0 \\ 0.2 \end{bmatrix}, \ E = [0.2 \ 0 \ 0.3], \ E_d = [0.4 \ 0 \ 0.1]$$

We first note that $\lambda(A) = \{0.0906, 0.31163, 0.5931\}$, $\lambda(A + A_d) = \{-0.1074, 0.1393, 0.468\}$ and hence both matrices A and $A + A_d$ are asymptotically stable. The result of implementing the iterative LMI-scheme is

$$Q = \begin{bmatrix} 0.6 & 0 & 0 \\ 0 & 0.6 & 0 \\ 0 & 0 & 0.6 \end{bmatrix}, \ \epsilon = 0.3001$$

$$P = \begin{bmatrix} 42.4291 & 0.2656 & 1.55095 \\ 0.2656 & 43.0743 & 1.7880 \\ 1.5095 & 1.7880 & 48.9937 \end{bmatrix}, \ \mu = 3.3396$$

This result shows clearly that the system is robustly asymptotically stable independent-of-delay.

2.3 Stability Tests Using $\mathcal{H}_\infty$-norm

It is well-known that one fundamental property of the *bounded real lemma* (see Appendix A) is that it relates the $\mathcal{H}_\infty$-norm of a linear dynamical system with an associated Riccati inequality (or equation , in the case of a stabilizing solution). We will make use of such a nice property in providing in the sequel a set of corollaries that characterize the asymptotic stability of some TDS using $\mathcal{H}_\infty$-setting.

   Corollary 2.3: *Subject to* **Assumption 2.1,** *the time delay system* (Σ_{dc}) *is globally asymptotically stable independent of delay if the following* $\mathcal{H}_\infty$-*norm constraint holds:*

$$||Q^{1/2}(sI - A)^{-1} A_d Q_\tau^{-1/2}||_\infty < 1 \qquad (2.86)$$

Corollary 2.4: *Consider the time-delay system* (Σ_{dc}) *satisfying* **Assumption 2.2.** *Then given a scalar* $\tau^* > 0$, *the system* (Σ_{dc}) *is globally asymptotically stable for any constant time-delay* τ *satisfying* $0 \leq \tau \leq \tau^*$ *if there exists scalars* $\epsilon > 0$ *and* $\alpha > 0$ *such that the following* $\mathcal{H}_\infty$-*norm constraint holds*:

$$\tau^* \| (\epsilon + \alpha)^{1/2} (sI - A - A_d)^{-1} [(\alpha)^{-1/2} A_d \quad (\epsilon)^{-1/2} A^t] \|_\infty < 1 \qquad (2.87)$$

Corollary 2.5: *Subject to* **Assumption 2.3,** *the time delay system* (Σ_{dd}) *is globally asymptotically stable independent of delay if the following* $\mathcal{H}_\infty$-*norm constraint holds*:

$$\| (zI - A)^{-1} A_d \|_\infty < 1 \qquad (2.88)$$

Remark 2.11: The interesting point to note is that inequalities (2.86)-(2.88) provide alternative computational methods for testing the stability of TDS based on continuous-time and discrete-time models. It is significant to observe that although (2.88) is obvious and straightforward, it has not been reported elsewhere in the literature about TDS. Also, when the delay is constant $(\tau^+ = 0)$, (2.86) reduces to

$$\| (sI - A)^{-1} A_d \|_\infty < 1$$

which has been reported before, see [49,53,70,106-107]. The reader should take note of the striking similarity between (2.86) and (2.88).

2.4   Stability of Time-Lag Systems

The class of time-delay systems in which the delay factor is a predominantly known constant is frequently called *time-lag systems* (TLS). The importance of TLS emerges from the fact that both time-domain and frequency-domain techniques are readily available for stability analysis. This section is devoted to the stability study of TLS using a different mathematical tool. The class of continuous-time TLS of interest has the form:

$$(\Sigma_{lc}): \quad \dot{x}(t) = (A + \Delta A)x(t) + (A_d + \Delta A_d)x(t - d) \qquad (2.89)$$

where $d > 0$ represents the amount of *lag* in the system and the uncertain matrices ΔA and ΔA_d satisfy:

Assumption 2.5: The uncertain matrices ΔA and ΔA_d are bounded in the form:

$$\|\Delta A\| \leq \beta, \quad \|\Delta A_d\| \leq \beta_d \tag{2.90}$$

where $\beta > 0, \beta_d > 0$ are known constants.

First, we suppress the uncertainties by setting $\Delta A = 0$ and $\Delta A_d = 0$. In this case, we obtain the nominal system:

$$(\Sigma_{lco}): \quad \dot{x}(t) = Ax(t) + A_d x(t - d) \tag{2.91}$$

which has the characteristic polynomial

$$\mathcal{F}(s) = sI - A - A_d e^{-ds} \tag{2.92}$$

This motivates the classical definition [218] that system (Σ_{lco}) is *asymptotically stable independent of delay* if and only if

$$det\left\{ sI - A - A_d e^{-ds} \right\} \neq 0, \quad \forall s \in \bar{C}_+, \quad \forall d \geq 0 \tag{2.93}$$

Since the eigenvalues of $\mathcal{F}(s) = 0$ are simply those of the matrix $(A - A_d e^{-ds})$, that is

$$s = \lambda_j(A + A_d e^{-ds}), \quad j = 1, \dots, n \tag{2.94}$$

Therefore, an equivalent statement to (2.93) would be that the roots of (2.94) lie in the open left half complex plane. There are several methods by which one can characterize the stability conditions for system (Σ_{lco}). One method is given below:

Lemma 2.4: System (Σ_{lco}) is asymptotically stable independent of delay if

$$\delta_1 := \mu(A) + \|A_d\| < 0 \tag{2.95}$$

Proof: Let $s = \alpha + j\omega$ be a root of (2.94) and assume $\alpha = Re\ s \geq 0$. By observing that $\mu(A) \leq \|A\|$ (see section 1.1.1), it follows that:

$$
\begin{aligned}
\alpha &= Re\ \lambda_j(A + A_d e^{-ds}) \\
&\leq \mu(A) + \mu(A_d e^{-jd\omega})e^{-d\alpha} \\
&\leq \mu(A) + \|(A_d e^{-jd\omega})\|e^{-d\alpha} \\
&\leq \mu(A) + \|A_d\| < 0 \tag{2.96}
\end{aligned}
$$

which contradicts the assumption $\alpha \geq 0$ and concludes the result.

It is readily seen that (2.95) is a very simple method to evaluate stability albeit its conservatism as it does not carry any information on the amount of lag d. This means that failing to satisfy (2.95) does not mean that system (2.91) is unstable. One way to relax this conservatism is to stretch the effect of the part $\mu(A_d e^{-jd\omega})e^{-d\alpha}$. By adopting another route of analysis, one seeks criteria that incorporates information on the amount of lag. Such a result can be obtained as follows [39-41].

Lemma 2.5: *Consider system* (Σ_{lco}) *such that* $\delta_1 > 0$. *Let* $\delta_2 := \mu(-jA) + \|A_d\|$. *Then the following condition*

$$\text{Re } \lambda_j(A + A_d e^{-ds}) < 0\,, \qquad j = 1, ..., n \tag{2.97}$$

assures the stability of system (Σ_{lco}) *in the region given by:*

$$s = j\omega\,, \quad 0 \leq \omega \leq \delta_2$$
$$s = \delta_1 + j\omega\,, \quad 0 \leq \omega \leq \delta_2$$
$$s = r + j\delta_2\,, \quad 0 \leq \omega \leq \delta_1$$

where $j := \sqrt{-1}$.

Observe that **Lemma 2.5** includes information on the size of the lag (delay) hence it represents some sort of delay-dependent stability condition. Eventually, the result amounts to enlarging the region for which system (Σ_{lco}) remains stable.

Recall that both **Lemma 2.4** and **Lemma 2.5** give sufficient conditions for stability. The following result rectifies this shortcoming. It relies on some known results from complex analysis [23], namely:

(1) The real and imaginary parts of an analytic function in some domain $\mathcal{D}_c$ are *harmonic functions* which are characterized by satisfying Laplace equations.
(2) The maximum value of a harmonic function on a closed set $\mathcal{D}_c$ is taken on the boundary of $\mathcal{D}_c$.

Theorem 2.5: *System (2.91) satisfying* **Assumption 2.1** *is asymptotically stable independent of delay if and only if*

(a) $\rho((j\omega I - A)^{-1}A_d) < 1\,, \quad \forall \omega > 0$

and either
(b) $\rho(A^{-1}A_d) < 1$, or
(c) $\rho(A^{-1}A_d) = 1$, $det(A + A_d) \neq 0$

Proof: ($\Rightarrow$) Suppose that **Assumption 2.1** holds and let $B_d(s) := (sI - A)^{-1}A_d$. It follows that $B_d(s)$ is analytic on $\bar{C}_+$ and so is $B_d(s)e^{-ds}$ $\forall d \geq 0$. Since $\rho(B_d(s)e^{-ds})$ is a subharmonic function on $\bar{C}_+$, it follows that

$$\sup_{s \in \bar{C}_+} \rho(B_d(s)e^{-ds}) = \sup_{\omega \geq 0} \rho(B_d(j\omega)e^{-jd\omega})$$

$$= \sup_{\omega \geq 0} \rho(B_d(j\omega))$$

with the maximum occuring on the boundary of $\bar{C}_+$. Thus conditions **(a)** and **(b)** assures that $\rho(B_d(s)e^{-ds}) < 1$ $\forall s \in \bar{C}_+$, $\forall d \geq 0$. In turn this leads to (2.93). On the other hand, conditions **(a)** and **(b)** imply that $\rho(B_d(s)e^{-ds}) < 1$ $\forall s \in \bar{C}_+$, $s \neq 0$, $\forall d \geq 0$ which again leads to (2.93) for all $d \geq 0$ and for all $s \in \bar{C}_+$, $s \neq 0$. Note that condition **(c)** ensures that (2.93) holds at $s = 0$. This means that system (2.91) is asymptotically stable.

($\Leftarrow$) Since the asymptotic stability independent of delay of system (2.91) implies that $\lambda(A) \in C^-$, it suffices to show that conditions **(a)**-**(c)** are necessary. Suppose that $\rho(B_d(j\omega)) = 1$ for some $\omega > 0$. This means that for some $\alpha \in [0, 2\pi]$, $\lambda(B_d(j\omega)) = e^{j\alpha}$. Take $d = \alpha/\omega$. It follows that $det(I - B_d(j\omega)e^{-jd\omega}) = 0$ which violates (2.93) at both $s = j\omega, d = \alpha/\omega$. This means that system (2.91) is not asymptotically stable independent of delay. In the same way, when $\rho(B_d(j\omega)) > 1$ for some $\omega > 1$ or $\rho(B_d(j\omega)) \leq 1$ for $\omega = 0$ it is easy to see that for some $\omega_o \in (\omega, \infty)$ that (2.93) is violated. Hence, the necessity of conditions **(a)**-**(c)** follows.

It is readily seen from the foregoing analysis that an alternative sufficient stability condition would be

$$\rho((j\omega I - A)^{-1}A_d) < 1, \quad \forall \omega \geq 0 \tag{2.98}$$

for which there exists efficient computational methods [3,4].

2.4.1   Example 2.9

Consider system (2.91) with

$$A = \begin{bmatrix} 0 & 1 & 0 & 0 \\ 0 & 0 & 1 & 0 \\ 0 & 0 & 0 & 0 \\ -2 & -3 & -5 & -2 \end{bmatrix}, \quad A_d = \begin{bmatrix} -0.05 & 0.005 & 0.25 & 0 \\ 0.005 & 0.005 & 0 & 0 \\ 0 & 0 & 0 & 0 \\ -1 & 0 & -0.5 & 0 \end{bmatrix}$$

A simple computation shows that A is asymptotically stable since $\lambda(A) = \{-0.6887 \pm j1.7636, 0.3113 \pm j0.6790\}$. However, it can easily verified that the system is not asymptotically stable independent of delay since $\mu(A) + \|A_d\| = 5.8290$ which contradicts **Lamma 2.4** and $\rho(A^{-1}A_d) = 1.2453$ which violates condition **(b)** of **Theorem 2.5**.

Next, we direct attention to the uncertain system (2.89). It has the characteristic polynomial

$$\mathcal{F}_\Delta(s) = sI - (A + \Delta A) - (A_d + \Delta A_d) e^{-ds}$$

Obviously s is a solution of $\mathcal{F}_\Delta(s) = 0$ if and only if $s = \lambda_j((A + \Delta A) + (A_d + \Delta A_d)e^{-ds})$. In this regard, we say that system (2.89) is *robustly asymptotically stable* if we consider the worst case in which

$$\delta_3 := \mu(A) + \|A_d\| + \beta + \beta_d < 0 \tag{2.99}$$

Again, (2.99) gives a delay-independent condition such that the stability is assured for any value of τ. Let us consider the worst case in which $\delta_3 > 0$. Extending on **Lemma 2.4** and **Lemma 2.5** we obtain the following results:

Lemma 2.6: *Consider system* (Σ_{lc}) *subject to* **Assumption 2.5** *such that* $\delta_3 > 0$. *Let* $\delta_4 := \mu(-jA) + \|A_d\| + \beta + \beta_d$. *If*

$$\det\left\{sI - (A + \Delta A) - (A_d + \Delta A_d)e^{-ds}\right\} \neq 0, \quad \forall s \in \bar{C}_+, \quad \forall d \geq 0 \tag{2.100}$$

for all admissible uncertainties satisfying (2.90), then system (2.89) is robustly asymptotically stable.

Proof: We start by the characteristic equation $s = \lambda_j((A + \Delta A) - (A_d + \Delta A_d)e^{-ds})$ and assume that it has a solution $s = \alpha + j\omega$ such that

$Re\ s = \alpha \geq 0$. We have

$$
\begin{aligned}
\omega &= Im\ \lambda_j((A + \Delta A) + (A_d + \Delta A_d)e^{-ds}) \\
&\leq \mu(-j(A + \Delta A)) + \mu(-j(A_d + \Delta A_d)e^{-ds}) \\
&\leq \mu(-jA) + \mu(-j\Delta A) + \| -j(A_d + \Delta A_d)e^{-d\alpha} \| \\
&\leq \mu(-jA) + \beta + \|A_d\| + \|\Delta A_d\|e^{-d\alpha} \\
&\leq \mu(A) + \beta + \|\Delta A_d\| + \beta_d \\
&:= \delta_4 > 0 \hspace{4cm} (2.101)
\end{aligned}
$$

$$
\begin{aligned}
\alpha &= Re\ \lambda_j((A + \Delta A) + (A_d + \Delta A_d)e^{-ds}) \\
&\leq \mu((A + \Delta A) + (A_d + \Delta A_d)e^{-ds}) \\
&\leq \mu((A + \Delta A)) + \mu((A_d + \Delta A_d)e^{-ds}) \\
&\leq \mu(A) + \mu(\Delta A) + \|(A_d + \Delta A_d)e^{-ds}\| \\
&\leq \mu(A) + \beta + \beta_d + \|A_d\| \\
&:= \delta_3 > 0 \hspace{4cm} (2.102)
\end{aligned}
$$

which is a contradiction of the initial claim and thus proves the lemma.

Lemma 2.7: *Consider system* (Σ_{lc}) *subject to* **Assumption 2.5** *such that* $\delta_3 > 0$. *Then the following condition*

$$\mu(A) + \beta + \beta_d + \mu(A_d e^{-ds}) < 0 \hspace{2cm} (2.103)$$

assures the robust stability of system (Σ_{lc}) *in the region* S *given by:*

$$
\begin{aligned}
s &= j\omega, \quad 0 \leq \omega \leq \delta_4, \\
s &= \delta_1 + j\omega, \quad 0 \leq \omega \leq \delta_4 \\
s &= r + j\delta_4, \quad 0 \leq \omega \leq \delta_1
\end{aligned}
$$

Proof: Since

$$
\begin{aligned}
Re\ \lambda_j((A + \Delta A) + (A_d + \Delta A_d)e^{-ds}) &\leq \\
\mu((A + \Delta A)) + \mu((A_d + \Delta A_d)e^{-ds}) &\leq \\
\mu(A) + \mu(\Delta A) + \mu(A_d e^{-ds}) + \mu((\Delta A_d)e^{-ds}) &\leq \\
\mu(A) + \beta + \beta_d + \mu(A_d e^{-ds}) &< 0 \hspace{1cm} (2.104)
\end{aligned}
$$

and from the properties of harmonic functions, it follows that the maximum value of $\lambda_j((A + \Delta A) + (A_d + \Delta A_d)e^{-ds})$ is located along the sides of the region S. By (2.104) and the fact that $Re\ s \geq 0$, it follows that $\lambda_j((A + \Delta A) + (A_d + \Delta A_d)e^{-ds})$ has no roots in S and in view of **Lemma 2.6** the proof is completed.

For sake of completeness, we develop some results for a class of discrete time-lag systems of the form:

$$(\Sigma_{ld}): \quad x(k+1) \;=\; (A + \Delta A)x(k) + (A_d + \Delta A_d)x(k-d) \quad (2.105)$$

$$\|\Delta A\| \;<\; \zeta \;,\; \|\Delta A_d\| \;<\; \zeta_d \qquad\qquad (2.106)$$

System (Σ_{ld}) has the pulse-transfer function:

$$\mathcal{F}_{d\Delta}(z) \;=\; zI \;-\; (A + \Delta A) \;-\; (A_d + \Delta A_d)z^{-d} \qquad (2.107)$$

The nominal discrete time-lag system is described by:

$$(\Sigma_{ldo}): \quad x(k+1) = Ax(k) + A_d x(k-d) \qquad (2.108)$$

and its nominal pulse-transfer function is given by:

$$\mathcal{F}_{d\Delta o}(z) \;=\; zI \;-\; A \;-\; A_d z^{-d} \qquad (2.109)$$

Unlike the continuous counterpart $\mathcal{F}_\Delta(s)$, $\mathcal{F}_{d\Delta}(z)$ and hence $\mathcal{F}_{d\Delta o}(z)$ represent finite polynomials for a given $d > 0$. System (Σ_{ldo}) is said to be *asymptotically stable* if and only if

$$|det\,(zI \;-\; A \;-\; A_d z^{-d})| \neq 0 \;,\quad \forall |z| \geq 1 \qquad (2.110)$$

Extending on this, we establish the following result:

Lemma 2.8: *System (Σ_{ld}) is robustly asymptotically stable for all admissible uncertainties satisfying (2.108) if*

$$|\mu(A)| \;+\; |\mu(A_d)| \;+\; \zeta + \zeta_d \;<\; 1 \qquad (2.111)$$

Proof: Assume that there exists a solution of $\mathcal{F}_{d\Delta}(z) = 0$ such that $|z| \geq 1$. We have

$$
\begin{aligned}
|z| \;&=\; |\lambda_j(A + \Delta A) + (A_d + \Delta A_d)z^{-d}| \\
&\leq\; |\lambda_j(A + \Delta A)| + |\lambda_j((A_d + \Delta A_d)z^{-d})| \\
&\leq\; |\mu(A + \Delta A)| + |\mu((A_d + \Delta A_d)z^{-d})| \\
&\leq\; |\mu(A)| + |\mu(\Delta A)| + \|(A_d + \Delta A_d)z^{-d}\| \\
&\leq\; |\mu(A)| + |\mu(\Delta A)| + \|A_d z^{-d}\| + \|\Delta A_d z^{-d}\| \\
&\leq\; |\mu(A)| + |\mu(\Delta A)| + \zeta + \zeta_d \;<\; 1 \qquad (2.112)
\end{aligned}
$$

which contradicts the claim that $|z| \geq 1$ and hence the lemma is proved.

By setting $\Delta A = 0, \Delta A_d = 0$, we specialize **Lemma 2.8** to:

Corollary 2.6: *System* (Σ_{ldo}) *is robustly asymptotically stable if*

$$|\mu(A)| + |\mu(A_d)| < 1 \tag{2.113}$$

Remark 2.12: An alternative result is given in [38].

2.5 Stability of Linear Neutral Systems

In this section, we direct attention to the stability of neutral functional differential equations (NFDE) with focus on linear systems. With reference to section 1.1, we consider the following class of linear NFDE with parametric uncertainties:

$$
\begin{aligned}
(\Sigma_{\Delta n}): \quad \dot{x}(t) - D\dot{x}(t - \tau) &= (A + \Delta A)x(t) + (A_d + \Delta A_d)x(t - \tau) \\
&= A_\Delta x(t) + A_{d\Delta} x(t - \tau) \tag{2.114} \\
x(t_o + \eta) &= \phi(\eta), \quad \forall \eta \in [-\tau, 0] \tag{2.115}
\end{aligned}
$$

where $x \in \Re^n$ is the state, $A \in \Re^{n \times n}$ and $A_d \in \Re^{n \times n}$ are known real constant matrices, $\tau > 0$ is a constant delay factor, $(t, \phi) \in \Re_+ \times \mathcal{C}_{n,-\tau}$ and $\Delta A \in \Re^{n \times n}$ and $\Delta A_d \in \Re^{n \times n}$ are matrices of uncertain parameters represented by the norm-bounded structure (2.45):

For system $(\Sigma_{\Delta n})$, we assume that:

Assumption 2.6: $|\lambda(D)| < 1$

Note that model (2.114) is continuous-time and **Assumption 2.6** gives a condition in the discrete-time sense. Now we establish the following stability result:

Theorem 2.6: *Subject to* **Assumption 2.1** *and* **Assumption 2.6**, *the linear neutral system* $(\Sigma_{\Delta n})$ *is robustly asymptotically stable independent of delay if the following conditions hold:*

(1) *There exist matrices* $0 < P = P^t \in \Re^{n \times n}$, $0 < S = S^t \in \Re^{n \times n}$ *and* $0 < R = R^t \in \Re^{n \times n}$ *and scalars* $\epsilon > 0, \rho > 0$ *satisfying the ARI*

$$PA + A^t P + (\epsilon + \rho)PHH^t P + \rho^{-1} EE^t +$$

$$[P(AD + A_d) + SD][R - \epsilon^{-1}(D^t E^t ED + E_d^t E_d)]^{-1}$$
$$[P(AD + A_d) + SD]^t + S \; < \; 0 \qquad (2.116)$$

(2) *There exist matrices* $0 < S = S^t \in \Re^{n \times n}$ *and* $0 < R = R^t \in \Re^{n \times n}$
satisfying the Lyapunov equation (LE)

$$D^t S D - S + R = 0 \qquad (2.117)$$

Proof: Introduce a Lyapunov-Krasovskii functional $V_7(x_t)$ of the form:

$$
\begin{aligned}
V_7(x_t) \;\; = \;\; & [x(t) - Dx(t - \tau)]^t P[x(t) - Dx(t - \tau)] \\
& + \int_{-\tau}^{0} x^t(t + \theta) S x(t + \theta) \; d\theta \qquad (2.118)
\end{aligned}
$$

Observe that $V_7(x_t)$ satisfy

$$\lambda_m(P) r^2 \; \leq \; V_7(r) \; \leq \; [\lambda_M(P) + \tau \lambda_M(S)] r^2 \qquad (2.119)$$

By differentiating $V_7(x_t)$ along the solutions of (2.114) and arranging terms, we get:

$$
\begin{aligned}
\dot{V_7}(x_t) \;\; = \;\; & [A_\Delta x(t) + A_{d\Delta} x(t - \tau)]^t P[x(t) - Dx(t - \tau)] \\
& + [x(t) - Dx(t - \tau)]^t P[A_\Delta x(t) + A_{d\Delta} x(t - \tau)] \\
& + x^t S x(t) - x(t - \tau) S x(t - \tau) \qquad (2.120)
\end{aligned}
$$

Algebraic manipulation of (2.120) using the difference operator $\mathcal{M}(x_t) := x(t) - Dx(t - \tau)$ and arranging terms, yields:

$$
\begin{aligned}
\dot{V_7}(x_t) \;\; = \;\; & \mathcal{M}^t(x_t)[PA_\Delta + A_\Delta^t P + S]\mathcal{M}(x_t) \\
& + \mathcal{M}^t(x_t)[PAD + SD + PA_{d\Delta}]x(t - \tau) \\
& + x^t(t - \tau)[D^t A^t P + D^t S + A_{d\Delta}^t P]\mathcal{M}(x_t) \\
& + x^t(t - \tau)[D^t S D - S]x(t - \tau) \qquad (2.121)
\end{aligned}
$$

In view of (2.117) and completing the squares, we get:

$$
\begin{aligned}
\dot{V_7}(x_t) \;\; = \;\; & \mathcal{M}^t(x_t)[PA_\Delta + A_\Delta^t P + S \\
& + (PA_\Delta D + SD + PA_{d\Delta})R^{-1}(PA_\Delta D + SD + PA_{d\Delta})^t]\mathcal{M}^t(x_t) \\
& - [(D^t A_\Delta^t P + A_{d\Delta}^t P + DS)\mathcal{M}(x_t) - Rx(t - \tau)]^t R^{-1} \\
& \quad [(D^t A_\Delta^t P + A_{d\Delta}^t P + DS)\mathcal{M}(x_t) - Rx(t - \tau)] \\
\leq \;\; & \mathcal{M}^t(x_t)[PA_\Delta + A_\Delta^t P + S \\
& + (PA_\Delta D + SD + PA_{d\Delta})R^{-1}(PA_\Delta D + SD + PA_{d\Delta})^t]\mathcal{M}^t(x_t) \\
& \qquad\qquad\qquad\qquad\qquad\qquad\qquad\qquad\qquad (2.122)
\end{aligned}
$$

By Lyapunov-Krasovskii theorem (see Appendix D), it is sufficient from (2.119) and (2.122) to conclude that system $(\Sigma_{\Delta n})$ is robustly asymptotically stable if:

$$PA_\Delta + A_\Delta^t P + S +$$
$$(PA_\Delta D + SD + PA_{d\Delta})R^{-1}(PA_\Delta D + SD + PA_{d\Delta})^t < 0$$

$$(2.123)$$

for all admissible uncertainties satisfying (2.45). Using **B.1.2, B.1.3** and (2.45), we get for some scalars $\epsilon > 0, \rho > 0$:

$$PA_\Delta + A_\Delta^t P \leq PA + A^t P + \rho PHH^t P + \rho^{-1} E^t E \qquad (2.124)$$

$$(PA_\Delta D + SD + PA_{d\Delta})R^{-1}(PA_\Delta D + SD + PA_{d\Delta})^t \leq \epsilon PHH^t P +$$
$$[P(AD + A_d) + SD][R - \epsilon^{-1}(D^t E^t ED + E_d^t E_d)]^{-1}[P(AD + A_d) + SD]^t$$

$$(2.125)$$

Substituting (2.124)-(2.125) into (2.122) yields ARI (2.116) such that S and R satisfy (2.117). This completes the proof.

By converting **Theorem 2.6** into an LMI feasibility problem using the Schur complements, we obtain the following result:

Corollary 2.7: *Subject to* **Assumption 2.1** *and* **Assumption 2.6,** *the linear neutral system* $(\Sigma_{\Delta n})$ *is asymptotically stable independent of delay if there exist matrices* $0 < P = P^t \in \Re^{n \times n}$, $0 < S = S^t \in \Re^{n \times n}$ *and scalars* $\epsilon > 0, \rho > 0$ *satisfying the LMIs*

$$\begin{bmatrix} PA + A^t P + \rho^{-1}EE^t & PH & P(AD + A_d) \\ +\epsilon PHH^t P + S & & +SD \\ H^t P & -\rho^{-1}I & 0 \\ (D^t A^t + A_d^t)P & 0 & -J \\ +D^t S & & \end{bmatrix} < 0$$

$$D^t S D - S < 0$$

$$\begin{bmatrix} D^t SD - S & D^t E^t & E_d^t \\ ED & -\epsilon I & 0 \\ E_d & 0 & -\epsilon I \end{bmatrix} < 0$$

$$(2.126)$$

where

$$J = D^t SD - S + \epsilon^{-1}(D^t E^t ED + E_d^t E_d)$$

On the other hand, by suppressing the uncertainties $E = 0, H = 0, E_d = 0$ in system (2.114), we get the system:

$$(\Sigma_{\Delta no}): \quad \dot{x}(t) - D\dot{x}(t - \tau) = Ax(t) + Ax(t - \tau) \qquad (2.127)$$
$$x(t_o + \eta) = \phi(\eta), \quad \forall \eta \in [-\tau, 0] \qquad (2.128)$$

for which the following hold:

Corollary 2.8: *Subject to* **Assumption 2.1** *and* **Assumption 2.6,** *the linear neutral system* $(\Sigma_{\Delta no})$ *is asymptotically stable independent of delay if the following conditions hold:*
(1) *There exist matrices* $0 < P = P^t \in \Re^{n \times n}$, $0 < S = S^t \in \Re^{n \times n}$ *and* $0 < R = R^t \in \Re^{n \times n}$ *satisfying the ARI*

$$PA + A^t P +$$
$$[P(AD + A_d) + SD]R^{-1}[P(AD + A_d) + SD]^t + S < 0 \quad (2.129)$$

(2) *There exist matrices* $0 < S = S^t \in \Re^{n \times n}$ *and* $0 < R = R^t \in \Re^{n \times n}$ *satisfying the Lyapunov equation (LE)*

$$D^t S D - S + R = 0 \qquad (2.130)$$

Corollary 2.9: *Subject to* **Assumption 2.1** *and* **Assumption 2.6,** *the linear neutral system* $(\Sigma_{\Delta no})$ *is asymptotically stable independent of delay if there exist matrices* $0 < P = P^t \in \Re^{n \times n}$ *and* $0 < S = S^t \in \Re^{n \times n}$ *satisfying the LMIs*

$$\begin{bmatrix} PA + A^t P + S & P(AD + A_d) + SD \\ (D^t A^t + A_d^t)P + D^t S & D^t SD - S \end{bmatrix} < 0$$
$$D^t S D - S < 0 \qquad (2.131)$$

Remark 2.13: It is important to mention that **Corollary 2.7** and **Corollary 2.8** recover the results of [122]. Interestingly enough, by deleting the matrix D in the linear neutral system (2.114) we recover the linear retarded system (2.43) and hence the results of **Theorem 2.5** and **Corollary 2.7** reduce to those of **Lemma 2.1**.

2.6 Stability of Multiple-Delay Systems

In this section, we extend some of the stability results of sections 2.1 and 2.2 to the class of uncertain time-delay systems containing several delay factors. For simplicity in exposition, we divide our efforts into multiple-delay continuous-time systems (MDCTS) and multiple-delay discrete-time systems (MDDTS). The class of MDCTS of interest is modeled by:

$$
\begin{aligned}
(\Sigma_{\Delta mc}): \quad \dot{x}(t) \;&=\; (A + \Delta A(t))x(t) \;+\; (A_{d1} + \Delta A_{d1}(t))x(t - \tau_1) \\
&\quad +\; \ldots + (A_{ds} + \Delta A_{ds}(t))x(t - \tau_s) \\
&=\; A_\Delta(t)x(t) + \sum_{j=1}^{s} A_{d\Delta j}x(t - \tau_j) \qquad (2.132)
\end{aligned}
$$

where $x \in \Re^n$ is the state, $A \in \Re^{n \times n}$, $A_{dj} \in \Re^{n \times n}; j = 1, .., s$ are real constant matrices and $\tau_j; j = 1, .., s$ are unknown time-varying delay factors satisfying

$$
0 \le \tau_j(t) \le \tau_j^o , \quad 0 \le \dot{\tau}_j(t) \le \tau_j^+ \le 1 \quad \forall j \in [1, s] \qquad (2.133)
$$

where $\tau_j^o, \tau_j^+, j = 1, .., s$ are known bounds. The uncertain matrices $\Delta A(t) \in \Re^{n \times n}$ and $\Delta A_{dj}(t) \in \Re^{n \times n}$ are represented by:

$$
\Delta A(t) = H\Delta(t)E , \quad \Delta A_{dj}(t) = H\Delta_j(t)E_{dj} , \quad \forall j \in [1, s] \qquad (2.134)
$$

where $E \in \Re^{\beta \times n}$, $E_{dj} \in \Re^{\beta \times n}$ and $H \in \Re^{n \times \alpha}$ are known constant matrices and $\Delta(t) \in \Re^{\alpha \times \beta}$ is an unknown matrix of uncertain parameters satisfying

$$
\Delta^t(t) \, \Delta(t) \le I , \quad \Delta_j^t(t) \, \Delta_j(t) \le I , \quad \forall t \; \forall j \in [1, s] \qquad (2.135)
$$

The corresponding MDDTS under consideration has the form:

$$
\begin{aligned}
(\Sigma_{\Delta md}): \quad x(k+1) \;&=\; A_\Delta x(k) + A_{d\Delta 1}x(k - \tau_1) + \ldots + A_{d\Delta s}x(k - \tau_s) \\
&=\; A_\Delta x(k) + \sum_{j=1}^{s} A_{d\Delta j}x(k - \tau_j) \qquad (2.136)
\end{aligned}
$$

where $x \in \Re^n$ is the state and τ_j are unknown integers representing the number of delay units in the state. The uncertain matrices $A_\Delta \in \Re^{n \times n}$ and $A_{d\Delta j} \in \Re^{p \times n}$ are given by:

$$
\begin{aligned}
A_\Delta \;&=\; A + H \, \Delta(k) \, E \\
A_{d\Delta j} \;&=\; A_{dj} + H \, \Delta_j(k) \, E_{dj} , \quad \forall j \in [1, s] \qquad (2.137)
\end{aligned}
$$

where $A \in \Re^{n \times n}$, $A_{dj} \in \Re^{n \times n}$, $H \in \Re^{n \times \alpha}$, $E \in \Re^{\beta \times n}$ and $E_{dj} \in \Re^{\beta \times n}$ are known real constant matrices and $\Delta(k) \in \Re^{\alpha \times \beta}$ is an unknown matrix satisfying

$$\Delta^t(k)\, \Delta(k) \leq I, \qquad \Delta_j^t(k)\, \Delta_j(k) \leq I, \qquad \forall k, \quad \forall j \in [1, s] \qquad (2.138)$$

Due to the strong similarity between models (2.44) and (2.132) in the continuous case and between models (2.74) and (2.136) in the discrete case, we are not going to repeat the stability analysis hereafter. Rather, we will present the main results in a sequence of corollaries without proof.

To study delay-independent stability of system (2.132), we consider the Lyapunov-Krasovskii functional $V_8(x_t)$:

$$V_8(x_t) = x^t(t)Px(t) + \sum_{j=1}^{s} \int_{t-\tau_j}^{t} x^t(\theta)Q_j x(\theta)\ d\theta \qquad (2.139)$$

where $0 < P = P^t \in \Re^{n \times n}$ and $0 < Q_j = Q_j^t \in \Re^{n \times n}; j = 1, ..., s$ are constant weighting matrices. It is easy to see that $V_8(x_t)$ is a natural generalization of $V_1(x_t)$.

Corollary 2.10: *Subject to* Assumption 2.1, *the time delay system* $(\Sigma_{\Delta c})$ *is robustly asymptotically stable independent of delay if there exist matrices* $0 < P = P^t \in \Re^{n \times n}$ *and* $0 < Q_j = Q_j^t \in \Re^{n \times n}; j = 1, ..., s$ *with* $Q_{\tau_j} = (1 - \tau_j^+)Q_j$ *and scalars* $\epsilon > 0, \nu_j > 0$ *such that* $Q_{\tau_j} - \nu_j^{-1}E_{dj}^t E_{dj} > 0; j = 1, ..., s$ *satisfying the ARI*

$$PA + A^tP + \sum_{j=1}^{s}Q_j + \epsilon^{-1}E^tE +$$

$$P\sum_{j=1}^{s}[(\epsilon + \nu_j)HH^t + A_{dj}(Q_{\tau_j} - \nu_j^{-1}E_{dj}^t E_{dj})^{-1}A_{dj}^t]P < 0$$

$$(2.140)$$

On the other hand, when studying delay-dependent stability we use the Lyapunov-Krasovskii functional $V_9(x_t)$ of the form:

$$V_9(x_t) = x^t(t)Px(t) + \sum_{j=1}^{s}W_j(x_t, t) \qquad (2.141)$$

$$W_j(x_t, t) = \int_- \tau_j \epsilon_{3j}(1 + \epsilon_{4j}^{-1}) \int_{t+\theta}^t \|A_\Delta(s)x(s)\|^2 ds d\theta$$

$$+ \int_- \tau_j \epsilon_{3j}(1 + \epsilon_{4j}) \int_{t+\theta}^t \|A_{d\Delta j}(s + \tau_j)x(s)\|^2 ds d\theta \quad (2.142)$$

where $0 < P = P^t \in \Re^{n \times n}$ and $\epsilon_{3j} > 0, \epsilon_{4j} > 0$ are scalars to be chosen. Again, (2.141)-(2.142) is a generalization of (2.59)-(2.60).

Corollary 2.11: *Consider the time-delay system* $(\Sigma_{\Delta c})$ *satisfying Assumption 2.2. Then given a scalar* $\tau^* > 0$, *the system* (Σ_{dc}) *is robustly asymptotically stable for any constant time-delay* $\tau_j; j = 1, .., s$ *satisfying* $0 \le \tau_j \le \tau_j^*$ *if there exist matrix* $0 < X = X^t \in \Re^{n \times n}$ *and scalars* $\epsilon_1 > 0, \epsilon_{3j} > 0, ..., \epsilon_{7j} > 0; j = 1, .., s$ *such that* $(I - \epsilon_{5j}HH^t) > 0$, $(I - \epsilon_{6j}HH^t) > 0$ *and* $(I - \epsilon_{7j}E_{dj}^t E_{dj}) > 0$ *satisfying the ARI*

$$\Xi(P, \epsilon, \tau_j^*) = P(A + \sum_{j=1}^s A_{dj}) + (A + \sum_{j=1}^s A_{dj})^t P + \epsilon_1 PHH^t P$$

$$+ \epsilon_1^{-1}(E + \sum_{j=1}^s E_{dj})^t(E + \sum_{j=1}^s E_{dj})$$

$$+ \tau_j^* \epsilon_3(1 + \epsilon_4^{-1})[\epsilon_5^{-1}E^t E + A^t(I - \epsilon_5 HH^t)^{-1}A]$$

$$+ \sum_{j=1}^s \left\{\tau_j^* \epsilon_{3j}(1 + \epsilon_{4j})[\epsilon_{6j}^{-1}E_{dj}^t E_{dj} + A_{dj}^t(I - \epsilon_{6j}HH^t)^{-1}A_{dj}]\right\}$$

$$+ \sum_{j=1}^s \left\{\tau_j^* \epsilon_{3j}^{-1} P[\epsilon_{7j}^{-1}HH^t + A_{dj}(I - \epsilon_{7j}E_{dj}^t E_{dj})^{-1}A_{dj}^t]P\right\}$$

$$< \quad 0 \quad (2.143)$$

In the discrete-time case, we use a discrete-type Lyapunov-Krasovskii functional $V_{10}(x_k)$ of the form:

$$V_{10}(x_k) = x^t(k)Px(k) + \sum_{j=1}^s \sum_{\theta=k-\tau_j}^{k-1} x^t(\theta)Q_j x(\theta) \quad (2.144)$$

where $0 < P = P^t \in \Re^{n \times n}$ and $0 < Q_j = Q_j^t \in \Re^{n \times n}; j = 1, ..., s$ are weighting matrices. It is needless to stress that (2.144) is just an extension of (2.79).

Corollary 2.12: *Subject to* **Assumption 2.3**, *the time delay system* $(\Sigma_{\Delta d})$ *is robustly asymptotically stable independent-of-delay if one of the following two equivalent conditions holds:*

(1) *There exist matrices* $0 < P = P^t \in \Re^{n \times n}$,
$0 < Q_j = Q_j^t \in \Re^{n \times n}; j = 1, ..., s$ *and scalars* $\epsilon_1 > 0, \epsilon_{2j} > 0; j = 1, ..., s$ *such that* $Q_j - \epsilon_{2j}^{-1} E_{dj}^t E_{dj} > 0; j = 1, ..., s$ *satisfying the ARI*

$$A^t \left\{ P^{-1} - \epsilon_1 H H^t - \sum_{j=1}^{s} [\epsilon_{2j} H H^t + A_{dj}(Q_j - \epsilon_2^{-1} E_{dj}^t E_{dj})^{-1} A_{dj}^t]^{-1} \right\} A$$
$$-P + \epsilon_1^{-1} E^t E + Q < 0 \qquad (2.145)$$

(2) *There exist matrices* $0 < P = P^t \in \Re^{n \times n}$,
$0 < Q_j = Q_j^t \in \Re^{n \times n}; j = 1, ..., s$ *and scalars* $\epsilon_1 > 0, \epsilon_{2j} > 0; j = 1, ..., s$ *such that*

$$\Lambda_j = [(\epsilon_1 + \sum_{j=1}^{s} \epsilon_{2j})^{1/2} H \quad \sum_{j=1}^{s} A_{dj}(Q_j - \epsilon_{2j}^{-1} E_{dj}^t E_{dj})^{-1/2}]$$

satisfying the LMIs

$$j=1,....,s$$

$$\begin{bmatrix} A^t P A - P + Q & A^t P \Lambda_j & E^t \\ \Lambda_j^t P A & -(I + \Lambda_j \Lambda_j^t) & 0 \\ E & 0 & -\epsilon_1 I \end{bmatrix} < 0$$
$$\epsilon_{2j}^{-1} E_{dj}^t E_{dj} - Q_j < 0 \qquad (2.146)$$

Remark 2.14: Despite the fact that **corollaries (2.9)-(2.11)** are basically stated for UTDS, it is a straightforward task to retrieve the results of TDS in section 2.1 by simply setting $H = 0, E = 0, E_{d1} = 0, ..., E_{ds} = 0$.

2.7   Stability Using Lyapunov-Razumikhin Theorem

As we mentioned at the beginning of this chapter, we can use the *Lyapunov-Razumikhin Theorem* as the main stability tool in the function space. Had we followed this route, we would have started with a Lyapunov-Razumikhin function of the form:

$$V_{11}(x(t)) = x^t(t) P x(t) , \qquad 0 < P = P^t \qquad (2.147)$$

where $V_{11}(x(t))$ possesses the properties stated in Appendix D. First a delay-independent stability result is summarized below:

Lemma 2.9: *Subject to* **Assumption 2.1**, *the time delay system (2.44)-(2.46) is robustly asymptotically stable independent of delay if one of the following two equivalent conditions holds:*

(1) *There exist a matrix* $0 < P = P^t \in \Re^{n \times n}$ *and scalars* $\nu > 0, \epsilon > 0, \alpha > 0$ *such that* $\alpha P - \nu^{-1} E_d^t E_d > 0$ *satisfying the ARI*

$$PA + A^t P + (\epsilon + \nu) PHH^t P + \epsilon^{-1} E^t E$$
$$PA_d(\alpha P - \nu^{-1} E_d^t E_d)^{-1} A_d^t P + \alpha P < 0 \qquad (2.148)$$

(2) $0 < P = P^t \in \Re^{n \times n}$ *and scalars* $\nu > 0, \epsilon > 0, \alpha > 0$ *satisfying the LMIs*

$$\begin{bmatrix} PA + A^t P + \alpha P + \epsilon^{-1} E^t E & PH & PH & PA_d \\ H^t P & -\epsilon^{-1} I & 0 & 0 \\ H^t P & 0 & -\nu^{-1} & 0 \\ A_d^t P & 0 & 0 & -(\alpha P - \nu^{-1} E_d^t E_d) \end{bmatrix} < 0$$

$$\nu^{-1} E_d^t E_d - \alpha P < 0 \qquad (2.149)$$

Proof: (1) By evaluating $\dot{V}_{11}$ along the solutions of (2.44) and applying the Lyapunov-Razumikhin theory (see Appendix D) for some $\alpha > 0$, we get:

$$\begin{aligned} \dot{V}_{11} &= x^t(t)[PA_\Delta + A_\Delta^t P]x(t) + x^t(t-\tau)A_{d\Delta}^t Px(t) + x^t(t)PA_{d\Delta}x(t-\tau) \\ &= x^t(t)[PA_\Delta + A_\Delta^t P]x(t) + x^t(t-\tau)A_{d\Delta}^t Px(t) + x^t(t)PA_{d\Delta}x(t-\tau) \\ &+ x^t(t-\tau)Px(t-\tau) - x^t(t-\tau)Px(t-\tau) \\ &\leq x^t(t)[PA_\Delta + A_\Delta^t P + \alpha P]x(t) + x^t(t-\tau)A_{d\Delta}^t Px(t) \\ &+ x^t(t)PA_{d\Delta}x(t-\tau) - x^t(t-\tau)Px(t-\tau) \end{aligned} \qquad (2.150)$$

Inequality (2.150) can be expressed in compact form as:

$$\begin{aligned} \dot{V}_{11} &= X^t(t)\Psi(P)X(t) \\ X(t) &= [x^t(t) \quad x^t(t-\tau)]^t \\ \Psi(P) &= \begin{bmatrix} PA_\Delta + A_\Delta^t P + \alpha P & PA_{d\Delta} \\ A_{d\Delta}^t P & -\alpha P \end{bmatrix} \end{aligned} \qquad (2.151)$$

Stability of system (2.44) follows if $\Psi(P) < 0$. Using the Schur complements, the latter condition corresponds to:

$$PA_\Delta + A_\Delta^t P + PA_{d\Delta}(\alpha P)^{-1} A_{d\Delta}^t P + \alpha P < 0 \qquad (2.152)$$

In line with **Lemma 2.1** and using inequalities (I.2) and (I.3) in (2.152), we get for some $\nu > 0, \epsilon > 0$:

$$PA + A^t P + (\epsilon + \nu)PHH^t P + \epsilon^{-1} E^t E$$
$$PA_d(\alpha P - \nu^{-1} E_d^t E_d)^{-1} A_d^t P + \alpha P < 0 \qquad (2.153)$$

which corresponds to (2.148).

(2) Follows from application of the Schur complements.

2.7.1 Example 2.10

In order to demonstrate **Lemma 2.9**, we consider the following data:

$$A = \begin{bmatrix} -5 & 0 \\ 0 & -2 \end{bmatrix}, \; A_d = \begin{bmatrix} -1 & 0 \\ -1 & -1 \end{bmatrix}, \; H = \begin{bmatrix} 1 \\ 0.5 \end{bmatrix}$$
$$E = [0.1 \; 0.2], \; E_d = [0.2 \; 0.1]$$

By properly manipulating the variables to guarantee the convexity requirement of the LMI solution, it is readily seen that we should deal with α, ν^{-1}, ϵ^{-1} as the unknown quantities in addition to P. The result of computer simulation using **LMI Control** software is then given by:

$$P = \begin{bmatrix} 8.2585 & 0.6234 \\ 0.6234 & 2.8243 \end{bmatrix}, \; \epsilon = 18.2805, \; \nu = 18.3221, \; \alpha = 0.6002$$

Corollary 2.13: *Subject to* **Assumption 2.1,** *the time delay system (2.1) is robustly asymptotically stable independent of delay if one of the following two equivalent conditions holds:*
(1) *There exist a matrix* $0 < P = P^t \in \Re^{n \times n}$ *and a scalar* $\alpha > 0$ *satisfying the ARI*

$$PA + A^t P + PA_d(\alpha P)^{-1} A_d^t P + \alpha P < 0 \qquad (2.154)$$

(2) *There exist matrices* $0 < P = P^t \in \Re^{n \times n}$, $0 < R = R^t \in \Re^{n \times n}$ *and a scalar* $\alpha > 0$ *satisfying the LMIs*

$$\begin{bmatrix} PA + A^t P + \alpha P & PA_d \\ A_d^t P & -R \end{bmatrix} < 0$$
$$R > 0, \quad R - \alpha P \leq 0 \qquad (2.155)$$

Remark 2.15: It is significant to observe the close similarity between ARIs (2.47) and (2.148). In particular, replacing the weighting matrix Q by

αP in (2.47) automatically yields (2.148). A simple interpretation for this is that the functional $V_1(x_t)$ has a penality term on the delayed state whereas the term αP is brought about to substitute the effect of the delayed state and hence meet the requirement imposed on $\dot{V}_1(x_t)$. Evidently there is an element of compromise here. One theorem allows for a treatment of the delayed state *a priori* and the other theorem deals with the delay *a posteriori*. In general, the application *Lyapunov-Krasovskii theorem* to TDS is quite direct but requires a lot of effort in constructing the functional. Alternatively, the function in *Lyapunov-Razumikhin theorem* is rather standard but some lengthy algebra is usually needed to satisfy its conditions.

Had we adopted the dynamic model (2.15) to represent the time-delay system, we would have achieved the following result:

Corollary 2.14: *Consider the time-delay system* (Σ_{dc}) *satisfying* **Assumption 2.2.** *Then given a scalar* $\tau^* > 0$, *the system* (Σ_{dc}) *is globally asymptotically stable for any constant time-delay* τ *satisfying* $0 \leq \tau \leq \tau^*$ *if one of the following two equivalent conditions holds:*
(1) *There exist matrix* $0 < X = X^t \in \Re^{n \times n}$ *and scalars* $\varepsilon > 0$ *and* $\alpha > 0$ *satisfying the ARI*

$$
(A + A_d)X + X(A + A_d)^t + \tau^* \varepsilon^{-1} A_d A X A^t A_d
$$
$$
+ \tau^* \alpha^{-1}(A_d)^2 X(A_d^t)^2 + \tau^*(\varepsilon + \alpha)X < 0 \tag{2.156}
$$

(2) *There exist matrix* $0 < X = X^t \in \Re^{n \times n}$ *and scalars* $\varepsilon > 0$ *and* $\alpha > 0$ *satisfying the LMI:*

$$
\begin{bmatrix}
(A + A_d)X + X(A + A_d)^t + \tau^*(\varepsilon + \alpha)X & \tau^* A_d A X & \tau^*(A_d)^2 X \\
\tau^* X A^t A_d^t & -(\tau^* \varepsilon)X & 0 \\
\tau^* X(A_d^t)^2 & 0 & -(\tau^* \alpha)X
\end{bmatrix} < 0
$$
$$\tag{2.157}$$

Remark 2.16: Extension of **Corollary 2.13** to the uncertain system (2.54)-(2.55) is straightforward and we will leave it as an excercise for the reader.

2.8 Stability Using Comparison Principle

Given a functional differential equation (Σ_{cs}) with known asymptotic behavior, a *comparison principle* provides a characterization of the set of conditions under which its asymptotic behavior implies the similar behavior of another system of the type (Σ_{dc}) or $(\Sigma_{\Delta c})$. In this case, system (Σ_{cs}) is called a *comparison system* for system (Σ_{dc}) or $(\Sigma_{\Delta c})$. Work on development comparison principles has been extensive, see [212,213]. Our purpose here is to shed some light on the application of the foregoing idea by examining the delay-dependent stability of TDS. For simplicity, we consider the delay factor $\tau = \tau^*$ as constant and provide the following results:

Lemma 2.10: *Consider the time-delay system* (Σ_{dc}) *satisfying Assumption 2.2 such that*

$$e^{(A+A_d)t} \leq c_d e^{-\eta_d t}, \; c_d > 1, \quad \eta_d > 0 \tag{2.158}$$

Given a scalar $\tau^* > 0$. *If the following inequality holds:*

$$\tau^* c_d \eta_d^{-1} \{\|A_d\|(\|A\| + \|A_d\|)\} < 1 \tag{2.159}$$

then system (Σ_{dc}) *is asymptotically stable for any constant time-delay* τ *satisfying* $0 \leq \tau \leq \tau^*$ *in the sense that*

$$\|x(t)\| \leq c_d \sup_{s \in [-2\tau^*, 0]} \|x(s)\| e^{-\sigma_d t} \tag{2.160}$$

where σ_d *is the solution of the transcendental equation*

$$1 - \sigma \eta_d^{-1} = c_d \|A_d\| \left\{ (\|A\| + \|A_d\| e^{\sigma \tau^*}) \right\} \frac{e^{\sigma \tau^*} - 1}{\sigma \eta_d} \tag{2.161}$$

Proof: Starting from (2.14), the solution is given by:

$$x(t) = e^{(A+A_d)t} x(0) - A_d \int_0^t \int_{s-\tau^*}^s e^{(A+A_d)(t-s)} A x(t+\theta) d\theta ds$$

$$\quad - A_d \int_0^t \int_{s-\tau^*}^s e^{(A+A_d)(t-s)} A_d x(t-\tau+\theta) d\theta ds \tag{2.162}$$

Using (2.158) and taking the norms of (2.162) with some standard manipulations, we get:

$$\|x(t)\| \leq x_d c_d e^{-\eta_d t} + \mathcal{N}(t, \|x(.)\|) \tag{2.163}$$

where

$$x_d = \sup_{s \in [-2\tau^*, 0]} \|x(s)\|$$

$$\mathcal{N}(t, \|x(.)\|) = \|A_d\| \int_0^t \int_{s-\tau^*}^s c_d e^{-\eta_d(t-s)} \chi(\theta, \tau) d\theta ds$$

$$\chi(\theta, \tau) = \|A\| \|x(\theta)\| + \|A_d\| \|x(\theta - \tau)\| \tag{2.164}$$

Now, let

$$w(t) = c_d x_d \, e^{-\sigma_d t} \,, \qquad t \geq -2\tau^* \tag{2.165}$$

where σ_d is the solution of (2.161). It can be readily verified that $w(t)$ satisfies:

$$w(t) = c_d x_d e^{-\sigma_d t} + ||A_d|| \int_0^t \int_{s-\tau^*}^s c_d e^{-\eta_d(t-s)} \chi(\theta,\tau) d\theta ds \,, \; t \geq -2\tau^* \tag{2.166}$$

From (2.164)-(2.166), it follows that the difference $v(t) = ||x(t)|| - w(t)$, $t \geq -2\tau^*$ satisfies $v(t) \leq x_d - c_d x_d \, e^{-\sigma_d t} < 0$, $t \geq -2\tau^*$. By applying the Bellamn-Gronwall lemma (**B.1.7**) over the successive time intervals $(0, 2\tau^*]$, $(2\tau^*, 4\tau^*]$, ..., and using (2.159), we conclude that $||x(t)|| \leq w(t) \forall t \geq 0$ and the lemma is proved.

Following parallel lines, we establish the following delay-independent result:

Corollary 2.15: *Consider the time-delay system* (Σ_{dc}) *satisfying Assumption 2.1 such that*

$$e^{At} \leq c_o \, e^{-\eta_o t} \,, c_o > 1 \,, \qquad \eta_o > 0 \tag{2.167}$$

If the following inequality holds:

$$c_o \eta_o^{-1} ||A_d|| \;< \; 1 \tag{2.168}$$

then system (Σ_{dc}) *is asymptotically stable independent of delay* τ *in the sense that*

$$||x(t)|| \; \leq \; c_o \sup_{s \in [-\tau^*, 0]} ||x(s)|| \, e^{-\sigma_o t} \tag{2.169}$$

where σ_o *is the solution of the transcendental equation*

$$1 \; - \; \sigma \eta_o^{-1} = c_o ||A_d|| \eta_o^{-1} e^{\sigma \tau^*} \tag{2.170}$$

Remark 2.17: Note that by writing (2.159) in the form

$$\tau^* \; < \; \frac{\eta_d}{c_d ||A_d|| (||A|| + ||A_d||)}$$

it provides a characterization of an interval of validity $[0, \tau^*)$ for the stability result. Note also that by setting $c_o = 1, \eta_o = -\mu(A)$ in (2.168), we recover the results of **Lemma 2.4**.

2.9 Notes and References

The literature on robust stability for TDS and UTDS abound. It is a heavy burden to undertake the task of categorizing an extensive literature on stability of systems with delays, whether they be TDS, UTDS or FDEs. Nevertheless, we provide some relevant notes on the published work. In addition to the edited volume [1] and the book chapter [2], the *Guided Tour* in [5] stands as a comprehensive, valuable source which is indispensable for researchers as it provides an overview of various topics and related references.

In the literature, approaches to delay-independent stability can be found in [6,11,12,16,53,57,70,72,74,77,83,106-108,110,113,119,124,135] for the case of norm-bounded uncertainties and in [7,8,14,15,21,99] for matched and/or mismatched uncertainties.

On the other hand, the references [34,46,78,100,131,137] contain different techniques on delay-dependent stability.

Frequency domain and related results can be found in [38-44,55,67-69,75]. Also in [76,80-82,86,87,90,91,91,98,101-105,125,133] whereas algebraic methods are available in [25,26,30,33,50,54,56,73,79,84,89,95-97,123,126-130] and also in [132,134,139].

The topic of α-stability is dealt with in [18,19,143] and D-stability is considered in [61,141-142]. Some results on discrete systems are presented in [112,115,117] and the topic of exponential stability are studied in [111, 114, 136]. Neutral systems are analyzed in [27,51,85,122] and some practical applications are contained in [88,93,94,116]. Results related to Razumkhin stability are presented in [144,145].

Despite the large number of problems tackled in the literature, there is still an ample amount of technical problems open for scientific research. This includes, but is not limited to, more refined results of delay-dependent stability and stabilization, exponential stability and discrete-time delay systems.

Chapter 3

Robust Stabilization

3.1 Introduction

There exists an extensive literature devoted to the control of delay systems. We can, in principle, distinguish between three broad approaches for designing stabilizing controllers:

Approach (1): In this approach [31,148], the delay is a known constant (frequently called time-lag) and the main design effort is to seek conditions for the existence of stabilizing controllers.

Approach (2): Here the controllers are designed independent of the size of the delay without destabilizing the closed-loop system and the delay may be arbitrarily large [7,8,14,15].

Approach (3): In this approach, Razumikhin-like theorems [4,6] are utilized to develop stabilization conditions in terms of an upper bound on the size of the delay.

In this chapter, we extend most of the robust stability results of Chapter 2 to the case of feedback systems. This is frequently called *robust stabilization* and focuses on the synthesis of feedback gains such that desirable stability behavior of the closed-loop system is preserved. In control engineering systems design, the primary objective is to construct feedback systems with better performance rates [146,147,149]. The H_∞-norm of the closed-loop system has been often considered an important performance index [214-219] and will be taken hereafter as the main design objective. For ease in exposition, we provide the stabilization results along two directions: one for time-delay systems (TDS) and the other for uncertain time-delay systems

(UTDS). Each of the developed results will be supplemented with a computational algorithm based on LMI format. It is interesting to report that the developed control methods might be of potential use in different engineering applications including: (1) modification of governor control action gas-turbine systems, (2) compensation of quality products in chemical reactors due to recycling, (3) enhancement of water pollution control systems, and (4) effective control of thickness in industrial mills.

3.2   Time-Delay Systems

In this section, we focus attention on time-delay systems without uncertainties. Later on we will incorporate parametric uncertainties.

3.2.1   Problem Description

Consider a class of time-delay of the form

$$\dot{x}(t) = Ax(t) + Bu(t) + A_d x(t - d) + B_h u(t - h) + Dw(t) \quad (3.1)$$

$$z(t) = Lx(t), \quad x(t) = \psi(t) \ \forall t \in [-max(d, h), 0] \quad (3.2)$$

where $t \in \Re$ is the time, $x(t) \in \Re^n$ is the state; $u(t) \in \Re^m$ is the control input; $w(t) \in \Re^q$ is the input disturbance which belongs to $\mathcal{L}_2[0, \infty)$; $z(t) \in \Re^p$ is the output and d, h represent the amount of delay in the state and at the input of the system, respectively. The matrices $A \in \Re^{n \times n}$, $B \in \Re^{n \times m}$ represent the nominal system without delay and the pair (A, B) is stabilizable; $A_d \in \Re^{n \times n}$, $B_h \in \Re^{n \times m}$, $D \in \Re^{n \times q}$ and $L \in \Re^{p \times n}$ are known matrices and $\psi(t) \in C[-max(d, h), 0]$ is a continuous vector valued initial function. In the sequel, the delay factors are taken to be different, that is $d \neq h$. Models of dynamical systems of the form (3.1)-(3.2) can be found in several engineering applications including river pollution control [220] and recycled continuous stirred-tank reactors [221]. The problem addressed in this chapter is that of designing a feedback controller of the form:

$$u(t) = \Phi[x(t)] \quad (3.3)$$

so that the closed-loop system is stabilized and the effect of disturbances is reduced to a pre-specified level. One of the popular forms of controller (3.3) is the linear constant-gain state-feedback in which

$$\Phi[x(t)] = Fx(t) \ ; \ F \in \Re^{m \times n} \quad (3.4)$$

Applying the control (3.4) to (3.1)-(3.2) yields the closed-loop system

$$\begin{aligned} \dot{x}(t) &= A_c x(t) + A_d x(t-d) + B_h F x(t-h) + D w(t) \\ z(t) &= E x(t), \quad A_c = A + BF \end{aligned} \tag{3.5}$$

for which the transfer function from the disturbance $w(t)$ to the output $z(t)$ is given by

$$\begin{aligned} T_{zw}(s) &= L[(sI - A_c) - (A_d e^{-ds} + B_h F e^{-hs})]^{-1} D \\ &= L[(sI - [A + BF]) - (A_d e^{-ds} + B_h F e^{-hs})]^{-1} D \tag{3.6} \end{aligned}$$

where s is the Laplace operator. Of particular interest in this chapter is the design of H_∞-controllers. Hence, the basis of designing controller (3.4) is to simultaneously stabilize (3.5) and to guarantee the H_∞ norm bound γ of the closed-loop transfer function T_{zw}, namely $\|T_{zw}\|_\infty \le \gamma$ and $\gamma > 0$.

3.2.2 State Feedback Synthesis

In the following, sufficient conditions are developed first for stabilizing the closed-loop system (3.5) and guaranteeing a desired H_∞ norm bound using Lyapunov's second method. Then sufficient conditions are established for H_∞-control synthesis by state-feedback.

Theorem 3.1: *The closed-loop system (3.5) is asymptotically stable with $w(t) = 0$ for $d, h \ge 0$ if one of the following equivalent conditions is satisfied:*

(1) *There exist matrices $0 < P^t = P \in \Re^{n \times n}$, $0 < Q_1^t = Q_1 \in \Re^{n \times n}$, $0 < Q_2^t = Q_2 \in \Re^{n \times n}$ satisfying ARI:*

$$\begin{aligned} P(A + BF) + (A + BF)^t P + P A_d Q_1^{-1} A_d^t P + \\ P B_h F Q_2^{-1} F^t B_h^t P + Q_1 + Q_2 < 0 \tag{3.7} \end{aligned}$$

(2) *There exist matrices $0 < P^t = P \in \Re^{n \times n}$, $0 < Q_1^t = Q_1 \in \Re^{n \times n}$, $0 < Q_2^t = Q_2 \in \Re^{n \times n}$ satisfying the LMI:*

$$W_o = \begin{bmatrix} P(A+BF) + (A+BF)^t P + Q_1 + Q_2 & PA_d & PB_hF \\ A_d^t P & -Q_1 & 0 \\ F^t B_h^t P & 0 & -Q_2 \end{bmatrix} < 0 \tag{3.8}$$

Proof: (1) Define a Lyapunov-Krasovskii functional $V_{12}(x_t)$ as

$$V_{12}(x_t) = x^t(t)Px(t) + \int_{t-d}^{t} x^t(v)Q_1x(v)dv + \int_{t-h}^{t} x^t(v)Q_2x(v)dv \quad (3.9)$$

where $P, Q_1, Q_2 > 0$. Observe that $V_{12}(x_t) > 0$, $x \neq 0$; $V_{12}(x_t) = 0$, $x = 0$. Evaluating the time derivative of (3.9) along the solutions of (3.5) results in:

$$
\begin{aligned}
\dot{V}_{12}(x_t) &= x^t(t)(PA_c + A_c^t P + Q_1 + Q_2)x(t) \\
&+ x^t(t)PA_dx(t-d) + x^t(t)PB_hFx(t-h) \\
&+ x^t(t-d)A_1^t Px(t) + x^t(t-h)F^t B_1^t Px(t) \\
&- x^t(t-d)Q_1x(t-d) - x^t(t-h)Q_2x(t-h) \\
&= Z_1^t(t)W_oZ_1(t) \quad (3.10)
\end{aligned}
$$

where the extended state vector $Z_1(t)$ is given by

$$Z_1(t) = \begin{bmatrix} x^t(t) & x^t(t-d) & x^t(t-h) \end{bmatrix}^t \quad (3.11)$$

and W_o is given by (3.8). The requirement of negative-definiteness of $\dot{V}_{12}(x_t)$ for stability is guaranteed by (3.7).

(2) Using **A.1**, one can easily obtain (3.8).

Therefore, we can conclude that the closed-loop system (3.5) is asymptotically stable for d, $h \geq 0$ as desired.

Remark 3.1: Despite its simplicity, **Theorem 3.1** provides a sort of delay-dependent stability criteria since condition (3.7) includes the delay factors d and h. However, it presumes the availability of the gain matrix F. It is of general form as it encompasses other available results as special cases.

Corollary 3.1: *The state-delay system*

$$
\begin{aligned}
\dot{x}(t) &= Ax(t) + Bu(t) + A_dx(t-d) \\
z(t) &= Lx(t), \quad x(t) = \psi(t) \; \forall t \in [-d, 0]
\end{aligned}
$$

can be stabilized by the controller $u(t) = Fx(t)$ *if there exist matrices* $0 < P^t = P \in \Re^{n \times n}$, $0 < Q_1^t = Q_1 \in \Re^{n \times n}$, *satisfying the LMI:*

$$W_{10} = \begin{bmatrix} P(A+BF) + (A+BF)^t P + Q_1 & PA_d \\ A_d^t P & -Q_1 \end{bmatrix} < 0 \quad (3.12)$$

or equivalently satisfying the ARI:

$$P(A+BF)+(A+BF)^tP+Q_1+PA_dQ_1^{-1}A_d^tP < 0 \qquad (3.13)$$

Proof: Set $B_h = 0$ in (3.7).

Corollary 3.2: *The free state-delay system*

$$\dot{x}(t) = Ax(t)+A_dx(t-d)$$
$$z(t) = Lx(t) , \quad x(t) = \psi(t) \ \forall t \in [-d, 0]$$

is asymptotically stable if there exist matrices $0 < P^t = P \in \Re^{n \times n}$, $0 < Q_1^t = Q_1 \in \Re^{n \times n}$ *satisfying the LMI:*

$$W_{20} = \begin{bmatrix} PA + A^tP + Q_1 & PA_d \\ A_d^tP & -Q_1 \end{bmatrix} < 0 \qquad (3.14)$$

or equivalently satisfying the ARI:

$$PA + A^tP + Q_1 + PA_dQ_1^{-1}A_d^tP < 0 \qquad (3.15)$$

Proof: Set $B = 0$ in (3.13).

Corollary 3.3: *The input-delay system*

$$\dot{x}(t) = Ax(t)+Bu(t)+B_hu(t-h)$$
$$z(t) = Lx(t) , \quad x(t) = \psi(t) \ \forall t \in [-h, 0]$$

can be stabilized by the controller $u(t) = Fx(t)$ *if there exist matrices* $0 < P^t = P \in \Re^{n \times n}$, $0 < Q_2^t = Q_2 \in \Re^{n \times n}$ *satisfying the LMI:*

$$W_{30} = \begin{bmatrix} P(A+BF) + (A+BF)^tP + Q_2 & PB_hF \\ F^tB_h^tP & -Q_2 \end{bmatrix} < 0 \qquad (3.16)$$

or equivalently satisfying the ARI:

$$P(A+BF)+(A+BF)^tP+Q_2+PB_hFQ_2^{-1}F^tB_h^tP < 0 \qquad (3.17)$$

Proof: Set $A_d = 0$ in (3.7).

Remark 3.2: It is important to note that **Corollaries 3.1-3.3** provide LMI-stability criteria for state-delay dynamical systems.

Since we plan to adopt $\mathcal{H}_\infty$-theory in the control synthesis, we now establish the conditions under which the controller (3.4) stabilizes (3.5) and guarantees the H_∞ norm bound γ of the closed-loop transfer function $||T_{zw}||$, namely $||T_{zw}||_\infty \leq \gamma;\ \gamma > 0$.

Theorem 3.2: *The closed-loop system (3.5) is asymptotically stable and* $||T_{zw}||_\infty \leq \gamma;\ \gamma > 0$ *for* $d, h \geq 0$ *if there exist matrices* $0 < P^t = P \in \Re^{n \times n}$, $0 < Q_1^t = Q_1 \in \Re^{n \times n}$, $0 < Q_2^t = Q_2 \in \Re^{n \times n}$ *satisfying the ARI:*

$$PA_c + A_c^t P + Q_1 + Q_2 + PA_d Q_1^{-1} A_d^t P + PB_h F Q_2^{-1} F^t B_h^t P$$
$$+ L^t L + \gamma^{-2} PDD^t P < 0 \tag{3.18}$$

Proof: By Theorem 3.1 and [45], the state-feedback controller (3.4) which satisfies inequality (3.18) stabilizes the time-delay system (3.1)-(3.2) for $d, h \geq 0$. Now, introduce the matrix:

$$-M := PA_c + A_c^t P + Q_1 + Q_2 + PA_d Q_1^{-1} A_d^t P$$
$$+ PB_h F Q_2^{-1} F^t B_h^t P + E^t E + \gamma^{-2} PDD^t P \tag{3.19}$$

so that

$$PA_c + A_c^t P + Q_1 + Q_2 + PA_d Q_1^{-1} A_d^t P + PB_h F Q_2^{-1} F^t B_h^t P$$
$$+ L^t L + \gamma^{-2} PDD^t P + M = 0 \tag{3.20}$$

Let $\omega \in \Re$ and construct the matrices

$$X(\beta, \omega) = [(\beta + j\omega)I - A_c - e^{-j\omega d} A_1 - e^{-j\omega h} B_h F]^{-1} \tag{3.21}$$

$$Y_1(\beta, \omega) = Q_1 + PA_d Q_1^{-1} A_d^t P - e^{j\omega d} A_1^t P e^{-\beta d}$$
$$- e^{-j\omega d} PA_d e^{\beta d} \tag{3.22}$$

$$Y_2(\beta, \omega) = Q_2 + PB_h F Q_2^{-1} F^t B_h^t P e^{-\beta h}$$
$$- e^{j\omega h} F^t B_h^t P e^{\beta h} - e^{-j\omega h} PB_h F \tag{3.23}$$

such that the use of (3.4) with $s = \beta + j\omega$, $\beta > 0$ ensures that

$$EX(\beta, \omega)D = T_{zw}(\beta + j\omega) \tag{3.24}$$

Note that $Y_1(\beta, \omega) > 0$ and $Y_2(\beta, \omega) > 0$. By adding and subtracting appropriate terms, we rewrite (3.20) using (3.21)-(3.23) as

$$PX^{-1}(\beta, \omega) + [X^t(\beta, -\omega)]^{-1} P - Y_1(\beta, \omega) - Y_2(\beta, \omega)$$
$$- E^t E - \gamma^{-2} PDD^t P - M = 0 \tag{3.25}$$

Premultiplying (3.25) by $X^t(\beta, -\omega)$, and postmultiplying by $X(\beta, \omega)$ and rearranging, we arrive at:

$$PX(\beta, \omega) + X^t(\beta, -\omega)P - \gamma^{-2}X^t(\beta, -\omega)PDD^tPX(\beta, \omega)$$
$$= X^t(\beta, -\omega)[Y_1(\beta, \omega) + Y_2(\beta, \omega) + L^tL + M]X(\beta, \omega) \qquad (3.26)$$

Further manipulation of (3.26) gives

$$D^tPX(\beta, \omega)D + D^tX^t(\beta, -\omega)PD$$
$$-\gamma^{-2}D^tX^t(\beta, -\omega)PDD^tPX(\beta, \omega)D - \gamma^2I$$
$$= -\gamma^2I + D^tX^t(\beta, -\omega)[Y_1(\beta, \omega) + Y_2(\beta, \omega)$$
$$+L^tL + M]X(\beta, \omega)D \qquad (3.27)$$

On completing the squares in (3.27), it follows that for all $\omega \in \Re$:

$$-[\gamma I - \gamma^{-1}D^tPX(\beta, -\omega)D]^t[\gamma I - \gamma^{-1}D^tPX(\beta, -\omega)D] =$$
$$[D^tX^t(\beta, -\omega)E^tEX(\beta, \omega)D]$$
$$-\gamma^2I + D^tX^t(\beta, -\omega)[Y_1(\beta, \omega) + Y_2(\beta, \omega) + M]X(\beta, \omega)D \qquad (3.28)$$

On observing that

$$-[\gamma I - \gamma^{-1}D^tPX(\beta, -\omega)D]^t[\gamma I - \gamma^{-1}D^tPX(\beta, -\omega)D] \le 0$$

then from (3.24) and (3.28), we get:

$$-\gamma^2I + D^tX^t(\beta, -\omega)[Y_1(\beta, \omega) + Y_2(\beta, \omega) + M]X(\beta, \omega)D$$
$$+T^t_{zw}(\beta + j\omega)T_{zw}(\beta + j\omega) \le 0 \qquad (3.29)$$

or equivalently,

$$T^t_{zw}(\beta + j\omega)T_{zw}(\beta + j\omega) \le$$
$$\gamma^2I - D^tX^t(\beta, -\omega)[Y_1(\beta, \omega) + Y_2(\beta, \omega) + M]X(\beta, \omega)D$$
$$\le \gamma^2I \qquad (3.30)$$

$\forall \beta > 0$, $\omega \in \Re$. We can then conclude that $||T_{zw}||_\infty \le \gamma$ as desired.

Remark 3.3: The usefulness of Theorem 3.2 lies in the fact that it provides condition (3.18) as a sufficient measure for the existence of a constant matrix F as the gain of an H_∞ controller.

The following theorem gives an LMI-based computational procedure to determine state-feedback controller with H_∞-norm constraint.

Theorem 3.3: *The closed-loop system (3.5) is asymptotically stable and* $\|T_{zw}\|_\infty \leq \gamma$; $\gamma > 0$ *for* $d, h \geq 0$ *if there exist matrices* $0 < Y^t = Y \in \Re^{n \times n}$, $0 < Q_t^t = Q_t \in \Re^{n \times n}$, $0 < Q_s^t = Q_s \in \Re^{n \times n}$, $S \in \Re^{m \times n}$, *satisfying the LMI:*

$$W_1 = \begin{bmatrix} AY + YA^t + BS \\ +S^tB^t + Q_t + Q_s & YE^t & B_hS & D & A_dY \\ LY & -I & 0 & 0 & 0 \\ S^tB_h^t & 0 & -Q_s & 0 & 0 \\ D^t & 0 & 0 & -\gamma^2 I & 0 \\ YA_d^t & 0 & 0 & 0 & -Q_t \end{bmatrix} < 0 \quad (3.31)$$

Moreover, the memoryless state-feedback controller is given by

$$u(t) = SY^{-1}x(t) \quad (3.32)$$

Proof: By **Theorem 3.2**, there exists a state-feedback controller with constant gain F such that the closed-loop system (3.5) is asymptotically stable and $\|T_{zw}\|_\infty \leq \gamma$; $\gamma > 0$ for $d, h \geq 0$. Now, letting $Y = P^{-1}$, $S = FY$, $Q_t = P^{-1}Q_1P^{-1}$, $Q_s = P^{-1}Q_2P^{-1}$, premultiplying (3.18) by P^{-1} and postmultiplying the result by P^{-1}, we get:

$$AY + YA^t + BS + S^tB^t + Q_t + Q_s + A_dYQ_t^{-1}YA_d^t + B_hSQ_s^{-1}S^tB_h^t + YL^tLY + \gamma^{-2}DD^t < 0 \quad (3.33)$$

which, in the light of **A.1**, can be conveniently arranged to yield the block form (3.31) as desired.

Remark 3.4: To implement such a controller, we can use the LMI-Control Toolbox to solve the following minimization problem:

$$\min_{Y, S, Q_s, Q_t} \gamma$$

$$s.t.\ -Y < 0, \quad -Q_s < 0,$$
$$-Q_t < 0, \quad W_1 < 0 \quad (3.34)$$

Two corollaries immediately follow:

Corollary 3.4: *The state-delay system*

$$\begin{aligned}
\dot{x}(t) &= Ax(t) + Bu(t) + A_d x(t-d) + Dw(t) \\
z(t) &= Lx(t) \ , \quad x(t) = \psi(t) \ \forall t \in [-d, 0]
\end{aligned} \tag{3.35}$$

with controller $u(t) = Fx(t)$ is asymptotically stable with disturbance atten-uation γ if there exist matrices $0 < Y^t = Y \in \Re^{n \times n}$, $0 < Q_t^t = Q_t \in \Re^{n \times n}$ and $S \in \Re^{m \times n}$, satisfying the LMI:

$$\begin{bmatrix}
AY + YA^t + BS + S^t B^t + Q_t & YL^t & D & A_d Y \\
LY & -I & 0 & 0 \\
D^t & 0 & -\gamma^2 I & 0 \\
YA_d^t & 0 & 0 & -Q_t
\end{bmatrix} < 0 \tag{3.36}$$

or equivalently satisfying the ARI:

$$\begin{aligned}
&AY + YA^t + BS + S^t B^t + Q_t \\
&+ A_d Y Q_t^{-1} Y A_d^t + YL^t LY + \gamma^{-2} DD^t < 0
\end{aligned} \tag{3.37}$$

The feedback gain is $F = SY^{-1}$.

Proof: Set $B_h = 0$ in (3.33).

Corollary 3.5: *The input-delay system*

$$\begin{aligned}
\dot{x}(t) &= Ax(t) + Bu(t) + B_h u(t-h) + Dw(t) \\
z(t) &= Lx(t) \ , \quad x(t) = \psi(t) \ \forall t \in [-h, 0]
\end{aligned} \tag{3.38}$$

with controller $u(t) = Fx(t)$ is asymptotically stable with disturbance atten-uation γ if there exist matrices $0 < Y^t = Y \in \Re^{n \times n}$, $0 < Q_s^t = Q_s \in \Re^{n \times n}$ and $S \in \Re^{m \times n}$, satisfying the LMI:

$$\begin{bmatrix}
AY + YA^t + BS + S^t B^t + Q_s & YL^t & B_h S & D \\
LY & -I & 0 & 0 \\
S^t B_h^t & 0 & -Q_s & 0 \\
D^t & 0 & 0 & -\gamma^2 I
\end{bmatrix} < 0 \tag{3.39}$$

or equivalently satisfying the ARI:

$$\begin{aligned}
&AY + YA^t + BS + S^t B^t + Q_s \\
&+ B_h S Q_s^{-1} S^t B_h^t + YL^t LY + \gamma^{-2} DD^t < 0
\end{aligned} \tag{3.40}$$

The feedback gain is $F = SY^{-1}$.

Proof: Set $B_h = 0$ in (3.33).

An alternative state-feedback controller is now presented.

Theorem 3.4: *The closed-loop system (3.5) is asymptotically stable and* $\|T_{zw}\|_\infty \leq \gamma; \gamma > 0$ *for $d, h \geq 0$ if there exist matrices $0 < Z^t = Z \in \Re^{n \times n}$, $0 < Q_t^t = Q_t \in \Re^{n \times n}$, $0 < Q_s^t = Q_s \in \Re^{n \times n}$ and a scalar $\rho > 0$ satisfying the LMI:*

$$W_2 =$$
$$\begin{bmatrix} AZ + ZA^t + Q_t + Q_s & A_dZ & \rho B & 1/2\rho^2 B_h B^t & ZL^t & D \\ ZA_d^t & -Q_t & 0 & 0 & 0 & 0 \\ \rho B^t & 0 & -I & 0 & 0 & 0 \\ 1/2\rho^2 BB_h^t & 0 & 0 & -Q_s & 0 & 0 \\ LZ & 0 & 0 & 0 & -I & 0 \\ D^t & 0 & 0 & 0 & 0 & -\gamma^2 I \end{bmatrix} < 0$$
$$(3.41)$$

Moreover, the gain of the memoryless state-feedback controller is given by

$$u(t) = (\rho^2/2)B^t Z^{-1} x(t) \qquad (3.42)$$

Proof: By Theorem 3.2, there exists a state-feedback controller with constant gain $F = \mu_o B^t P$, $\mu_o > 0$ such that the closed-loop system (3.5) is asymptotically stable and $\|T_{zw}\|_\infty \leq \gamma; \gamma > 0$ for $d, h \geq 0$. This implies that

$$PA + A^t P + Q_1 + Q_2 + PA_d Q_1^{-1} A_d^t P + \gamma^{-2} PLL^t P$$
$$+\mu_o^2 PB_h B^t PQ_2^{-1} PBB_h^t P + L^t L + 2\mu_o PBB^t P < 0 \qquad (3.43)$$

Now, letting $Z = P^{-1}$, $\mu_o = \rho^2/2B^t$, $Q_t = P^{-1}Q_1 P^{-1}$, $Q_s = P^{-1}Q_2 P^{-1}$, premultiplying the above inequality by P^{-1} and postmultiplying the result by P^{-1}, we get:

$$AZ + ZA^t + Q_t + Q_s + ZL^t LZ + \gamma^{-2} DD^t$$
$$+A_d ZQ_t^{-1} ZA_d^t + 1/4\rho^4 B_h B^t Q_s^{-1} BB_h^t + \rho^2 BB^t < 0 \qquad (3.44)$$

which, in the light of **A.1**, can be conveniently arranged to yield the block form (3.41) as desired.

Remark 3.5: To implement controller (3.42), one has to solve the minimization problem

$$\min_{Z, Q_s, Q_t, \rho} \gamma$$

$$s.\,t. \;\; -Z < 0, \;\; -Q_s < 0,$$
$$-Q_t < 0, \;\; -\rho < 0, \;\; W_2 < 0 \tag{3.45}$$

which is amenable to standard LMI-format of the **LMI-Control Toolbox**.

Remark 3.6: In comparing between the state-feedback controllers (3.32) and (3.42), one should observe that the gain of (3.32) has two matrices S and Y to be determined using the LMI solver whilst the controller (3.42) has one matrix Z and a scalar ρ. This implies that not only the computational load of the latter would be less than that of the former but also less in the degrees of freedom (one versus $m \times n$). On the other hand, the former would seem to encompass the latter since (3.42) would correspond to (3.32) in the case $S \to (\rho^2/2)B^t$.

3.2.3 Two-Term Feedback Synthesis

As a departure from the memoryless state feedback, we now propose the controller

$$u(t) = Fx(t) + Kx(t - d) \tag{3.46}$$

which combines the effect of the instantaneous as well as the delay states. In some sense, it can be called a DP controller since it provides two degrees of freedom: one by the proportional (P) term (Fx) and the other through the delay (D) term $Kx(t - d)$. Note that we did not use the term PD to avoid confusion with the standard proportional-plus-derivative controller. Now, applying controller (3.46) to system (3.1)-(3.2) gives the following closed-loop system:

$$
\begin{aligned}
\dot{x}(t) &= A_c x(t) + A_d x(t - d) + B_h F x(t - h) \\
&+ B_h K x(t - d - h) + D w(t) \\
z(t) &= L x(t) \\
A_c &= A + BF, \;\; A_h = A_d + BK
\end{aligned}
\tag{3.47}
$$

The transfer function from the disturbance $w(t)$ to the output $z(t)$ is given by

$$T_{zw}(s) = E[(sI - A_c) - (A_d e^{-ds} + B_1 F e^{-hs} + B_1 K e^{-(d+h)s})]^{-1} D \quad (3.48)$$

To study the stability behavior in this case, we define the quadratic Lyapunov-Krasovskii functional $V_{13}(x_t)$ as

$$V_{13}(x_t) = x^t(t)Px(t) + \int_{t-d}^{t} x^t(v)Q_1 x(v)dv + \int_{t-h}^{t} x^t(v)Q_2 x(v)dv$$
$$+ \int_{t-d-h}^{t} x^t(v)Q_3 x(v)dv \quad (3.49)$$

Theorem 3.5: *The closed-loop system (3.47) is asymptotically stable for d, $h \geq 0$ if one of the following equivalent conditions is satisfied:*

(1) *There exist matrices $0 < P^t = P \in \Re^{n \times n}$, $0 < Q_1^t = Q_1 \in \Re^{n \times n}$, $0 < Q_2^t = Q_2 \in \Re^{n \times n}$ and $0 < Q_3^t = Q_3 \in \Re^{n \times n}$ satisfying ARI:*

$$PA_c + A_c^t P + Q_1 + Q_2 + Q_3 + PA_h Q_1^{-1} A_h^t P$$
$$+ PB_h F Q_2^{-1} F^t B_h^t P + PB_h K Q_3^{-1} K^t B_h^t P < 0 \quad (3.50)$$

(2) *There exist matrices $0 < P^t = P \in \Re^{n \times n}$, $0 < Q_1^t = Q_1 \in \Re^{n \times n}$, $0 < Q_2^t = Q_2 \in \Re^{n \times n}$ and $0 < Q_3^t = Q_3 \in \Re^{n \times n}$ satisfying the LMI:*

$$W_3 = \begin{bmatrix} PA_c + A_c^t P + Q_1 + Q_2 + Q_3 & PA_h & PB_h F & PB_h K \\ A_h^t P & -Q_1 & 0 & 0 \\ F^t B_h^t P & 0 & -Q_2 & 0 \\ K^t B_h^t P & 0 & 0 & -Q_3 \end{bmatrix} < 0$$
$$(3.51)$$

Proof: (1) First, observe that $V_{13}(x_t) > 0$, $x \neq 0$; $V_{13}(x_t) = 0$, $x = 0$. The Lyapunov derivative $\dot{V}_{13}(x_t)$ evaluated along the solutions of system (3.47) is given by:

$$\dot{V}_{13}(x_t) = x^t(t)(PA_c + A_c^t P + Q_1 + Q_2 + Q_3)x(t)$$
$$+ x^t(t)PA_h x(t - d) + x^t(t)PB_h F x(t - h)$$
$$+ x^t(t)PB_h K x(t - d - h) + x^t(t - d)A_h^t Px(t)$$
$$+ x^t(t - h)F^t B_h^t Px(t) + x^t(t - d - h)K^t B_h^t Px(t)$$
$$- x^t(t - d)Q_1 x(t - d) - x^t(t - h)Q_2 x(t - h)$$
$$- x^t(t - d - h)Q_3 x(t - d - h)$$
$$= Z_2^t(t)W_3 Z_2(t) \quad (3.52)$$

where the extended state vector

$$Z_2(t) = \left[x^t(t) \quad x^t(t-d) \quad x^t(t-h) \quad x^t(t-d-h) \right]^t \qquad (3.53)$$

where W_3 is given by (3.51). The requirement of negative-definiteness of $\dot{V}_{13}(x_t)$ for stability is implied by (3.50).

(2) Given (3.50) and using **A.1**, one can easily obtain (3.51).

Therefore, we can conclude that the closed-loop system (3.47) is asymptotically stable for d, $h \geq 0$ as desired.

Remark 3.7: In a similar way, **Theorem 3.5** provides a delay-dependent stability criteria since condition (3.50) includes the delay factors d and h. However, the gain matrices F and K are needed for practical implementation.

Corollary 3.6: *The state-delay system (3.11) is stabilizable by the controller (3.46) if there exist matrices $0 < P^t = P \in \Re^{n \times n}$, $0 < Q_1^t = Q_1 \in \Re^{n \times n}$, satisfying the LMI:*

$$W_{30} = \begin{bmatrix} PA_c + A_c^t P + Q_1 & PA_h \\ A_h^t P & -Q_1 \end{bmatrix} < 0 \qquad (3.54)$$

or equivalently satisfying the ARI:

$$PA_c + A_c^t P + Q_1 + PA_h Q_1^{-1} A_h^t P < 0 \qquad (3.55)$$

Remark 3.7: For the case of input-delay systems (3.16), it is meaningless to use a two-term controller and a one-term controller would be sufficient.

Theorem 3.6: *The closed-loop system (3.47) is asymptotically stable and $\|T_{zw}\|_\infty \leq \gamma$; $\gamma > 0$ for d, $h \geq 0$ if there exist matrices $0 < Y^t = Y \in \Re^{n \times n}$, $0 < Q_t^t = Q_t \in \Re^{n \times n}$, $0 < Q_s^t = Q_s \in \Re^{n \times n}$, $0 < Q_r^t = Q_r \in \Re^{n \times n}$, S, $V \in \Re^{m \times n}$ and positive scalars σ, κ satisfying the LMI:*

$$W_4 = \begin{bmatrix} \Phi(Y) & B_c S & N & M \\ S^t B_c^t & -J_t & 0 & 0 \\ N & 0 & -J_n & 0 \\ M^t & 0 & 0 & -J_d \end{bmatrix} < 0 \qquad (3.56)$$

$$\Phi(Y) = {}' AY + YA^t + BS + S^t B^t + Q_t + Q_s + Q_r$$

$$
\begin{aligned}
B_c &= [\kappa B \quad B_h], \quad J_t = [\kappa Q_t \quad Q_s] \\
N &= [\sigma A_d Y \quad B_h V] \; , \quad J_n = [\sigma Q_t \quad Q_r] \\
M &= [Y L^t \quad D], \quad J_d = [I \quad \gamma^2 I]
\end{aligned}
\tag{3.57}
$$

Moreover, the delayed state-feedback controller is given by

$$
u(t) = SY^{-1}x(t) + VY^{-1}x(t-d)
\tag{3.58}
$$

Proof: By **Theorem 3.2**, there exists a delayed state-feedback controller with constant gains $[F \quad K]$ such that the closed-loop system (3.50) is asymptotically stable and $\|T_{zw}\|_\infty \leq \gamma;\; \gamma > 0$ for $d, h \geq 0$. Now, letting $Y = P^{-1}$, $S = FY$, $V = KY$, $Q_t = P^{-1}Q_1 P^{-1}$, $Q_s = P^{-1}Q_2 P^{-1}$, $Q_r = P^{-1}Q_3 P^{-1}$, premultiplying (3.50) by P^{-1} and postmultiplying the result by P^{-1}, we get:

$$
\begin{aligned}
&AY + YA^t + BS + S^t B^t + Q_t + Q_s + Q_r + (A_d + BF)Q_1^{-1}(A_d + BF)^t \\
&+B_h SQ_s^{-1}S^t B_h^t + B_h V Q_r^{-1}V^t B_h^t + Y L^t L Y + \gamma^{-2}DD^t < 0
\end{aligned}
\tag{3.59}
$$

Using **B.1.2** in the term $(A_d + BF)Q_1^{-1}(A_d + BF)^t$ and manipulating, it becomes:

$$
\begin{aligned}
(A_d + BF)Q_1^{-1}(A_d + BF)^t &= \\
A_d Q_1^{-1}A_d^t + BFQ_1^{-1}F^t B^t &+ BFQ_1^{-1}A_d^t + A_d Q_1^{-1}F^t B^t \\
&\leq (1+\alpha)A_d Q_1^{-1}A_d^t + (1+\alpha^{-1})BFQ_1^{-1}F^t B^t \\
&\leq \sigma A_d Y Q_t^{-1}Y A_d^t + \kappa BSQ_t^{-1}S^t B^t \qquad \forall \sigma, \kappa > 0
\end{aligned}
\tag{3.60}
$$

so that (3.60) can be expressed as

$$
\begin{aligned}
&AY + YA^t + BS + S^t B^t + Q_t + Q_s + Q_r + \\
&\sigma A_d Y Q_t^{-1}Y A_d^t + \kappa BSQ_t^{-1}S^t B^t + B_h SQ_s^{-1}S^t B_h^t + \\
&B_h V Q_r^{-1}V^t B_h^t + Y L^t L Y + \gamma^{-2}DD^t < 0
\end{aligned}
\tag{3.61}
$$

which, in the light of **A.1**, can be conveniently arranged to yield the block form (3.57) as desired.

Remark 3.9: **Theorem 3.6** provides a delay-dependent condition for a two-term H_∞-controller which guarantees the norm-bound γ of the transfer function T_{zw}. To determine the gains of such a controller, one has to solve the following minimization problem

$$\min_{Y, S, V, Q_s, Q_t, Q_r, \sigma, \kappa} \gamma$$

$$s.\,t. \; -Y < 0, \;\; -Q_s < 0, -Q_t < 0,$$
$$-Q_r < 0, -\sigma < 0, \;\; -\kappa < 0, \;\; W_4 < 0 \qquad (3.62)$$

iteratively using the following procedure:

Step 1: Select initial values for σ, κ and choose arbitrary
$0 < Y^* = Y^{*^t} \in \Re^{n \times n}$.
Set the iteration index $j = 1$.
Step 2: Solve problem (3.64) and let $Y = Y^{(j)}$.
Step 3: If $||Y^* - Y^{(j)}|| < \delta$, a predetermined tolerance, then **STOP** and
record $Y^{(j)}$ as the desired feasible solution.
Otherwise, set $Y^* = Y^{(j)}$, $\sigma = \sigma + \Delta\sigma$, $\kappa = \kappa + \Delta\kappa$, $j = j + 1$ and go to
Step 2 where Δ is a predetermined increment.

Remark 3.10: It is significant to note that the minimization problem addressed in **Remark 3.9** has the form of a generalized eigenvalue problem which is known to be solved numerically very efficiently using interior-point methods [3,4]. The software **LMI-Control Toolbox** provides an efficient tool for computer implementation. Experience has indicated that only a few number of iterations are usually required to converge to an acceptable, feasible solution.

3.2.4 Static Output Feedback Synthesis

Now, we consider system (3.1)-(3.2) when a limited number of states are accessible for measurement. In this regard, we recall the output measurement

$$y(t) = Cx(t) \qquad (3.63)$$

where $y \in \Re^s$ is the measured output and $C \in \Re^{s \times n}$ is a constant matrix such that the pair (A, C) is detectable. Our purpose is to develop an output feedback control for system (3.1)-(3.2) and (3.63) of the form $u(t) = \Phi[y(t)]$. In this section, the control law is given by:

$$u(t) = Gy(t) = GCx(t) \qquad (3.64)$$

where G is a static gain matrix to be determined. It should be emphasized that controller (3.63) can be considered as a replica of the state-feedback controller (3.4) and as such no generality is claimed at this point. In [158], it was clearly stated that results pertaining to output-feedback synthesis for dynamical systems are few and restrictive. The results of [166] are appealing in this regard. The only way to resolve this problem is to impose some condition on G. An initial attempt to alleviate this restriction for systems with state-delay would be to invoke the strict positive realness condition [167,206] by considering that the nominal transfer function

$$T_o(s) = GC[sI - A]^{-1}B \tag{3.65}$$

is strictly positive real (SPR). It is known that condition (3.67) corresponds to

$$GC = B^t P \tag{3.66}$$

where P is a Lyapunov matrix for the free delayless version of (3.1)-(3.2). To relax the restrictive condition (3.68), we replace it here by

$$GC = \omega B^t P + \Omega \ , \ \ \omega > 0 \tag{3.67}$$

The closed-loop system of (3.1)-(3.2), and (3.63)-(3.64) is given by:

$$\begin{aligned}
\dot{x}(t) &= (A + BGC)x(t) + A_d x(t - d) \\
&+ \ B_h GC x(t - h) + Dw(t) \\
z(t) &= Lx(t)
\end{aligned} \tag{3.68}$$

The following theorem summarizes the desired result:

Theorem 3.7: *The closed-loop system (3.68) is asymptotically stable and $||T_{zw}||_\infty \le \gamma$; $\gamma > 0$ for $d, h \ge 0$ if there exist matrices $0 < Y^t = Y \in \Re^{n \times n}$, $0 < Q_t^t = Q_t \in \Re^{n \times n}$, $0 < Q_s^t = Q_s \in \Re^{n \times n}$, $\Gamma \in \Re^{m \times n}$ and scalar ω satisfying the LMI:*

$$W_5 \ = \ \begin{bmatrix} \Psi(Y) & M_d & \omega B_h B^t & M \\ M_d^t & -J_m & 0 & 0 \\ \omega BB_h^t & 0 & -Q_s & 0 \\ M^t & 0 & 0 & -J_d \end{bmatrix} < 0 \tag{3.69}$$

where

$$\begin{aligned}
\Psi(Y) &= AY + YA^t + Q_t + Q_s + B\Gamma + \Gamma^t B^t + 2\omega BB^t \\
&+ \ \omega B_h \Gamma Q_s^{-1} BB_h 1^t + \omega B_h B^t Q_s^{-1} \Gamma^t B_h^t \\
M &= [YL^t \ \ D] \ , \ \ J_d = [I \ \ \gamma^2 I] \\
M_d &= [A_d Y \ \ B_h \Gamma] \ , \ \ J_m = [Q_t \ \ Q_s]
\end{aligned} \tag{3.70}$$

Moreover, the feedback controller is given by

$$u(t) = (\omega B^t Y^{-1} + \Gamma Y^{-1}) x(t) \qquad (3.71)$$

Proof: By Theorem 3.2, there exists a memoryless feedback controller with constant gain $F = GC = \omega B^t P + \Omega$ such that the closed-loop system (3.68) is asymptotically stable and $\|T_{zw}\|_\infty \leq \gamma$; $\gamma > 0$ for $d, h \geq 0$. The stabilizing controller satisfies inequality (3.18) such that:

$$P(A + BGC) + (A + BGC)^t P + Q_1 + Q_2 + PA_d Q_1^{-1} A_d^t P$$
$$+ PB_h GCQ_2^{-1} C^t G^t B_h^t P + L^t L + \gamma^{-2} PDD^t P < 0 \qquad (3.72)$$

Now, letting $Y = P^{-1}$, $\Omega = \Gamma P$, $Q_t = P^{-1} Q_1 P^{-1}$, $Q_s = P^{-1} Q_2 P^{-1}$, premultiplying (3.72) by P^{-1} and postmultiplying the result by P^{-1}, we get:

$$AY + YA^t + Q_t + Q_s + B\Gamma + \Gamma^t B^t + 2\omega BB^t A_d Y Q_t^{-1} Y A_d^t +$$
$$\omega^2 B_h B^t Q_s^{-1} BB_h^t + B_h \Gamma Q_s^{-1} \Gamma^t B_h^t + Y L^t L Y +$$
$$\gamma^{-2} DD^t + \omega B_h \Gamma Q_s^{-1} BB_h 1^t + \omega B_h B^t Q_s^{-1} \Gamma^t B_h^t < 0 \qquad (3.73)$$

which, in the light of **A.1**, can be conveniently arranged to yield the block form (3.70) as desired.

Remark 3.11: To determine the gain factors ω, Γ, Y one has to solve the following minimization problem:

$$\min_{Y, \Gamma, Q_s, Q_t, \omega} \gamma$$

$$s.t. \ -Y < 0, \ -Q_s < 0,$$
$$-Q_t < 0, \ -\omega < 0, \ W_5 < 0 \qquad (3.74)$$

Remark 3.12: It is significant to observe that when specializing the gain of the output-feedback controller (3.71) to the case $\omega = 0$, it reduces to the constant gain state-feedback controller (3.32) with $S \rightarrow \Gamma$.

3.2.5 Dynamic Output Feedback Synthesis

Given the time-delay system (3.1)-(3.2) with the output measurement (3.63), we now consider the problem of output feedback control by using a state observer-based control scheme. Let the state-observer be described by:

$$
\begin{aligned}
\dot{\xi}(t) &= A_o\xi(t) + B_o[y(t) - C\xi(t)] \\
&+ C_d\xi(t - d) + C_h\xi(t - h) \\
u(t) &= G_o\xi(t)
\end{aligned} \tag{3.75}
$$

where the matrices A_o, B_o, G_o, C_h, C_d will be specified shortly. Introducing the variables

$$
e(t) := \xi(t) - x(t) \; ; \quad x_a(t) := [x^t(t) \; e^t(t)]^t \tag{3.76}
$$

then the closed-loop system corresponding to (3.1)-(3.2), (3.63) and (3.75)-(3.76) is given by the state model:

$$
\begin{aligned}
\dot{x}_a(t) &= A_a x_a(t) + B_a x_a(t - d) + C_a x_a(t - h) + D_a w(t) \\
z(t) &= L_a x_a(t)
\end{aligned} \tag{3.77}
$$

with

$$
T_{zw}(s) = E_a[(sI - A_a) - (B_a e^{-ds} + C_a e^{-hs})]^{-1} D_a \tag{3.78}
$$

Let the matrices A_o, C_d, C_h be defined by

$$
\begin{aligned}
A_o &= A + BG_o + \gamma^{-2}DD^tP - L^{-1}G_o^t B^t P \\
C_d &= A_d \; ; \; C_h = B_h g_o
\end{aligned} \tag{3.79}
$$

such that

$$
A_a = \begin{bmatrix} A + BG_o & BG_o \\ A_o - BG_o - A & A_o - BG_o - B_oC \end{bmatrix}; \; B_a = \begin{bmatrix} A_d & 0 \\ 0 & A_d \end{bmatrix} \tag{3.80}
$$

$$
C_a = \begin{bmatrix} B_hG_o & B_hG_o \\ 0 & 0 \end{bmatrix}; \; D_a = \begin{bmatrix} D \\ -D \end{bmatrix}; \; E_a = [L \quad 0] \tag{3.81}
$$

which describes a free time-delay system. The following theorem summarizes the desired result:

Theorem 3.8: *The closed-loop system (3.78) is asymptotically stable and $\|T_{zw}\|_\infty \leq \gamma; \gamma > 0$ for $d, h \geq 0$ if there exist matrices $0 < Y^t = Y \in \Re^{n \times n}$, $0 < X^t = X \in \Re^{n \times n}$ $0 < Q_{dd}^t = Q_{dd} \in \Re^{n \times n}$, $0 < Q_{hh}^t = Q_{hh} \in$*

$\Re^{n\times n}$, $0 < Q_{dh}^t = Q_{dh} \in \Re^{n\times n}$, $0 < Q_{hd}^t = Q_{hd} \in \Re^{n\times n}$, $0 < Q_{ll}^t = Q_{ll} \in \Re^{n\times n}$, $S_o \in \Re^{m\times n}$, $M_o \in \Re^{n\times s}$ and scalar $\varphi > 0$ satisfying the LMIs:

$$W_6 = \begin{bmatrix} \Theta_1(Y) & A_d Y & \bar{B}_h S_o & M \\ Y A_d^t & -Q_{dd} & 0 & 0 \\ S_o^t \bar{B}_h^t & 0 & -\bar{Q}_{dh} & 0 \\ M^t & 0 & 0 & -J_d \end{bmatrix} < 0 \qquad (3.82)$$

and,

$$W_7 = \begin{bmatrix} \Theta_2(X) & X N_o & A_d X & \bar{D} \\ N_o^t X & -I & 0 & 0 \\ X A_d^t & 0 & -Q_{ll} & 0 \\ \bar{D} & 0 & 0 & -U \end{bmatrix} < 0 \qquad (3.83)$$

where

$$\begin{aligned}
\Theta_1(Y) &= AY + YA^t + BS_o + S_o^t B^t + Q_{dd} + Q_{hd} \\
A_e &= A_o + \gamma^{-2} DD^t Y^{-1} \\
\Theta_2(X) &= A_e X + X A_e^t + Q_{ll} + Q_{dd} \\
\bar{B}_h &= [B_h \quad B_h], \quad \bar{Q}_{dh} = [Q_{dh} \quad Q_{hd}], \quad \bar{D} = [D \quad C^t M_o^t] \\
N_o N_o^t &= [\varphi I - Y^{-1}(BS_o + S_o^t B^t)Y^{-1}], \quad U = [\gamma^2 I \quad \varphi^{-1} I] \quad (3.84)
\end{aligned}$$

Moreover, the observer-based feedback controller is given by

$$\begin{aligned}
\dot{\xi} &= [A + BS_o Y^{-1} + \gamma^{-2} DD^t Y^{-1} - X G_o^t B^t Y^{-1}]\xi(t) \\
&\quad - M_o[y(t) - C\xi(t)] + A_d \xi(t-d) + B_h S_o Y^{-1}\xi(t-h) \quad (3.85) \\
u(t) &= S_o Y^{-1}\xi(t) \qquad (3.86)
\end{aligned}$$

Proof: By **Theorem 3.1**, the closed-loop time-delay system (3.77) is asymptotically stable for $d, h \geq 0$ and satisfies the inequality

$$\begin{aligned}
W_a &= P_a A_a + A_a^t P_a + Q_d + Q_h + E_a^t E_a + P_a B_a Q_d^{-1} B_a^t P_a \\
&\quad + P_a C_a Q_h^{-1} C_a^t P_a + \gamma^{-2} P_a D_a D_a^t P_a < 0 \qquad (3.87)
\end{aligned}$$

where $0 < P_a^t = P_a \in \Re^{2n\times 2n}$, $0 < Q_d^t = Q_d \in \Re^{2n\times 2n}$ and $0 < Q_h^t = Q_h \in \Re^{2n\times 2n}$. Introducing

$$P_a = \begin{bmatrix} P_1 & 0 \\ 0 & P_2 \end{bmatrix}; \quad Q_d = \begin{bmatrix} Q_{d1} & 0 \\ 0 & Q_{d2} \end{bmatrix}; \quad Q_h = \begin{bmatrix} Q_{h1} & 0 \\ 0 & Q_{h2} \end{bmatrix} \qquad (3.88)$$

Expansion of W_a in (3.86) yields the form

$$W_a = \begin{bmatrix} W_{a1} & W_{a2} \\ W_{a2}^t & W_{a3} \end{bmatrix} \tag{3.89}$$

where

$$
\begin{aligned}
W_{a1} &= P_1(A + BG_o) + (A + BG_o)^t P_1 + Q_{d1} + Q_{h1} + P_1 A_d Q_{d1}^{-1} A_d^t P_1 \\
&\quad + P_1 B_h G_o (Q_{h1}^{-1} + Q_{h2}^{-1}) G_o^t B_h^t P_1 + L^t L + \gamma^{-2} P_1 D D^t P_1 \tag{3.90} \\
W_{a2} &= P_1 B G_o + (A_o - BG_o - A)^t P_2 - \gamma^{-2} P_1 D D^t P_2 \tag{3.91} \\
W_{a3} &= P_2 (A_o - BG_o - B_o C) + (A_o - BG_o - B_o C)^t P_2 + Q_{d2} \\
&\quad + Q_{h2} + P_2 A_d Q_{d2}^{-1} A_d^t P_2 + \gamma^{-2} P_2 D D^t P_2 \tag{3.92}
\end{aligned}
$$

It is readily seen in view of (3.80)-(3.81) that $W_{a2} = 0$. Letting $Y = P_1^{-1}$, $X = P_2^{-1}$, $S_o = G_o Y$, $Q_{dd} = P_1^{-1} Q_{d1} P_1^{-1}$, $Q_{hd} = P_1^{-1} Q_{h1} P_1^{-1}$, $Q_{dh} = P_1^{-1} Q_{h2} P_1^{-1}$, $Q_{hh} = P_2^{-1} Q_{h2} P_2^{-1}$, $Q_{ll} = P_2^{-1} Q_{d2} P_2^{-1}$, and $M_o = -b_o$. We premultiply (3.90) by P_1^{-1} and (3.92) by P_2^{-1}; then postmultiply the results by P_1^{-1} and P_2^{-1} respectively, and manipulating with the aid of **B.1.2** to get the conditions:

$$
\begin{aligned}
&AY + YA^t + BS_o + S_o^t B^t + Q_{dd} + Q_{hd} + YL^t LY + \gamma^{-2} DD^t + \\
&A_d Y Q_{dd}^{-1} Y A_d^t + B_h S_o Q_{dh}^{-1} S_o^t B_h^t + B_h S_o Q_{hd}^{-1} S_o^t B_h^t < 0 \tag{3.93}
\end{aligned}
$$

and,

$$
\begin{aligned}
&(A + \gamma^{-2} DD^t Y^{-1}) X + X(A + \gamma^{-2} DD^t Y^{-1})^t + Q_{ll} + Q_{hh} + \\
&X N_o N_o^t X + \gamma^{-2} DD^t + A_d X Q_{ll}^{-1} X A_d^t + \varphi^{-1} C^t M_o^t M_o C < 0 \tag{3.94}
\end{aligned}
$$

where $M_o = -B_o, \varphi > 0$ and $N_o N_o^t$ is given in (3.84).

In the light of **A.1**, inequalities (3.94)-(3.95) can be conveniently arranged to yield the block forms (3.83) and (3.83), respectively, as desired.

Remark 3.13: To determine the gain factors S_o, Y, X, M_o one has to solve the following problems sequentially:

Problem P1:

$$\min_{Y, S_o, Q_{dd}, Q_{hd}} \gamma$$

$$\text{s. t. } -Y < 0,$$

$$-Q_{dd} < 0, \quad -Q_{hd} < 0, \quad W_6 < 0 \tag{3.95}$$

Problem P2:

$$\min_{X, M_o, Q_{dh}, Q_{hh}} \varphi$$

$$given \ Y, \ S_o, \ \gamma$$

$$s.\,t. \ -X < 0, \ -Q_{dh} < 0, \ -Q_{hh} < 0, \ W_7 < 0 \qquad (3.96)$$

3.3 Simulation Examples

In the following, we present several examples to illustrate the control synthesis methods. The examples differ both in structure and in the associated data information in order to examine the impact of delay factors.

3.3.1 Example 3.1

Consider a dynamical system of form (3.1)-(3.2) with

$$A = \begin{bmatrix} -1 & 0 \\ 0 & -2 \end{bmatrix} ; \ A_d = \begin{bmatrix} -0.2 & 0 \\ -1 & -1 \end{bmatrix} ; \ B = \begin{bmatrix} -1 \\ 1 \end{bmatrix}$$

$$B_h = \begin{bmatrix} -0.1 \\ 0.2 \end{bmatrix} ; \ D = \begin{bmatrix} 0.45 \\ 0.25 \end{bmatrix}$$

$$L = [0.45 \ \ 0.65] ; \ d = 0.1 ; \ h = 0.2$$

In view of Chapter 2, we note that the homogenous part $\dot{x}(t) = A_o x(t) + A_1 x(t - d)$ is unstable since $\mu(A_o) + \|A_1\| = 0.4213 > 0$. Note also that d, h are of the same order of magnitude and the input delay is greater than the state delay; that is $h > d$. Now to determine the gain of state-feedback controller (3.32), we solve problem (3.34) using the software LMI-Control Toolbox [4]. The result is

$$S = [1.1092 \ \ 0.6407]; \ \ Y = \begin{bmatrix} 0.8953 & -0.7290 \\ -0.7290 & 2.4292 \end{bmatrix} ; \gamma_{min} = 0.6031$$

so that the state-feedback control (3.32) takes the form

$$u(t) = Fx(t) = [1.9237 \ \ 0.8410]x(t) \ ; \ \ \|F\| = 2.0996$$

On the other hand, the LMI solution for problem (3.45) has the form

$$Z = \begin{bmatrix} 2.8401 & -3.5612 \\ -3.5612 & 8.0650 \end{bmatrix}; \; \rho = 0.4282 \; ; \; \gamma_{min} = 4.5035$$

$$F = [-0.0404 \quad -0.0065]; \quad ||F|| = 0.0409$$

It should be observed that the gain of controller (3.42) is smaller in magnitude than the gain of controller (3.32).

3.3.2   Example 3.2

Consider the third-order system of the form (3.1)-(3.2) with the matrices

$$A = \begin{bmatrix} -3 & 2 & 0 \\ 1 & -2 & -1 \\ 3 & 0 & -6 \end{bmatrix}; \; A_d = \begin{bmatrix} -1 & 0 & 1 \\ 1 & 1 & 0 \\ 0 & 0 & 2 \end{bmatrix}; \; B = \begin{bmatrix} 1 \\ 4 \\ 2 \end{bmatrix}$$

$$B_h = \begin{bmatrix} 0.1 \\ 0.2 \\ 0.6 \end{bmatrix}; \; D = \begin{bmatrix} 0.2 \\ 0.4 \\ 1 \end{bmatrix}$$

$$L = [1 \quad 0 \quad 1]; \quad d = 0.1, \quad h = 0.2$$

Note that $\mu(A_o) + ||A_1|| = 1.4386 > 0$ but the pair (A, B) is stabilizable. Here the amount of delays d, h are of the same order of magnitude. Using the weighting matrices, $Q_t = 0.2I_3$, $Q_s = 0.4I_3$, the LMI solution results of problem (3.34) are

$$Y = \begin{bmatrix} 0.6364 & -0.2164 & -0.0295 \\ -0.2164 & 1.3851 & -0.1771 \\ -0.0295 & -0.1771 & 0.6224 \end{bmatrix}; \; \gamma_{min} = 0.4756$$

$$S = [-1.0521 \quad -0.2056 \quad 0.0248]$$

so that the state-feedback control (3.32) takes the form

$$u(t) = Fx(t) = [-1.8158 \quad -0.4546 \quad -0.1754]x(t); \quad ||F|| = 1.8801$$

Alternatively, the LMI solution results of problem (3.45) are

$$Z = \begin{bmatrix} 2.2626 & -1.1204 & -1.1657 \\ -1.1204 & 0.9080 & 0.9243 \\ -1.1657 & 0.9243 & 3.2323 \end{bmatrix}; \; \gamma_{min} = 331.7427$$

with

$$
\begin{aligned}
u(t) &= Fx(t) \\
&= [0.1161 \quad 0.2339 \quad -0.0143] \times 10^{-3} x(t) \\
\|F\| &= 2.6151 \, 10^{-4}; \quad \rho = 0.0059
\end{aligned}
$$

In the following two examples, the iterative LMI procedure is used starting from $\sigma = 1$, $\kappa = 1$.

3.3.3 Example 3.3

Consider the dynamical system of **Example 3.1** with $Q_r = 0.1I_2$. Now, to determine the gains of state-delayed feedback controller (3.58) we apply the iterative LMI procedure to problem (3.62) with $\epsilon = 10^{-6}$. The result is obtained when $\sigma = 3$, $\kappa = 5$ as

$$
S = [0.0496 \quad 0.1348]; Y = \begin{bmatrix} 0.5712 & -0.0518 \\ -0.0518 & 0.5644 \end{bmatrix}
$$

$$
V = 10^{-5}[-0.4669 \quad -0.3767]; \gamma_{min} = 1.0209
$$

so that the state-delayed feedback controller (3.58) takes the form

$$
\begin{aligned}
u(t) &= Fx(t) + Kx(t-d) = [0.1094 \quad 0.2490]X(t) \\
&+ 10^{-5}[-0.8854 \quad -0.7487]x(t-d) \\
\|F\| &= 0.2720; \quad \|K\| = 1.1595 \times 10^{-5}
\end{aligned}
$$

3.3.4 Example 3.4

The iterative LMI solution results for the third-order of **Example 3.2** with $Q_r = 0.1I_3$ are summarized for $\sigma = 4$, $\kappa = 7$ by

$$
Y = \begin{bmatrix} 0.2612 & 0.0202 & 0.0778 \\ 0.0778 & 0.3389 & 0.0051 \\ 0.0778 & 0.0051 & 0.2503 \end{bmatrix}; \quad \gamma_{min} = 1.1374
$$

$$
\begin{aligned}
S &= [-0.1125 \quad -0.0873 \quad -0.0287] \\
V &= 10^{-6}[-0.7684 \quad -0.6715 \quad -0.3588]
\end{aligned}
$$

so that the state-feedback control takes the form

$$\begin{aligned}
u(t) &= [-0.4187 \quad -0.2329 \quad 0.0203]x(t) \\
&+ \quad 10^{-5}[-0.2628 \quad -0.1816 \quad -0.0580]x(t-d) \\
\|F\| &= \quad 0.4795; \quad \|K\| = 3.2465 \ 10^{-6}
\end{aligned}$$

3.3.5 Example 3.5

Consider the dynamical system of **Example 3.1** with $C = [0.1 \quad 0]$. To determine the gains of static output-feedback controller (3.71), we solve problem (3.74) using the software LMI-Control Toolbox using the same weighting matrices. The result is

$$\begin{aligned}
\omega &= \quad 0.1441 \ ; \ \Gamma = [1.2111 \quad 0.9522]; \\
Y &= \begin{bmatrix} 2.5603 & -2.9813 \\ -2.9813 & 4.0871 \end{bmatrix}; \ \gamma_{min} = 142.956
\end{aligned}$$

so that the static output-feedback control law (3.71) takes the form

$$u(t) = GCx(t) = [4.8406 \quad 3.7991]x(t)$$

$$\|GC\| = 6.1535$$

3.3.6 Example 3.6

Considering the system treated in Example 3.2, the LMI solution results in

$$Y = \begin{bmatrix} 0.3676 & -0.1086 & 0.0499 \\ -0.1086 & 0.5754 & -0.0156 \\ 0.0499 & -0.0156 & 0.2305 \end{bmatrix}; \ \omega = 0.0522$$

$$[-0.3527 \quad -0.4238 \quad -0.2218]; \ \gamma_{min} = 2.4771$$

so that the static output-feedback control (3.71) takes the form

$$u(t) = Fx(t) = [-0.9359 \quad -0.5594 \quad -0.3442]x(t); \ \|F\| = 1.1434$$

3.3.7 Example 3.7

Consider the second-order system of **Example 3.1** with $C = [0.1 \quad 0]$ and $Q_{dd} = 0.2I_2$, $Q_{hd} = 0.4I_2$, $Q_{dh} = 0.1I_2$ and $Q_{hh} = 0.3I_2$. To determine the gains of observer-based controller (3.85)-(3.86), we solve problem (3.95) first to get:

$$Y = \begin{bmatrix} 2.6931 & -2.8697 \\ -2.8697 & 4.2178 \end{bmatrix}; \; S_o = [1.4705 \quad 0.5254]; \; \gamma_{min} = 3.1770$$

Then we proceed to solve problem (3.96)

$$X = \begin{bmatrix} 0.5639 & -0.1372 \\ -0.1372 & 1.1490 \end{bmatrix}; \; M_o = \begin{bmatrix} -0.0100 \\ -0.0027 \end{bmatrix}; \; \varphi_{min} = 13.5151$$

3.3.8 Example 3.8

Consider the third-order system of **Example 3.2** with $Q_{dd} = 0.2I_3$, $Q_{hd} = 0.4I_3$, $Q_{dh} = 0.1I_3$ and $Q_{hh} = 0.6I_3$. The LMI solution results of problems (3.95) and (3.96) are summarized by

$$Y = 10^3 \begin{bmatrix} 4.2547 & 1.0915 & 0.4948 \\ 1.0915 & 8.9179 & 1.2371 \\ 0.4948 & 1.2371 & 3.3279 \end{bmatrix}; \; \gamma_{min} = 10.8489$$

$$S_o = [1.3709 \quad 20.6155 \quad 4.5595]$$

$$X = \begin{bmatrix} 8.0285 & -1.6558 & -2.1686 \\ -1.6558 & 34.1855 & 18.6520 \\ -2.1686 & 18.6520 & 20.9717 \end{bmatrix}; \; M_o = \begin{bmatrix} -0.0100 \\ 0.0010 \\ -0.0033 \end{bmatrix}; \; \varphi_{min} = 42693$$

3.4 Uncertain Time-Delay Systems

In this section, we address the problems of robust performance and state feedback control synthesis for a class of nominally linear systems with state and input delays as well as time-varying parametric uncertainties. Here, we consider that the delays are time-varying and unknown-but-bounded with known bounds. In order to bring together the robust stabilization results of uncertain time-delay systems into one framework, we consider three classes of uncertainties: matched, mismatched and norm-bounded. We restrict attention on state-feedback and develop sufficient conditions for robust stability and performance for asymptotically-convergent closed-loop controlled systems. These conditions are basically delay-dependent with focus on H_∞-control synthesis schemes and thereby generalizing the available results in the literature.

3.4.1 Problem Statement and Definitions

Consider a class of uncertain time-delay systems of the form:

$$
\begin{aligned}
(\Sigma_\Delta): \quad \dot{x}(t) &= [A + \Delta A(t)]x(t) + [B + \Delta B(t)]u(t) + Dw(t) \\
&+ [A_d + \Delta D(t)]x(t - \tau(t)) + [B_h + \Delta E(t)]u(t - \eta(t)) \\
x(t) &= \phi(t) \quad \forall t \in [-max(\tau, \eta), 0] \\
z(t) &= Lx(t)
\end{aligned}
\tag{3.97}
$$

where $t \in \Re$ is the time, $x \in \Re^n$ is the instantaneous state; $u \in \Re^m$ is the control input; $w(t) \in \Re^q$ is the input disturbance which belongs to $\mathcal{L}_2[0, \infty)$; $z(t) \in \Re^p$ is the controlled output; $\phi(t)$ is a continuous vector valued initial function, and $\tau(t), \eta(t)$ stands for the amount of delay in the state and at the input of the system, respectively, satisfying

$$
\begin{aligned}
0 \le \tau(t) \le \tau^* < \infty, \quad \dot{\tau}(t) \le \tau^+ &< 1 \\
0 \le \eta(t) \le \eta^* < \infty, \quad \dot{\eta}(t) \le \eta^+ &< 1
\end{aligned}
\tag{3.98}
$$

with the bounds τ^*, η^* are known otherwise $\tau(t), \eta(t)$ are unknown. In (3.97) the matrices A, B represent the nominal system and the triplet (A, B, L) is stabilizable and detectable; A_d, B_h, L and D are known constant matrices. Models of dynamical systems of the type (3.97) can be found in several engineering applications [2,220,221]. The problem addressed in this work is that of designing a feedback controller $u(t) = \Psi[x(t)]$ so that the closed-loop system is stabilized in the presence of uncertainties and disturbance is reduced to a prespecified level. Specifically, the objective is to achieve a desirable performance in H_∞-setting [214-219]. In this regard, the H_∞-control problem of interest is to choose a feedback control law $u(t) = Kx(t)$ such that

$$
\begin{aligned}
\max_{0 \ne w \in \mathcal{L}_2} \int_0^\infty & \quad [z^t(t)z(t) - \gamma^2 w^t w(t)]dt \le 0 \\
\dot{x}(t) &= Ax(t) + Bu(t) + A_d x(t - \tau(t)) \\
&+ B_h u(t - \eta(t)) + Dw(t), \quad x(t) = 0 \\
z(t) &= Lx(t)
\end{aligned}
\tag{3.99}
$$

Distinct from (3.97) are the following systems:

$$
\begin{aligned}
(\Sigma_{\Delta o}): \quad \dot{x}(t) &= Ax(t) + Bu(t) + A_d x(t - \tau(t)) + B_h u(t - \eta(t)) \\
z(t) &= Lx(t) \tag{3.100} \\
(\Sigma_{\Delta w}): \quad \dot{x}(t) &= Ax(t) + Bu(t) + A_d x(t - \tau(t)) + B_h u(t - \eta(t)) \\
&\quad + Dw(t) \\
z(t) &= Lx(t) \tag{3.101} \\
(\Sigma_{\Delta wo}): \quad \dot{x}(t) &= [A + \Delta A(t)]x(t) + [B + \Delta B(t)]u(t) \\
&\quad + [A_d + \Delta D(t)]x(t - \tau(t))[B_h + \Delta E(t)]u(t - \eta(t)) \\
z(t) &= Lx(t) \tag{3.102}
\end{aligned}
$$

We will focus attention on system $(\Sigma_{\Delta wo})$ since it includes systems $(\Sigma_{\Delta o})$ and $(\Sigma_{\Delta w})$ as special cases.

3.4.2   Closed-Loop System Stability

Consider system $(\Sigma_{\Delta wo})$ subject to (3.98) and the state-feedback control $u(t) = Kx(t)$. The following theorem provides stability conditon of the closed-loop system:

$$
\begin{aligned}
\dot{x}(t) &= \{[A + \Delta A(t)] + [B + \Delta B(t)]K\}x(t) \\
&\quad + [A_d + \Delta D(t)]x(t - \tau(t)) + [B_h + \Delta E(t)]Kx(t - \eta(t)) \\
z(t) &= Lx(t) \tag{3.103}
\end{aligned}
$$

Theorem 3.9: *The closed-loop system (3.103) is asymptotically stable for delays $\tau(t)$, $\eta(t)$ satisfying (3.99) and given $0 < Q_c^t = Q_c \in \Re^{n \times n}, 0 < Q_u^t = Q_u \in \Re^{n \times n}$ if one of the following conditions is satisfied:*
(1) *There exists a matrix $0 < P^t = P \in \Re^{n \times n}$ solving the LMI:*

$$
W_1 =
\begin{bmatrix}
\Pi(P) & PD_o + P\Delta D(t) & PE_oK + P\Delta E(t)K \\
D_o^t P + \Delta D^t(t)P & -R_c & 0 \\
K^t E_o^t P + K^t \Delta E^t(t)P & 0 & -R_u
\end{bmatrix}
< 0
\tag{3.104}
$$

$$
\begin{aligned}
\Pi(P) &= PA_c + A_c^t P + Q_c + Q_u \\
&\quad + \Delta A^t(t)P + P\Delta A(t) + K^t \Delta B^t(t)P + P\Delta B(t)K \\
A_c &= A + BK \;\; ; \;\; R_c = Q_c(1 - \tau^+) \;, \;\; R_u = Q_u(1 - \eta^+) \tag{3.105}
\end{aligned}
$$

(2) *There exist matrices* $0 < P^t = P \in \Re^{n \times n}$ *satisfying the ARI:*

$$PA_c + A_c^t P + Q_c + Q_u + \Delta A^t(t)P + P\Delta A(t) +$$
$$P(D_o + \Delta D(t))R_c^{-1}(D_o^t + \Delta D^t(t))P + K^t \Delta B^t(t)P + P\Delta B(t)K +$$
$$P(E_o K + \Delta E(t)K)R_u^{-1}(K^t E_o^t + K^t \Delta E^t(t))P < 0 \qquad (3.106)$$

Proof: (1) Define a Lyapunov-Krasovskii functional $V_{13}(x_t)$ as

$$V_{13}(x_t) = x^t(t)Px(t) + \int_{t-\tau(t)}^{t} x^t(v)Q_c x(v)dv$$

$$+ \int_{t-\eta(t)}^{t} x^t(v)Q_u x(v)dv \qquad (3.107)$$

where $0 < P = P^t$, $0 < Q_c = Q_c^t$, $0 < Q_u = Q_u^t$. Observe that $V_{13}(x_t) > 0$, $x \neq 0$; $V_{13}(x_t) = 0$, $x = 0$. Evaluating $\dot{V}_{13}(x_t)$ along the solutions of (3.103) using (3.105) and with some manipulations in view of (3.98) we get:

$$\dot{V}_{13}(x_t) < x^t(t)[PA_c + A_c^t P + P\Delta A + \Delta A^t P + K^t \Delta B^t P + P\Delta BK]x(t)$$
$$+x^t(t)[PD_o + P\Delta D]x(t-\tau) + x^t(t-\tau)[D_o^t P + \Delta D^t P]x(t)$$
$$-x^t(t-\tau)R_c x(t-\tau) - x^t(t-\eta)R_u x(t-\eta)$$
$$+x^t(t)[PE_o K + P\Delta EK]x(t-\eta)$$
$$+x^t(t-\eta)[K^t E_o^t P + K^t \Delta E^t P]x(t)$$
$$+x^t(t)[Q_c + Q_u]x(t)$$
$$= Z_3^t(t)W Z_3(t) \qquad (3.108)$$

where $Z_3(t) = [x^t(t) \ x^t(t-\tau) \ x^t(t-\eta)]^t$ and W_1 is given by (3.104). For a given realization $\Delta A(t)$, $\Delta B(t)$, $\Delta D(t)$, $\Delta E(t)$ and a state-feedback gain K if $\dot{V}(x,t) < 0$ when $x \neq 0$, then $x \to 0$ as $t \to \infty$ and the asymptotic stability is guaranteed. This condition follows from (3.104). Therefore, we conclude that the controlled system (3.103) is stable for τ, η satisfying (3.98).

(2) By **A.1**, ARI (3.106) is equivalent to LMI (3.104).

Remark 3.14: It is significant to observe that the problem of determining the stability of the uncertain time-delay system (3.103) is converted to an LMI feasibility problem. This problem is convex but if it turns out to be infeasible, it means that system (3.103) cannot be stabilized. Indeed, neither form (3.104) nor (3.106) is directly amenable for direct computation due to the presence of uncertainty. However, their usefulness lies in their general format which will serve as the cornerstone in the subsequent analysis and design. Finally, note that the existence of K is guaranteed by the stabilizability-detectability of the triplet (A, B, L).

Corollary 3.7: *System $(\Sigma_{\Delta o})$ is asymptotically stable via state-feedback $u(t) = Kx(t)$ for τ, η satisfying (3.98) if given $0 < Q_c^t = Q_c \in \Re^{n \times n}$, $0 < Q_u^t = Q_u \in \Re^{n \times n}$ there exists matrix $0 < P^t = P \in \Re^{n \times n}$ solving the LMI:*

$$
W_{12} = \begin{bmatrix} PA_c + A_c^t P + Q_c + Q_u & PA_d & PB_h K \\ A_d^t P & -R_c & 0 \\ K^t B_h^t P & 0 & -R_u \end{bmatrix} < 0 \qquad (3.109)
$$

Proof: Setting $\Delta A \equiv 0, \Delta B \equiv 0, \Delta D \equiv 0$ and $\Delta E \equiv 0$ in (3.104).

3.5 Nominal Control Synthesis

We first provide a sufficient condition for the asymptotic stability of the nominal time-delay system (6).

Theorem 3.10: *System $(\Sigma_{\Delta w})$ is asymptotically stable with disturbance attenuation γ via a memoryless state-feedback controller for τ, η satisfying (3.98) if given matrices $0 < Q_t^t = Q_t \in \Re^{n \times n}$ and $0 < Q_s^t = Q_s \in \Re^{n \times n}$ there exist matrices $0 < Y^t = Y \in \Re^{n \times n}$ and $S \in \Re^{m \times n}$ satisfying the LMI:*

$$
W_2 = \begin{bmatrix} AY + YA^t + Q_t + Q_s & YL^t & B_h S & A_d & D \\ +BS + S^t B^t & & & & \\ LY & -I & 0 & 0 & 0 \\ S^t B_h^t & 0 & -R_s & 0 & 0 \\ A_d^t & 0 & 0 & -R_c & 0 \\ D^t & 0 & 0 & 0 & -\gamma^2 I \end{bmatrix} < 0 \quad (3.110)
$$

Moreover, the gain of the memoryless state-feedback controller is given by

$$
K = SY^{-1} \qquad (3.111)
$$

Proof: We choose $V_{13}(x_t)$ as in (3.108). From $\mathcal{H}_\infty$ theory [214-219], it is known that the $\mathcal{L}_2$-induced norm from $v(t) = \gamma w(t)$ to $z(t)$ is less than unity if the Hamiltonian

$$
H(x, v, t) = \dot{V}(x, t) + [z^t z - v^t v] < 0 \qquad (3.112)
$$

A straightforward computation of $H(x,v,t)$ and taking into consideration (3.98) yields:

$$
\begin{aligned}
H(x,v,t) \;<\; & \Big(x^t(t)[PA + A^tP + Q_c + Q_u + PBK + K^tB^tP + L^tL]x(t)\Big) \\
+ & \Big(x^t(t)PA_d x(t-\tau) + x^t(t-\tau)A_d^t Px(t)\Big) \\
+ & \Big(\gamma^{-1}x^t(t)PDv(t) + \gamma^{-1}v^t(t)D^t Px(t) - v^t(t)v(t)\Big) \\
+ & \Big(x^t(t)PB_h Kx(t-\eta) + x^t(t-\eta)K^tB_h^t Px(t)\Big) \\
- & \Big(x^t(t-\tau)R_c x(t-\tau) + x^t(t-\eta)R_u x(t-\eta)\Big) \\
= \; & Z_2^t(t)W_\gamma Z_2(t) \hspace{4cm} (3.113)
\end{aligned}
$$

where $Z_4(t) = [x^t(t)\; x^t(t-\tau)\; x^t(t-\eta)\; v^t(t)]^t$ and

$$
W_\gamma = \begin{bmatrix}
\begin{matrix} PA + A^tP + Q_c + Q_u \\ + L^tL + K^tB^tP + PBK \end{matrix} & PA_d & PB_h K & \gamma^{-1}PD \\
A_d^t P & -R_c & 0 & 0 \\
K^tB_h^t P & 0 & -R_u & 0 \\
\gamma^{-1}D^t P & 0 & 0 & -I
\end{bmatrix}
\hspace{1cm} (3.114)
$$

Note that $H(x,v,t) < 0$ corresponds to $W_\gamma < 0$. Using **A.1**, we get

$$
\begin{aligned}
& PA + A^tP + Q_c + Q_u + L^tL + K^tB^tP + PBK + \\
& PA_d R_c^{-1} A_d^t P + PB_h K R_u^{-1} K^t B_h^t P + \gamma^{-2}PDD^t P < 0 \quad (3.115)
\end{aligned}
$$

Inequality (3.115) is not convex in P and K, however, with the substitutions of $Y = P^{-1}$, $S = KY$, $Q_t = P^{-1}Q_c P^{-1}$, $Q_s = P^{-1}Q_u P^{-1}$, $R_s = Q_s(1-\eta^+)$, pre- and postmultiplication by P^{-1}, we get:

$$
\begin{aligned}
& AY + YA^t + Q_t + Q_s + L^tL + S^tB^t + BS + A_d R_c^{-1} A_d^t + \\
& B_h S R_s^{-1} S^t B_h^t + \gamma^{-2}HH^t < 0 \hspace{3cm} (3.116)
\end{aligned}
$$

which is now convex in both Y and S. By **A.1**, (3.116) is equivalent to (3.110).

Remark 3.15: Theorem 3.10 provides a sufficient delay-dependent condition for a memoryless H_∞-controller guaranteeing the norm bound γ. It is expressed in the easily computable LMI format. To implement such a controller, one has to solve the following minimization problem using the LMI-Control Toolbox:

$$\underset{Y, S, Q_s, Q_t}{Min} \quad \gamma$$

$$subject\ to\ -Y < 0, \quad -Q_s < 0,$$
$$-Q_t < 0\ , W_2 < 0$$

3.5.1 Example 3.9

Consider a fourth-order system modeled in the format (3.97) with nominal data

$$A = \begin{bmatrix} -6 & -2.236 & 0 & 0 \\ 2.236 & 0 & 0 & 0 \\ 0 & 0 & -0.6 & -0.2236 \\ 0 & 0 & 0.2236 & 0 \end{bmatrix} ; \ B = \begin{bmatrix} 1.0 & 0 \\ 0.5 & 0 \\ 0 & 0.2 \\ 0 & 1.0 \end{bmatrix}$$

$$A_d = \begin{bmatrix} 0.1 & 0.05 & 0 & 0 \\ 0 & 0 & 0.3 & 0.02 \\ 0.1 & 0 & -0.6 & -0.2236 \\ 0 & 0 & 0.2 & 0.01 \end{bmatrix} ; \ B_h = \begin{bmatrix} 0.2 & 0 \\ 1.0 & 0 \\ 0 & 0.2 \\ 0 & 1 \end{bmatrix}$$

$$D = \begin{bmatrix} 0.1 \\ 0.05 \\ 0 \\ 0 \end{bmatrix}, \ L = \begin{bmatrix} 1.0 & 0.5 & 0.2 & 0.1 \end{bmatrix}$$

In order to demonstrate **Theorem 3.10**, we use the following weighting factors: $Q_t = diag(0.2, 0.2, 0.2, 0.2)$ and $Q_s = diag(0.4, 0.4, 0.4, 0.4)$. Hence we get

$$R_c = diag(6.5518, 6.5518, 6.5518, 6.5518)$$

$$R_s = diag(0.8867, 0.8867, 0.8867, 0.8867)$$

Using the LMI-Control Toolbox, the solution is given by:

$$Y = \begin{bmatrix} 1.6317 & -1.6317 & -0.0766 & 0.0186 \\ -1.6173 & 5.3262 & -0.3098 & -0.3298 \\ -0.0766 & -0.3098 & 6.6404 & -5.4220 \\ 0.0186 & -0.3298 & -5.4220 & 13.4592 \end{bmatrix}, \ \gamma_{min} = 2.8029$$

$$S = \begin{bmatrix} -0.3458 & -0.2620 & -0.0156 & -0.1749 \\ -0.0534 & -0.0442 & -0.1243 & -0.1085 \end{bmatrix}$$

$$K = \begin{bmatrix} -0.3817 & -0.1696 & -0.0421 & -0.0336 \\ -0.0667 & -0.0326 & -0.0420 & -0.0257 \end{bmatrix}, \quad \|K\| = 0.4287$$

3.6 Uncertainty Structures

The uncertainties within system (3.97) are represented by the real-valued matrix functions $\Delta A(t)$, $\Delta B(t)$, $\Delta D(t)$, $\Delta E(t)$. Characterization of these functions for state-space models reveals the nature of uncertainty structure. In what follows we provide some of these structures:

Class I: (Matched Uncertainties)
The matrix functions $\Delta A(t)$, $\Delta B(t)$, $\Delta D(t)$, $\Delta E(t)$ are assumed to have the form

$$[\Delta A(t) \quad \Delta B(t) \quad \Delta D(t) \quad \Delta E(t)] = B[A_1(t) \quad B_1(t) \quad D_1(t) \quad E_1(t)];$$
$$\forall t \in \Re \tag{3.117}$$

in which the uncertainties are restricted to lie in the range-space of the input matrix B.

Class II: (Mismatched Uncertainties)
Here, the matrix functions $\Delta A(t)$, $\Delta B(t)$, $\Delta D(t)$, $\Delta E(t)$, are assumed to have the form

$$[\Delta A(t) \quad \Delta B(t) \quad \Delta D(t) \quad \Delta E(t)] = B[A_1(t) \quad B_1(t) \quad D_1(t) \quad E_1(t)]$$
$$+[A_2(t) \quad B_2(t) \quad D_2(t) \quad E_2(t)]; \quad \forall t \in \Re \tag{3.118}$$

which consists of two parts: a matched part (e.g., BA_1) and a mismatched part (e.g., A_2).

Class III: (Norm-Bounded Uncertainties)
Let the matrix functions $\Delta A(t)$, $\Delta B(t)$, $\Delta D(t)$, $\Delta E(t)$ be expressed $\forall t \in \Re$

$$[\Delta A(t) \quad \Delta B(t)] = H\Delta(t)[H \quad E_b], \quad \Delta^t(t)\Delta(t) \leq \sigma_1 I$$
$$\Delta E(t) = H_e F_e(t) E_e, \quad \Delta_e^t(t)\Delta_e(t) \leq \sigma_2 I,$$
$$\Delta D(t) = H_d F_d(t) E_d, \quad \Delta_d^t(t)\Delta_d(t) \leq \sigma_3 I$$
$$0 < \sigma_1, \quad \sigma_2, \quad \sigma_3 < 1 \tag{3.119}$$

where the elements of $\Delta_{ij}(t)$, $(\Delta_e(t))_{ij}$, $(\Delta_d(t))_{ij}$ are Lebsegue measurable $\forall i, j$; $\Delta(t) \in \Re^{\alpha_f \times \beta_f}$, $\Delta_e \in \Re^{\alpha_e \times \beta_e}$, $\Delta_d \in \Re^{\alpha_d \times \beta_d}$, and $H, E, E_b, H_d, H_e, E_e, E_d$ are constant matrices.

3.6.1 Control Synthesis for Matched Uncertainties

By applying the state-feedback $u(t) = Kx(t)$ to system (3.97) subject to (3.117), we get the closed-loop system

$$
\begin{aligned}
\dot{x}(t) &= [A + BK + B\{A_1(t)K + B_1(t)K\}]x(t) \\
&+ [A_d + BD_1(t)]x(t - \tau(t)) + [B_h + BE_1(t)]Kx(t - \eta(t)) \\
&+ Dw(t) \\
z(t) &= Lx(t) \tag{3.120}
\end{aligned}
$$

Introducing

$$
\epsilon_a = \sup_t \lambda_M[A_1(t)A_1^t(t)]; \quad \epsilon_b = \sup_t \lambda_M[B_1(t)B_1^t(t)]
$$

$$
\epsilon_d = \sup_t \lambda_M[D_1(t)D_1^t(t)]; \quad \epsilon_e = \sup_t \lambda_M[E_1(t)E_1^t(t)] \tag{3.121}
$$

Theorem 3.11: *The closed-loop system (3.120) is stable with distur-bance attenuation γ via a memoryless state-feedback controller for τ, η sat-isfying (3.98) if given matrices $0 < Q_t^t = Q_t \in \Re^{n \times n}$, $0 < Q_s^t = Q_s \in \Re^{n \times n}$, and scalars $\alpha_1 > 0$, $\alpha_2 > 0$, $\alpha_3 > 0$, $\alpha_4 > 0$, α_5 such that $\alpha_3^{-1}B^t B < I$ there exist matrices $0 < Y^t = Y \in \Re^{n \times n}$ and $S \in \Re^{m \times n}$ satisfying the LMI:*

$$
W_{\gamma 1} = \begin{bmatrix} AY + YA^t + Q_t + Q_s \\ +BS + S^t B^t + \alpha_3 \epsilon_d R_c^{-1} & \bar{S} & \bar{B} \\ \bar{S}^t & -J_s & 0 \\ \bar{B}^t & 0 & -J_h \end{bmatrix} < 0 \tag{3.122}
$$

$$
\bar{S} = [Y \ S^t \ A_d] \ , \quad \bar{H} = [B \ B_h S \ \gamma^{-1}D]
$$
$$
R_s = Q_s(1 - \eta^+) \quad , R_{cc}^{-1} = R_c^{-1/2}[I - \alpha_3^{-1}B^t B]^{-1}(R_c^{-1/2})^t
$$
$$
L = L^t L + \alpha_1^{-1} I \quad , R_{ss}^{-1} = (R_s^{-1/2})^t[I - \alpha_4 B^t B](R_c^{-1/2})
$$
$$
J_s = diag[L^{-1} \ \alpha_2 I \ R_{cc}] \ , \quad J_h = diag[\alpha_t I \ R_{ss} \ I]
$$
$$
\alpha_t = \epsilon_a \alpha_1 + \epsilon_b \alpha_2 + \epsilon_e \alpha_5 \tag{3.123}
$$

Moreover, the gain of the memoryless state-feedback controller is given by

$$
K = S Y^{-1} \tag{3.124}
$$

Proof: We start by **Theorem 3.10** and make the following changes

$$
A_c \rightarrow A + BK + B(A_1 + B_1 K) \ ; \quad A_d \rightarrow A_d + BD_1 \ ; \quad B_h K \rightarrow B_h K + BE_1 K \tag{3.125}
$$

we obtain the Hamiltonian

$$H(x, v, t) < Z_2^t(t) W_I Z_2(t)$$

$$Z_2(t) = \begin{bmatrix} x^t(t) & x^t(t-\tau) & x^t(t-\eta) & v^t(t) \end{bmatrix}^t \qquad (3.126)$$

where

$$W_I =$$

$$\begin{bmatrix}
\Omega(P) & PA_d + PBD_1 & PB_hK + PBE_1K & \gamma^{-1}PD \\
A_d^tP + D_1^tB^tP & -R_c & 0 & 0 \\
K^tA_d^tP + K^tE_1^tB^tP & 0 & -R_u & 0 \\
\gamma^{-1}D^tP & 0 & 0 & -I
\end{bmatrix}$$

$$\Omega(P) = PA + A^tP + Q_c + Q_u + L^tL + PBA_1 +$$
$$A_1^tB^tP + K^tB^tP + PBK + PBB_1K + K^tB_1^tB^tP \qquad (3.127)$$

Evidently the sufficient stability condition $H(x, v, t) < 0$ corresponds to $W_I < 0$. Now applying **A.1**, the latter condition is equivalent to the nonstandard ARI:

$$\Omega(P) + \gamma^{-2}PDD^tP + (PA_d + PBD_1)R_c^{-1}(A_d^tP + D_1^tB^tP)$$
$$+ (PA_dK + PBE_1K)R_u^{-1}(K^tA_d^tP + K^tE_1^tB^tP) < 0 \qquad (3.128)$$

which can be shown to be non-convex in P and K. To remove this constraint, we substitute $Y = P^{-1}$, $S = KY$, $Q_t = P^{-1}Q_cP^{-1}$, $Q_s = P^{-1}Q_uP^{-1}$, $R_s = Q_s(1 - \eta^+)$, $R_{ss} = R_s - \alpha_4 S^tS$, then pre- and postmultiplying by P^{-1}, we get the matrix inequality:

$$AY + YA^t + Q_t + Q_s + (A_d + BD_1)R_c^{-1}(A_d^t + D_1^tB^t) +$$
$$YL^tLY + \gamma^{-2}DD^t + YA_1^tB^t + BA_1Y$$
$$+ BB_1S + S^tB_1^tB^t +$$
$$(B_h + BE_1)SR_s^{-1}S^t(B_h^t + E_1^tB^t) + S^tB^t + BS < 0 \qquad (3.129)$$

which is now convex in both Y and S. Using **B.1.2**, **B.1.3**, (3.122) and selecting $\alpha_1 > 0$, $\alpha_2 > 0$, $\alpha_3 > 0$, $\alpha_4 > 0$, $\alpha_5 = \alpha_4^{-1}$ such that $\alpha_3^{-1}B^tB < I$, we obtain the following bounds:

$$BA_1Y + YA_1^tB^t \leq B_o(\alpha_1\epsilon_a I)B^t + Y(\alpha_1^{-1}I)Y$$
$$BB_1S + S^tB_1^tB^t \leq B(\alpha_2\epsilon_b I)B^t + S^t(\alpha_2^{-1}I)S$$
$$(A_d + BD_1)R_c^{-1}(A_d^t + D_1^tB^t) \leq \alpha_3R_c^{-1}(\epsilon_d I) + A_dR_{cc}^{-1}A_d^t$$
$$(B_h + BE_1)SR_s^{-1}S^t(B_h^t + E_1^tB^t) \leq \alpha_5E_1SR_s^{-1}S^tE_1^t + B_hSR_{ss}^{-1}S^tB_h^t$$
$$\qquad (3.130)$$

where $R_{cc}^{-1} = R_c^{-1/2}[I - \alpha_3^{-1}B^t B]^{-1}(R_c^{-1/2})^t$; $\alpha_3^{-1}B^t B < I$. By substituting inequalities (3.130) into (3.129) and grouping similar terms, we obtain:

$$AY + YA^t + Q_t + Q_s + A_d R_{cc}^{-1} A_d^t + YLY +$$
$$S^t(\alpha_2^{-1}I)S + \alpha_3 \epsilon_d R_c^{-1} +$$
$$\gamma^{-2}DD^t + B(\alpha_t I)B^t +$$
$$B_h S R_{ss}^{-1} S^t B_h^t + S^t B_o^t + B_o S < 0 \qquad (3.131)$$

where $\alpha_3^{-1}B_o^t B_o < I$. Note that for a given ϵ_a, ϵ_b, ϵ_e then α_t then is affinely linear in $\alpha_1, \alpha_2, \alpha_5$. Simple rearrangement of (3.131) using A.1 yields the LMI (3.122).

Remark 3.16: Theorem 3.11 provides a sufficient delay-dependent condition for a memoryless H_∞-controller guaranteeing the norm bound γ and it is expressed in the easily computable LMI format. To implement such a controller one has to solve the following minimization problem:

$$\underset{Y, S, Q_s, Q_t, \alpha_1, \ldots, \alpha_4, \alpha_5}{Min} \quad \gamma$$

$$s.t. \ to \ -Y < 0, \ -Q_s < 0,$$
$$-Q_t < 0, \ -R_{ss} < 0, \ W_{\gamma 1} < 0$$

3.6.2 Example 3.10

Here, we consider **Example 3.9** in addition to the set of uncertainties $\{\Delta A(t), \Delta B(t), \Delta D(t), \Delta E(t)\}$ satisfying the matching condition (3.117) with

$$A_1(t) = \begin{bmatrix} 0.1 \ sin(3t) & -0.001 & 0 & 0 \\ 0 & 0 & 0.05 & -0.02 \ cos(3t) \end{bmatrix}$$

$$B_1(t) = \begin{bmatrix} 0.2 \ cos(2t) & 0 \\ 0.1 & 0 \end{bmatrix}; \quad E_1(t) = \begin{bmatrix} 0.01 & 0.02 \\ 0.1 \ sin(5t) & 0.1 \end{bmatrix}$$

$$D_1(t) = \begin{bmatrix} 0.05 \ cos(t) & 0 & 0 & 0 \\ 0 & 0 & 0.05 & 0.5 \end{bmatrix}$$

First, we evaluate the norms in (3.121) over the period $[-2, 15]$ to give $\epsilon_a = 0.0098$, $\epsilon_b = 0.0366$, $\epsilon_d = 0.25$, $\epsilon_e = 0.011$. Selecting the same weighting factors of **Example 3.9** plus $\alpha_1 = 8$, $\alpha_2 = 4$, $\alpha_3 = 2$, $\alpha_4 = 2$, $R_{ss} = diag(5, 5, 5, 5)$. This selection yields

$$R_{cc} = \begin{bmatrix} 3.2759 & -1.6379 & 0 & 0 \\ -1.6379 & 5.7328 & 0 & 0 \\ 0 & 0 & 3.2759 & -1.6379 \\ 0 & 0 & -1.6379 & 5.7328 \end{bmatrix}$$

Finally using the LMI-Control Toolbox, we obtain

$$Y = \begin{bmatrix} 8.6962 & -5.9864 & -0.6429 & 0.0491 \\ -5.8964 & 8.9217 & -0.0073 & -0.0024 \\ -0.6429 & -0.0073 & 3.2644 & -0.5508 \\ 0.0491 & -0.0024 & -0.5508 & 0.7626 \end{bmatrix}, \quad \gamma_{min} = 0.1055$$

$$S = \begin{bmatrix} -0.8505 & -0.15421 & -0.4296 & 0.0887 \\ -0.0394 & -0.0045 & 0.7347 & -2.0856 \end{bmatrix}$$

$$K = \begin{bmatrix} -0.4333 & -0.4638 & -0.2208 & -0.0167 \\ -0.0863 & -0.06 & -0.8002 & -3.3075 \end{bmatrix}, \quad \|K\| = 3.4057$$

To examine the sensitivity of the obtained results to the set of initial data, the computational algorithm I is executed iteratively to obtain feasible solutions while changing the set $\{\alpha_1, \alpha_2, \alpha_3, \alpha_4\}$ around the base value $8, 4, 8, 2$ and observing the variation in gain K as measured by $\|K\|$. It has been found that:

(1) Varying the factor α_1 only over the range (8-40) results in changing $\|K\|$ from (3.4057) to (1.0728), that is, as α_1 is increased by 5 times, $\|K\|$ decreases by about 62.55 percent.

(2) Varying the factor α_2 only over the range (4-20) results in changing $\|K\|$ from (3.4057) to (2.2910), that is, as α_2 is increased by 5 times, $\|K\|$ decreases by about 28 percent.

(3) Varying the factor α_3 only over the range (2-8) results in changing $\|K\|$ from (0.9071) to (3.4057), that is, as α_3 is increased by four times, $\|K\|$ increases by more than 260 percent.

(4) Varying the factor α_4 only over the range (2-10) results in changing $\|K\|$ from (3.4057) to (3.3876), that is, as α_4 is increased by 5 times, $\|K\|$ remains almost constant.

3.6.3 Control Synthesis for Mismatched Uncertainties

In this case, we use (3.118) to get the closed-loop controlled system

$$
\begin{aligned}
\dot{x}(t) &= [A + B\{A_1(t) + B_1(t)K\} + \{A_2(t) + B_2(t)K\}]x(t) \\
&+ [A_d + BD_1(t) + D_2(t)]x(t - \tau(t)) \\
&+ [B_h + BE_1(t) + E_2(t)]Kx(t - \eta(t)) + Dw(t) \\
z(t) &= Lx(t) \qquad\qquad\qquad\qquad\qquad\qquad (3.132)
\end{aligned}
$$

Introducing

$$
\epsilon_a = \sup_t \lambda_M[A_1(t)A_1^t(t)], \quad \epsilon_b = \sup_t \lambda_M[B_1(t)B_1^t(t)]
$$

$$
\epsilon_c = \sup_t \lambda_M[A_2(t)A_2^t(t)], \quad \epsilon_d = \sup_t \lambda_M[B_2(t)B_2^t(t)]
$$

$$
\epsilon_e = \sup_t \lambda_M[D_1(t)R_c^{-1}D_1^t(t)], \quad \epsilon_f = \sup_t \lambda_M[D_2(t)R_c^{-1}D_2^t(t)]
$$

$$
\epsilon_g = \sup_t \lambda_M[E_1(t)E_1^t(t)], \quad \epsilon_h = \sup_t \lambda_M[E_2(t)R_k^{-1}E_2^t(t)] \qquad (3.133)
$$

Theorem 3.12: *The closed-loop system (3.132) is asymptotically stable with disturbance attenuation γ via a memoryless state-feedback controller for τ, η satisfying (3.98) if given matrices $0 < Q_t^t = Q_t \in \Re^{n \times n}$, $0 < Q_s^t = Q_s \in \Re^{n \times n}$, $0 < R_k = R_k^t \in \Re^{m \times m}$ and scalars $\delta_1 > 0, ..., \delta_{10} > 0$ and $\varphi_1 > 0,..., \varphi_{10} > 0$ there exist matrices $0 < Y^t = Y \in \Re^{n \times n}$ and $S \in \Re^{m \times n}$ satisfying the LMI:*

$$
W_{\gamma 2} = \begin{bmatrix} \Lambda_1(Y) & \Lambda_2 & \Lambda_3 \\ \Lambda_2^t & -J_b & 0 \\ \Lambda_3^t & 0 & -J_d \end{bmatrix} < 0 \qquad (3.134)
$$

where

$$
\Lambda_1(Y) = AY + YA^t + Q_t + Q_s + BS + S^t B^t + \delta_v
$$

$$
\Lambda_2 = [Y \quad S^t \quad B] \; ; \; J_b = [L_g \quad \delta_p I \quad \delta_q I]
$$

$$
\Lambda_3 = [B_h \quad A_d \quad D] \; ; \; J_d = [R_{km} \quad R_{cd} \quad \gamma^2 I]
$$

$$
R_{km}^{-1} = \delta_m R_k^{-1}; \quad \delta_v = \epsilon_c \delta_2 + \epsilon_d \delta_4 + \epsilon_f \delta_e + \epsilon_h \delta_n
$$

$$
\delta_p = \varphi_2 + \varphi_4; \quad L_g = L^t L + (\varphi_1 + \varphi_2)I;
$$

$$
\delta_t = 1 + \delta_5 + \delta_6; \quad \delta_e = 1 + \delta_7 + \varphi_5; \quad \delta_r = 1 + \varphi_9 + \varphi_{10};
$$

$$
\delta_s = 1 + \varphi_6 + \varphi_7; \quad \delta_m = 1 + \delta_8 + \delta_9; \quad \delta_n = 1 + \delta_{10} + \varphi_8 \qquad (3.135)
$$

Moreover, the gain of the memoryless state-feedback controller is given by

$$
K = SY^{-1} \qquad (3.136)
$$

Proof: Applying **Theorem 3.10** along with the changes

$$A_c \to A + BK + B(A_1 + B_1K) + A_2 + B_2K \; ; \; A_d \to A_d + BD_1 + D_2 \; ;$$
$$B_hK \to B_hK + BE_1K + E_2K \tag{3.137}$$

we obtain the Hamiltonian in the form:

$$H(x,v,t) < Z_3^t(t)W_{II}Z_3(t)$$

$$Z_3(t) = \begin{bmatrix} x^t(t) & x^t(t-\tau) & x^t(t-\eta) & v^t(t) \end{bmatrix}^t$$

$$W_{II} = \begin{bmatrix} W_{II,1} & W_{II,2} & W_{II,3} & \gamma^{-1}PD \\ W_{II,2}^t & -R_c & 0 & 0 \\ W_{II,3}^t & 0 & -R_u & 0 \\ \gamma^{-1}D^tP & 0 & 0 & -I \end{bmatrix}$$

$$\begin{aligned}
W_{II,1} &= PA + A^tP + Q_c + Q_u + PBK + K^tB^tP \\
&+ P(BA_1 + A_2) + (A_1^tB^t + A_2^t)P + L^tL \\
&+ P(BB_1 + B_2)K + K^t(B_1^tB^t + B_2^t)P \\
W_{II,2} &= PA_d + PBD_1 + PD_2 \\
W_{II,3} &= PB_hK + PBE_1K + PE_2K
\end{aligned} \tag{3.138}$$

From (3.138), the stability sufficient condition $H(x,v,t) < 0$ corresponds to $W_{II} < 0$. By applying **A.1**, this condition is equivalent to:

$$W_{II,1} + W_{II,2}R_c^{-1}W_{II,2}^t + W_{II,3}R_u^{-1}W_{II,3}^t + \gamma^{-2}PDD^tP < 0 \tag{3.139}$$

To convexify (3.139), we first substitute $Y = P^{-1}, S = KY, Q_t = P^{-1}Q_cP^{-1}$, $Q_s = P^{-1}Q_uP^{-1}, R_k^{-1} = SPR_u^{-1}PS^t$, then pre- and postmultiplying by P^{-1}, to get :

$$\begin{aligned}
&AY + YA^t + Q_t + Q_s + YA_1^tB^t + BA_1Y + YL^tLY + \\
&B_2S + S^tB_2^t + (A_d + BD_1 + D_2)R_c^{-1}(A_d^t + D_1^tB^t + D_2^t) + \gamma^{-2}DD^t + \\
&+BS + S^tB^t + (B_h + BE_1 + E_2)R_k^{-1}(B_h^t + E_1^tB^t + E_2^t) \\
&+YA_2^t + A_2Y + BB_1S + S^tB_1^tB^t < 0
\end{aligned} \tag{3.140}$$

Using **B.I.2**, (3.135) and by selecting scalars $\delta_1 > 0$, ..., $\delta_{10} > 0$ and $\varphi_1 > 0,..., \varphi_{10} > 0$ such that $\varphi_j = \delta_j^{-1}; \; j = 1, ..., 10$, with some algebraic manipulations, we obtain the following bounds:

$$BA_1Y + YA_1^tB^t \leq B(\delta_1\epsilon_a I)B^t + Y(\varphi_1 I)Y$$

$$
\begin{aligned}
A_2 Y + Y A_2^t &\leq (\delta_1 \epsilon_a I) + Y(\varphi_2 I) Y \\
B B_1 S + S^t B_1^t B^t &\leq B(\delta_3 \epsilon_b I) B^t + S^t(\varphi_3 I) S \\
B_2 S + S^t B_2^t &\leq (\delta_4 \epsilon_d I) + S^t(\varphi_4 I) S \\
(A_d + B D_1 + D_2) R_c^{-1}(A_d^t + D_1^t B^t + D_2^t) &\leq \delta_t A_d R_c^{-1} A_d^t + (\epsilon_f \delta_e I) \\
&\quad + B(\epsilon_e \delta_s I) B^t \\
(B_h + B E_1 + E_2) R_k^{-1}(B_h^t + E_1^t B^t + E_2^t) &\leq \delta_m B_h R_k^{-1} B_h^t + (\epsilon_h \delta_n I) \\
&\quad + B(\epsilon_r \delta_g I) B^t
\end{aligned}
$$

$$(3.141)$$

By substituting inequalities (3.141) into (3.138) and grouping similar terms, we obtain:

$$
\begin{aligned}
AY + Y A^t + Q_t + Q_s + A_d R_{cd}^{-1} A_d^t + Y L_g Y + S^t(\delta_p I) S & \\
+ \gamma^{-2} D D^t + B(\delta_q I) B^t + B_h R_{km}^{-1} B_h^t + (\delta_v I) + S^t B^t + B S &< 0
\end{aligned}
$$

$$(3.142)$$

where

$$
\begin{aligned}
\delta_v &= \epsilon_c \delta_2 + \epsilon_d \delta_4 + \epsilon_f \delta_e + \epsilon_h \delta_n; \quad L_g = L_o^t L_o + (\varphi_1 + \varphi_2) I; \\
\delta_p &= \varphi_2 + \varphi_4; \quad R_{cd}^{-1} = \delta_t R_c^{-1}; \\
R_{km}^{-1} &= \delta_m R_k^{-1}; \quad \delta_q = \epsilon_a \delta_1 + \epsilon_b \delta_3 + \epsilon_e \delta_s + \epsilon_g \delta_r
\end{aligned}
$$

$$(3.143)$$

We observe that for a given $\epsilon_a, ..., \epsilon_f$ then δ_t, δ_v, δ_p, δ_q, δ_m, δ_n are affinely linear in $\delta_1, ..., \delta_{10}$, $\varphi_1, ..., \varphi_{10}$. Simple rearrangement of (3.142) using **A.1** yields the LMI (3.134) as desired.

Remark 3.17: Theorem 3.12 provides a sufficient LMI-based delay-dependent condition for a memoryless H_∞-controller guaranteeing the norm bound γ with mismatched uncertainties. To implement such a controller, one solves the following minimization problem:

$$
\underset{Y, S, Q_s, Q_t, \delta_1, ..., \delta_{10}, \varphi_1, ..., \varphi_{10}}{Min} \quad \gamma
$$

$$
s.t. \ -Y < 0 \ , \ -Q_s < 0,
$$

$$
-Q_t < 0 \ , \ -R_k < 0 \ , \ W_{\gamma 3} < 0
$$

3.6.4   Example 3.11

We consider **Example 3.9** plus the uncertainties $\Delta A(t), \Delta B(t), \Delta D(t)$ and $\Delta E(t)$ satisfying the mismatching condition (3.118) with

$$A_1(t) = \begin{bmatrix} 0.1\ sin(3t) & -0.001 & 0 & 0 \\ 0 & 0 & 0.05 & -0.02\ cos(3t) \end{bmatrix}$$

$$B_1(t) = \begin{bmatrix} 0.4\ cos(2t) & 0 \\ 0.1 & 0 \end{bmatrix}$$

$$D_1(t) = \begin{bmatrix} 0.1\ cos(t) & 0.1 & 0 & 0 \\ 0 & 0 & 0.1 & 0.1 \end{bmatrix} ;\ E_1(t) = \begin{bmatrix} 0.01 & 0 \\ 0 & 0.1\ sin(5t) \end{bmatrix}$$

$$A_2(t) = \begin{bmatrix} 0.1 & 0 & 0 & 0 \\ 0 & 0.2\ sin(4t) & 0 & 0 \\ 0 & 0 & -0.3 & 0 \\ 0 & 0 & 0 & 0.4 \end{bmatrix} ;\ B_2(t) = \begin{bmatrix} 0.1 & 0 \\ -0.1 & 0 \\ 0 & 0.2\ sin(5t) \\ 0 & 0.1 \end{bmatrix}$$

$$D_2(t) = \begin{bmatrix} 0.05\ sin(t) & 0 & 0 & 0 \\ 0 & 0.05\ cos(2t) & 0 & 0 \\ 0 & 0 & 0 & 0 \\ 0 & 0 & 0 & 0 \end{bmatrix} ;\ E_2(t) = \begin{bmatrix} 0.05 & 0 \\ 0.1 & 0 \\ 0 & 0.05 \\ 0 & 0.1\ sin(3t) \end{bmatrix}$$

We evaluate the norms in (3.132) over the period [-2,15] to give $\epsilon_a = 0.0098$, $\epsilon_b = 0.0366$, $\epsilon_d = 0.16$ $\epsilon_e = 0.02$, $\epsilon_e = 0.0031$, $\epsilon_f = 0.0003$, $\epsilon_g = 0.0011$ $\epsilon_h = 0.0208$.

Next, we select the different weighting factors: $Q_s = diag(0.2, 0.2, 0.2, 0.2)$, $Q_c = diag(0.4, 0.4, 0.4, 0.4)$, $Q_t = diag(0.3, 0.3, 0.3, 0.3)$, $R_k = diag(0.6, 0.6)$, $\tau^* = 0.4$, $\tau^+ = 0.7$, $\eta^* = 0.2$, $\eta^+ = 0.4$, $\beta = 5$, $\delta_1 = 5, \delta_2 = 4$, $\delta_3 = 8, \delta_4 = 2, \delta_5 = 2, \delta_6 = 2, \delta_7 = 2, \delta_8 = 2, \delta_9 = 2, \delta_{10} = 2, \varphi_1 = 0.2, \varphi_2 = 0.25, \varphi_3 = 0.125, \varphi_4 = 0.5, \varphi_5 = 0.5, \varphi_6 = 0.5, \varphi_7 = 0.5, \varphi_8 = 0.5, \varphi_9 = 0.5, \varphi_{10} = 0.5$. Letting R_{ss}, R_s, R_{cc} and using the LMI-Control Toolbox , we obtain

$$Y = \begin{bmatrix} 2.3994 & -1.5430 & -0.8531 & 0.1339 \\ -1.5430 & 3.7746 & -3.1093 & -1.5613 \\ -0.8531 & -3.2570 & 12.0809 & -1.2782 \\ -0.1339 & -1.5613 & -1.2314 & 11.0809 \end{bmatrix} ,\quad \gamma_{min} = 0.2890$$

$$S = \begin{bmatrix} -1.6206 & -0.7654 & -0.0079 & -0.0024 \\ -0.0088 & -0.0022 & -1.5657 & -0.7856 \end{bmatrix}$$

$$K = \begin{bmatrix} -2.0476 & -1.7441 & -0.6507 & -0.3453 \\ -0.5270 & -0.6024 & -0.3507 & -0.2026 \end{bmatrix}, \quad \|K\| = 2.9222$$

To examine the sensitivity of the obtained results to the set of initial data, the computational algorithm in **Remark 3.17** is executed iteratively to obtain feasible solutions while changing the set $\{\delta_1, ..., \delta_{10}, \varphi_1, ..., \varphi_{10}\}$ around the base value $\{8, 4, 2, 2\}$ and observing the variation in gain K as measured by $\|K\|$. It has been found that varying $(\delta_6, ... \delta_{10})$ yields small changes in $\|K\|$ whereas variations in $(\delta_1, ... \delta_3)$ results in decreasing $\|K\|$ and in (δ_4) causes $\|K\|$ to increase.

3.6.5 Control Synthesis for Norm-Bounded Uncertainties

With the norm-bounded structure (3.119), the closed-loop controlled system has the form:

$$\begin{aligned} \dot{x}(t) &= [A + BK + H\Delta(t)\{E + +LK\}]x(t) + Dw(t) \\ &\quad + [A_d + H_d\Delta_d(t)E_d]x(t - \tau(t)) + [B_h + H_e\Delta_e(t)E_e]Kx(t - \eta(t)) \\ z(t) &= Lx(t) \end{aligned} \tag{3.144}$$

Theorem 3.13: *The closed-loop system (3.144) is stable with disturbance attenuation γ via a memoryless state-feedback controller for τ, η satisfying (3.98) if given matrices $0 < Q_t^t = Q_t \in \Re^{n \times n}$, $0 < Q_s^t = Q_s \in \Re^{n \times n}$ and scalars $\alpha_1 > 0$, $\alpha_2 > 0$, $\alpha_3 > 0$, $\alpha_4 > 0$ such that $\alpha_3 E_d^t E_d < I$ and $\alpha_4 H_e^t H_e < I$ there exist matrices $0 < Y^t = Y \in \Re^{n \times n}$ and $S \in \Re^{m \times n}$ satisfying the LMI:*

$$W_{\gamma 3} = \begin{bmatrix} \Upsilon_1(Y) & \Upsilon_2 & \Upsilon_3 \\ \Upsilon_2^t & -J_\alpha & 0 \\ \Upsilon_3^t & 0 & -J_\beta \end{bmatrix} < 0 \tag{3.145}$$

where

$$\begin{aligned} \Upsilon_1(Y) &= AY + YA^t + Q_t + Q_s + BS + S^t B^t + H(\alpha_m I)H^t \\ \alpha_m &= \sigma_1 \alpha_1 + \sigma_1 \alpha_2; \quad L_t = L^t L + \alpha_1^{-1} E^t E; \quad R_{sd} = \alpha_3 \sigma_3^{-1} R_c; \\ R_{se} &= \alpha_4 \sigma_2^{-1} R_s; R_{ee}^{-1} = R_s^{-1/2}[I - \alpha_4^{-1} H_e^t H_e]^{-1}(R_s^{-1/2})^t; \\ R_{dd}^{-1} &= R_c^{-1/2}[I - \alpha_4^{-1} E_d^t E_d]^{-1}(R_c^{-1/2})^t; \quad L_s = L^t L \\ \Upsilon_2 &= [Y \;\; S^t \;\; A_d], \quad J_\alpha = [L_t \;\; \alpha_2 L_s \;\; R_{dd}^{-1}] \\ \Upsilon_3 &= [E_d \;\; E_e S \;\; S \;\; D], \quad J_\beta = [R_{sd}^{-1} \;\; R_{se}^{-1} \;\; R_{22} \;\; \gamma^2 I] \end{aligned} \tag{3.146}$$

Moreover, the gain of the memoryless state-feedback controller is given by

$$K = SY^{-1} \tag{3.147}$$

Proof: In a similar way, we apply **Theorem 3.10** with the changes

$$A_c \rightarrow A + BK + H\Delta(t)(E + LK) \;;\quad A_d \rightarrow A_d + H_d\Delta_d(t)E_d \;,$$
$$B_h K \rightarrow B_h K + H_e \Delta_e(t) E_e K$$

$$\tag{3.148}$$

to obtain the Hamiltonian

$$H(x,v,t) \;<\; Z_4^t(t) W_{III} Z_4(t) \tag{3.149}$$

$$Z_4(t) \;=\; \begin{bmatrix} x^t(t) & x^t(t-\tau) & x^t(t-\eta) & v^t(t) \end{bmatrix}^t$$

$$W_{III} \;=\; \begin{bmatrix} W_{III_{11}} & W_{III_{12}} & W_{III_{13}} & \gamma^{-1}PD \\ W_{III_{12}}^t & -R_c & 0 & 0 \\ W_{III_{13}}^t & 0 & -R_u & 0 \\ \gamma^{-1}D^t P & 0 & 0 & -I \end{bmatrix} \tag{3.150}$$

where

$$\begin{aligned}
W_{III_{11}} \;=\;& PA + A^t P + Q_c + Q_u + L^t L + K^t B^t P + PBK \\
&+\; PH\Delta E + E^t \Delta^t H^t P + PH\Delta LK + K^t L^t \Delta^t H^t P \tag{3.151} \\
W_{III_{12}} \;=\;& PD + H_d \Delta_d E_d \tag{3.152} \\
W_{III_{13}} \;=\;& PA_d K + PH_e \Delta_e E_d K \tag{3.153}
\end{aligned}$$

Again $H(x,v,t) < 0$ is implied by $W_{III} < 0$. Now applying **A.1**, we get the inequality:

$$W_{III_{11}} + W_{III_{12}} R_c^{-1} W_{III_{12}}^t + W_{III_{13}} R_u^{-1} W_{III_{13}}^t \gamma^{-2} PDD^t P < 0 \tag{3.154}$$

Using (3.151)-(3.153), letting $Y = P^{-1}$, $S = KY$, $Q_t = P^{-1}Q_c P^{-1}$, $Q_s = P^{-1}Q_u P^{-1}$, $R_s = Q_s(1 - \eta^+)$, then pre- and postmultiplying by P^{-1}, we get:

$$\begin{aligned}
& AY + YA^t + Q_t + Q_s + S^t B^t + BS + YL^t LY + \\
& H\Delta EY + YE^t \Delta^t H^t + H\Delta LS + S^t L^t \Delta^t H^t + \\
& (A_d + H_d \Delta_d E_d) R_c^{-1} (A_d + H_d \Delta_d E_d)^t + \\
& \gamma^{-2} DD^t + (B_h + H_e \Delta_e E_e) SR_s^{-1} S^t (B_h + H_e \Delta_e E_e)^t < 0
\end{aligned}$$

$$\tag{3.155}$$

Using **B.1.2,B.1.3** and by selecting $\alpha_1 > 0$, $\alpha_2 > 0$, $\alpha_3 > 0$, $\alpha_4 > 0$ such that $\alpha_3 E_d^t E_d < I$, $\alpha_4 H_e^t H_e < I$, we obtain the following bounds:

$$
\begin{aligned}
H \Delta E Y + Y E^t \Delta^t H^t &\leq H(\alpha_1 \sigma_1 I) H^t + Y(\alpha_1^{-1} E^t E) Y \\
H \Delta L S + S^t L^t \Delta^t H^t &\leq H(\alpha_2 \sigma_2 I) H^t + S^t(\alpha_2^{-1} L^t L) S \\
(A_d + H_d \Delta_d E_d) R_c^{-1} (A_d + H_d \Delta_d E_d)^t &\leq E_d(\sigma_3 \alpha_3^{-1} R_c^{-1} E_d^t + A_d R_{dd}^{-1} A_d^t) \\
(B_h + H_e \Delta_e E_e) S R_s^{-1} S^t (B_h + H_e \Delta_e E_e)^t &\leq E_e S(\sigma_2 \alpha_4^{-1} R_s^{-1} S^t E_e^t \\
&+ B_h S R_{ee}^{-1} S^t B_h^t)
\end{aligned}
$$

$$(3.156)$$

By substituting inequalities (3.156) into (3.155) and grouping similar terms, we obtain:

$$
\begin{aligned}
&AY + YA^t + Q_t + Q_s + A_d R_{dd}^{-1} A_d^t + Y L_t Y + S^t(\alpha_2^{-1} L^t L) S + \\
&H(\alpha_m I) H^t + BS + S^t B^t + \gamma^{-2} DD^t + E_d(\sigma_3 \alpha_3^{-1} R_c^{-1} E_d^t) \\
&+ E_e S(\sigma_2 \alpha_4^{-1} R_s^{-1} S^t E_e^t + E_o S R_{ee}^{-1} S^t E_o^t) \leq 0
\end{aligned}
$$

$$(3.157)$$

where

$$\alpha_m = \sigma_1 \alpha_1 + \sigma_2 \alpha_2; \quad L_t = L_o^t L_o + \alpha_1^{-1} N^t N; \quad \alpha_3 E_d^t E_d < I, \quad \alpha_4 E_e^t E_e < I$$

Note in (3.156) that for a given σ_1, σ_2; then α_m is affinely linear in α_1, α_2. Simple rearrangement of (3.156) using **A.1** yields the LMI (3.145) as desired.

Remark 3.18: The main result of **Theorem 3.13** is that it establishes a sufficient delay-dependent condition for a memoryless H_∞-controller guaranteeing the norm-bound γ for systems with norm-bounded uncertainties and it casts the condition into the easily computable LMI format. In this regard, it generalizes existing results to the case of unknown state and input delays. To implement such controller, one has to solve the following minimization problem:

$$
\underset{Y, S, Q_s, Q_t, \alpha_1, ..., \alpha_4,}{Min} \quad \gamma
$$

$$\text{subject to } -Y < 0, \quad -Q_s < 0 \quad -Q_t < 0, \quad W_{\gamma 3} < 0$$

3.6.6   Example 3.12

Consider Example 3.9 together with $\{\Delta A(t),\ \Delta B(t),\ \Delta D(t),\ \Delta E(t)\}$ with

$$E = \begin{bmatrix} 0.4 & 0 & -0.2 & 0 \\ 0 & 0.4 & 0 & 0.1 \end{bmatrix} ; \quad L = \begin{bmatrix} 0.1 & 0 \\ 0 & 0.1 \end{bmatrix} ; \quad E_e = \begin{bmatrix} 0.1 & 0 \\ 0 & 0.05 \\ -0.2 & 0 \\ 0 & 0.15 \end{bmatrix}$$

$$H = \begin{bmatrix} -0.5 & 0 \\ 0 & -0.6 \\ 0.3 & 0 \\ 0 & 0.5 \end{bmatrix} ; \quad H_d(t) = \begin{bmatrix} 0.1 \\ -0.05 \\ 0.5 \\ -0.2 \end{bmatrix} ; \quad H_e = \begin{bmatrix} -0.04 \\ 0.12 \\ 0.05 \\ 0.10 \end{bmatrix}$$

$$E_d(t) = \begin{bmatrix} 0.02 & 0.3 & 0 & 0 \\ 0 & 0 & 0.5 & 0.2 \\ 0.1 & 0.2 & 0 & 0 \\ 0 & 0 & 0.4 & 0.8 \end{bmatrix} ; \quad \Delta(t) = \begin{bmatrix} 0.22\,cos(2t) & 0.05 \\ 0.18 & -0.05\,sin(2t) \end{bmatrix}$$

$$\Delta_d(t) = [0.1 \ -0.15sin(3t) \ 0.05 \ 0.01], \quad \Delta_e = [0.2cos(5t) \ 0 \ 0 \ 0]$$

Simple computation gives $R_{dd} = diag(3.5789, 3.5789, 3.5789, 3.5789),\ L_s = diag(0.01, 0.01),\ R_{ee} = diag(0.8488, 0.8488, 0.8488, 0.8488)$ and

$$L_t = \begin{bmatrix} 1.032 & 0.5 & 0.184 & 0.1 \\ 0.5 & 0.282 & 0.1 & 0.058 \\ 0.184 & 0.1 & 0.048 & 0.02 \\ 0.1 & 0.058 & 0.02 & 0.012 \end{bmatrix}$$

Next, computing the scalars in (3.119) yields $\sigma_1 = 0.0408$, $\sigma_2 = 0.0372$, $\sigma_3 = 0.03466$. By selecting the different weighting factors as in **Example 3.11**, we evaluate the respective matrices and then proceed to solve the underlying minimization problem to obtain:

$$Y = \begin{bmatrix} 180 & -2350 & 160 & 5330 \\ -1350 & 17330 & -2710 & -70730 \\ 160 & -2710 & 1030 & 10360 \\ 5330 & -70730 & 10360 & 291240 \end{bmatrix}, \quad \gamma_{min} = 0.0229$$

$$S = \begin{bmatrix} -2.4185 & -10.3692 & 5.1672 & 2.5054 \\ 4.2873 & -46.2425 & -23.1474 & -10.9651 \end{bmatrix}$$

$$K = \begin{bmatrix} 0.2365 & 0.3093 & 0.1009 & 0.0669 \\ -1.2767 & -1.5419 & -0.5491 & -0.3316 \end{bmatrix}, \quad \|K\| = 2.1417$$

3.7 Notes and References

Robust stabilization schemes based on state-feedback can be found in [7,8,14, 15,28,35-36,64-66,77,99,111,118] for matched and/or mismatched uncertainties and in [9-12,16,20,37,46,71,78,82,92,108-109, 120-121, 135,137-138] for norm-bounded uncertainties. Results based on classical and algebraic methods are available in [24,31,52-53,58-60,67,140,147-149,158-164,166,173].

Chapter 4

Robust $\mathcal{H}_\infty$ Control

We recall that the previous two chapters have a common denominator. Chapter 2 was concerned with the internal stability of open-loop systems and Chapter 3 dealt with closed-loop system stability under state-feedback. Therefore Chapter 3 was a natural extension of Chapter 2. We also adopted, in Chapter 3, $\mathcal{H}_\infty$-bound as one tangible design criterion in synthesizing a feedback controller and considered uncertain systems with state and input delay. In this chapter, we expand this philosophy further and examine the robust $\mathcal{H}_\infty$-control for different classes of uncertain time-delay systems. The results presented hereafter complement those of Chapter 3, but also add results on nonlinear, discrete-time as well as multiple-delay systems. It should be emphasized that in systems theory $\mathcal{H}_\infty$ attenuation has been proven to be a very useful performance measure. Loosely speaking, if $w(t) \in \mathcal{L}_2[0, \infty)$ is the exogenous input and $z(t)$ is the controlled output then the γ-disturbance attenuation problem is to choose a state-feedback law $u(t) = K_s x(t)$ which guarantees the closed-loop internal stability and ensures that

$$\max_{0 \neq w \in \mathcal{L}_{\in[r,\infty)}} (\|z\|_2^2 - \gamma^2 \|w\|_2^2) \leq 0 \,, \gamma > 0$$

The last decade has witnessed major advances in $\mathcal{H}_\infty$ control theory [214-219] of linear dynamical systems. It seems, however, that little results are available so far on $\mathcal{H}_\infty$ control of time-delay systems. In [32], a frequency-domain approach is used to design an $\mathcal{H}_\infty$ controller when the time-delay is constant. A Lyapunov approach is adopted in [47] for the design of state and dynamic output feedback controllers for a class of time-varying delay systems.

121

Here, we consider the robust $\mathcal{H}_\infty$ control problem for a class of time-delay systems with norm-bounded uncertainties and unknown constant state-delay. Specifically, our objective is to guarantee that the internal stability of the closed-loop feedback system with a prescribed $\mathcal{H}_\infty$-norm bound constraint on disturbance attentuation for all admissible uncertainties and unknown state delay. The main thrust for solving the foregoing problem stems from a Lyapunov functional approach which eventually leads to finite-dimensional Riccati equations that can be effectively handled by existing software.

4.1 Linear Uncertain Systems

4.1.1 Problem Statement and Preliminaries

We consider a class of uncertain time-delay systems represented by:

$$(\Sigma_\Delta): \quad \dot{x}(t) = [A + \Delta A(t)]x(t) + [A_d + \Delta A_d(t)]x(t-\tau)$$
$$+ [B + \Delta B(t)]u(t) + Rw(t)$$
$$= A_\Delta(t)x(t) + A_{d\Delta}(t)x(t-\tau) + B_\Delta(t)u(t) + Rw(t)$$
$$z(t) = Lx(t) + Du(t) \quad (4.1)$$

where $x(t) \in \Re^n$ is the state, $u(t) \in \Re^p$ is the control input and $z(t) \in \Re^r$ is the controlled output. In (4.1), $A \in \Re^{n\times n}$, $B \in \Re^{n\times p}$, $A_d \in \Re^{n\times n}$, $L \in \Re^{m\times n}$ and $D \in \Re^{m\times p}$ are real constant matrices representing the nominal plant. Here, τ is an unknown constant scalar representing the amount of delay in the state. The matrices $\Delta A(t)$, $\Delta B(t)$ and $\Delta A_d(t)$ represent time-varying parameteric uncertainties which are of the form:

$$[\Delta A(t) \ \Delta B(t)] = H\Delta_1(t)[E \ E_b]; \quad \Delta A_d(t) = H_d\Delta_2(t)E_d \quad (4.2)$$

where $H \in \Re^{n\times\alpha}$, $H_d \in \Re^{n\times\sigma}$, $E \in \Re^{\beta\times n}$, $E_b \in \Re^{\beta\times p}$ and $E_d \in \Re^{\omega\times p}$ are known real constant matrices and $\Delta_1(t) \in \Re^{\alpha\times\beta}$, $\Delta_2(t) \in \Re^{\sigma\times\omega}$ are unknown matrices with Lebsegue measurable elements satisfying

$$\Delta_1^t(t)\Delta_1(t) \leq I; \quad \Delta_2^t(t)\Delta_2(t) \leq I, \ \forall \ t \quad (4.3)$$

The initial condition is specified as $\langle x(t_o), x(s)\rangle = \langle x_o, \phi(s)\rangle$, where $\phi(\cdot) \in \mathcal{L}_2[-\tau, t_o]$ and $w(t) \in \mathcal{L}_2[0, \infty)$ is an input disturbance signal.

Distinct from system (Σ_Δ) are three systems:

$$
\begin{aligned}
(\Sigma_{\Delta o}): \quad \dot{x}(t) &= A_\Delta(t)x(t) + E_\Delta(t)x(t-\tau) \\
(\Sigma_{\Delta w}): \quad \dot{x}(t) &= A_\Delta(t)x(t) + E_\Delta(t)x(t-\tau) + R_o w(t) \\
z(t) &= Hx(t) \\
(\Sigma_{\Delta u}): \quad \dot{x}(t) &= A_\Delta(t)x(t) + E_\Delta(t)x(t-\tau) + B_\Delta(t)u(t) \quad (4.4)
\end{aligned}
$$

We note that systems $(\Sigma_{\Delta o})$, $(\Sigma_{\Delta w})$ and $(\Sigma_{\Delta u})$ represent, respectively, the unforced disturbance-free portion, the disturbance-free portion and the unforced portion of system (Σ_Δ).

We recall from Chapter 2 that system $(\Sigma_{\Delta o})$ with uncertainties satisfying (4.2) is robustly stable independent of delay if there exist scalars $(\mu > 0, \sigma > 0)$ and a matrix $0 < W = W^t \in \Re^{n\times n}$ such that the algebraic Ricatti equation (ARE):

$$
\bar{P}A + A^t\bar{P} + \bar{P}B(\mu,\sigma)B^t(\mu,\sigma)\bar{P} + \mu^{-1}E^tE + W = 0 \qquad (4.5)
$$

has a stabilizing solution $0 < \bar{P} = \bar{P}^t \in \Re^{n\times n}$, with

$$
B(\mu,\sigma)B^t(\mu,\sigma) = \mu HH^t + \sigma H_d H_d^t + A_d(W - \sigma^{-1}E_d^t E_d)^{-1}A_d^t \qquad (4.6)
$$

Based on this, we establish the following result for system $(\Sigma_{\Delta u})$.

Theorem 4.1: *System $(\Sigma_{\Delta u})$ is robustly stable via memoryless state feedback if there exist scalars $(\mu > 0, \sigma > 0)$ and a matrix $0 < W = W^t \in \Re^{n\times n}$ such that ARE:*

$$
\bar{P}A + A^t\bar{P} + \bar{P}\bar{B}(\mu,\sigma)\bar{B}^t(\mu,\sigma)\bar{P} + \mu^{-1}E^t\{I - E_b(E_b^t E_b)^{-1}E_b^t\}E + W = 0 \qquad (4.7)
$$

has a stabilizing solution $0 < \bar{P} = \bar{P}^t \in \Re^{n\times n}$, where

$$
\bar{B}(\mu,\sigma)\bar{B}^t(\mu,\sigma) = B(\mu,\sigma)B^t(\mu,\sigma) - \mu B\{E_b^t E_b\}^{-1}B^t \qquad (4.8)
$$

Furthermore, the stabilizing control law is given by:

$$
\begin{aligned}
u(t) &= K_s\, x(t) \\
K_s &= -\mu(E_b^t E_b)^{-1}(B^t P + \mu^{-1}E_b^t E) \qquad (4.9)
\end{aligned}
$$

Proof: System $(\Sigma_{\Delta u})$ with the memoryless feedback control law $u(t) = K_s x(t)$ becomes:

$$(\Sigma_{\Delta u}): \quad \dot{x}(t) = A_{\Delta c}(t)x(t) + A_{d\Delta}(t)x(t - \tau)$$
$$= [A_c + H\Delta_1(t)M_c]x(t) + A_{d\Delta}(t)x(t - \tau)$$
$$A_c = A + BK_s, \quad M_c = E + E_b K_s \qquad (4.10)$$

By **Lemma 2.1**, system (4.10) is robustly stable independent of delay if:

$$PA_{\Delta c} + A_{\Delta c}^t P + PA_{d\Delta}W^{-1}A_{d\Delta}^t P + W < 0 \qquad (4.11)$$

Applying **B.1.3** and **A.2** and using K_s from (4.9), it is readily seen that (4.11) with the help of (4.6) and (4.8) reduces to:

$$PA + A^t P + P\left\{\bar{B}(\mu,\sigma)\bar{B}^t(\mu,\sigma)\right\}P + \mu^{-1}E^t\{I - E_b(E_b^t E_b)^{-1}E_b^t\}E + W < 0 \qquad (4.12)$$

Finally, by **A.3.1**, it follows that the existence of a matrix $0 < P = P^t \in \Re^{n \times n}$ satisfying inequality (4.12) is equivalent to the existence of a stabilizing solution $0 < \bar{P} = \bar{P}^t \in \Re^{n \times n}$ to the ARE (4.7).

4.1.2   Robust $\mathcal{H}_\infty$ Control

Now, we proceed to examine closely the robust $\mathcal{H}_\infty$ control problem. First, we consider the robust $\mathcal{H}_\infty$ performance analysis for system $(\Sigma_{\Delta w})$.

Theorem 4.2: *System $(\Sigma_{\Delta w})$ is robustly stable with a disturbance attenuation γ if there exist scalars $(\mu > 0, \sigma > 0)$ and a matrix $0 < W = W^t \in \Re^{n \times n}$ such that the ARE:*

$$\bar{P}A + A^t\bar{P} + \bar{P}\hat{B}(\mu,\sigma,\gamma)\hat{B}^t(\mu,\sigma,\gamma)\bar{P} + \mu^{-1}E^t E + W + L^t L = 0 \qquad (4.13)$$

has a stabilizing solution $0 < \bar{P} = \bar{P}^t \in \Re^{n \times n}$, with

$$\hat{B}(\mu,\sigma,\gamma)\hat{B}^t(\mu,\sigma,\gamma) = \mu HH^t + \sigma H_d H_d^t + \gamma^{-2}RR^t + A_d(W - \sigma^{-1}E_d^t E_d)^{-1}A_d^t \qquad (4.14)$$

Proof: In order to show that system $(\Sigma_{\Delta w})$ is robustly stable with a disturbance attenuation γ, it is required that the associated Hamiltonian

$$H(x,w,t) = \dot{V}_1(x_t) + z^t(t)z(t) - \gamma^2 w^t(t)w(t) < 0$$

where $V_1(x_t)$ is given by (2.5). Little algebra shows that:

$$
\begin{aligned}
H(x,w,t) &= \xi^t(t)\,\Psi(P)\,\xi(t) \\
\Psi(P) &= \begin{bmatrix} A_\Delta^t P + PA_\Delta + W + L^t L & PA_{d\Delta} & PR \\ A_{d\Delta}^t P & -W & 0 \\ R^t P & 0 & -\gamma^{-2}I \end{bmatrix} \\
\xi(t) &= [x^t(t) \quad w^t(t) \quad x^t(t-\tau)]^t \qquad (4.15)
\end{aligned}
$$

The requirement $H(x,w,t) < 0$, $\forall \xi(t) \neq 0$ is implied by $\Psi(P) < 0$ which, in turn, by A.1 implies that:

$$ PA_\Delta + A_\Delta^t P + PA_{d\Delta} W^{-1} A_{d\Delta}^t P + \gamma^{-2} PRR^t P + W + L^t L < 0 \quad (4.16) $$

Using B.1.2 and B.1.3, it follows for some $\mu > 0, \sigma > 0$ that

$$
\begin{aligned}
& PA + A^t P + \mu^{-1} E^t E + W + L^t L + \\
& P\left\{ \mu HH^t + \sigma H_d H_d^t + \gamma^{-2} RR^t + A_d(W - \sigma^{-1} E_d^t E_d)^{-1} A_d^t \right\} P < 0
\end{aligned}
\quad (4.17)
$$

By A.1, it follows that the existence of a matrix $0 < P = P^t \in \Re^{n \times n}$ satisfying inequality (4.17) is equivalent to the existence of a stabilizing solution $0 < \bar{P} = \bar{P}^t \in \Re^{n \times n}$ to the ARE (4.13).

By considering the robust synthesis problem for system (Σ_Δ), we establish the following result.

Theorem 4.3: *System (Σ_Δ) is robustly stable with a disturbance attenuation γ via memoryless state feedback if there exist scalars $(\mu > 0, \sigma > 0)$ and a matrix $0 < W = W^t \in \Re^{n \times n}$ such that the algebraic Riccati equation (ARE):*

$$
\begin{aligned}
& \bar{P}A + A^t \bar{P} + \bar{P}\hat{W}(\mu,\sigma,\gamma)\hat{W}^t(\mu,\sigma,\gamma)\bar{P} + \mu^{-1} E^t E + L^t L + W \\
& - \{\bar{P}B + \mu^{-1} E^t E_b + L^t D\}\{D^t D + \mu^{-1} E_b^t E_b\}^{-1} \\
& \{B^t \bar{P} + \mu^{-1} E_b^t E + D^t L\} = 0
\end{aligned}
\quad (4.18)
$$

has a stabilizing solution $0 < \bar{P} = \bar{P}^t \in \Re^{n \times n}$, where the stabilizing control law is given by:

$$
\begin{aligned}
u(t) &= K_* \, x(t) \\
K_* &= -\{D^t D + \mu^{-1} E_b^t E_b\}^{-1}\{B^t \bar{P} + \mu^{-1} E_b^t E + D^t L\} \qquad (4.19)
\end{aligned}
$$

Proof: System (Σ_Δ) subject to the control law $u(t) = K_* x(t)$ has the form:

$$
\begin{aligned}
\dot{x}(t) &= A_{\Delta c}(t)x(t) + A_{d\Delta}(t)x(t-\tau) + Rw(t) \\
&= [A_c + E\Delta_1(t)M_c]x(t) + Rw(t) \\
z(t) &= [L + DK_*]x(t)
\end{aligned}
\tag{4.20}
$$

where

$$
A_c = A + BK_*, \qquad M_c = M_1 + M_2 K_*
\tag{4.21}
$$

Using $V_1(x_t)$ of (2.5), it is easy to see with the aid of (4.20) that:

$$
\begin{aligned}
H(x,w,t) &= \xi^t(t)\,\Psi_*(P)\,\xi(t) \\
\Psi_*(P) &= \begin{bmatrix} \begin{array}{c} A_{\Delta c}^t P + P A_{\Delta c} + W \\ +[L+DK_*]^t[L+DK_*] \\ A_{d\Delta}^t P \\ R^t P \end{array} & \begin{array}{c} PA_{d\Delta} \\ -W \\ 0 \end{array} & \begin{array}{c} PR \\ 0 \\ -\gamma^{-2}I \end{array} \end{bmatrix} \\
\xi(t) &= [x^t(t) \ \ w^t(t) \ \ x^t(t-\tau)]^t
\end{aligned}
\tag{4.22}
$$

By similarity to **Theorem 4.2**, we use **A.1**, **B.1.2** and **B.1.3** in the condition $\Psi_*(P) < 0$ to obtain for some $\mu > 0, \sigma > 0$:

$$
\begin{aligned}
&PA_c + A_c^t P + Q + [H + DK_*]^t[H + DK_*] + \mu^{-1}M_c^t M_c + \\
&P\hat{W}(\mu,\sigma,\gamma)\hat{W}^t(\mu,\sigma,\gamma)P < 0
\end{aligned}
\tag{4.23}
$$

Using K_* from (4.19) and manipulating, we get:

$$
\begin{aligned}
&PA + A^t P + P\hat{W}(\mu,\sigma,\gamma)\hat{W}^t(\mu,\sigma,\gamma)P + \mu^{-1}E^t E + L^t L + W \\
&-\{PB + \mu^{-1}E^t E_b + L^t D\}\{D^t D + \mu^{-1}E_b^t E_b\}^{-1} \\
&\{B^t P + \mu^{-1}E_b^t E + D^t L\} < 0
\end{aligned}
\tag{4.24}
$$

By **A.1**, it follows that the existence of a matrix $0 < P = P^t \in \Re^{n \times n}$ satisfying inequality (4.24) is equivalent to the existence of a stabilizing solution $0 < \bar{P} = \bar{P}^t \in \Re^{n \times n}$ to the ARE (4.18).

4.2 Nonlinear Systems

Now, we move one step further and consider the robust $\mathcal{H}_\infty$ control problem for a class of uncertain nonlinear time-delay systems. The parametric uncertainties are real time-varying and norm-bounded and the nonlinearities are state-dependent and cone-bounded. The delays are time-varying and bounded both in the state and at the input.

4.2.1 Problem Statement and Preliminaries

Consider a class of nonlinear dynamical systems with state and input delays as well as uncertain parameters of the form:

$$
\begin{aligned}
(\Sigma_\Delta): \ \dot{x}(t) &= [A + \Delta A(t)]x(t) + [G + \Delta G(t)]g[x(t)] \\
&+ [B + \Delta B(t)]u(t) + [A_d + \Delta A_d(t)]x(t - \tau(t)) \\
&+ [B_h + \Delta B_h(t)]u(t - \eta(t)) + Rw(t) \\
&= A_\Delta(t)x(t) + G_\Delta(t)g[x(t)] + B_\Delta(t)u(t) \\
&+ A_{d\Delta}(t)x(t - \tau(t)) + B_{h\Delta}(t)u(t - \eta(t)) + Rw(t) \\
y(t) &= [C + \Delta C(t)]x(t) + [H_h + \Delta H_h(t)]h[x(t)] \\
&+ [F + \Delta F(t)]u(t) + Tw(t) \\
&= C_\Delta(t)x(t) + H_{h\Delta}(t)h[x(t)] + F_\Delta u(t) + Tw(t) \\
z(t) &= Lx(t) + Ju(t)
\end{aligned}
\tag{4.25}
$$

where $x \in \Re^n$ is the state; $u \in \Re^m$ is the control input; $y \in \Re^q$ is the measured output; $z \in \Re^p$ is the controlled output and $w(t) \in \Re^p$ is an input disturbance signal which belongs to $\mathcal{L}_2[0, \infty)$. Here, τ, η stand for the amount of delay in the state and at the input of the system, respectively, with the following properties:

$$
\begin{aligned}
0 \leq \tau(t) \leq \tau^* < \infty, \quad \dot{\tau}(t) \leq \tau^+ < 1 \\
0 \leq \eta(t) \leq \eta^* < \infty, \quad \dot{\eta}(t) \leq \eta^+ < 1
\end{aligned}
\tag{4.26}
$$

such that the bounds τ^*, η^*, τ^+ and η^+ are known. In (4.25) the matrices $A \in \Re^{n \times n}$, $B \in \Re^{n \times m}$ and $C \in \Re^{q \times n}$, represent the nominal system without delay and uncertainties and $A_d \in \Re^{n \times n}$, $H_h \in \Re^{n \times m}$, $G \in \Re^{n \times n}$, $L_o \in \Re^{p \times n}$, $J \in \Re^{p \times m}$ and $H_h \in \Re^{q \times n}$ are known constant matrices. The uncertain matrices $\Delta A(t)$, $\Delta B(t)$, $\Delta C(t)$, $\Delta A_d(t)$, $\Delta B_h(t)$, $\Delta F(t)$, $\Delta G(t)$ and $\Delta H_h(t)$ are given by:

$$
\begin{aligned}
\begin{bmatrix} \Delta A(t) & \Delta B(t) \\ \Delta C(t) & \Delta F(t) \end{bmatrix} &= \begin{bmatrix} H \\ H_c \end{bmatrix} \Delta(t) \begin{bmatrix} E & E_b \end{bmatrix} \\
\Delta A_d(t) &= H_d \Delta_1(t) E_d, \quad \Delta H_h(t) = H_t \Delta_2(t) E_t \\
\Delta G(t) &= H_g \Delta_g(t) E_g, \quad \Delta H(t) = H_n \Delta_h(t) E_n
\end{aligned}
\tag{4.27}
$$

where $\Delta(t) \in \Re^{\alpha_1 \times \alpha_2}$, $\Delta_1(t) \in \Re^{\alpha_3 \times \alpha_4}$, $\Delta_2(t) \in \Re^{\alpha_5 \times \alpha_6}$, $\Delta_g(t) \in \Re^{\alpha_7 \times \alpha_8}$, $\Delta_h(t) \in \Re^{\alpha_9 \times \alpha_{10}}$ are matrices of uncertain parameters satisfying the bounds

$$
\begin{aligned}
\Delta^t(t)\Delta(t) \leq I, \quad \Delta_1(t)\Delta_1^t(t) \leq I, \quad \Delta_2(t)\Delta_2^t(t) \leq I \\
\Delta_g(t)\Delta_g^t(t) \leq I, \quad \Delta_h(t)\Delta_h^t(t) \leq I
\end{aligned}
\tag{4.28}
$$

and $H \in \Re^{n \times \alpha_1}$, $H_c \in \Re^{p \times \alpha_1}$, $H_d \in \Re^{n \times \alpha_3}$, $H_t \in \Re^{m \times \alpha_5}$, $H_g \in \Re^{n \times \alpha_7}$, $H_n \in \Re^{m \times \alpha_9}$, $E \in \Re^{\alpha_2 \times n}$, $E_b \in \Re^{\alpha_2 \times m}$, $E_d \in \Re^{\alpha_4 \times n}$, $E_t \in \Re^{\alpha_6 \times m}$, $E_g \in \Re^{\alpha_8 \times n}$ and $E_h \in \Re^{\alpha_{10} \times m}$ are known real constant matrices. The unknown vector functions $g(.)$ and $h(.)$ are assumed to satisfy the following boundedness conditions:

Assumption 4.1: There exist known scalars $k_1 > 0$ and $k_2 > 0$ and matrices W_g, W_h such that for all $x \in \Re^n$

$$\|g(x)\| \le k_1 \|W_g x\|, \quad \|h(x)\| \le k_2 \|W_h x\| \tag{4.29}$$

Finally, the initial functions of system (4.25) are specified as $x_0 \in C([-\tau, 0]; \Re^n)$ and $u_0 \in C([-\eta, 0]; \Re^m)$.

Remark 4.1: System (Σ_Δ) represents a general state-space framework encompassing models of many physical systems which include water pollution systems, chemical reactors with recycling and hot stripping mills. The structure of the uncertainties in (4.27-4.28) is known to belong to the class of norm-bounded uncertainties.

Distinct from system (Σ_Δ) are the following systems:

$$
\begin{aligned}
(\Sigma_{\Delta o}): \quad \dot{x}(t) &= A_\Delta(t)x(t) + G_\Delta(t)g[x(t)] + A_{d\Delta}(t)x(t-\tau) \\
(\Sigma_{\Delta w}): \quad \dot{x}(t) &= A_\Delta(t)x(t) + G_\Delta(t)g[x(t)] + A_{d\Delta}(t)x(t-\tau) + Rw(t) \\
z(t) &= Lx(t) \\
(\Sigma_{\Delta u}): \quad \dot{x}(t) &= A_\Delta(t)x(t) + G_\Delta(t)g[x(t)] + A_{d\Delta}(t)x(t-\tau) + B_\Delta(t)u(t) \\
&+ B_{h\Delta}(t)u(t-\eta) \\
z(t) &= Lx(t) + Ju(t)
\end{aligned}
\tag{4.30}
$$

which represent, respectively, the unforced disturbance-free portion, the unforced portion of system and the disturbance-free portion of system (Σ_Δ). As usual, we examine properties of systems $(\Sigma_{\Delta o})$, $(\Sigma_{\Delta w})$ and $(\Sigma_{\Delta u})$ for the determination of the behavior of system (Σ_Δ).

We consider the problem of robust stabilization of the uncertain time-delay system (Σ_Δ) using linear dynamic output-feedback. We will develop our results based on the properties of systems $(\Sigma_{\Delta o})$ and $(\Sigma_{\Delta w})$. In the sequel, we will establish an interconnection between the robust stability of $(\Sigma_{\Delta o})$ and the disturbance attenuation property of the following parameterized linear time invariant system.

$$(\Sigma_\xi): \quad \dot{\xi}(t) = A\xi(t) + \tilde{B}(\rho_1, \rho_2, \rho_3)\tilde{\psi}(t) \tag{4.31}$$
$$\tilde{z}(t) = \tilde{C}(\rho_1, \rho_2, \rho_3)\xi(t) \tag{4.32}$$

where $\xi(t) \in \Re^n$ is the state, $\tilde{\psi}(t) \in \Re^{n_\psi}$ is the disturbance input which belongs to $\mathcal{L}_2[0, \infty)$, $\tilde{z}(t) \in \Re^{p_\xi}$ is the controlled output and the matrix functions $\tilde{B}(\rho_1, \rho_2, \rho_3)$ and $\tilde{C}(\rho_1, \rho_2, \rho_3)$ satisfy

$$\tilde{B}(\rho_1, \rho_2, \rho_3)\tilde{B}^t(\rho_1, \rho_2, \rho_3) = \rho_1^2 HH^t + \rho_2^2 H_g H_g^t + G(I - \rho_2^{-2} E_g^t E_g)^{-1} G^t$$
$$+ \rho_3^2 H_d H_d^t + A_d(\bar{W}_1 - \rho_3^{-2} E_d^t E_d)^{-1} A_d^t \tag{4.33}$$
$$\tilde{C}(\rho_1, \rho_2, \rho_3)\tilde{C}^t(\rho_1, \rho_2, \rho_3) = k_1^t W_g^t W_g + \rho_1^{-2} E^t E + W_1 \tag{4.34}$$

where $\bar{W}_1 = (1 - \tau^+)W_1$, and the scaling parameters $\rho_1 > 0$, $\rho_2 > 0$ and $\rho_3 > 0$ are such that $\rho_2^{-2} M_g^t M_g < I$ and $\rho_3^{-2} M_3^t M_3 < \bar{W}_1$.

The next theorem establishes a robust stability result of system (Σ_Δ) using the Riccati equation approach.

Theorem 4.4: *Consider system $(\Sigma_{\Delta o})$ satisfying* **Assumption 1.** *Then this system is robustly stable if there exist a matrix $0 < W_1 = W_1^t$ and scaling parameters $\rho_1 > 0$, $\rho_2 > 0$ and $\rho_3 > 0$ such that: $(1)\rho_2^{-2} M_g^t M_g < I$ and $\rho_3^{-2} M_3^t M_3 < (1 - \tau^+)W_1$; (2) system (Σ_ξ) has unitary disturbance attenuation.*

Proof: Introduce a Lyapunov-Krasovskii functional $V_n(x_t)$

$$V_n(x_t) = x^t(t)Px(t) + k_1^2 \int_0^t (W_g x(v))^t W_g x(v) dv$$
$$- \int_0^t g^t(x(v))g(x(v))dv + \int_{t-\tau(t)}^t x^t(v)W_1 x(v)dv \tag{4.35}$$

where $0 < P = P^t$ and $0 < W_1 = W_1^t$ are weighting matrices. Observe that $V_n(x_t) > 0$, $x \neq 0$, and $V_n(x_t) = 0$, $x = 0$. By evaluating the time-derivative of $\dot{V}_n(x_t)$ along the solutions of $(\Sigma_{\Delta o})$, we obtain:

$$\dot{V}_n(x_t) = x^t(t)[A_\Delta^t P + PA_\Delta]x(t) + g^t(x)G_\Delta^t Px(t) + x^t(t)PG_\Delta g(x)$$
$$+ x^t(t - \tau)D_\Delta^t Px(t) + x^t(t)PD_\Delta x(t - \tau) + k_1^2 x^t(t)W_g^t W_g x(t)$$
$$- g^t(x)g(x) + x^t(t)W_1 x(t) - x^t(t - \tau)W_{11}x(t - \tau) \tag{4.36}$$

where $W_{11} = (1 - \dot{\tau})W_1$. On completing the squares in (4.36) and using (4.26), we get:

$$
\begin{aligned}
\dot{V}_1(x,t) &= x^t(t)\left\{A_\Delta^t P + PA_\Delta x(t) + P(G_\Delta G_\Delta^t + D_\Delta W_{11}^{-1} D_\Delta^t)P\right\}x(t) \\
&+ x^t(t)\left\{k_1^2 W_g^t W_g + W_1\right\}x(t) \\
&- [g(x) - G_\Delta^t Px(t)]^t[g(x) - G_\Delta^t Px(t)] \\
&- [x(t-\tau) - W_{11}^{-1} D_\Delta^t Px(t)]^t W_{11}[x(t-\tau) - W_{11}^{-1} D_\Delta^t Px(t)] \\
&< x^t(t)\left\{A_\Delta^t P + PA_\Delta x(t) + P(G_\Delta G_\Delta^t + D_\Delta \bar{W}_1^{-1} D_\Delta^t)P\right\}x(t) \\
&+ x^t(t)\left\{k_1^2 W_g^t W_g + W_1\right\}x(t)
\end{aligned}
\tag{4.37}
$$

For internal stability, $\dot{V}_n(x_t) < 0$ for $x(t) \neq 0$. This holds if:

$$
PA_\Delta + A_\Delta^t P + PD_\Delta \bar{W}_1^{-1} D_\Delta^t P + PG_\Delta G_\Delta^t P + k_1^2 W_g^t W_g + W_1 < 0 \tag{4.38}
$$

On using B.1.2, we get

$$
\begin{aligned}
PA_\Delta + A_\Delta^t P &= PA + A^t P + PH\Delta E + E^t \Delta^t E^t P \\
&\leq PA + A^t P + \rho_1^2 PHH^t P + \rho_1^{-2} E^t E
\end{aligned}
\tag{4.39}
$$

By B.1.3, we have

$$
\begin{aligned}
PD_\Delta \bar{W}_1^{-1} D_\Delta^t P &= P(A_d + H_d \Delta_1 E_d)\bar{W}_1^{-1}(A_d + H_d \Delta_1 E_d)^t P \\
&\leq P[\rho_3^2 H_d H_d^t + A_d(\bar{W}_1 - \rho_3^{-2} E_d^t E_d)^{-1} A_d^t]P
\end{aligned}
\tag{4.40}
$$

$$
\begin{aligned}
PG_\Delta G_\Delta^t P &= P(G + H_g \Delta_g E_g)(G + H_g \Delta_g E_g)^t P \\
&\leq P[\rho_2^2 H_g H_g^t + G(I - \rho_2^{-2} E_g^t E_g)^{-1} G^t]P
\end{aligned}
\tag{4.41}
$$

Substituting (4.39-4.41) into (4.37) yields the ARI

$$
\begin{aligned}
&PA + A^t P + \rho_1^{-2} E^t E + P(\rho_1^2 HH^t + \rho_2^2 H_g H_g^t + \rho_3^2 H_d H_d^t)P \\
&+ k_1^2 W_g^t W_g + PA_d(\bar{W}_1 - \rho_3^{-2} E_d^t E_d)^{-1} A_d^t P \\
&+ PG(I - \rho_2^{-2} E_g^t E_g)^{-1} G^t P + W_1 < 0
\end{aligned}
\tag{4.42}
$$

On the other hand, by A.3.1, system (Σ_ξ) is robustly stable with unitary disturbance attenuation if there exist matrices $0 < P = P^t$ and $0 < W_1 = W_1^t$ such that

$$
\begin{aligned}
&A^t P + PA + P\tilde{B}(\rho_1, \rho_2, \rho_3)\tilde{B}^t(\rho_1, \rho_2, \rho_3)P + \\
&\tilde{C}^t(\rho_1, \rho_2, \rho_3)\tilde{C}(\rho_1, \rho_2, \rho_3) < 0
\end{aligned}
\tag{4.43}
$$

A comparison of (4.42)-(4.43) in the light of (4.33)-(4.34) shows that $(\Sigma_{\Delta o})$ is globally asymptotically stable for all admissible uncertainties.

Remark 4.2: The key point disclosed by **Theorem 4.4** is that the robust stability problem of a class of nonlinear systems with norm-bounded uncertainties and time-varying state-delay of the form (Σ_Δ) can be effectively converted into a "parameterized" H_∞ analysis for a linear time-invariant system without parameter uncertainties and without state-delay.

Corollary 4.1: *System* $(\Sigma_{\Delta o})$ *satisfying* **Assumption 4.1** *is robustly stable if there exist a matrix* $0 < W_1 = W_1^t$ *and positive scaling parameters* ρ_1, ρ_2 *and* ρ_3 *such that:* (1)$\rho_2^{-2}E_g^t E_g < I$ *and* $\rho_3^{-2}E_d^t E_d < (1-\tau^+)W_1$; (2) *the ARE*

$$A^t\bar{P} + \bar{P}A + \bar{P}\tilde{B}(\rho_1,\rho_2,\rho_3)\tilde{B}^t(\rho_1,\rho_2,\rho_3)\bar{P} +$$
$$\tilde{C}^t(\rho_1,\rho_2,\rho_3)\tilde{C}(\rho_1,\rho_2,\rho_3) + W_1 = 0 \tag{4.44}$$

has a stabilizing solution $0 \leq \bar{P} = \bar{P}^t$.

Proof: By **Theorem 4.4** and applying **A.1**, it follows that the existence of a matrix $0 < P = P^t \in \Re^{n \times n}$ satisfying inequality (4.42) is equivalent to the existence of a stabilizing solution $0 \leq \bar{P} = \bar{P}^t \in \Re^{n \times n}$ to the ARE (4.44).

Corollary 4.2: *System* $(\Sigma_{\Delta o})$ *satisfying* **Assumption 4.1** *is robustly stable if*
(1) *A is stable matrix, and*
(2) *The following H_∞ norm bound is satisfied*

$$\left\| \begin{bmatrix} \rho_1^{-1}E \\ k_1 W_g \\ W_1^{1/2} \end{bmatrix} [sI - A]^{-1} [\bar{L} \ \ \bar{D}] \right\|_\infty < 1$$
$$\bar{L} = [\rho_1 H \ \ \rho_2 H_g \ \ \rho_3 H_d]$$
$$\bar{D} = [A_d(\bar{W}_1 - \rho_3^{-2}H_d^t H_d)^{-1/2} \ \ G(I - \rho_2^{-2}E_g^t E_g)^{-1/2}]$$

Remark 4.3: The combined results of **Theorem 4.4**, **Corollary 4.1** and **Corollary 4.2** provide alternative ways to assess the robust stability of system $(\Sigma_{\Delta o})$. For practical implementation, a convenient way would be to convert (4.42) into the LMI

$$
\begin{bmatrix}
PA + A^t P + W_1 \\
+ k_1^t W_g^t W_g & E^t & P\check{L} & P\check{D} \\
E & -\rho_1^2 I & 0 & 0 \\
\check{L}^t P & 0 & -\check{J}_1 & 0 \\
\check{D}^t P & 0 & 0 & -\check{J}_2
\end{bmatrix} < 0
$$

where

$$
\begin{aligned}
\check{L} &= [H \ H_g \ H_d] \ , \quad \check{J}_1 = [\rho_1^{-2} I \ \rho_2^{-2} I \ \rho_3^{-2} I] \\
D_{oo} &= D_o(\bar{W}_1 - \rho_3^{-2} E_d^t E_d)^{-1/2} \\
G_{oo} &= G_o(I - \rho_2^{-2} M_g^t M_g)^{-1/2} \\
\check{D} &= [PD_{oo} \ PG_{oo}] \ , \quad \check{J}_2 = [I \ I]
\end{aligned}
$$

In this regard, the software LMI-Toolbox [4] provides a numerically efficient method for computing the matrix P as well as the scaling parameters ρ_1, ρ_2 and ρ_3.

Now, we consider the robust stabilization of system (Σ_Δ). In this regard, we introduce the following parameterized linear time-invariant system :

$$
\begin{aligned}
(\Sigma_u) \quad \dot{\zeta} &= A\zeta(t) + \mathcal{B}(\epsilon_1, \epsilon_2, \epsilon_4, \epsilon_5)\tilde{w}(t) + Bu(t) & (4.45) \\
\tilde{z} &= \begin{bmatrix} \epsilon_1^{-1} E \\ W_{gh} \end{bmatrix} \zeta(t) + \begin{bmatrix} \epsilon_1^{-1} E_b \\ 0 \end{bmatrix} u(t) & (4.46) \\
y(t) &= C\zeta(t) + \mathcal{H}(\epsilon_1, \epsilon_2, \epsilon_3)\tilde{w}(t) + Fu(t) & (4.47)
\end{aligned}
$$

where $\zeta(t) \in \Re^{n_\zeta}$ is the state, $u(t) \in \Re^m$ is the input, $\tilde{w}(t) \in \Re^{w_\zeta}$ is the disturbance input, $y(t) \in \Re^r$ is the measured output, $\tilde{z}(t) \in \Re^r$ is the controlled output, $W_{gh}^t W_{gh} = k_1^2 W_g^t W_g + k_2^2 W_h^t W_h$ and

$$
\begin{aligned}
\mathcal{B}(\epsilon_1, \epsilon_2, \epsilon_4, \epsilon_5) &= [\mathcal{B}_1(\epsilon_1, \epsilon_2) \ \mathcal{B}_2(\epsilon_4, \epsilon_5)] \ , \\
\mathcal{B}_1(\epsilon_1, \epsilon_2) &= [\epsilon_1 H, \ G(I - \epsilon_2^{-2} E_g^t E_g)^{-1/2}, \ \epsilon_2 H_g, \ 0] \ , \\
\mathcal{B}_2(\epsilon_4, \epsilon_5) &= [\mathcal{B}_{21}(\epsilon_4, \epsilon_5) \ \mathcal{B}_{22}(\epsilon_4, \epsilon_5)] \ , \\
\mathcal{B}_{21}(\epsilon_4, \epsilon_5) &= [\epsilon_4 H_d, \ D_o(\bar{W}_1 - \epsilon_4^{-2} E_d^t E_d)^{-1/2}] \\
\mathcal{B}_{22}(\epsilon_4, \epsilon_5) &= [\epsilon_5 H_t, \ B_h(I - \epsilon_5^{-2} E_t^t E_t)^{-1/2}] & (4.48) \\
\mathcal{H}(\epsilon_1, \epsilon_2, \epsilon_3) &= [\mathcal{H}_1(\epsilon_1, \epsilon_2, \epsilon_3), \ 0, 0, 0, 0] \\
\mathcal{H}_1(\epsilon_1, \epsilon_2, \epsilon_3) &= [\epsilon_1 H_c, \ 0, \ H_h(I - \epsilon_3^{-2} E_n^t E_n)^{-1/2}, \ \epsilon_3 H_n] & (4.49)
\end{aligned}
$$

The scaling parameters $\epsilon_1 > 0$, $\epsilon_2 > 0$, $\epsilon_3 > 0$, $\epsilon_4 > 0$ and $\epsilon_5 > 0$ are such that $\epsilon_2^{-2} M_g^t M_g < I$, $\epsilon_4^{-2} M_3^t M_3 < \bar{W}_1$ and $\epsilon_3^{-2} M_h^t M_h < I$.

The next theorem provides an interconnection between the robust stabilization of system $(\Sigma_{\Delta u})$ and the H_∞-control of system (Σ_u) using a linear, strictly proper, output-feedback controller $G_u(s)$ which has the following realization:

$$(G_u(s)): \quad \dot{\zeta}(t) = A_f \zeta(t) + B_f y(t); \quad \zeta(0) = 0 \qquad (4.50)$$
$$u(t) = K_f \zeta(t) \qquad\qquad\qquad\qquad\qquad (4.51)$$

where the dimension of the controller n_f and the matrices A_f, B_f and K_f are to be chosen.

Theorem 4.5: *Consider system $(\Sigma_{\Delta u})$ satisfying* **Assumption 4.1.** *Then this system is robustly stabilizable via controller $G_u(s)$ if there exist scaling parameters $\epsilon_1 > 0, \epsilon_2 > 0$ and $\epsilon_3 > 0$ such that: (1) $\epsilon_2^{-2} E_g^t E_g < I$, $\epsilon_4^{-2} E_d^t E_d < \bar{W}_1$ and $\epsilon_3^{-2} E_n^t E_n < I$; (2) the closed-loop system formed by system (Σ_u) and controller $G_u(s)$ has a unitary disturbance attenuation.*

Proof: We proceed by augmenting system $(\Sigma_{\Delta u})$ and the controller $G_u(s)$ to get the closed-loop system:

$$\dot{\xi}(t) = [A_a + L_a \Delta(t) M_a]\xi(t) + [G_c + L_c \Delta_c(t) M_c]f[\xi]$$
$$+ \ [D_d + L_d \Delta_d(t) M_d]\xi(t - \tau) + [E_e + L_e \Delta_e(t) M_e]\xi(t - \eta)$$
$$(4.52)$$

where

$$\xi(t) = \begin{bmatrix} \zeta(t) \\ x_f(t) \end{bmatrix}, \ A_a = \begin{bmatrix} A & BK_f \\ B_f C & A_f + B_f F K_f \end{bmatrix}, \ L_a = \begin{bmatrix} \epsilon_1 H \\ \epsilon_1 B_f H_c \end{bmatrix}$$

$$M_a = [\epsilon_1^{-1} E \quad \epsilon_1^{-1} E_b K_f], \ G_c = \begin{bmatrix} G & 0 \\ 0 & B_f H_h \end{bmatrix}, \ \Delta_c(t) = \begin{bmatrix} \Delta_2(t) & 0 \\ 0 & 0 \end{bmatrix}$$

$$L_c = \begin{bmatrix} \epsilon_2 H_g & 0 \\ 0 & \epsilon_3 B_f H_n \end{bmatrix}, \ f[\xi] = \begin{bmatrix} g[x] \\ h[x] \end{bmatrix}, \ M_e = \begin{bmatrix} \epsilon_5^{-1} E_t & 0 \\ 0 & 0 \end{bmatrix},$$

$$M_c = \begin{bmatrix} \epsilon_2^{-1} E_g & 0 \\ 0 & \epsilon_3^{-1} E_n \end{bmatrix}, \ D_d = \begin{bmatrix} A_d & 0 \\ 0 & 0 \end{bmatrix}, \ L_d = \begin{bmatrix} \epsilon_4 H_d & 0 \\ 0 & 0 \end{bmatrix},$$

$$M_d = \begin{bmatrix} \epsilon_4^{-1} E_d & 0 \\ 0 & 0 \end{bmatrix}, \quad E_e = \begin{bmatrix} B_h & 0 \\ 0 & 0 \end{bmatrix}, \quad L_e = \begin{bmatrix} \epsilon_5 H_t & 0 \\ 0 & 0 \end{bmatrix},$$

$$\Delta_c(t) = \begin{bmatrix} \Delta_g(t) & 0 \\ 0 & \Delta_h(t) \end{bmatrix}, \quad \Delta_d(t) = \begin{bmatrix} \Delta_1(t) & 0 \\ 0 & 0 \end{bmatrix} \tag{4.53}$$

Observe from (4.53) that $\Delta_c(t)\Delta_c^t(t) \leq I \ \forall t$ and $\forall \xi \in \Re^{n+n_f}, \|f[\xi]\| \leq k_{12}\|[I, \ 0]\xi\|$, $k_{12} = \|W_{gh}^t W_{gh}\|$.

Alternatively, system (Σ_u) under the action of controller $G_u(s)$ yields the closed-loop system:

$$\begin{aligned} \dot{\xi}(t) &= A_a \xi(t) + B_a \tilde{w}(t) \\ \tilde{z}(t) &= C_a \xi(t) \end{aligned} \tag{4.54}$$

where

$$B_a = \begin{bmatrix} B(\epsilon_1, \epsilon_2, \epsilon_4, \epsilon_5) \\ B_f \mathcal{H}(\epsilon_1, \epsilon_2 \epsilon_3) \end{bmatrix}, \quad C_a = \begin{bmatrix} M_a \\ \bar{W}_{gh} \end{bmatrix} \tag{4.55}$$

with $\bar{W}_{gh} = [W_{gh}^t W_{gh} \ \ 0]$ and the remaining matrices are as in (4.53). From (4.48)-(4.49)and (4.53), it is a simple task to verify that

$$\begin{aligned} B_a B_a^t &= L_a L_a^t + G_c (I - M_c^t M_c)^{-1} G_c^t + L_c L_c^t \\ &+ L_d L_d^t + D_d (\bar{W}_1 - M_d^t M_d)^{-1} D_d^t + L_e L_e^t + E_e (I - M_e^t M_e)^{-1} E_e^t \\ C_a^t C_a &= M_a^t M_a + W_{gh}^t W_{gh} [I \ 0]^t [I \ 0] \end{aligned} \tag{4.56}$$

Finally, by applying **Theorem 4.4** to systems (4.52) and (4.54)-(4.56), the desired result follows immediately.

Corollary 4.3: *Consider system (Σ_Δ) satisfying* **Assumption 4.1** *with all the state variables being measurable. Then this system is robustly stabilizable via the state-feedback controller $u(t) = K_s \zeta(t)$ where $K_s \in \Re^{m \times n}$ is a constant gain matrix, if there exist scaling parameters $\epsilon_1, \epsilon_2, \epsilon_3$ such that: (1) $\epsilon_2^{-2} E_g^t E_g < I$, $\epsilon_4^{-2} E_d^t E_d < \bar{W}_1$ and $\epsilon_3^{-2} E_n^t E_n < I$; (2) system (Σ_u) under the control action $u(t) = K_s \zeta(t)$ has a unitary disturbance attenuation.*

Remark 4.4: The results of **Theorem 4.5, Corollary 4.3** generalize previous results and show that the robust stabilization problem can be conveniently converted to a parameterized $\mathcal{H}_\infty$-control problem which does not involve parametric uncertainties, unknown nonlinearities or unknown delays. The solution to the $\mathcal{H}_\infty$-control problem is by now quite standard and can be obtained via algebraic Riccati equations (AREs), see [214-219].

4.2.2   Robust $\mathcal{H}_\infty$− Performance Results

We now examine the problem of robust performance with $\mathcal{H}_\infty$-bound. For this purpose we consider system (Σ_Δ) and address the following problem:

Given a scalar $\gamma > 0$, design a linear time-invariant feedback controller $u(t) = G_c(s)y(t)$ such that the closed-loop system is robustly stable and guarantees that under zero initial conditions, $||z||_2 < \gamma ||w||_2$ for all non-zero $w \in \mathcal{L}_2[0, \infty)$ and for all admissible uncertainties satisfying (4.3) and for unknown state-delay. In this case, we say that system (Σ_Δ) is robustly stabilizable with disturbance attenuation γ and the closed-loop system of (4.1)-(4.3) with $u(t) = G_c(s)y(t)$ is robustly stable with disturbance attenuation γ.

Toward our goal, we now focus on system $(\Sigma_{\Delta w})$. In line of the analytical development of the previous section, we define the following parameterized linear time-invariant system

$$(\Sigma_{\xi w}): \quad \dot{\xi}(t) \;=\; A_o\xi(t) + [R_o \quad \gamma\tilde{B}(\rho_1, \rho_2, \rho_3)]\bar{\psi}(t) \qquad (4.57)$$

$$\bar{z}(t) \;=\; \begin{bmatrix} L_o \\ \tilde{C}(\rho_1, \rho_2, \rho_3) \end{bmatrix} \xi(t) \qquad (4.58)$$

where $\xi(t) \in \Re^n$ is the state, $\bar{\psi}(t) \in \Re^{q_\xi}$ is the disturbance input which belongs to $\mathcal{L}_2[0, \infty)$, $\bar{z}(t) \in \Re^{\nu_\xi}$ is the controlled output , $\tilde{B}(\rho_1, \rho_2, \rho_3)$, $\tilde{C}(\rho_1, \rho_2, \rho_3)$ are given by (4.33)-(4.34), and A, B, L are the same as in (4.25).

Theorem 4.6: *Consider system $(\Sigma_{\Delta w})$ satisfying* Assumption 1. *Given a scalar $\gamma > 0$, this system is robustly stable with disturbance attenuation if there exist a matrix $0 < W_1 = W_1^t$ and scaling parameters $\rho_1 > 0$, $\rho_2 > 0$ and $\rho_3 > 0$ such that:* 1)$\rho_2^{-2}E_g^t E_g < I$ *and* $\rho_3^{-2}E_d^t E_d < (1 - \tau^+)W_1$; 2) *system $(\Sigma_{\xi w})$ has unitary disturbance attenuation.*

Proof: By A.1 applied to system $(\Sigma_{\Delta w})$, it follows that there exist matrices $0 < P = P^t$ and $0 < W_1 = W_1^t$ such that

$$A^t P + PA + P\tilde{B}(\rho_1, \rho_2, \rho_3)\tilde{B}^t(\rho_1, \rho_2, \rho_3)P + \gamma^{-2}PRR^t P +$$
$$L^t L + \tilde{C}^t(\rho_1, \rho_2, \rho_3)\tilde{C}(\rho_1, \rho_2, \rho_3) < 0 \qquad (4.59)$$

On considering (4.33)-(4.34), inequality (4.59) reduces to:

$$A^t P + PA + P\{\rho_1^2 HH^t + \rho_2^2 H_g H_g^t + G(I - \rho_2^{-2}E_g^t E_g)^{-1}G^t +$$

$$\rho_3^2 H_d H_d^t + A_d (\bar{W}_1 - \rho_3^{-2} E_d^t E_d)^{-1} A_d^t \} P +$$
$$\gamma^{-2} P R R^t P + L^t L + k_1^2 W_g^t W_g + \rho_1^{-2} E^t E + W_1 < 0 \qquad (4.60)$$

Using **B.1.2** and **B.1.3**, inequality (4.60) implies:

$$PA_\Delta + A_\Delta^t P + P D_\Delta \bar{W}_1^{-1} D_\Delta^t P + P G_\Delta G_\Delta^t P + \gamma^{-2} P R R^t P +$$
$$k_1^2 W_g^t W_g + L^t L + W_1 < 0 \qquad (4.61)$$

which means that system $(\Sigma_{\Delta w})$ is robustly stable.

Next, to establish that $||z||_2 < ||w||_2$ whenever $||w||_2 \neq 0$, we introduce

$$J = \int_0^\infty (z^t z - \gamma^2 w^t w) \, dt \qquad (4.62)$$

In view of the asymptotic stability of system $(\Sigma_{\Delta w})$ and that $w \in L_2[0, \infty)$, it is readily seen that J is bounded. Using (4.1) with $x_o \equiv 0$, it follows from (4.62) that:

$$
\begin{aligned}
J &= \int_0^\infty \left\{ x^t L_o^t L_o x + \frac{d}{dt}(x^t P x) - \gamma^2 w^t w \right\} dt \\
&= \int_0^\infty \{ P A_\Delta + A_\Delta^t P + P D_\Delta \bar{W}_1^{-1} D_\Delta^t P + P G_\Delta G_\Delta^t P + \gamma^{-2} P R R^t P \\
&\quad + k_1^2 W_g^t W_g + L^t L + W_1 \} dt \\
&\quad - \int_0^\infty \left\{ [g(x) - G_\Delta^t P x(t)]^t [g(x) - G_\Delta^t P x(t)] \right\} dt \\
&\quad - \int_0^\infty \left\{ [x(t - \tau) - W_{11}^{-1} D_\Delta^t P x(t)]^t W_{11} [x(t - \tau) - W_{11}^{-1} D_\Delta^t P x(t)] \right\} dt \\
&\quad - \int_0^\infty \gamma^2 \left\{ [w - \gamma^{-2} R^t P x]^t [w - \gamma^{-2} R^t P x] \right\} dt \\
&\quad - \int_0^\infty \left\{ x^t(t - \tau) W_{11} x(t - \tau) - g^t(x) g(x) + k_1^2 (W_g x)^t (W_g x) \right\} dt \quad (4.63)
\end{aligned}
$$

By **Assumption 4.1** and inequality (4.61), it is easy to see that $J < 0$. This means that $||z||_2 < \gamma ||w||_2 \; \forall 0 \neq w(t) \in \mathcal{L}_2[0, \infty)$ and for all admissible uncertainties.

We are now in a position to attend to the problem of robust $\mathcal{H}_\infty$ control of system (Σ_Δ) by converting it into a parameterized $\mathcal{H}_\infty$ control problem.

For this purpose, we introduce the following system:

$$(\Sigma_{\xi u}): \quad \dot{\xi}(t) = A\xi(t) + [R \quad \gamma B(\epsilon_1, \epsilon_2, \epsilon_4, \epsilon_5)]\bar{\psi}(t) + Bu(t) \quad (4.64)$$

$$\bar{z}(t) = \begin{bmatrix} L \\ \epsilon_1^{-1}E \\ k_{12}I \end{bmatrix} \xi(t) + \begin{bmatrix} J \\ \epsilon_1^{-1}E_b \\ 0 \end{bmatrix} u(t) \quad (4.65)$$

$$y(t) = C\xi(t) + [T \quad \mathcal{H}(\epsilon_1, \epsilon_2, \epsilon_3)]\bar{\psi}(t) + Fu(t) \quad (4.66)$$

where $\xi(t) \in \Re^n$ is the state, $\bar{\psi}(t) \in \Re^{q_\epsilon}$ is the disturbance input which belongs to $\mathcal{L}_2[0, \infty)$ and $\bar{z}(t) \in \Re^{\nu_\epsilon}$ is the controlled output, $\gamma > 0$ is the desired disturbance attenuation level for system (Σ_Δ), the matrices $B(\epsilon_1, \epsilon_2, \epsilon_4, \epsilon_5)$, $\mathcal{H}(\epsilon_1, \epsilon_2, \epsilon_3)$ are given by (4.48)-(4.49) and the matrices (A, B, C, R, L, J, F, T) are the same as in (4.25).

The next theorem summarizes the main result.

Theorem 4.7: *Consider system* (Σ_Δ) *satisfying* **Assumption 1.** *Then this system is robustly stabilizable with disturbance attenuation* γ *via a linear dynamic output-feedback controller* $G_u(s)$ *if there exist scaling parameters* $\epsilon_1 > 0, \epsilon_2 > 0$ *and* $\epsilon_3 > 0$ *such that:* (1) $\epsilon_2^{-2}E_g^t E_g < I$, $\epsilon_4^{-2}E_d^t E_d < \bar{W}_1$ *and* $\epsilon_3^{-2}E_n^t E_n < I$; (2) *the closed-loop system formed by system* $(\Sigma_{\xi u})$ *and controller* $G_u(s)$ *has a unitary disturbance attenuation.*

Proof: Follows directly by applying **Theorem 4.6** to the closed-loop system of (4.25) with the controller (4.50)-(4.51) on one hand and to the closed-loop system of (4.64)-(4.66) with the same controller (4.50)-(4.51) on the other hand.

Remark 4.5: The result disclosed by **Theorem 4.7** indicates that the robust stabilization problem with $\mathcal{H}_\infty-$ performance for a class of uncertain nonlinear time-delay systems of the type (4.25) can be converted into a parameterized $\mathcal{H}_\infty-$control problem for linear time-invariant systems without uncertainties and delay terms. The solution of the latter problem can be obtained by standard methods [214-219].

4.3 Discrete-Time Systems

This section considers a class of discrete-time systems with norm-bounded uncertainty and unknown constant state-delay. We investigate conditions of robust state feedback stabilization guaranteeing a prescribed $\mathcal{H}_\infty$ performance. Using a Lyapunov functional approach, we express these conditions in terms of finite-dimensional Riccati equations.

4.3.1 Problem Description and Preliminaries

We consider a class of uncertain time-delay systems represented by:

$$
\begin{aligned}
(\Sigma_\Delta): \quad x(k+1) &= [A + \Delta A(k)]x(k) + [B + \Delta B(k)]u(k) \\
&\quad + A_d x(k-\tau) + Rw(k) \\
&= A_\Delta(k)x(k) + B_\Delta(k)u(k) + A_d x(k-\tau) + Rw(k) \\
z(k) &= Lx(k) \qquad\qquad\qquad\qquad\qquad\qquad (4.67)
\end{aligned}
$$

where $x(k) \in \Re^n$ is the state, $u(k) \in \Re^m$ is the control input, $w(k) \in \Re^p$ is the disturbance input which belongs to $\ell_2[0,\infty)$ with a weighting matrix $R \in \Re^{n\times p}$, $z(k) \in \Re^q$ is the controlled output and the matrices $A \in \Re^{n\times n}$, $B \in \Re^{n\times m}$, $A_d \in \Re^{n\times n}$ and $L \in \Re^{q\times n}$ are real constant matrices representing the nominal plant. Here, τ is unknown constant scalar representing the amount of delay in the state. For all practical purposes, we consider $\tau \leq \tau^*$ with τ^* being known. The matrices $\Delta A(k)$ and $\Delta B(k)$ represent parameteric uncertainties which are represented by:

$$
[\Delta A(k) \quad \Delta B(k)] = H\,\Delta(k)\,[E \quad E_b] \qquad\qquad (4.68)
$$

where $L \in \Re^{n\times\alpha}$, $E \in \Re^{\beta\times n}$ and $E_b \in \Re^{\beta\times m}$ are known constant matrices which characterize how the uncertainties affect the nominal system and we assume that the matrix $E_b^t E_b$ is nonsingular. The matrix $\Delta(k) \in \Re^{\alpha\times\beta}$ is unknown but bounded in the form:

$$
\Delta^t(k)\,\Delta(k) \leq I \quad \forall k \qquad\qquad (4.69)
$$

The initial condition is specified as $\langle x(0), x(s)\rangle = \langle x_o, \phi(s)\rangle$, where $\phi(.) \in \ell_2[-\tau, 0]$.

Distinct from system (Σ_Δ) are the following systems:

$$
\begin{aligned}
(\Sigma_D): \quad x(k+1) &= A_\Delta(k)x(k) + A_d x(k-\tau) \\
(\Sigma_w): \quad x(k+1) &= A_\Delta(k)x(k) + A_d x(k-\tau) + Rw(k) \\
z(k) &= Lx(k) \\
(\Sigma_u): \quad x(k+1) &= A_\Delta(k)x(k) + A_d x(k-\tau) + B_\Delta(k)u(k) \qquad (4.70)
\end{aligned}
$$

We learned from **Lemma 2.3** that system (Σ_D) is robustly stable independent of delay (RSID) if there exists matrices $0 < P = P^t \in \Re^{n \times n}$ and $0 < W = W^t \in \Re^{n \times n}$ satisfying the ARI:

$$A_\Delta^t(k)PA_\Delta(k) - P + A_\Delta^t(k)PA_d[W - A_d^tPA_d]^{-1}A_d^tPA_\Delta(k) + \\ W < 0 \quad \forall \Delta : \Delta^t(k)\,\Delta(k) \leq I \, , \quad \forall k \tag{4.71}$$

Algebraic manipulation of (4.71) using **A.2**, **B.1.2** and **B.1.3** shows that system (Σ_D) is (RSID) if and only if there exists matrices $0 < P = P^t \in \Re^{n \times n}$ and $0 < W = W^t \in \Re^{n \times n}$ satisfying the ARI:

$$A^t \left\{ P^{-1} - A_dW^{-1}A_d^t - \mu HH^t \right\} A - P + \mu^{-1}E^tE + W = 0 \tag{4.72}$$

Extending on this, we have the following result:

Theorem 4.8: *Given a scalar $\gamma > 0$, system (Σ_w) is robustly stable with a disturbance attenuation γ if there exists a scalar $\mu > 0$ and a matrix $0 < W = W^t \in \Re^{n \times n}$ such that the following ARE*

$$A^t \left\{ \hat{P}^{-1} - \mu HH^t - \gamma^{-2}RR^t - A_dW^{-1}A_d^t \right\}^{-1} A - \hat{P} + \mu^{-1}E^tE + L^tL + W = 0 \tag{4.73}$$

has a stabilizing solution $0 \leq \hat{P} = \hat{P}^t$.

Proof: In order to show that (Σ_w) is robustly stable with a disturbance attenuation γ, it is required that the associated Hamiltonian $H(x, w, k) = \Delta V_6(x_k) + z^t(k)z(k) - \gamma^2 w^t(k)w(k) < 0$, where $V_6(x_k)$ is given by (2.78). Standard matrix manipulations produce:

$$H(x, w, k) = \xi^t(k)\Omega(P)\xi(k)$$

$$\Omega(P) =$$

$$\begin{bmatrix} A_\Delta^t PA_\Delta - P + \\ W + L^tL & A_\Delta^t PR & A_\Delta^t PA_d \\ R^t PA_\Delta & \begin{matrix} -\gamma^2 I + \\ R^t PR \end{matrix} & R^t PA_d \\ A_d^t PA_\Delta & A_d^t PR & \begin{matrix} -W + \\ A_d^t PA_d \end{matrix} \end{bmatrix}$$

$$\xi(k) = [x^t(k) \quad w^t(k) \quad x^t(k - \tau)]^t \tag{4.74}$$

The requirement $H(x, w, k) < 0$, $\forall\, \xi(k) \neq 0$ is implied by $\Omega(P) < 0$. By A.1, it is expressed as:

$$
\begin{bmatrix}
-P + W + L^t L & 0 & 0 & A_\Delta^t \\
0 & -\gamma^2 I & 0 & R^t \\
0 & 0 & -W & A_d^t \\
A_\Delta & R & A_d & -P^{-1}
\end{bmatrix} < 0 \qquad (4.75)
$$

Upon expansion using (4.75), it becomes:

$$
\begin{bmatrix}
-P + W + L^t L & 0 & 0 & A^t \\
0 & -\gamma^2 I & 0 & R^t \\
0 & 0 & -W & A_d^t \\
A & R & A_d & -P^{-1}
\end{bmatrix} +
$$

$$
\begin{bmatrix} 0 \\ 0 \\ 0 \\ H \end{bmatrix} \Delta(k) [E\ \ 0\ \ 0\ \ 0] +
\begin{bmatrix} E^t \\ 0 \\ 0 \\ 0 \end{bmatrix} \Delta^t(k)\, [0\ \ 0\ \ 0\ \ H^t] < 0 \quad (4.76)
$$

By B.1.4, (4.76) holds if and only if

$$
\begin{bmatrix}
-P + W + L^t L & 0 & 0 & A^t \\
0 & -\gamma^2 I & 0 & R^t \\
0 & 0 & -W & A_d^t \\
A & R & A_d & -P^{-1}
\end{bmatrix} +
$$

$$
\begin{bmatrix} 0 \\ 0 \\ 0 \\ \sqrt{\mu} H \end{bmatrix} [0\ \ 0\ \ 0\ \ \sqrt{\mu} H^t] +
\begin{bmatrix} \frac{1}{\sqrt{\mu}} E^t \\ 0 \\ 0 \\ 0 \end{bmatrix} [\tfrac{1}{\sqrt{\mu}} E\ \ 0\ \ 0\ \ 0] =
$$

$$
\begin{bmatrix}
-P + W + L^t L \\ \quad + \mu^{-1} E^t E & 0 & 0 & A^t \\
0 & -\gamma^2 I & 0 & R^t \\
0 & 0 & -W & A_d^t \\
A & R & A_d & \begin{matrix} -P^{-1} + \\ \mu H H^t \end{matrix}
\end{bmatrix} < 0 \qquad (4.77)
$$

Define $S = [P^{-1} - \mu HH^t]^{-1}$ and $Q = [W - A_d^t SA_d]^{-1}$. Then by repeatedly applying **A.1**, it can be shown that the LMI (4.77) is equivalent to the ARI:

$$A^t SA - P + W + L^t L + \mu^{-1} E^t E + A^t SA_d QA_d^t SA +$$
$$\left\{ A^t SR + A^t SA_d QA_d^t SR \right\} \left\{ \gamma^2 I - R^t SR - R^t SA_d QA_d^t SR \right\}^{-1}$$
$$\left\{ R^t SA + R^t SA_d QA_d^t SA \right\} < 0 \tag{4.78}$$

Using **A.2**, it can easily be shown that

$$S + SA_d QA_d^t S = [P^{-1} - \mu LL^t - A_d W^{-1} A_d^t]^{-1}$$
$$= \Pi(P) \tag{4.79}$$

The use of (4.79) into (4.78) yields

$$A^t \Pi(P) A - P + W + L^t L + \mu^{-1} E^t E +$$
$$\{ A^t \Pi(P) R \} \{ \gamma^2 I - R^t \Pi(P) R \}^{-1} \{ R^t \Pi(P) A \} < 0 \tag{4.80}$$

Once again, application of **A.2** produces:

$$A^t \left\{ P^{-1} - \mu LL^t - \gamma^{-2} RR^t - A_d W^{-1} A_d^t \right\}^{-1} A - P +$$
$$\mu^{-1} E^t E + H^t H + W < 0 \tag{4.81}$$

By **A.3.2**, it follows that the existence of a matrix $0 < P = P^t$ satisfying inequality (4.80) is equivalent to the existence of a stabilizing solution $0 \leq \hat{P} = \hat{P}^t$ to (4.72).

4.3.2 Robust $\mathcal{H}_\infty$ Control

Now, we present the results on robust control synthesis.

Theorem 4.9: *System* (Σ_u) *is robustly stabilizable via memoryless state feedback* $u(t) = K_s x(t)$ *if there exists matrices* $0 < P = P^t \in \Re^{n \times n}$ *and* $0 < W = W^t \in \Re^{n \times n}$ *and a scalar* $\mu > 0$ *such that the ARE*

$$A^t \left\{ \bar{P}^{-1} - A_d W^{-1} A_d^t - \mu LL^t - \mu B(E_b^t E_b)^{-1} B^t \right\}^{-1} A - \bar{P} +$$
$$\mu^{-1} H^t H + W = 0 \tag{4.82}$$

has a stabilizing solution $0 \leq \bar{P} = \bar{P}^t$ *where the stabilizing control law is given by*

$$
\begin{array}{rcl}
u_s(k) & = & K_s\, x(k) \\
K_s & = & -\left\{B^t\Theta B + \mu^{-1}E_b^t E_b\right\}B^t\Theta A \\
\Theta & = & [P^{-1} - A_d W^{-1}A_d^t - \mu L L^t]^{-1}
\end{array}
\tag{4.83}
$$

Proof: System (Σ_u) with the memoryless feedback control $u(t) = K_s x(t)$ becomes:

$$
\begin{array}{rcl}
x(k+1) & = & A_{\Delta c}(k)x(k) + A_d x(k-\tau) \\
& = & [A_c + H\Delta M_c]x(k) + A_d x(k-\tau)
\end{array}
\tag{4.84}
$$

where

$$
A_c = A + BK_s, \quad M_c = E + E_b K_s
\tag{4.85}
$$

It follows that system (4.84)-(4.85) is robustly stable if

$$
A_{\Delta c}^t(k)[P^{-1} - A_d W^{-1}A_d^t]^{-1}A_{\Delta c}(k) - P + W < 0
\tag{4.86}
$$
$$
\Delta^t(k)\,\Delta(k) \leq I, \quad \forall k
$$

Applying **B.1.5** to (4.86) and then manipulating using **A.2**, we obtain:

$$
A_c^t\Theta A_c - P + \mu^{-1}M_c^t M_c + W < 0
\tag{4.87}
$$

Expanding (4.87) and using K_s from (4.83), we get:

$$
A^t\Theta A - P + \mu^{-1}H^t H + W - A^t\Theta B[B^t\Theta B + \mu^{-1}E_b^t E_b]^{-1}B^t\Theta A < 0
\tag{4.88}
$$

Using **A.2** once again:

$$
A^t\left\{P^{-1} - A_d W^{-1}A_d^t - \mu H H^t - \mu B(E_b^t E_b)^{-1}B^t\right\}^{-1} A
$$
$$
- P + \mu^{-1}E^t E + W < 0
\tag{4.89}
$$

Finally, from **A.3.2**, it follows that the existence of a matrix $0 < P = P^t$ satisfying inequality (4.89) solution is equivalent to the existence of a stabilizing solution $0 \leq \bar{P} = \bar{P}^t$ to (4.82).

Theorem 4.10: *Given a scalar* $\gamma > 0$, *system* (Σ_Δ) *is robustly stable with a disturbance attenuation* γ *via memoryless state-feedback controller if* there exists a scalar $\mu > 0$ and a matrix $0 < W = W^t \in \Re^{n \times n}$ such that the ARE

$$
A^t \left\{ \hat{P}^{-1} - A_d W^{-1} A_d^t - \mu H H^t - \mu B (E_b^t E_b)^{-1} B^t - \gamma^{-2} R R^t \right\}^{-1} A
$$
$$
-\hat{P} + \mu^{-1} E^t E + L^t L + W = 0 \tag{4.90}
$$

has a stabilizing solution $0 \leq \hat{P} = \hat{P}^t$. Moreover the stabilizing controller is given by:

$$
\begin{aligned}
u(k) &= K_* \, x(k) \\
K_* &= - \left\{ B \Psi B + \mu^{-1} E_b^t E_b \right\}^{-1} B^t \Psi A \\
\Psi &= \left\{ \hat{P}^{-1} - A_d W^{-1} A_d^t - \mu H H^t - \mu B (E_b^t E_b)^{-1} B^t \right\}^{-1} \tag{4.91}
\end{aligned}
$$

Proof: System (Σ_Δ) subject to the control law $u(k) = K_* \, x(k)$ has the form:

$$
\begin{aligned}
x(k+1) &= A_{\Delta c}(k) x(k) + A_d x(k - \tau) + R w(k) \\
&= [A_c + H \Delta M_c] x(k) + A_d x(k - \tau) + R w(k) \\
z(k) &= L x(k) \tag{4.92}
\end{aligned}
$$

where $A_c = A + B K_*$ and $M_c = E + E_b K_*$. By **Theorem 4.8**, system (4.92) is robustly stable with a disturbance attenuation γ if there exists a stabilizing solution $0 \leq \hat{P} = \hat{P}^t$ to the following ARE

$$
A_c^t \left\{ \hat{P}^{-1} - \mu H H^t - \gamma^{-2} R R^t - A_d W^{-1} A_d^t \right\}^{-1} A_c - \hat{P} +
$$
$$
\mu^{-1} M_c^t M_c + L^t L + W = 0 \tag{4.93}
$$

Algebraic manipulation of (4.93) using Ψ and K_* from (4.91) leads to:

$$
A^t \Psi A - \hat{P} + \mu^{-1} E^t E + W -
$$
$$
A^t \Psi B [B^t \Psi B + \mu^{-1} E_b^t E_b]^{-1} B^t \Psi A + L^t L = 0 \tag{4.94}
$$

By **A.2**, inequality (4.94) reduces to (4.90).

4.4 Multiple-Delay Systems

In this section, we deal with the $\mathcal{H}_\infty$ feedback control synthesis problem of systems with multiple state and input time-invariant delays. We address initially the case of certain systems with focus on the stability conditions and the development of H_∞-control scheme by state-feedback. Then the results are extended to norm-bounded uncertainties. All the results obtained in this work are conveniently expressed in the framework of LMIs. For ease in exposition, the following short-hand notations will be used.

Given a set of constant matrices, $A_1,...,A_p \in \Re^{n\times n}$, $B_1,...,B_q \in \Re^{n\times m}$, $\mathcal{A}_t \triangleq [A_1 .... A_p] \in \Re^{n\times np}$, $\mathcal{B}_t \triangleq [B_1 .... B_q] \in \Re^{n\times mq}$. For some positive constants $(a_1,...,a_p)$ with vectors $x(t-a_p) \in \Re^n$, $\forall j \in J_p$, and some positive constants $(b_1,...,b_q)$ with vectors $u(t-b_k) \in \Re^m$, $\forall k \in J_q$, we let

$$\chi(t,a,p) \triangleq \begin{bmatrix} x(t-a_1) \\ x(t-a_2) \\ \vdots \\ x(t-a_p) \end{bmatrix} \in \Re^{np} \ , \ \ \nu(t,b,q) \triangleq \begin{bmatrix} u(t-b_1) \\ u(t-b_2) \\ \vdots \\ u(t-b_q) \end{bmatrix} \in \Re^{mq}$$

$$E(s,a,p) \triangleq \begin{bmatrix} e^{-sa_1} \\ e^{-sa_2} \\ \vdots \\ e^{-sa_p} \end{bmatrix}$$

where t stands for the time and s is the Laplace variable.

4.4.1   Problem Description

Consider a class of linear systems with multiple state and input delays:

$$\begin{aligned} \dot{x}(t) &= Ax(t) + Bu(t) + \sum_{j=1}^{p} A_j x(t-d_j) + \sum_{k=1}^{q} B_k u(t-h_k) + Dw(t) \\ &:= Ax(t) + Bu(t) + \mathcal{A}_t\,\chi(t,d,p) + \mathcal{B}_t\,\nu(t,h,q) + Dw(t) \\ z(t) &= Lx(t) \hspace{5cm} (4.95) \\ x(t) &= \psi(t) \ \ \ \forall t \in [-max(d_j,h_k) \ (j \in J_p \ , \ k \in J_q) \, , \, 0] \end{aligned}$$

where $t \in \Re$ is the time, $x \in \Re^n$ is the state, $u \in \Re^m$ is the control input; $w \in \Re^s$ is the input disturbance which belongs to $L_2[0,\,\infty)$; $z \in \Re^s$ is the output, and $d_j,\,h_k \geq 0$ $(j \in J_p, k \in J_q)$ are constant scalars representing the amount of delays in the state and at the input of the system, respectively. The matrices $A \in \Re^{n\times n}$ and $B \in \Re^{n\times m}$ are real constant matrices representing the nominal plant with the pair (A, B) being controllable. The matrices

$D \in \Re^{n \times s}$ and $L \in \Re^{r \times n}$ are real distribution matrices at the input and output respectively. We shall focus from now onwards on the case of linear constant-gain state-feedback in which,

$$u[x(t)] = Kx(t) \quad ; \quad K \in \Re^{m \times n} \tag{4.96}$$

The closed-loop system consisting of (4.95) and (4.96) is given by:

$$\dot{x}(t) = A_c x(t) + A_t \chi(t, d, p) + B_t K_d \chi(t, h, q) + Dw(t) \tag{4.97}$$

$$z(t) = Lx(t) \quad ; \quad A_c = A + BK$$

where $K_d = diag(K, K, ..., K) \in \Re^{mq \times nq}$.

The transfer function from the disturbance $w(t)$ to the output $z(t)$ has the form:

$$
\begin{aligned}
T_{zw}(s) &= L[(sI - A_c) - A_t E(s, d, p) - B_t K_d E(s, h, q)]^{-1} D \\
&= L[(sI - A - BK) - A_t E(s, d, p) - B_t K_d E(s, h, q)]^{-1} D
\end{aligned}
\tag{4.98}
$$

From [214-219], the basis of designing H_∞-controller is to simultaneously stabilize (4.97) and to guarantee the H_∞-norm bound γ of the closed-loop transfer-function T_{zw}; namely $\|T_{zw}\| \le \gamma, \gamma > 0$.

Distinct from (4.97) is the disturbance-free system ($w = 0$) given by

$$\dot{x}(t) = A_c x(t) + A_t \chi(t, d, p) + B_t K_d \chi(t, h, q) \tag{4.99}$$

$$z(t) = Lx(t)$$

for which we will start our design.

4.4.2 State Feedback $\mathcal{H}_\infty$-Control

In this section, basic results for H_∞-control and stabilization by state-feedback for the closed-loop time-delay systems (4.97) and (4.99) are developed. First, a general result will be provided.

Theorem 4.11: *The closed-loop time-delay system (4.99) is asymptotically stable $\forall d_j, h_k \ge 0$ ($j \in J_p, k \in J_q$) if one of the following equivalent conditions is satisfied:*

(1) *There exist matrices* $0 < P = P^t \in \Re^{n \times n}$, $0 < Q_j = Q_j^t \in \Re^{n \times n}$
$(j \in J_p)$, $0 < S_k = S_k^t \in \Re^{n \times n}$ $(k \in J_q)$ *solving the LMI:*

$$
W_o = \begin{bmatrix} PA_c + A_c^t P + \\ \sum_{j=1}^p Q_j + \sum_{k=1}^q S_k & PA_t & PB_t K_d \\ A_t^t P & -\Omega(d,p) & 0 \\ K_d^t B_t^t P & 0 & -\Gamma(h,q) \end{bmatrix} < 0 \qquad (4.100)
$$

where

$$
\Omega(d,p) = block - diag[Q_1, ..., Q_p] \ ; \ \ \Gamma(h,q) = block - diag[S_1, ..., S_q] \tag{4.101}
$$

(2) *There exist matrices* $0 < P = P^t \in \Re^{n \times n}$ *and* $0 < Q_j = Q_j^t \in \Re^{n \times n}$
$(j \in J_p)$, $0 < S_k = S_k^t \in \Re^{n \times n}$ $(k \in J_q)$ *solving the ARI:*

$$
PA_c + A_c^t P + \sum_{j=1}^p Q_j + \sum_{k=1}^q S_k + PA_t \Omega^{-1} A_t^t P + PB_t K_d \Gamma^{-1} K_d^t B_t^t P < 0 \tag{4.102}
$$

Proof: (1) Define a Lyapunov function candidate $V(x_t)$

$$
\begin{aligned}
V(x_t) &= x^t(t) P x(t) + \sum_{j=1}^p \int_{t-d_j}^t x^t(\tau_j) Q_j x(\tau_j) d\tau_j \\
&+ \sum_{k=1}^q \int_{t-h_k}^t x^t(\tau_k) S_k x(\tau_k) d\tau_k
\end{aligned} \tag{4.103}
$$

where $0 < P = P^t \in \Re^{n \times n}$, $0 < Q_j = Q_j^t \in \Re^{n \times n}$ $(j \in J_p)$, $0 < S_k = S_k^t \in \Re^{n \times n}$ $(k \in J_q)$. Observe that $V(x_t) > 0$, $x \neq 0$; $V(x_t) = 0$, $x = 0$. It is a straightforward task to show that

$$
\begin{aligned}
\dot{V}(x,t) &= Z^t(t) W_o Z(t) \\
Z^t(t) &= \begin{bmatrix} x^t(t) & \chi^t(t,d,p) & \chi^t(t,h,q) \end{bmatrix}
\end{aligned} \tag{4.104}
$$

where W_o is given by (4.100). That $\dot{V}(x_t) < 0$ is implied by $W_o < 0$ as desired.

(2) Follows from application of **A.1**.

Corollary 4.4: *Consider the state-delay system*

$$
\dot{x}(t) = Ax(t) + Bu(t) + \sum_{j=1}^p A_j x(t - d_j) \tag{4.105}
$$

under state-feedback $u(t) = Kx(t)$. Then the closed-loop system is asymptotically stable if there exist matrices $0 < P = P^t \in \Re^{n \times n}$ and $0 < Q_j = Q_j^t \in \Re^{n \times n}$ ($j \in J_p$) solving the LMI:

$$W_{1o} = \begin{bmatrix} PA_c + A_c^t P + \sum_{j=1}^{p} Q_j & PA_t \\ A_t^t P & -\Omega(d, p) \end{bmatrix} < 0 \qquad (4.106)$$

or equivalently solving the ARI:

$$PA_c + A_c^t P + \sum_{j=1}^{p} Q_j + PA_t \Omega^{-1} A_t^t P < 0 \qquad (4.107)$$

Corollary 4.5: Consider the input-delay system

$$\dot{x}(t) = Ax(t) + Bu(t) + \sum_{k=1}^{q} B_k u(t - h_k) \qquad (4.108)$$

under state-feedback $u(t) = Kx(t)$. The controlled system is then asymptotically stable if there exist matrices $0 < P = P^t \in \Re^{n \times n}$ and $0 < S_k = S_k^t \in \Re^{n \times n}$ ($k \in J_q$) solving the LMI:

$$W_{2o} = \begin{bmatrix} PA_c + A_c^t P + \sum_{k=1}^{q} S_k & PB_t K_d \\ K_d^t B_t^t P & -\Gamma(h, q) \end{bmatrix} < 0 \qquad (4.109)$$

or equivalently solving the ARI:

$$PA_c + A_c^t P + \sum_{k=1}^{q} S_k + PB_t K_d \Gamma^{-1} K_d^t B_t^t P < 0 \qquad (4.110)$$

Remark 4.6: It is interesting to note that **Theorem 4.11, Corollary 4.4** and **Corollary 4.5** generalize the results of Chapters 2 on multiple-delay systems and Chapter 3 on stabilization of single-delay systems. The existence of the gain matrix K is guaranteed by the controllability of the pair (A, B).

Now, we proceed to establish the conditions under which the controller (4.96) stabilizes the time-delay system (4.97) and guarantees the H_∞-norm bound γ of T_{zw}.

Theorem 4.12: The controlled system (4.97) is asymptotically stable and $\|T_{zw}\|_\infty \leq \gamma$ ($\gamma > 0$) $\forall d_j, h_k \geq 0$ ($j \in J_p$, $k \in J_q$), if there exist matrices $0 < P = P^t \in \Re^{n \times n}$, $0 < Q_j = Q_j^t \in \Re^{n \times n}$ $(j \in J_p)$, $0 < S_k = S_k^t \in \Re^{n \times n}$, $(k \in J_q)$ solving the ARI:

$$PA_c + A_c^t P + \sum_{j=1}^{p} Q_j + \sum_{k=1}^{q} S_k + PA_t \Omega^{-1} A_t^t P$$
$$+ PB_t K_d \Gamma^{-1} K_d^t B_t^t P + L^t L + \gamma^{-2} PDD^t P < 0 \quad (4.111)$$

Proof: It follows from **Theorem 4.11** and [44] that the controller (4.96) which satisfies inequality (4.111) stabilizes the time-delay system (4.95) $\forall\, d_j, h_k \geq 0$ $(j \in J_p, k \in J_q)$. Consider the matrix

$$-M \equiv PA_c + A_c^t P + \sum_{j=1}^{p} Q_j + \sum_{k=1}^{q} S_k + PA_t \Omega^{-1} A_t^t P$$
$$+ PB_t K_d \Gamma^{-1} K_d^t B_t^t P + L^t L + \gamma^{-2} PDD^t P \quad (4.112)$$

so that

$$PA_c + A_c^t P + \sum_{j=1}^{p} Q_j + \sum_{k=1}^{q} S_k + PA_t \Omega^{-1} A_t^t P$$
$$+ PB_t K_d \Gamma^{-1} K_d^t B_t^t P + L^t L + \gamma^{-2} PDD^t P + M = 0 \quad (4.113)$$

Let $s = \beta + j\omega$, $\beta > 0$, $\omega \in \Re$, and construct the matrices

$$
\begin{aligned}
Y_1(\beta, \omega) &= [(\beta + j\omega)I - A_c - A_t E(\beta + j\omega, d, p) \\
&\quad - B_t K_d E(\beta + j\omega, h, q)]^{-1} \quad (4.114)
\end{aligned}
$$

$$
\begin{aligned}
Y_2(\beta, \omega) &= \sum_{j=1}^{p} Q_j + PA_t \Omega^{-1} A_t^t P - PA_t E(\beta + j\omega, d, p) \\
&\quad - \sum_{j=1}^{p} A_t^t PE(\beta - j\omega, d, p) \quad (4.115)
\end{aligned}
$$

$$
\begin{aligned}
Y_3(\beta, \omega) &= \sum_{k=1}^{q} S_k + PB_t K_d \Gamma^{-1} K_d^t B_t^t P - PB_t K_d E(\beta + j\omega, h, q) \\
&\quad - \sum_{k=1}^{q} K_d^t B_t^t PE(\beta - j\omega, h, q) \quad (4.116)
\end{aligned}
$$

such that the use of (4.97) ensures that

$$LY_1(\beta, \omega)D = T_{zw}(\beta + j\omega) \quad (4.117)$$

Note that $Y_2(\beta, \omega) \geq 0$ and $Y_3(\beta, \omega) \geq 0$. Manipulating (4.114)-(4.116), we get:

$$D^t PY_1(\beta, \omega)D - \gamma^2 I +$$
$$D^t Y_1^t(\beta, -\omega)PD - \gamma^{-2}D^t Y_1^t(\beta, -\omega)PD - D^t PY_1(\beta, \omega)D =$$
$$-\gamma^2 I + D^t Y_1^t(\beta, -\omega)[Y_2(\beta, \omega) + Y_3(\beta, \omega) + L^t L + M]Y_1(\beta, \omega)D$$
$$(4.118)$$

It then follows that for all $\omega \in \Re$:

$$
\begin{aligned}
- \quad & [\gamma I - \gamma^{-1}D^t PY_1(\beta, -\omega)D]^t[\gamma I - \gamma^{-1}D^t PY_1(\beta, -\omega)D] \\
= \quad & -\gamma^2 I + D^t Y_1^t(\beta, -\omega)[Y_2(\beta, \omega) + Y_3(\beta, \omega) + M]Y_1(\beta, \omega)D \\
+ \quad & [D^t Y_1^t(\beta, -\omega)E_o^t E_o Y_1(\beta, \omega)D]
\end{aligned}
$$

On noting that

$$-[\gamma I - \gamma^{-1}D^t PY_1(\beta, -\omega)D]^t[\gamma I - \gamma^{-1}D^t PY_1(\beta, -\omega)D] \leq 0$$

we finally obtain from (4.119):

$$
\begin{aligned}
T_{zw}^t(\beta + j\omega)T_{zw}(\beta + j\omega) \quad & \leq \\
\gamma^2 I - D^t Y_1^t(\beta, -\omega)[Y_2(\beta, \omega) + Y_3(\beta, \omega) + M]Y_1(\beta, \omega)D \quad & \leq \\
\gamma^2 I \quad \forall \beta > 0, \omega \in \Re \quad & (4.119)
\end{aligned}
$$

We can conclude that $\|T_{zw}\|_\infty \leq \gamma$ as desired.

It is significant to observe that **Theorem 4.12** provides a sufficient measure, inequality (4.111), for the existence of a constant matrix K as the constant gain of an H_∞-controller.

We now move ahead to establish the H_∞-feedback control synthesis result.

Theorem 4.13: *There exists a memoryless state-feedback controller such that the closed-loop time-delay system (4.97) is asymptotically stable and $\|T_{zw}\|_\infty \leq \gamma$ $(\gamma > 0)$ $\forall d_j, h_k \geq 0$ $(j \in J_p, k \in J_q)$, if there exist matrices $0 < Y = Y^t \in \Re^{n \times n}$, $0 < Q_{t_j} = Q_{t_j}^t \in \Re^{n \times n}$ $(j \in J_p)$, $0 < Q_{s_k} = Q_{s_k}^t \in$*

$\Re^{n\times n}$ $(k \in J_q)$, $N \in \Re^{m\times n}$ solving the LMI:

$$
W_\gamma = \begin{bmatrix}
AY + YA^t + \\ BN + N^t B^t + \\ \sum_{j=1}^{p} Q_{t_j} + \sum_{k=1}^{q} Q_{s_k} & A_t L_d & B_t N_{od} & LE_o^t & D \\
L_d A_t^t & -\Omega_t(d,p) & 0 & 0 & 0 \\
N_{od}^t B_t^t & 0 & -\Gamma_t(h,q) & 0 & 0 \\
E_o L & 0 & 0 & -I & 0 \\
D^t & 0 & 0 & 0 & -\gamma^2 I
\end{bmatrix} < 0 \tag{4.120}
$$

where

$$
\begin{aligned}
\Omega_t(d,p) &= block - diag[Q_{t_1}, ..., Q_{t_p}] \\
\Gamma_t(h,q) &= block - diag[Q_{s_1}, ..., Q_{s_q}] \\
N_{od} &= block - diag[N, N ..., N] \in \Re^{mq\times nq} \\
L_d &= block - diag[L, L ..., L] \in \Re^{np\times np}
\end{aligned} \tag{4.121}
$$

The gain of the memoryless state-feedback controller is given by

$$
K = NY^{-1} \tag{4.122}
$$

Proof: By Theorem 4.12, there exists a state-feedback controller with constant gain K such that the closed-loop system (4.97) is asymptotically stable and $\|T_{zw}\|_\infty \le \gamma$ $(\gamma > 0)$ $\forall$ $d_j, h_k \ge 0$ $(j \in J_p, k \in J_q)$. We note that inequality (4.111) is not convex in P and K. However with the substitutions of $Y = P^{-1}$, $N = KY$, $Q_{t_j} = P^{-1}Q_jP^{-1}$, $Q_{s_k} = P^{-1}S_kP^{-1}$, the premultiplication by P^{-1} and postmultiplication of the result by P^{-1} yields:

$$
AY + YA^t + BN + N^t B^t + \sum_{j=1}^{p} Q_{t_j} + \sum_{k=1}^{q} Q_{s_k} + YL^t LY +
$$
$$
\gamma^{-2}DD^t + A_t L_d \Omega_t^{-1} L_d A_t^t + B_t N_{od}\Gamma_t^{-1} N_{od}^t B_t^t < 0 \tag{4.123}
$$

where $\Omega_t(d,p)$, $\Gamma_t(h,q)$ are given by (4.121). Application of A.1 to (4.123) puts it directly to form (4.120) as required.

Remark 4.7: Theorem 4.13 provides an LMI-based delay-dependent condition for a memoryless H_∞-controller which guarantees the norm bound

γ of the transfer function T_{zw}. To implement such a controller, one has to solve the following minimization problem:

$$\min_{L, W_\gamma, Q_{t_1}, \dots, Q_{t_p}, Q_{s_1}, \dots, Q_{s_q}} \gamma \qquad (4.124)$$

$$subject \ to \ Y > 0, W_\gamma < 0$$

Corollary 4.6: *For the case of state-delay systems, there exist matrices* $0 < Y = Y^t \in \Re^{n \times n}, 0 < Q_{t_j} = Q_{t_j}^t \in \Re^{n \times n}$ $(j \in J_p), N \in \Re^{m \times n}$ *solving the LMI:*

$$W_{\gamma_1} = \begin{bmatrix} AY + YA^t + \\ BN + N^t B^t + & A_t L_d & YL^t & D \\ \sum_{j=1}^p Q_{t_j} \\ L_d A_t^t & -\Omega_t(d,p) & 0 & 0 \\ LY & 0 & -I & 0 \\ D^t & 0 & 0 & -\gamma^2 I \end{bmatrix} < 0 \qquad (4.125)$$

or equivalently solving the ARI:

$$AY + YA^t + BN + N^t B^t + \sum_{j=1}^p Q_{t_j} + A_t L_d \Omega_t^{-1} L_d A_t^t +$$

$$YL^t LY + \gamma^{-2} DD^t < 0 \qquad (4.126)$$

Corollary 4.7: *For the case of input-delay systems, there exists matrices* $0 < Y = Y^t \in \Re^{n \times n}, 0 < Q_{s_k} = Q_{s_k}^t \in \Re^{n \times n}$ $(k \in J_q),$ *and* $N \in \Re^{m \times n}$ *solving the LMI:*

$$W_{\gamma_2} = \begin{bmatrix} AY + YA^t + \\ BN + N^t B^t + & B_t N_{od} & YL^t & D \\ \sum_{k=1}^q Q_{s_k} \\ N_{od}^t B_t^t & -\Gamma_t(h,q) & 0 & 0 \\ LY & 0 & -I & 0 \\ D^t & 0 & 0 & -\gamma^2 I \end{bmatrix} < 0 \qquad (4.127)$$

or equivalently solving the ARI:

$$AY + YA^t + BN + N^t B^t + \sum_{k=1}^q Q_{s_k} + B_t N_{od} \Gamma_t^{-1} N_{od}^t B_t^t +$$

$$YL^t LY + \gamma^{-2} DD^t < 0 \qquad (4.128)$$

4.4.3 Example 4.1

Extending on the water-quality dynamic model [35] with two states and two inputs ($n = 2$, $m = 2$), a more realistic system with three state delays, and two input delays ($p = 3$, $q = 2$) is considered here. The system can be written in the form (4.95) with

$$A = \begin{bmatrix} -1 & 1 \\ -2 & -3 \end{bmatrix}, B = \begin{bmatrix} 1 & 0 \\ 0 & 0.5 \end{bmatrix}, L = \begin{bmatrix} 0.5 & 0 \\ 0 & 0.5 \end{bmatrix}$$

$$A_1 = \begin{bmatrix} 0 & -0.1 \\ 0.5 & 1 \end{bmatrix}, A_2 = \begin{bmatrix} 0.1 & 0 \\ 0.3 & 0.8 \end{bmatrix}, A_3 = \begin{bmatrix} 0 & 0.1 \\ 0.1 & 0.5 \end{bmatrix}$$

$$B_1 = \begin{bmatrix} 0.7 & 0 \\ 0.1 & 0.3 \end{bmatrix}, B_2 = \begin{bmatrix} 0.3 & 0.1 \\ 0 & 0.2 \end{bmatrix}, D = \begin{bmatrix} 0.1 & 0 \\ 0 & 0.1 \end{bmatrix}$$

In this case, $x = [x_1 \quad x_2]^t \in \Re^2$, where the state variables x_1 and x_2 are the concentrations of pollutants A and B. The matrices Q_{t_1}, Q_{t_2}, Q_{t_3}, Q_{s_1} and Q_{s_2} are chosen such that, $Q_{t_1} = Q_{t_2} = Q_{t_3} = 0.1I$, and $Q_{s_1} = Q_{s_2} = 0.01I$.

With the help of the **LMI Toolbox**, it was found that

$$Y = \begin{bmatrix} 0.2311 & -0.0717 \\ -0.0717 & 0.1784 \end{bmatrix}, N = \begin{bmatrix} -0.0033 & 0.0092 \\ -0.0028 & -0.0259 \end{bmatrix}, \gamma = 3.29.$$

The gain of the memoryless state-feedback controller is

$$K = \begin{bmatrix} 0.0019 & 0.0522 \\ -0.0655 & -0.1717 \end{bmatrix}$$

4.4.4 Problem Description with Uncertainties

Extending on system (4.95), we now consider a class of linear systems with multiple state and input delays as well as uncertain parameters of the form

$$\begin{aligned}
\dot{x}(t) &= [A + \Delta A(t)]x(t) + [B + \Delta B(t)]u(t) \\
&+ \sum_{j=1}^{p}[A_j + \Delta A_j(t)]x(t - d_j) + \sum_{k=1}^{q}[B_k \\
&+ \Delta B_k(t)]u(t - h_k) + Dw(t) \\
&:= [A + \Delta A(t)]x(t) + [B + \Delta B(t)]u(t) \\
&+ [\mathcal{A}_t + \Delta \mathcal{A}_t(t)] \chi(t, d, p) + [\mathcal{B}_t + \Delta \mathcal{B}_t(t)] \nu(t, h, q) + Dw(t)
\end{aligned}$$

$$z(t) = Lx(t) \tag{4.129}$$
$$x(t) = \psi(t) \quad \forall t \in [-max(d_j, h_k) \ (j \in J_p, \ k \in J_q), \ 0]$$

where

$$\Delta \mathcal{A}_t(t) = [\Delta A_1(t) .... \Delta A_p(t)] \in \Re^{n \times np}$$
$$\Delta \mathcal{B}_t(t) = [\Delta B_1(t) .... \Delta B_q(t)] \in \Re^{n \times mq} \tag{4.130}$$

The real-valued functions $\{\Delta A(t), \ \Delta B(t), \ \Delta \mathcal{A}_t(t), \ \Delta \mathcal{B}_t(t)\}$ represent the uncertainty structure. In the sequel, we consider that this uncertainty structure belongs to the class of norm-bounded uncertainties represented $\forall t \in \Re$ by:

$$[\Delta A(t) \ \ \Delta B(t)] = H\Delta(t)[E \ \ E_b]$$
$$\Delta^t(t)\Delta(t) \le \sigma I, \quad 0 < \sigma < 1$$
$$\Delta \mathcal{A}_t(t) = H_a F_a(t) E_a, \quad \Delta_a^t(t)\Delta_a(t) \le \sigma_a I, \quad 0 < \sigma_a < 1$$
$$\Delta \mathcal{B}_t(t) = H_b F_b(t) E_b, \quad \Delta_b^t(t)\Delta_b(t) \le \sigma_b I, \quad 0 < \sigma_b < 1 \tag{4.131}$$

where the elements $\Delta_{ij}(t)$, $(\Delta_a(t))_{ij}$, $(\Delta_b(t))_{ij}$ are Lebsegue measurable $\forall i, j$; $H \in \Re^{n \times t}$, $\Delta(t) \in \Re^{t \times \eta}$, $E \in \Re^{\eta \times n}$ and $E_b \in \Re^{\eta \times m}$. $H_a \in \Re^{n \times t_a}$, $\Delta_a(t) \in \Re^{t_a \times \eta_a}$ and $G_a \triangleq [G_1 .... G_p] \in \Re^{\eta_a \times np}$. $H_b \in \Re^{n \times t_b}$, $F_b(t) \in \Re^{t_b \times \eta_b}$ and $T_b \triangleq [T_1 .... T_q] \in \Re^{\eta_b \times mq}$.

Under the state feedback control (4.96) and using (4.130)-(4.131), the closed-loop system is given by:

$$\dot{x}(t) = [A + BK + H\Delta(t)(E + E_b K)]x(t) + [\mathcal{A}_t + H_a \Delta_a(t)G_a] \chi(t, d, p)$$
$$+ \ [\mathcal{B}_t + H_b F_b(t)T_b] K_d \chi(t, h, q) + Dw(t)$$
$$z(t) = Lx(t) \tag{4.132}$$

For system (4.132), it turns out [195,196] that the L_2-induced norm from $v(t) = \gamma w(t)$ to $z(t)$ is less than unity if the associated Hamiltonian

$$H(x, v, t) = \dot{V}(x_t) + (z^t z - v^t v) \tag{4.133}$$

is negative definite $\forall \ x(t)$ and $v(t)$.

The following theorem establishes the main result.

Theorem 4.14: *There exists a memoryless state-feedback controller such that the closed-loop system (4.132) is asymptotically stable and $\|T_{zw}\|_\infty \leq \gamma$ ($\gamma > 0$) $\forall \, d_j, h_k \geq 0$ $(j \in J_p, \, k \in J_q)$, if there exist matrices $0 < Y = Y^t \in \Re^{n \times n}$, $0 < Q_{t_j} = Q_{t_j}^t \in \Re^{n \times n}$ $(j \in J_p)$, $0 < Q_{s_k} = Q_{s_k}^t \in \Re^{n \times n}$ $(k \in J_q)$, $N \in \Re^{m \times n}$ and scalars $\alpha_1, \alpha_2, \alpha_a, \alpha_b > 0$, $\alpha_m = \sigma(\alpha_1 + \alpha_2)$ such that $\alpha_a M_a^t M_a < I$ and $\alpha_b M_b^t M_b < I$ solving the LMI:*

$$W_{\gamma u} = \begin{bmatrix} \Gamma(Y) & \bar{L} & \bar{G} & \bar{N} \\ \bar{L}^t & -J_l & 0 & 0 \\ \bar{G} & 0 & -J_g & 0 \\ \bar{N} & 0 & 0 & -J_n \end{bmatrix} < 0 \qquad (4.134)$$

where

$$
\begin{aligned}
\Gamma(Y) &= AY + YA^t + BN + N^t B^t + \sum_{j=1}^p Q_{t_j} + \sum_{k=1}^q Q_{s_k} + H(\alpha_m I) H^t \\
E_t &= L^t L + \alpha_1^{-1} E^t E, \quad T_s = \alpha_2^{-1} E_b^t E_b \\
\Omega_e &= \Omega_t^{-1/2}[I - \alpha_a H_a^t H_a]^{-1} \Omega_t^{-1/2}, \quad \Omega_s = \sigma_a \alpha_a^{-1} \Omega_t^{-1} \\
\Gamma_e &= \Gamma_t^{-1/2}[I - \alpha_b H_b^t H_b]^{-1} \Gamma_t^{-1/2}, \quad \Gamma_s = \sigma_b \alpha_b^{-1} \Gamma_t^{-1} \\
\bar{L} &= [L \;\; N^t \;\; \gamma^{-1} D], \quad J_l = diag[E_t \;\; T_s \;\; I], \quad \bar{N} = [\mathcal{B}_t N_{od} \;\; T_b N_{od}] \\
\bar{G} &= [A_t L_d \;\; G_a L_d], \quad J_g = diag[\Omega_e \;\; \Omega_s], \quad J_n = [\Gamma_e \;\; \Gamma_s] \qquad (4.135)
\end{aligned}
$$

Proof: By differentiating (4.103) along the solutions of (4.132) and manipulating, we get:

$$
\begin{aligned}
H(x, v, t) &= Z_*^t(t) W_u Z_*(t) \\
Z_*^t(t) &= \begin{bmatrix} x^t(t) & \chi^t(t, d, p) & \chi^t(t, h, q) & v^t \end{bmatrix} \\
W_u &= \begin{bmatrix} W_{u,1} & W_{u,2} & W_{u,3} & \gamma^{-1} PD \\ W_{u,2}^t & -\Omega(d, p) & 0 & 0 \\ W_{u,3}^t & 0 & -\Gamma(h, q) & 0 \\ \gamma^{-1} D^t P & 0 & 0 & -I \end{bmatrix} \qquad (4.136)
\end{aligned}
$$

where

$$
\begin{aligned}
W_{u,1} &= P(A + BK) + (A + BK)^t P + L^t L + \sum_{j=1}^p Q_j + \sum_{k=1}^q S_k + PH\Delta E \\
&+ E^t \Delta^t H^t P + PH\Delta E_b K + K^t E_b^t \Delta^t H^t P
\end{aligned}
$$

$$W_{u,2} = P(\mathcal{A}_t + H_a \Delta_a G_a)$$
$$W_{u,3} = P(\mathcal{B}_t + H_b F_b T_b) K_d$$

$$(4.137)$$

A sufficient condition for stability is that $W_u < 0$. Expanding this condition using **A. 1**, we get:

$$W_{u,1} + W_{u,2} \Omega(d,p)^{-1} W_{u,2}^t + W_{u,3} \Gamma(h,q)^{-1} W_{u,3}^t + \gamma^{-2} PDD^t P < 0$$

$$(4.138)$$

By employing the convexification procedure of **Theorem 4.13** and introducing $Y = P^{-1}$, $N = KY$, $Q_{t_j} = P^{-1} Q_j P^{-1}$, $Q_{s_k} = P^{-1} S_k P^{-1}$, $Y_d = diag(Y, ..., Y)$ with (4.137), then premultiplying (4.138) by P^{-1} and postmultiplying the result by P^{-1}, we obtain:

$$AY + YA^t + BN + N^t B^t + (H\Delta EY + YH^t \Delta^t H^t) +$$
$$(H\Delta E_b N + N^t E_b^t \Delta^t H^t) + YL^t LY + \gamma^{-2} DD^t +$$
$$(\mathcal{A}_t + H_a \Delta_a G_a) Y_d \Omega_t^{-1} Y_d (\mathcal{A}_t{}^t + G_a^t \Delta_a^t H_a^t) +$$
$$(\mathcal{B}_t + H_b \Delta_b T_b) N_{od} \Gamma_t^{-1} N_{od}^t (\mathcal{B}_t{}^t + E_b^t \Delta_t^t T_b^t) + \sum_{j=1}^{p} Q_{t_j} + \sum_{k=1}^{q} Q_{s_k} < 0$$

$$(4.139)$$

Applying **B.1.2, B.1.3** repeatedly and by selecting scalars $\alpha_1 > 0$, $\alpha_2 > 0$, $\alpha_a > 0$, $\alpha_b > 0$ such that $\alpha_a H_a^t H_a < I$, $\alpha_b H_b^t H_b < I$, we obtain the following bounds:

$$H\Delta EY + YH^t \Delta^t H^t \leq H(\alpha_1 \sigma I)H^t + Y(\alpha_1^{-1} E^t E)H \qquad (4.140)$$
$$H\Delta E_b N + N^t E_b^t \Delta^t H^t \leq H(\alpha_2 \sigma I)H^t + N^t(\alpha_2^{-1} E_b^t E_b)N \quad (4.141)$$
$$(\mathcal{A}_t + H_a \Delta_a G_a) Y_d \Omega_t^{-1} Y_d (\mathcal{A}_t{}^t + E_a^t \Delta_a^t G_a^t)$$
$$\leq G_a Y_d (\sigma_a \alpha_a^{-1} \Omega_t^{-1}) Y_d G_a{}^t + \mathcal{A}_t Y_d \Omega_e Y_d \mathcal{A}_t{}^t$$
$$= G_a Y_d \Omega_s Y_d G_a{}^t + \mathcal{A}_t Y_d \Omega_e Y_d \mathcal{A}_t{}^t \qquad (4.142)$$
$$(\mathcal{B}_t + H_b \Delta_b T_b) N_{od} \Gamma_t^{-1} N_{od}^t (\mathcal{B}_t{}^t + T_b^t \Delta_t^t H_b^t)$$
$$\leq T_b N_{od} (\sigma_b \alpha_b^{-1} \Gamma_t^{-1}) N_{od}^t T_b^t + \mathcal{B}_t N_{od} \Gamma_e N_{od}^t \mathcal{B}_t{}^t$$
$$= T_b N_{od} \Gamma_s N_{od}^t T_b^t + \mathcal{B}_t N_{od} \Gamma_e N_{od}^t \mathcal{B}_t{}^t \qquad (4.143)$$

By substituting inequalities (4.140)-(4.143) and grouping similar terms, we get,

$$AY + YA^t + BN + N^tB^t + \sum_{j=1}^{p} Q_{t_j} + \sum_{k=1}^{q} Q_{s_k} + H(\alpha_m I)H^t + YE_tY +$$
$$N^tT_sN + A_tY_d\Omega_eY_dA_t{}^t + E_aY_d\Omega_sY_dE_a{}^t + \mathcal{B}_tN_{od}\Gamma_e N_{od}^t\mathcal{B}_t{}^t +$$
$$E_bN_{od}\Gamma_sN_{od}^tE_b^t + \gamma^{-2}DD^t \; \leq \; 0 \tag{4.144}$$

Simple rearrangement of (4.144) using **A.1** yields the block form (4.134) as desired.

Remark 4.7: Theorem 4.14 provides a necessary and sufficient delay-dependent condition for a memoryless H_∞-controller which guarantees the norm bound γ of the transfer function T_{zw}. To implement such a controller, one has to solve the following minimization problem:

$$\min_{L, W_{\gamma u}, Q_{t_1}, ..., Q_{t_p}, Q_{s_1}, ..., Q_{s_q}} \gamma$$
$$s.t. \; to \; L > 0, W_{\gamma u} \leq 0 \tag{4.145}$$

Corollary 4.8: For the case of multiple state-delay systems with parameter uncertainty, there exist matrices $0 < Y = Y^t \in \Re^{n \times n}$, $0 < Q_{t_j} = Q_{t_j}^t \in \Re^{n \times n}$ $(j \in J_p)$, $N \in \Re^{m \times n}$ solving the LMI:

$$W_{\gamma u 1} = \begin{bmatrix} AY + YA^t + BN + N^tB^t \\ + \sum_{j=1}^{p} Q_{t_j} + H(\alpha_m I)H^t & \tilde{L} & \tilde{G} \\ \tilde{L}^t & -\tilde{J}_1^{-1} & 0 \\ \tilde{G}^t & 0 & -\tilde{J}_2^{-1} \end{bmatrix} \leq 0 \tag{4.146}$$

where

$$\tilde{L} = [L \; N^t \; \gamma^{-1}D] \; , \; \tilde{J}_1 = [E_t \; T_s \; I]$$
$$\tilde{G} = [A_tY_d \; G_aY_d] \; , \; \tilde{J}_2 = [\Omega_e \; \Omega_s]$$

It is interesting to note that **Corollary 4.8** recovers the results of [165].

Corollary 4.9: For the case of multiple input-delay systems with parameter uncertainty, there exists matrices $0 < Y = Y^t \in \Re^{n \times n}$, $0 < Q_{s_k} =$

$Q_{s_k}^t \in \Re^{n \times n}$ $(k \in J_q)$, and $N \in \Re^{m \times n}$ solving the LMI:

$$W_{\gamma u 2} = \begin{bmatrix} AY + YA^t + BN + N^t B^t & & \\ + \sum_{k=1}^q Q_{s_k} + H(\alpha_m I) H^t & \tilde{L} & \tilde{B} \\ \tilde{L}^t & -\tilde{J}_1^{-1} & 0 \\ \tilde{B}^t & 0 & -\tilde{J}_3^{-1} \end{bmatrix} \leq 0 \quad (4.147)$$

where

$$\tilde{B} = [B_t N_{od} \; T_b N_{od}], \quad \tilde{J}_3 = [\Gamma_e \; \Gamma_s]$$

4.4.5 Example 4.2

The nominal data of the water-quality dynamic model given in **Example 4.1** will be used here. The matrices representing parameter uncertainties on the system are given by

$$H = \begin{bmatrix} 0.6 & -0.2 \\ 0 & 0.8 \end{bmatrix}, \; \Delta(t) = \begin{bmatrix} 0.7 \sin(t) & 0 \\ 0 & 0.3 \sin(3t) \end{bmatrix}, \; E = \begin{bmatrix} 0 & 0.4 \\ 0.2 & 0.2 \end{bmatrix}$$

$$E_b = \begin{bmatrix} 0.6 & 0.4 \\ 0 & -0.4 \end{bmatrix}, \; H_a = \begin{bmatrix} 0.25 & 0 & 0 & 0 & 0 & 0 \\ 0 & 0 & 0 & 0 & 0 & 0.2 \end{bmatrix}$$

$$H_b = \begin{bmatrix} 0.15 & 0 & 0 & 0 \\ 0 & 0 & 0 & 0.25 \end{bmatrix}, \; \Delta_a^t = \begin{bmatrix} 0.5\sin(2t) & 0 & 0 & 0 & 0 & 0 \\ 0 & 0 & 0 & 0 & 0 & 0.4\sin(t) \end{bmatrix}$$

$$\Delta_b^t = \begin{bmatrix} 0.4 \sin(2t) & 0 & 0 & 0 \\ 0 & 0 & 0 & 0.6 \sin(t) \end{bmatrix}, \; E_1 = \begin{bmatrix} 0.4 & 0 \\ -0.2 & 0.6 \end{bmatrix}$$

$$E_2 = \begin{bmatrix} -0.2 & 0.4 \\ 0 & 0.2 \end{bmatrix}, \; E_3 = \begin{bmatrix} 0.1 & -0.4 \\ 0.2 & 0 \end{bmatrix}$$

$$E_{b1} = \begin{bmatrix} 0.2 & -0.8 \\ 0 & 0.2 \end{bmatrix}, \; E_{b2} = \begin{bmatrix} -0.4 & 0 \\ 0.2 & 0.8 \end{bmatrix}$$

The matrices $Q_{t_1}, Q_{t_2}, Q_{t_3}, Q_{s_1}$ and Q_{s_2} are chosen such that, $Q_{t_1} = Q_{t_2} = Q_{t_3} = 0.001I$, and $Q_{s_1} = Q_{s_2} = 0.0001I$. The other design parameters are $\alpha_1 = \alpha_2 = \alpha_m = 0.001$, $\alpha_a = \alpha_b = 5$, $\sigma = 0.5$, $\sigma_a = 0.5$ and $\sigma_b = 0.4$. Using the **LMI** toolbox, it was found that $\gamma = 3.2394$ and

$$Y = \begin{bmatrix} 0.0014 & -0.0004 \\ -0.0004 & 0.0015 \end{bmatrix}, \; N = \begin{bmatrix} -0.1429e - 3 & 0.0602e - 3 \\ 0.0381e - 3 & -0.2073e - 3 \end{bmatrix}$$

The gain of the memoryless state-feedback controller is

$$K = \begin{bmatrix} -0.0975 & 0.0174 \\ -0.0083 & -0.1413 \end{bmatrix}$$

4.5   Linear Neutral Systems

In Chapter 2, we considered a class of neutral functional differential equations (NFDE) described by a linear model with parameteric uncertainties:

$$
\begin{aligned}
(\Sigma_{\Delta n}): \quad \dot{x}(t) - D\dot{x}(t - \tau) &= (A + \Delta A)x(t) + (A_d + \Delta A_d)x(t - \tau) \\
&= A_\Delta x(t) + A_{d\Delta}x(t - \tau) \quad\quad (4.148) \\
x(t_o + \eta) &= \phi(\eta), \quad \forall \eta \in [-\tau, 0] \quad\quad (4.149)
\end{aligned}
$$

where $x \in \Re^n$ is the state, $A \in \Re^{n \times n}$ and $A_d \in \Re^{n \times n}$ are known real constant matrices, $\tau > 0$ is an unknown constant delay factor and $\Delta A \in \Re^{n \times n}$ and $\Delta A_d \in \Re^{n \times n}$ are matrices of uncertain parameters represented by:

$$[\Delta A(t) \ \ \Delta A_d(t)] = H\Delta(t)[E \ \ E_d], \quad \Delta^t(t)\Delta(t) \leq I ; \quad \forall \, t \quad (4.150)$$

where $H \in \Re^{n \times \alpha}$, $E \in \Re^{\beta \times n}$, $E_d \in \Re^{\beta \times n}$ are known real constant matrices and $\Delta(t) \in \Re^{\alpha \times \beta}$ is an unknown matrix with Lebsegue measurable elements. The initial condition is specified as $\langle x(t_o), x(s) \rangle = \langle x_o, \phi(s) \rangle$, where $\phi(\cdot) \in \mathcal{L}_2[-\tau, t_o]$. Note that when $\Delta A \equiv 0, \Delta A_d \equiv 0$, system $(\Sigma_{\Delta n})$ reduces to the standard linear neutral systems [122]. We also proved that subject to **Assumption 2.1** and **Assumption 2.6**, the neutral system $(\Sigma_{\Delta n})$ is robustly asymptotically stable independent of delay if the following conditions hold:

(1) There exist matrices $0 < P = P^t \in \Re^{n \times n}$, $0 < S = S^t \in \Re^{n \times n}$ and $0 < R = R^t \in \Re^{n \times n}$ and scalars $\epsilon > 0, \rho > 0$ satisfying the ARI:

$$
\begin{aligned}
&PA + A^t P + (\epsilon + \rho)PHH^t P + \rho^{-1}EE^t + S + \\
&[P(AD + A_d) + SD][R - \epsilon^{-1}(D^t E^t ED + E_d^t E_d)]^{-1} \\
&[P(AD + A_d) + SD]^t < 0 \quad\quad (4.151)
\end{aligned}
$$

(2) There exist matrices $0 < S = S^t \in \Re^{n \times n}$ and $0 < R = R^t \in \Re^{n \times n}$ satisfying the Lyapunov equation (LE)

$$D^t S D - S + R = 0 \quad\quad (4.152)$$

Our task in the following sections is to develop robust $\mathcal{H}_\infty$ control results.

4.5.1 Robust Stabilization

Here, we consider a controlled-form of system $(\Sigma_{\Delta n})$ represented by:

$$(\Sigma_{\Delta nu}): \quad \dot{x}(t) - D\dot{x}(t-\tau) = (A + \Delta A)x(t) + (A_d + \Delta A_d)x(t - \tau)$$
$$+ [B + \Delta B(t)]u(t)$$
$$= A_\Delta x(t) + A_{d\Delta}x(t - \tau) + B_\Delta u(t) \quad (4.153)$$
$$x(t_o + \eta) = \phi(\eta), \quad \forall \eta \in [-\tau, 0] \quad (4.154)$$

where $u(t) \in \Re^p$ is the control input, $B \in \Re^{n \times p}$ is a real matrix and $\Delta B(t)$ represents time-varying parameteric uncertainties at the input which is of the form:

$$\Delta B(t) = H\Delta(t)E_b \quad (4.155)$$

and $E_b \in \Re^{\beta \times p}$ is a known constant matrix. The remaining matrices are as in (4.148)-(4.150). We restrict ourselves in the robust stabilization problem of the uncertain neutral system $(\Sigma_{\Delta nu})$ on using a linear memoryless state-feedback $u(t) = K_s x(t)$ and establish the following result.

Theorem 4.15: *System $(\Sigma_{\Delta nu})$ is robustly stable via memoryless state feedback $u(t) = K_s x(t)$ if there exist scalars $(\epsilon > 0, \rho > 0)$, matrices $0 < Y = Y^t \in \Re^{n \times n}, 0 < R = R^t \in \Re^{n \times n}, 0 < S = S^t \in \Re^{n \times n}$ and $X \in \Re^{m \times n}$ satisfying the LMIs:*

$$\begin{bmatrix} W(X,Y,S) & YE^t + X^t E_b^t & G(X,Y,S) \\ EY + E_b X & \rho I & 0 \\ G^t(x,y,s) & 0 & -J_s \end{bmatrix} < 0$$
$$D^t SD - S < 0$$
$$(E_d + [E + E_b XY^{-1}]D)^t(E_d + [E + E_b XY^{-1}]D) - \epsilon R < 0$$
$$(4.156)$$

Moreover, the feedback gain is given by:

$$K_s = X Y^{-1} \quad (4.157)$$

where

$$W(X,Y,S) = YA^t + AY + (\epsilon + \rho)HH^t + YSY + BX + X^t B^t$$
$$G(X,Y,S) = AD + A_d + BX^{-1}YD + YSD$$
$$J_s = R - \epsilon^{-1}(E_d + [E + E_b XY^{-1}]D)^t(E_d + [E + E_b XY^{-1}]D)$$
$$(4.158)$$

Proof: System ($\Sigma_{\Delta nu}$) with the memoryless feedback control law $u(t) = K_s x(t)$ becomes:

$$
(\Sigma_{\Delta nu}): \quad \dot{x}(t) - D\dot{x}(t-\tau) = A_{\Delta c}(t)x(t) + A_{d\Delta}(t)x(t-\tau)
$$
$$
= [A_c + H\Delta(t)M_c]x(t) + A_{d\Delta}(t)x(t-\tau)
$$
$$(4.159)$$

where

$$
A_c = A + BK_s, \qquad M_c = E + E_b K_s \qquad (4.160)
$$

By **Theorem 2.6**, system (4.159)-(4.160) is robustly, asymptotically stable if:

$$
PA_{\Delta c} + A_{\Delta c}^t P + S +
$$
$$
(PA_{\Delta c}D + SD + PA_{d\Delta})R^{-1}(PA_{\Delta c}D + SD + PA_{d\Delta})^t < 0
$$
$$(4.161)$$

for all admissible uncertainties satisfying (4.150). Applying **B.1.2** and **B.1.3**, it can be shown that (4.160) reduces for some ($\epsilon > 0$, $\rho > 0$) to:

$$
PA + A^t P + (\epsilon + \rho)PHH^t P + S + PBK_s + K_s^t B^t P +
$$
$$
\rho^{-1}(E + E_b K_s)^t(E + E_b K_s) + \{P[(A + BK_s)D + A_d] + SD\}
$$
$$
\left\{R - \epsilon^{-1}(E_d + [E + E_b K_s]D)^t(E_d + [E + E_b K_s]D)\right\}^{-1}
$$
$$
\{P[(A + BK_s)D + A_d] + SD\}^t < 0 \qquad (4.162)
$$

Premultiplying and postmultiplying (4.162) by P^{-1}, letting $Y = P^{-1}$ and using (4.157), we get:

$$
AY + YA^t + (\epsilon + \rho)HH^t + YSY + \rho^{-1}(EY + E_b X)^t(EY + E_b X) +
$$
$$
BX + X^t B^t + \left\{(A + BX^{-1}Y)D + A_d + YSD\right\}
$$
$$
\left\{R - \epsilon^{-1}(E_d + [E + E_b XY^{-1}]D)^t(E_d + [E + E_b XY^{-1}]D)\right\}
$$
$$
\left\{(A + BX^{-1}Y)D + A_d + YSD\right\}^t < 0 \qquad (4.163)
$$

Finally by **A.1**, the LMI (4.156) follows from the ARE (4.163).

Corollary 4.10: *System (Σ_n) with $\Delta A \equiv 0, \Delta A_d \equiv 0$ is robustly stable via memoryless state feedback $u(t) = K_s x(t)$ if there exist matrices $0 < Y =$*

$Y^t \in \Re^{n\times n}$, $0 < R = R^t \in \Re^{n\times n}$, $0 < S = S^t \in \Re^{n\times n}$ and $X \in \Re^{m\times n}$ satisfying the LMIs:

$$
\begin{bmatrix}
YA^t + AY + & AD + A_d + \\
YSY + BX + X^tB^t & BX^{-1}YD + YSD \\
D^tS^tY + D^tYX^{-t}B^t & \\
+A_d^t + D^tA^t & -R
\end{bmatrix}
< 0
$$

$$D^tSD - S < 0 \quad (4.164)$$

Moreover, the feedback gain is given by:

$$K_s = X Y^{-1} \qquad (4.165)$$

Proof: Set $E = 0, H = 0, E_b = 0$ in (4.163).

4.5.2 Robust $\mathcal{H}_\infty$ Performance

We proceed further and extend the robust stabilization results developed in the previous section to the case of robust $\mathcal{H}_\infty$ performance problem. For this purpose, we consider the following system:

$$
\begin{aligned}
(\Sigma_{\Delta nw}): \quad \dot{x}(t) - D\dot{x}(t-\tau) &= A_\Delta x(t) + A_{d\Delta} x(t-\tau) \\
&+ Nw(t) & (4.166) \\
z(t) &= Cx(t) & (4.167) \\
x(t_o + \eta) &= \phi(\eta), \quad \forall \eta \in [-\tau, 0] & (4.168)
\end{aligned}
$$

where $z \in \Re^p$ is the controlled output, $N \in \Re^{n\times p}$, $C \in \Re^{p\times n}$ are known real constant matrices and $w(t) \in \mathcal{L}_2[0,\infty)$ is the external input.

Theorem 4.16: *Subject to* **Assumption 2.1** *and* **Assumption 2.6**, *the neutral system* $(\Sigma_{\Delta nw})$ *is robustly asymptotically stable independent of delay with disturbance attenuation* γ *if the following conditions hold:*

(1) *There exist matrices* $0 < P = P^t \in \Re^{n\times n}$, $0 < S = S^t \in \Re^{n\times n}$ *and* $0 < R = R^t \in \Re^{n\times n}$ *and scalars* $\epsilon > 0, \rho > 0$ *satisfying the ARI:*

$$
\begin{aligned}
&PA + A^tP + P[(\epsilon + \rho)HH^t + \gamma^{-2}NN^t]P + S + C^tC + \rho^{-1}EE^t + \\
&[P(AD + A_d) + (S + C^tC)D][R_u - \epsilon^{-1}(D^tE^tED + E_d^tE_d)]^{-1} \\
&[P(AD + A_d) + (S + C^tC)D]^t < 0 \qquad (4.169)
\end{aligned}
$$

(2) *There exist matrices $0 < S = S^t \in \Re^{n\times n}$ and $0 < R = R^t \in \Re^{n\times n}$ such that $R_u = R + C^tC$ satisfying the LE*

$$D^t\,(S + C^tC)\,D - (S + C^tC) + R_u = 0 \qquad (4.170)$$

Proof: In order to show that system $(\Sigma_{\Delta w})$ is robustly stable with a disturbance attenuation γ , it is required that the associated Hamiltonian $H(x, w, t)$ satisfies [218]:

$$H(x, w, t) = \dot{V}_7(x_t) + z^t(t)z(t) - \gamma^2 w^t(t)w(t) < 0$$

where $V_7(x_t)$ is given by (2.118). By differentiating (2.118) along the trajectories of (4.166)-(4.167), it yields:

$$
\begin{aligned}
H(x, w, t) &= [A_\Delta x(t) + A_{d\Delta}x(t-\tau)]^t P[x(t) - Dx(t-\tau)] \\
&+ [x(t) - Dx(t-\tau)]^t P[A_\Delta x(t) + A_{d\Delta}x(t-\tau)] \\
&+ x^t Sx(t) - x^t(t-\tau)Sx(t-\tau) + x^t C^tCx - \gamma^{-2}w^tw \\
&+ w^t N^t P[x(t) - Dx(t-\tau)] + [x(t) - Dx(t-\tau)]^t PNw
\end{aligned}
$$
$$(4.171)$$

In terms of $\mathcal{M}$, we manipulate (4.171) to reach:

$$
\begin{aligned}
H(x, w, t) &= \mathcal{M}^t(x_t)[PA_\Delta + A_\Delta^t P + S + C^tC]\mathcal{M}(x_t) \\
&+ \mathcal{M}^t(x_t)[PA_\Delta D + (S + C^tC)D + PA_{d\Delta}]x(t-\tau) \\
&+ x^t(t-\tau)[D^t A_\Delta^t P + D^t(S + C^tC) + A_{d\Delta}^t P]\mathcal{M}(x_t) \\
&+ x^t(t-\tau)[D^t(S + C^tC)D - S]x(t-\tau) \\
&+ w^t N^t P\mathcal{M}(x_t) + \mathcal{M}^t(x_t)PNw - \gamma^{-2}w^tw
\end{aligned}
$$
$$(4.172)$$

Using $R_u = R + C^tC$, completing the squares in (4.172) and arranging terms we reach:

$$
\begin{aligned}
H(x, w, t) &\leq \mathcal{M}^t(x_t)[PA_\Delta + A_\Delta^t P + S + C^tC + \gamma^{-2}PNN^t P]\mathcal{M}(x_t) \\
&+ \mathcal{M}^t(x_t)[PA_\Delta D + (S + C^tC)D + PA_{d\Delta}]R_u^{-1} \\
&\quad [PA_\Delta D + (S + C^tC)D + PA_{d\Delta}]^t \mathcal{M}(x_t)
\end{aligned}
$$
$$(4.173)$$

For asymptotic stability of system $(\Sigma_{\Delta nw})$, it is sufficient that

$$
\begin{aligned}
&PA_\Delta + A_\Delta^t P + S + C^tC + \gamma^{-2}PNN^t P + \\
&[PA_\Delta D + (S + C^tC)D + PA_{d\Delta}]R_u^{-1} \\
&[PA_\Delta D + (S + C^tC)D + PA_{d\Delta}] < 0
\end{aligned}
$$
$$(4.174)$$

Using **B.1.2** and **B.1.3** in (4.174), it follows for some $\mu > 0, \sigma > 0$ that

$$PA + A^t P + P[(\epsilon + \rho)HH^t + \gamma^{-2}NN^t]P + S + C^tC + \rho^{-1}EE^t +$$
$$[P(AD + A_d) + (S + C^tC)D][R_u - \epsilon^{-1}(D^tE^tED + E_d^tE_d)]^{-1}$$
$$[P(AD + A_d) + (S + C^tC)D]^t < 0 \qquad (4.175)$$

Finally, ARI (4.175) corresponds to (4.169) such that S and R satisfy (4.170).

Corollary 4.11: *Subject to* **Assumption 2.1** *and* **Assumption 2.6,** *the neutral system* $(\Sigma_{\Delta nw})$ *is asymptotically stable independent of delay if there exist matrices* $0 < Q = Q^t \in \Re^{n \times n}$, $0 < S = S^t \in \Re^{n \times n}$ *and scalars* $\epsilon > 0, \rho > 0$ *satisfying the LMIs*

$$\begin{bmatrix} AQ + QA^t + \epsilon HH^t \\ +Q(\rho^{-1}EE^t + S + C^tC)Q & H & \begin{matrix} AD + A_d \\ +QSD \end{matrix} & \gamma^{-1}N \\ H^t & -\rho^{-1}I & 0 & 0 \\ \begin{matrix} D^tA^t + A_d^t \\ +D^tSQ \end{matrix} & 0 & -J_w & 0 \\ \gamma^{-1}N^t & 0 & 0 & -I \end{bmatrix} < 0$$

$$D^t(S + C^tC)D - (S + C^tC) < 0$$
$$\epsilon[D^t(S + C^tC)D - (S + C^tC)] + [D^tE^tED + E_d^tE_d] < 0 \qquad (4.176)$$

where

$$J_w = D^t(S + C^tC)D - (S + C^tC) + \epsilon^{-1}(D^tE^tED + E_d^tE_d)$$

Proof: By **A.1**, ARI (4.169) and (4.170) with $Q = P^{-1}$ are equivalent to the LMI (4.176).

We now consider the robust synthesis problem for system $(\Sigma_{\Delta nwu})$:

$$(\Sigma_{\Delta nwu}): \quad \dot{x}(t) - D\dot{x}(t - \tau) = A_\Delta x(t) + A_{d\Delta}x(t - \tau)$$
$$+ B_\Delta u(t) + Nw(t) \qquad (4.177)$$
$$x(t_o + \eta) = \phi(\eta), \quad \forall \eta \in [-\tau, 0] \qquad (4.178)$$

The following theorem establishes the main result.

Theorem 4.17: *System* $(\Sigma_{\Delta nwu})$ *is robustly stable with a disturbance attenuation* γ *via memoryless state feedback if there exist scalars* $(\epsilon > 0, \rho >$

0), matrices $0 < Y = Y^t \in \Re^{n \times n}$, $0 < R = R^t \in \Re^{n \times n}$, $0 < S = S^t \in \Re^{n \times n}$ and $X \in \Re^{m \times n}$ satisfying the LMIs:

$$
\begin{bmatrix}
W_*(X,Y,S) & YE^t + X^t E_b^t & G_*(X,Y,S) \\
EY + E_b X & \rho I & 0 \\
G_*^t(X,Y,S) & 0 & -J_*
\end{bmatrix} < 0
$$

$$D^t(S + C^t C)D - (S + C^t C) < 0$$

$$(E_d + [E + E_b XY^{-1}]D)^t(E_d + [E + E_b XY^{-1}]D) - \epsilon(R + C^t C) < 0$$

$$(4.179)$$

Moreover, the feedback gain is given by:

$$K_s = X Y^{-1} \tag{4.180}$$

where

$$
\begin{aligned}
W_*(X,Y,S) &= YA^t + AY + [(\epsilon + \rho)HH^t + \gamma^{-2}NN^t] + YSY \\
&+ BX + X^t B^t + Y(S + C^t C + \rho^{-1} EE^t)Y \\
G_*(X,Y,S) &= (A + BXY^{-1})D + A_d + Y(S + C^t C)D \\
J_* &= (R + C^t C) - \epsilon^{-1}(E_d + [E + E_b XY^{-1}]D)^t \\
&\quad (E_d + [E + E_b XY^{-1}]D)
\end{aligned} \tag{4.181}
$$

Proof: System $(\Sigma_{\Delta n w u})$ subject to the control law $u(t) = K_* x(t)$ has the form:

$$
\begin{aligned}
\dot{x}(t) - D\dot{x}(t - \tau) &= A_{\Delta *}(t)x(t) + A_{d\Delta}(t)x(t - \tau) + Nw(t) \\
&= [A_* + H\Delta(t)M_*]x(t) + Nw(t) \tag{4.182} \\
z(t) &= Cx(t) \tag{4.183}
\end{aligned}
$$

where

$$A_* = A + B K_*, \qquad M_* = E + E_b K_* \tag{4.184}$$

In terms of $V_7(x_t)$ in (2.118), we evaluate the Hamiltonian $H(x, w, t)$ associated with (4.182)-(4.183) in the manner of **Theorem 4.16** to yield:

$$
\begin{aligned}
H(x, w, t) &= [A_{\Delta *}x(t) + A_{d\Delta}x(t - \tau)]^t P[x(t) - Dx(t - \tau)] \\
&+ [x(t) - Dx(t - \tau)]^t P[A_{\Delta *}x(t) + A_{d\Delta}x(t - \tau)] \\
&+ x^t Sx(t) - x^t(t - \tau)Sx(t - \tau) + x^t C^t Cx - \gamma^{-2}w^t w \\
&+ w^t N^t P[x(t) - Dx(t - \tau)] + [x(t) - Dx(t - \tau)]^t PNw
\end{aligned} \tag{4.185}
$$

In terms of $\mathcal{M}$, we manipulate (4.185) to reach:

$$H(x, w, t) \leq \mathcal{M}^t(x_t)[PA_{\Delta*} + A_{\Delta}^t P + S + C^t C + \gamma^{-2} PNN^t P]\mathcal{M}(x_t)$$
$$+ \mathcal{M}^t(x_t)[PA_{\Delta*}D + (S + C^t C)D + PA_{d\Delta}]R_u^{-1}$$
$$[PA_{\Delta}D + (S + C^t C)D + PA_{d\Delta}]^t \mathcal{M}(x_t) \qquad (4.186)$$

from which it is sufficient for asymptotic stability that

$$PA_{\Delta*} + A_{\Delta*}^t P + S + C^t C + \gamma^{-2} PNN^t P +$$
$$[PA_{\Delta*}D + (S + C^t C)D + PA_{d\Delta}]R_u^{-1}$$
$$[PA_{\Delta*}D + (S + C^t C)D + PA_{d\Delta}] < 0 \qquad (4.187)$$

Using B.1.2 and B.1.3 in (4.187), it follows for some $\mu > 0, \sigma > 0$ that

$$PA + A^t P + PBK_* + K_*^t B^t P + P[(\epsilon + \rho)HH^t + \gamma^{-2}NN^t]P +$$
$$S + C^t C + \rho^{-1}EE^t + \rho^{-1}(E + E_b K_*)^t (E + E_b K_*)$$
$$[P(A + BK_*)D + PA_d + (S + C^t C)D]$$
$$[R_u - \epsilon^{-1}(E_d + [E + E_b K_*]D)^t (E_d + [E + E_b K_*]D)]^{-1}$$
$$[P(A + BK_*)D + PA_d + (S + C^t C)D]^t < 0 \qquad (4.188)$$

Finally, premultiplying (4.188) and postmultiplying by P^{-1}, letting $Y = P^{-1}$ and using (4.180), the LMIs (4.179) follow.

4.6 Notes and References

For basic results on $\mathcal{H}_\infty$ control of time-delay systems, the reader can consult [13,17,19,22,32,47,191,222]. What we have attempted in this chapter is to present general results on robust $\mathcal{H}_\infty$ control for wide classes of uncertain time-delay systems. Despite this effort, there are ample interesting problems to be solved. These include, but are not limited to, $\mathcal{H}_\infty$ control incorporating delay-dependent internal stability, $\mathcal{H}_\infty$ control of other classes of nonlinear time-delay systems and discrete-time systems.

Chapter 5

Guaranteed Cost Control

In Chapter 3, we addressed the problem of designing stabilizing feedback controllers using a standard state-feedback approach. Then in Chapter 4, we discussed a second approach to the same problem based on $\mathcal{H}_\infty$ theory. Here, we move another step further and examine a third design approach called *guaranteed cost control*. As we shall see in the sequel, this appraoch uses a fixed Lyapunov functional to establish an upper bound on the closed-loop value of a quadratic cost function. Keeping up with our objective throughout the book, we will start by treating continuous-time systems and then deal with discrete-time systems.

5.1 Continuous-Time Systems

5.1.1 Uncertain State-Delay Systems

We consider a class of uncertain time-delay systems represented by:

$$
\begin{aligned}
(\Sigma_\Delta): \quad \dot{x}(t) &= [A + \Delta A(t)]x(t) + [B + \Delta B(t)]u(t) \\
&+ [A_d + \Delta A_d(t)]x(t - \tau) \\
&= A_\Delta(t)x(t) + B_\Delta(t)u(t) + A_{d\Delta}(t)x(t - \tau) \quad (5.1)
\end{aligned}
$$

where $x(t) \in \Re^n$ is the state, $u(t) \in \Re^m$ is the control input and the matrices $A \in \Re^{n \times n}$, $B \in \Re^{n \times m}$ and $A_d \in \Re^{n \times n}$ are real constant matrices representing the nominal plant. Here, τ is an unknown constant integer representing the number of delay units in the state. For all practical purposes, we consider $0 \leq \tau \leq \tau^*$ with τ^* being known. The matrices $\Delta A(t)$, $\Delta B(t)$ and $\Delta D(t)$

represent parameteric uncertainties which are of the form:

$$[\Delta A(t) \quad \Delta B(t) \quad \Delta A_d(t)] \quad = \quad H \, \Delta(t) \, [E \quad E_b \quad E_d] \qquad (5.2)$$

where $H \in \Re^{n \times \alpha}$, $E \in \Re^{\beta \times n}$, $E_b \in \Re^{\beta \times m}$ and $E_d \in \Re^{\beta \times n}$ are known constant matrices and $\Delta(t) \in \Re^{\alpha \times \beta}$ is an unknown matrix satisfying

$$\Delta^t(t) \, \Delta(t) \quad \leq \quad I \quad \forall t \qquad (5.3)$$

The initial condition is specified as $\langle x(0), x(s) \rangle = \langle x_o, \phi(s) \rangle$, where $\phi(.) \in \mathcal{L}_2[-\tau, 0]$.

Associated with the uncertain system (Σ_Δ) is the cost function:

$$J = \int_0^\infty [\, x^t(t) Q x(t) + u^t(t) R u(t) \,] \; dt \qquad (5.4)$$

where $0 < Q = Q^t \in \Re^{n \times n}$, $0 < R = R^t \in \Re^{m \times m}$ are given state and control weighting matrices.

Distinct from system (Σ_Δ) is the free-system

$$(\Sigma_D): \quad \dot{x}(t) \quad = \quad [A + \Delta A(t)] x(t) + [A_d + \Delta A_d(t)] x(t - \tau)$$
$$= \quad A_\Delta(t) x(t) + A_{d\Delta}(t) x(t - \tau) \qquad (5.5)$$

for which we associate the cost function:

$$J_o = \int_0^\infty [\, x^t(t) Q x(t) \,] \; dt \qquad (5.6)$$

In the sequel, we consider the problem of designing a robust state-feedback control that renders the closed-loop system robustly stable and guarantees a prescribed level of performance.

5.1.2   Robust Performance Analysis I

Since the stability of system (Σ_D) is crucial to the development of the guaranteed cost control for (Σ_Δ), we adopt hereafter the notion of robust stability independent of delay which was discussed in Chapter 2. Recall from **Lemma 2.1** with Lyapunov-Krasovskii functional (2.5) that system (Σ_D) is *robustly stable* (RS) independent of delay if there exist matrices $0 < P = P^t \in \Re^{n \times n}$ and $0 < W = W^t \in \Re^{n \times n}$ satisfying the ARI:

$$A_\Delta^t(t) P + P A_\Delta(t) + P A_{d\Delta}(t) W^{-1} A_{d\Delta}^t(t) P + W < 0 \qquad (5.7)$$
$$\forall \Delta : \Delta^t(t) \, \Delta(t) \quad \leq \quad I \; \forall t$$

Or equivalently, there exist matrices $0 < P = P^t \in \Re^{n \times n}$ and $0 < W = W^t \in \Re^{n \times n}$ satisfying the LMI

$$\begin{bmatrix} PA_\Delta(t) + A_\Delta^t(t)P + W & PA_{d\Delta}(t) \\ A_{d\Delta}^t(t)P & -W \end{bmatrix} < 0 \qquad (5.8)$$

$$\forall \Delta : \quad \Delta^t \Delta \leq I \ \forall t$$

Based on this, we have the following definition:

Definition 5.1: *System (Σ_D) with cost function (6) is said to be robustly stable (RS) independent of delay with a quadratic cost matrix (QCM) $0 < P = P^t \in \Re^{n \times n}$ if there exists a matrix $0 < W = W^t \in \Re^{n \times n}$ such that*

$$A_\Delta^t(t)P + PA_\Delta(t) + PA_{d\Delta}(t)W^{-1}A_{d\Delta}^t(t)P + W + Q < 0$$
$$\forall \Delta : \Delta^t(t)\,\Delta(t) \leq I \ \forall t$$

Then, we have the following result:

Theorem 5.1: *Consider system (Σ_D) and cost function (6). If $0 < P = P^t$ is a QCM, then (Σ_D) is RS and the cost function satisfies the bound*

$$J_o \leq x_o^t P x_o + \int_{-\tau}^0 x^t(\alpha) W x(\alpha) d\alpha \qquad (5.9)$$

Conversely, if system (Σ_D) is RS then there will be a QCM for this system and cost function (5.6).

Proof:$(\Longrightarrow)$ Let $0 < P = P^t$ be a QCM for system (Σ_D) and cost function (5.6). It follows from **Definition 5.1** that there exists a matrix $0 < W = W^t \in \Re^{n \times n}$ such that

$$X^t(t)\begin{bmatrix} PA_\Delta + A_\Delta^t P + W + Q & PA_{d\Delta} \\ A_{d\Delta}^t P & -W \end{bmatrix}X(t) < 0 \quad (5.10)$$

$$\forall \Delta : \quad \Delta^t \Delta \leq I \ , \quad X(k) \neq [0\ 0]$$

where

$$X(t) = [x^t(t) \ \ x^t(t - \tau)]^t \qquad (5.11)$$

Therefore, system (Σ_D) is RS.

Now by evaluating the derivative $\dot{V}_1(x_t)$ of the functional (2.5), we obtain:

$$\dot{V}_1(x_t) = X^t(t) \begin{bmatrix} PA_\Delta + A_\Delta^t P + W & PA_{d\Delta} \\ A_{d\Delta}^t P & -W \end{bmatrix} X(t)$$

$$\leq -x^t(k)Qx(k) \tag{5.12}$$

from which we conclude that

$$x^t(k)Qx(k) \leq -\dot{V}_1(x_t) \tag{5.13}$$

integrating (5.13) over the period $t \in [0, \infty]$ and using (5.6), we get:

$$J_o \leq V_1(x_o) - V_1(x(\infty)) \tag{5.14}$$

By (5.10), system (Σ_D) is RS leading to $V(x(t) \rightarrow 0$ as $t \rightarrow \infty$ and therefore (5.14) reduces to

$$J_o \leq x_o^t P x_o + \int_{-\tau}^0 x^t(\alpha)Wx(\alpha)d\alpha$$

($\Longleftarrow$) Let system (Σ_D) be RS. It follows that there exist $0 < P = P^t$ and $0 < W = W^t$ such that

$$A_\Delta^t(t)P + PA_\Delta(t) + PD_\Delta(t)W^{-1}D_\Delta^t(t)P + W < 0 \tag{5.15}$$
$$\forall \Delta : \quad \Delta^t \; \Delta \; \leq \; I \; \; \forall t$$

Hence, one can find some $\rho > 0$ such that the following inequality holds:

$$\rho^{-1}[A_\Delta^t(t)P + PA_\Delta(t) + PA_{d\Delta}(t)W^{-1}A_{d\Delta}^t(t)P + W] + Q =$$
$$[A_\Delta^t(t)\breve{P} + \breve{P}A_\Delta(t) + \breve{P}A_{d\Delta}(t)\breve{W}^{-1}A_{d\Delta}^t(t)\breve{P} + \breve{W}] + Q < 0 \tag{5.16}$$
$$\forall \Delta : \quad \Delta^t \; \Delta \; \leq \; I \; \; \forall t$$

The above inequality implies that there exists a matrix $\breve{W} = \rho^{-1}W$ such that the matrix $\breve{P} = \rho^{-1}P$ is a QCM for system (Σ_D).

Remark 5.1: In applying **Theorem 5.1** to a particular example we will need the bound τ^* on the time-delay factor τ to evaluate the upper bound on J_o .

Theorem 5.2: A matrix $0 < P = P^t \in \Re^{n \times n}$ is a QCM for system (Σ_D) and cost function (5.6) if and only if there exists a matrix $0 < W = W^t \in$

$\Re^{n \times n}$ and a scalar $\mu > 0$ such that $W_d = W - \mu^{-1} E_d^t E_d > 0$ and satisfying the LMIs

$$
\begin{bmatrix}
PA + A^t P + W + \\
Q + \mu^{-1} E^t E & PH & PA_d + \\
& & \mu^{-1} E^t E_d \\
H^t P & -\mu^{-1} I & 0 \\
A_d^t P + \\
\mu^{-1} E_d^t E_d & 0 & -W_d
\end{bmatrix} < 0
$$

$$\mu^{-1} E^t E_d - W < 0 \qquad (5.17)$$

Proof: By Definition 5.1 and A.1, system (Σ_D) with cost function (5.6) is QCM which implies that

$$
\begin{bmatrix}
PA_\Delta(t) + A_\Delta^t(t)P + W + Q & PA_{d\Delta}(t) \\
A_{d\Delta}^t(t)P & -W
\end{bmatrix} =
$$

$$
\begin{bmatrix}
PA + A^t P + W + Q & PA_d + \\
PH\Delta(t)E + E^t \Delta^t(t) H^t P & PH\Delta(t) E_d \\
A_d^t P + & \\
E_d^t \Delta^t(t) H^t P & -W
\end{bmatrix} < 0 \qquad (5.18)
$$

$$\forall \Delta : \quad \Delta^t \Delta \leq I \quad \forall t$$

Inequality (5.18) holds if and only if

$$
\begin{bmatrix}
PA + A^t P + W + Q & PA_d \\
A_d^t P & -W
\end{bmatrix} +
\begin{bmatrix}
PH \\
0
\end{bmatrix} \Delta(t) [E \quad E_d]
$$

$$
\begin{bmatrix}
E^t \\
E_d^t
\end{bmatrix} \Delta^t(t) [L^t P \quad 0] < 0 \qquad (5.19)
$$

$$\forall \Delta : \quad \Delta^t(t) \Delta(t) \leq I \quad \forall t$$

By **B.1.4**, inequality (5.19) for some $\mu > 0$ is equivalent to

$$
\begin{bmatrix}
PA + A^t P + W + Q & PA_d \\
A_d^t P & -W
\end{bmatrix} + \mu
\begin{bmatrix}
PH \\
0
\end{bmatrix} [H^t P \quad 0] +
$$

$$
\mu^{-1}
\begin{bmatrix}
E^t \\
E_d^t
\end{bmatrix} [E \quad E_d] =
$$

$$
\begin{bmatrix}
PA + A^t P + W + \\
Q + \mu PHH^t P + \mu^{-1} E^t E & PA_d + \mu^{-1} E^t E_d \\
A_d^t P + \mu^{-1} E_d^t E & -(W - \mu^{-1} E_d^t E_d)
\end{bmatrix} < 0 \quad (5.20)
$$

Simple rearrangement of (5.20) yields LMIs (5.17).

For convenience, we define the following matrix expressions:

$$\bar{W}_d = I + \mu^{-1} E_d W^{-1} E_d^t , \quad \bar{A} = A + \mu^{-1} A_d W_d^{-1} E_d^t E$$
$$\Omega^t \Omega = E^t \bar{W}_d E, \quad \Gamma \Gamma^t = A_d W_d^{-1} A_d^t \tag{5.21}$$

Corollary 5.1: *A matrix* $0 < P = P^t \in \Re^{n \times n}$ *is a QCM for system* (Σ_D) *and cost function (5.6) if and only if there exists a matrix* $0 < W = W^t \in \Re^{n \times n}$ *and a scalar* $\mu > 0$ *such that* $W_d = W - \mu^{-1} A_d^t A_d > 0$ *and satisfying the ARI*

$$P\bar{A} + \bar{A}^t P + P[\mu H H^t + \Gamma \Gamma^t] P + W + Q + \mu^{-1} \Omega^t \Omega < 0 \tag{5.22}$$

Proof: Follows from application of **A.1** (5.21) using **A.2** and (5.21).

Recalling **A.3.1**, it is possible to cast **Theorem 5.2** into an H_∞−norm bound setting in the following sense. There exists a matrix $0 < P = P^t \in \Re^{n \times n}$ such that inequality (5.17) is satisfied if and only if the following conditions hold:

(a) A is Hurwitz
(b) The following H_∞−norm bound is satisfied

$$\left\| \begin{bmatrix} \mu^{-1/2} \Omega \\ W^{1/2} \\ Q^{1/2} \end{bmatrix} [sI - \bar{A}]^{-1} [\mu^{1/2} H \quad \Gamma] \right\|_\infty < 1 \tag{5.23}$$

An important point to observe is that the set $\Upsilon = \{(\mu, W) : \mu > 0, 0 < W = W^t\}$ over which inequality (5.17) has a solution $0 < P = P^t$ can be determined by finding those values of $\mu > 0$, $0 < W = W^t$ such that (5.23) is satisfied. More importantly, for any such μ, W the following ARE

$$\breve{P}\bar{A} + \bar{A}^t \breve{P} + \breve{P}[\mu H H^t + \Gamma \Gamma^t]\breve{P} + W + Q + \mu^{-1} \Omega^t \Omega = 0 \tag{5.24}$$

has a stabilizing solution $\breve{P} \geq 0$.

Corollary 5.2: *Consider the time-delay system*

$$(\Sigma_{Do}): \quad \dot{x}(t) = A_\Delta(t)x(t) + A_d x(t - \tau) \tag{5.25}$$

where the matrix A is Hurwitz. A matrix $0 < P = P^t \in \Re^{n \times n}$ is a QCM for system (Σ_{Do}) and cost function (5.6) if and only if any one of the following equivalent conditions hold:

(1) There exists a matrix $0 < W = W^t \in \Re^{n \times n}$ and a scalar $\mu > 0$ satisfying the LMI

$$\begin{bmatrix} PA + A^t P + W + Q + \mu^{-1} E^t E & PH & PA_d \\ H^t P & -\mu^{-1} I & 0 \\ A_d^t P & 0 & -W \end{bmatrix} < 0 \qquad (5.26)$$

(2) There exists a matrix $0 < W = W^t \in \Re^{n \times n}$ and a scalar $\mu > 0$ satisfying the ARI

$$PA + A^t P + P[\mu HH^t + A_d W^{-1} A_d^t]P + W + Q + \mu^{-1} E^t E < 0 \qquad (5.27)$$

(3) There exists a matrix $0 < W = W^t \in \Re^{n \times n}$ and a scalar $\mu > 0$ such that the ARE

$$\check{P} A + A^t \check{P} + \check{P}[\mu HH^t + A_d W^{-1} A_d^t]\check{P} + W + Q + \mu^{-1} E^t E = 0 \qquad (5.28)$$

has a stabilizing solution $\check{P} < P$.
(4) There exists a matrix $0 < W = W^t \in \Re^{n \times n}$ and a scalar $\mu > 0$ satisfying the following H_∞−norm bound

$$\left\| \begin{bmatrix} \mu^{-1/2} E \\ W^{1/2} \\ Q^{1/2} \end{bmatrix} [sI - A]^{-1} [\mu^{1/2} H \quad A - dW^{-1/2}] \right\|_\infty < 1 \qquad (5.29)$$

Proof: Follows easily from **Theorem 5.2** and **Corollary 5.1** by setting $E_d = 0$.

5.1.3 Robust Performance Analysis II

To complete our work, we deal here with robust performance analysis based on the delay-dependent robust stability. This was previously discussed in **Lemma 2.2**. In the following, reference is made to the delay system

$$\begin{aligned} (\Sigma_D:) \quad \dot{x}(t) &= (A + A_d)x(t) \\ &- A_d \left\{ \int_{-\tau}^0 Ax(t+\theta)d\theta + \int_{-\tau}^0 A_d x(t - \tau + \theta)d\theta \right\} \end{aligned} \qquad (5.30)$$

which represents a functional differential equation with initial conditions over the interval $[-2\tau^*, 0]$. To deal with system (Σ_D), we introduce a Lyapunov-Krasovskii functional $V_5(x_t)$ of the form:

$$
\begin{aligned}
V_5(x_t) &= x^t(t)Px(t) + \int_{-\tau}^{0}\int_{t+\theta}^{t} r_1[x^t(s)A_{\Delta}^t A_{\Delta}x(s)]ds d\theta \\
&+ \int_{-\tau}^{0}\int_{t-\tau+\theta}^{t} r_2[x^t(s)A_{d\Delta}^t A_{d\Delta}x(s)]ds d\theta
\end{aligned}
\tag{5.31}
$$

where $0 < P = P^t \in \Re^{n \times n}$ and $r_1 > 0$, $r_2 > 0$ are weighting factors. Note that the second and third terms are constructed to take care of the delayed state. Recall that $V_5(x_t)$ here is slightly modified from (2.59) in Chapter 2.

Definition 5.2: *System (Σ_D) satisfying* **Assumption 2.2** *is said to be robustly stable for any constant time-delay τ satisfying $0 \leq \tau \leq \tau^*$ if, given $\tau^* > 0$, there exists matrix $0 < P = P^t \in \Re^{n \times n}$ and scalars $r_1 > 0$ and $r_2 > 0$ satisfying the ARI*

$$
\begin{aligned}
&P(A_{\Delta} + A_{d\Delta}) + (A_{\Delta} + A_{d\Delta})^t P + \tau^* P A_{d\Delta} A_{d\Delta}^t P \\
&+\tau^* r_1 A_{\Delta}^t A_{\Delta} + \tau^* r_2 A_{d\Delta}^t A_{d\Delta} < 0 \qquad \forall \, \Delta : \, \Delta^t(t)\,\Delta(t) \leq I \quad \forall t
\end{aligned}
$$

Definition 5.3: *System (Σ_D) satisfying* **Assumption 2.2** *with cost function (5.6) is said to be robustly stable with a quadratic cost matrix (QCM) $0 < P = P^t \in \Re^{n \times n}$ for any constant time-delay τ satisfying $0 \leq \tau \leq \tau^*$ if, given $\tau^* > 0$, there exist matrix $0 < P = P^t \in \Re^{n \times n}$ and scalars $r_1 > 0$ and $r_2 > 0$ satisfying the ARI*

$$
\begin{aligned}
&P(A_{\Delta} + A_{d\Delta}) + (A_{\Delta} + A_{d\Delta})^t P + \tau^* P A_{d\Delta} A_{d\Delta}^t P + \tau^* r_1 A_{\Delta}^t A_{\Delta} \\
&+\tau^* r_2 A_{d\Delta}^t A_{d\Delta} + Q < 0 \qquad \forall \, \Delta : \, \Delta^t(t)\,\Delta(t) \leq I \quad \forall t
\end{aligned}
$$

The following theorem derives an upper bound on the cost fucntion J_o in the case of robust delay-dependent stability.

Theorem 5.3: *Consider system (Σ_D) and cost function (5.6). Given scalars $\tau^* > 0, \sigma > 0, \mu > 0, r_1 > 0, r_2 > 0$, if $0 < P = P^t$ is a QCM then (Σ_D) is robustly stable for any constant time-delay τ satisfying $0 \leq \tau \leq \tau^*$ and the cost function satisfies the bound*

$$
\begin{aligned}
J_o &\leq x_o^t P x_o + \int_{-\tau}^{0} r_1 x^t(\alpha)\left(\sigma E^t E + A^t(I - \sigma HH^t)A\right)x(\alpha)d\alpha \\
&+ \int_{-\tau}^{0} r_2 x^t(\alpha)\left(\mu E_d^t E_d + A_d^t(I - \mu HH^t)A_d\right)x(\alpha)d\alpha
\end{aligned}
\tag{5.32}
$$

Conversely, if system (Σ_D) is robustly stable for any constant time-delay τ satisfying $0 \leq \tau \leq \tau^*$ then there will be a QCM for this system and cost function (5.6).

Proof:$(\Longrightarrow)$ Let $0 < P = P^t$ be a QCM for system (Σ_D) and cost function (5.6). It follows from **Definition 5.3** that there exist scalars $\tau^* > 0, r_1 > 0, r_2 > 0$ such that

$$X^t(t) \begin{bmatrix} P(A_\Delta + A_{d\Delta}) + \\ (A_\Delta + A_{d\Delta})^t P + Q & \tau^* P A_{d\Delta} & \tau^* A_\Delta^t & \tau^* A_{d\Delta}^t \\ \tau^* A_{d\Delta}^t P & -\tau^* I & 0 & 0 \\ \tau^* A_\Delta^t & 0 & -\tau^* r_1^{-1} I & 0 \\ \tau^* A_{d\Delta}^t & 0 & 0 & -\tau^* r_2^{-1} I \end{bmatrix} X(t)$$

$$< 0$$

$$\forall \Delta: \quad \Delta^t \Delta \leq I, \quad X(k) \neq [0\ 0\ 0\ 0]$$
$$X(t) = [x^t(t)\ x^t(t-\tau)\ x^t(t)\ x^t(t-\tau)]^t \tag{5.33}$$

Note that the matrix in (5.33) is continuously dependent on τ^*. Therefore, system (Σ_D) is robustly stable for any constant time-delay τ satisfying $0 \leq \tau \leq \tau^*$.

Now by evaluating the derivative $\dot{V}_5(x(t))$ of the functional (5.31), we obtain:

$$\dot{V}_5(x_t) =$$
$$X^t(t) \begin{bmatrix} P(A_\Delta + A_{d\Delta}) + \\ (A_\Delta + A_{d\Delta})^t P + Q & \tau^* P A_{d\Delta} & \tau^* A_\Delta^t & \tau^* A_{d\Delta}^t \\ \tau^* A_{d\Delta}^t P & -\tau^* I & 0 & 0 \\ \tau^* A_\Delta^t & 0 & -\tau^* r_1^{-1} I & 0 \\ \tau^* A_{d\Delta}^t & 0 & 0 & -\tau^* r_2^{-1} I \end{bmatrix} X(t)$$

$$\leq -x^t(k) Q x(k) \tag{5.34}$$

from which we conclude that

$$x^t(k) Q x(k) \leq -\dot{V}_5(x_t) \tag{5.35}$$

Integrating (5.35) over the period $t \in [0, \infty]$ and using (5.6), we get:

$$J_o \leq V_5(x_o) - V_5(x_\infty) \tag{5.36}$$

By (5.33), system (Σ_D) is robustly stable for any constant time-delay τ satisfying $0 \leq \tau \leq \tau^*$. This leads to $V_5(x_t) \to 0$ as $t \to \infty$. With the help of B.1.2 for some scalars $\sigma > 0$, $\mu > 0$ (5.36) reduces to

$$J_o \leq x_o^t P x_o + \int_{-\tau}^0 r_1 x^t(\alpha) \left(\sigma E^t E + A^t(I - \sigma H H^t) A \right) x(\alpha) d\alpha$$
$$+ \int_{-\tau}^0 r_2 x^t(\alpha) \left(\mu E_d^t E_d + A_d^t(I - \mu H H^t) A_d \right) x(\alpha) d\alpha$$

($\Longleftarrow$) Let system (Σ_D) be robustly stable for any constant time-delay τ satisfying $0 \leq \tau \leq \tau^*$. It follows that there exist matrix $0 < P = P^t$ and scalars $r_1 > 0, r_2 > 0$ such that

$$P(A_\Delta + A_{d\Delta}) + (A_\Delta + A_{d\Delta})^t P + \tau^* P A_{d\Delta} A_{d\Delta}^t P$$
$$+\tau^* r_1 A_\Delta^t A_\Delta + \tau^* r_2 A_{d\Delta}^t A_{d\Delta} < 0$$
$$\forall \Delta: \quad \Delta^t(t) \Delta(t) \leq I \ \forall t \tag{5.37}$$

Hence, one can find some $\rho > 0$ such that the following inequality holds:

$$\vartheta^{-1}[P(A_\Delta + A_{d\Delta}) + (A_\Delta + A_{d\Delta})^t P + \tau^* P A_{d\Delta} A_{d\Delta}^t P$$
$$+\tau^* r_1 A_\Delta^t A_\Delta + \tau^* r_2 A_{d\Delta}^t A_{d\Delta}] + Q \ =$$
$$[(A_\Delta + A_{d\Delta})^t \breve{P} + \breve{P} A_\Delta + A_{d\Delta}) + \tau^o \breve{P} A_{d\Delta} A_{d\Delta}^t \breve{P}$$
$$+\tau^o \breve{r}_1 A_\Delta^t A_\Delta + \tau^o \breve{r}_2 A_{d\Delta}^t A_{d\Delta}] + Q \ < \ 0 \tag{5.38}$$
$$\forall \Delta: \quad \Delta^t(t) \Delta(t) \leq I \ \forall t$$

The above inequality implies that there exist scalars $\tau^o > 0, \breve{r}_1 > 0, \breve{r}_2 > 0$ such that the matrix $\breve{P} = \vartheta^{-1} P$ is a QCM for system (Σ_D).

5.1.4   Synthesis of Guaranteed Cost Control I

In this section, we focus attention on the problem of optimal guaranteed cost control based on state-feedback for the uncertain delay system (Σ_Δ) with uncertainties satisfying (5.3) and based on delay-independent robust stability. Here the cost function is given by (5.4). To proceed further, we provide the following definition:

Definition 5.4: *A state-feedback controller $u(t) = K_s x(t)$ is said to define a quadratic guaranteed cost control (QGCC) associated with cost* matrix $0 < P = P^t \in \Re^{n \times n}$ for system (Σ_Δ) and cost function (5.4) if there exists a matrix $0 < W = W^t \in \Re^{n \times n}$ such that

$$P(A_c + H\Delta(t)E_c) + (A_c + H\Delta(t)E_c)^t P + W + Q + K_s^t R K_s +$$
$$P(A_d + H\Delta(t)E_d)W^{-1}(A_d + H\Delta(t)E_d)^t P < 0 \qquad (5.39)$$
$$\forall \Delta: \quad \Delta^t(t)\,\Delta(t) \leq I \quad \forall t$$

where

$$A_c = A + BK_s, \quad E_c = E + E_b K_s \qquad (5.40)$$

The following theorem establishes that the problem of determining a QGCC for system (Σ_Δ) and cost function (5.4) can be recast to an algebraic matrix inequality (AMI) feasibility problem.

Theorem 5.4: *Suppose that there exist a scalar $\mu > 0$ and a matrix $0 < W = W^t$ such that the μ–dependent ARE*

$$P(A + \mu^{-1}A_d W_d^{-1} E_d^t E) + (A + \mu^{-1}A_d W_d^{-1} E_d^t E)^t P + + W + Q$$
$$+ P[\mu HH^t + A_d W_d^{-1} A_d^t]P + \mu^{-1}E^t \bar{W}_d^{-1} E$$
$$- \left\{ PB + \mu^{-1}(E^t \bar{W}_d^{-1} + PA_d W_d^{-1} E_d^t)E_b \right\} (R + \mu^{-1}E_b^t \bar{W}_d^{-1} E_b)^{-1}$$
$$\left\{ B^t P + \mu^{-1}E_b^t(\bar{W}_d^{-1} E + E_d W_d^{-1} A_d^t P) \right\} = 0 \qquad (5.41)$$

has a stabilizing solution $0 < P = P^t$. In this case, the state-feedback controller

$$u(k) = K_s x(k)$$
$$K_s = -(R + \mu^{-1}E_b^t \bar{W}_d^{-1} E_b)^{-1} \left\{ B^t P + \mu^{-1}E_b^t(\bar{W}_d^{-1} E + E_d W_d^{-1} A_d^t P) \right\} \qquad (5.42)$$

is a QGCC for system Σ_Δ with cost matrix $\check{P}$ which satisfies $P < \check{P} < P + \rho I$ for any $\rho > 0$.

Conversely given any QGCC with cost matrix $0 < \check{P} = \check{P}^t$, there exists a scalar $\mu > 0$ and a matrix $0 < W = W^t$ such that the ARE (5.41) has a stabilizing solution $0 < P_* = P_*^t$ where $P_* < \check{P}$.

Proof: $(\Longrightarrow)$ Let the control law $u(t)$ be defined by (5.42). By substituting (5.40) and (5.42) into (5.41) and manipulating using **A.2**, it can be shown that (5.41) is equivalent to:

$$PA_c + A_c^t P + \mu PHH^t P + \mu^{-1} E_c^t E_c + W + Q + K_s^t RK_s +$$
$$(PA_d + \mu^{-1} E_c^t E_d)W_d^{-1}(A_d^t P + \mu^{-1} E_d^t E_c) = 0 \qquad (5.43)$$

By B.1.1, it follows that there exists a matrix $0 < \breve{P} = \breve{P}^t$ such that

$$\breve{P}A_c + A_c^t \breve{P} + \mu \breve{P}HH^t \breve{P} + \mu^{-1} E_c^t E_c + W + Q + K_s^t RK_s +$$
$$(\breve{P}A_d + \mu^{-1} E_c^t E_d)W_d^{-1}(A_d^t \breve{P} + \mu^{-1} E_d^t E_c) < 0 \qquad (5.44)$$

which implies that there exists a matrix $\Phi > 0$ such that

$$\breve{P}A_c + A_c^t \breve{P} + \mu \breve{P}HH^t \breve{P} + \mu^{-1} E_c^t E_c + W + Q + K_s^t RK_s +$$
$$(\breve{P}A_d + \mu^{-1} E_c^t E_d)W_d^{-1}(A_d^t \breve{P} + \mu^{-1} E_d^t E_c) + \Phi = 0 \qquad (5.45)$$

Given $\sigma \in (0,1)$, it follows from (5.45) and the properties of the ARE that

$$\hat{P}A_c + A_c^t \hat{P} + \mu \hat{P}HH^t \hat{P} + \mu^{-1} E_c^t E_c + W + Q + K_s^t RK_s +$$
$$(\hat{P}A_d + \mu^{-1} E_c^t E_d)W_d^{-1}(A_d^t \hat{P} + \mu^{-1} E_d^t E_c) + \sigma \Phi = 0 \qquad (5.46)$$

has a stabilizing solution $0 < \hat{P} = \hat{P}^t$. In addition, $\hat{P} > P$ and as $\sigma \to 0$, $\hat{P} \to P$. Therefore, given any $\rho > 0$, we can find a $\sigma > 0$ such that $P < \breve{P} = \hat{P} < P + \rho I$.

($\Longleftarrow$) Suppose that $u(k) = K_s x(k)$ is a QGCC with a cost matrix $\breve{P}$. By **Theorem 5.2**, it follows that there exists a scalar $\mu > 0$ and a matrix $0 < W = W^t$ such that

$$PA_c + A_c^t P + \mu PHH^t P + \mu^{-1} E_c^t E_c + W + Q + K_s^t RK_s +$$
$$(PA_d + \mu^{-1} E_c^t E_d)W_d^{-1}(A_d^t P + \mu^{-1} E_d^t E_c) < 0 \qquad (5.47)$$

In terms of (5.40), inequality (5.47) is equivalent to:

$$P(A + BK_s) + (A + BK_s)^t P + \mu PHH^t P +$$
$$\mu^{-1}(E + E_b K_s)^t (E + E_b K_s) + W + Q + K_s^t RK_s +$$
$$\left\{ PA_d + \mu^{-1}(E + E_b K_s)^t E_d \right\} W_d^{-1} \left\{ A_d^t P + \mu^{-1} E_d^t (E + E_b K_s) \right\}$$
$$< 0 \qquad (5.48)$$

Define

$$\bar{K}_s = \mu^{-1/2} K_s, \qquad \bar{R} = \mu R$$
$$\bar{B} = B + \mu^{-1} A_d W_d^{-1} E_d^t E_b \qquad (5.49)$$

By substituting (5.21), (5.49) into (5.48), it follows that there exists $\bar{P} > 0$ satisfying

$$\bar{P}(\bar{A} + \mu^{1/2}\bar{B}\bar{K}_s) + (\bar{A} + \mu^{1/2}\bar{B}\bar{K}_s)^t\bar{P} + \bar{P}(\mu H H^t +$$
$$A_d W_d^{-1} A_d^t)\bar{P} + W + Q + \bar{K}_s^t\bar{R}\bar{K}_s +$$
$$(\mu^{-1/2}\bar{W}_d^{-1/2} E + \bar{W}_d^{-1/2} E_b\bar{K}_s)^t(\mu^{-1/2}\hat{W}_d^{-1/2} E + \bar{W}_d^{-1/2} E_b\bar{K}_s) < 0$$
$$(5.50)$$

Now, consider the state feedback $\mathcal{H}_\infty$ control problem of the system:

$$\dot{x} = \bar{A} + [\mu^{1/2}H \quad A_d W_d^{-1/2}]w + (\mu^{1/2}\bar{B})u$$
$$z = \begin{bmatrix} \mu^{-1/2}\bar{W}_d^{-1/2} E \\ 0 \\ (W + Q)^{1/2} \end{bmatrix} x + \begin{bmatrix} E_b\bar{W}_d^{-1/2} \\ \mu^{1/2}R^{1/2} \\ 0 \end{bmatrix} u \qquad (5.51)$$

It follows that from [182] that system (5.42) with the state feedback $u(t) = \bar{K}_s x(t)$ has the μ-dependent ARE (5.41). Moreover, it has a stabilizing solution $P^* \geq 0$ such that $P^* < \bar{P}$. Since $Q > 0, W > 0$, it follows that $P^* > 0$.

Corollary 5.3: *Consider the system*

$$(\Sigma_{\Delta o}): \quad \dot{x}(t) = [A + \Delta A(t)]x(t) + [B + \Delta B(t)]u(t) + A_d x(t - \tau) \quad (5.52)$$

which is obtained from system (Σ_Δ) *by setting* $E_d = 0$; *In this case, the ARE (5.41) and controller (5.42) reduce to:*

$$PA + A^t P + P[\mu H H^t + A_d W^{-1} A_d^t]P +$$
$$W + Q + \mu^{-1}E^t E -$$
$$\left\{PB + \mu^{-1}E^t E_b\right\}(R + \mu^{-1}E_b^t E_b)^{-1}\left\{B^t P + \mu^{-1}E_b^t E\right\}$$
$$= 0 \qquad (5.53)$$

$$u(t) = K_s x(t)$$
$$K_s = -(R + \mu^{-1}E_b^t E_b)^{-1}\left\{B^t P + \mu^{-1}E_b^t E\right\} \qquad (5.54)$$

Corollary 5.4: *Consider the system*

$$(\Sigma_{\Delta f}): \quad \dot{x}(t) = Ax(t) + Bu(t) + A_d x(t - \tau) \qquad (5.55)$$

which is obtained from system (Σ_Δ) by setting $E_d = 0, H = 0, E_b = 0$ corresponding to the case of delay systems without uncertainties. In such case, the ARE (5.51) and controller (5.54) reduce to:

$$PA + A^tP + W + Q + P(A_dW^{-1}A_d^t - BR^{-1}B^t)P = 0 \quad (5.56)$$
$$u(t) = -R^{-1}B^tPx(t) \quad (5.57)$$

which provides a guaranteed cost control for linear continuous-time systems with state-delay.

5.1.5   Synthesis of Guaranteed Cost Control II

Here, we deal with the problem of optimal guaranteed cost control based on state-feedback for system (Σ_D) and adopting the notion of delay-dependent stability. Here the cost function is given by (5.4). The following definition is given:

Definition 5.5: *A state-feedback controller $u(t) = K_s x(t)$ is said to define a quadratic guaranteed cost control (QGCC) associated with cost matrix $0 < P = P^t \in \Re^{n \times n}$ for system (Σ_Δ) and cost function (5.4) for any constant time-delay τ satisfying $0 \leq \tau \leq \tau^*$ if, given $\tau^* > 0$, there exist matrix $0 < P = P^t \in \Re^{n \times n}$ and scalars $r_1 > 0$ and $r_2 > 0$ satisfying the ARI*

$$PA_{cd\Delta} + A_{cd\Delta}^t P + \tau^* PA_{d\Delta} A_{d\Delta}^t P + K_s^t RK_s$$
$$+\tau^* r_1 A_{c\Delta}^t A_{c\Delta} + \tau^* r_2 A_{d\Delta}^t A_{d\Delta} < 0$$
$$\forall \Delta : \quad \Delta^t(t) \Delta(t) \leq I \quad \forall t \quad (5.58)$$

where

$$\begin{aligned}
A_{cd\Delta} &= A_{cd} + H\Delta(t)E_{cd} \\
&= (A + A_d + BK_s) + H\Delta(t)(E + E_d + E_bK_s) \\
A_{c\Delta} &= A_c + H\Delta(t)E_c \\
&= (A + BK_s) + H\Delta(t)(E + E_bK_s) \quad (5.59)
\end{aligned}$$

The following two theorems establish that the problem of determining a QGCC for system (Σ_D) and cost function (5.4) can be recast to an ARI or LMI-feasibility problem.

Theorem 5.5: *Given system (Σ_Δ) and cost function (5.4). Suppose that there exist matrix $0 < P = P^t$, scalars $r_1 > 0, r_2 > 0, \mu_1 > 0, \mu_2 >$*

$0, \mu_3 > 0, \mu_4 > 0$ and $\tau^* > 0$ such that for any constant time-delay τ such that $0 \leq \tau \leq \tau^*$ satisfying the ARI

$$P(A + A_d) + (A + A_d)^t P + P[\mu_1 H H^t + \tau^* M_1 M_1^t]P +$$
$$[\mu_1^{-1}(E + E_d)^t(E + E_d) + \tau^* M_2^t M_2]$$
$$- \left\{ PB + \mu_1^{-1}(E + E_d)^t E_b + \tau^* N_1^t N_1 \right\} \left\{ R + \mu_1^{-1} E_b^t E_b + \tau^* N_2^t N_2 \right\}^{-1}$$
$$\left\{ B^t P + \mu_1^{-1} E_b^t(E + E_d) + \tau^* N_1^t N_1 \right\}^t < 0 \qquad (5.60)$$

with

$$\begin{aligned}
M_1 M_1^t &= \mu_4^{-1} H H^t + A_d(I - \mu_4 E_d^t E_d)^{-1} A_d^t \\
M_2^t M_2 &= r_1 \mu_2^{-1} E^t E + r_2 \mu_3^{-1} E_d^t E_d \\
&\quad + r_1 A^t(I - \mu_2 H H^t)^{-1} A + r_2 A_d^t(I - \mu_3 H H^t)^{-1} A_d \\
N_1^t N_1 &= r_1 \left(A^t(I - \mu_2 H H^t)^{-1} B + \mu_2^{-1} E^t E_b \right) \\
N_2^t N_2 &= r_1 \left(\mu_2^{-1} E_b^t E_b + B^t(I - \mu_2 H H^t)^{-1} B \right) \qquad (5.61)
\end{aligned}$$

Then, the state-feedback controller

$$\begin{aligned}
u(t) &= K_s x(t) \\
K_s &= -\left\{ R + \mu_1^{-1} E_b^t E_b + \tau^* N_2^t N_2 \right\}^{-1} \\
&\quad \left\{ B^t P + \mu_1^{-1} E_b^t(E + E_d) + \tau^* N_1^t N_1 \right\} \qquad (5.62)
\end{aligned}$$

is a QGCC for system Σ_Δ with cost matrix $0 < P = P^t$.

Proof: From (5.59) and **A.1, B.1.2**, we get:

$$\begin{aligned}
PA_{cd\Delta} + A_{cd\Delta}^t P &= P(A + A_d + BK_s) + (A + A_d + BK_s)^t P + \\
&\quad PH\Delta(E + E_d + E_b K_s) + (E + E_d + E_b K_s)^t \Delta^t H^t P \\
&\leq P(A + A_d + BK_s) + (A + A_d + BK_s)^t P + \\
&\quad \mu_1 PHH^t P + \mu_1^{-1}(E + E_d + E_b K_s)^t(E + E_d + E_b K_s) \qquad (5.63)
\end{aligned}$$

$$\begin{aligned}
A_{c\Delta}^t A_{c\Delta} &= \\
&[(A + BK_s) + H\Delta(t)(E + E_b K_s)]^t[(A + BK_s) + H\Delta(t)(E + E_b K_s)] \\
&\leq \mu_2^{-1}(E + E_b K_s)^t(E + E_b K_s) + (A + BK_s)^t(I - \mu_2 H H^t)^{-1}(A + BK_s) \qquad (5.64)
\end{aligned}$$

$$(A_d + H\Delta E_d)^t(A_d + H\Delta E_d) \leq \mu_3^{-1}E_d^t E_d + A_d^t(I - \mu_3 HH^t)^{-1}A_d \quad (5.65)$$

$$P(A_d + H\Delta E_d)(A_d + H\Delta E_d)^t P \leq \mu_4^{-1}PHH^t P$$
$$+ \quad P[A_d(I - \mu_4 E_d^t E_d)^{-1}A_d^t]P$$
$$(5.66)$$

for any scalars $\mu_1 > 0, ..., \mu_4 > 0$ satisfying $(I - \mu_2 HH^t) > 0, (I - \mu_3 HH^t) > 0, (I - \mu_4 E_d^t E_d) > 0$. It follows from **Definition 5.3** , (5.61) and (5.63)-(5.66) with some arrangement that

$$P(A + A_d + BK_s) + (A + A_d + BK_s)^t P + P[\mu_1 HH^t + \tau M_1 M_1^t]P +$$
$$[\mu_1^{-1}(E + E_d)^t(E + E_d) + \tau M_2^t M_2] +$$
$$K_s^t\left\{B^t P + \mu_1^{-1}E_b^t(E + E_d) + \tau N_1^t N_1\right\} +$$
$$\left\{PB + \mu_1^{-1}(E + E_d)^t E_b + \tau N_1^t N_1\right\}K_s +$$
$$K_s^t\left\{R + \mu_1^{-1}E_b^t E_b + \tau N_2^t N_2\right\}K_s \quad < \quad 0 \qquad (5.67)$$

Observe that (5.67) is continuously dependent on τ. On completing the squares in (5.67) with respect to K_s and arranging terms, one obtains the control law (5.62) such that $0 < P = P^t$ satisfies ARE (5.60).

Theorem 5.6: *Given system* (Σ_Δ) *and cost function (5.4), suppose that there exist matrices* $0 < Y = Y^t$, $0 < S = S^t$, *scalars* $r_1 > 0, r_2 > 0, \mu_1 > 0, \mu_2 > 0, \mu_3 > 0, \mu_4 > 0$ *and* $\tau^* > 0$ *such that for any constant time-delay* τ *such that* $0 \leq \tau \leq \tau^*$ *satisfying the LMIs*

$$\begin{bmatrix} \Gamma(Y,\tau^*,\mu) & \tau^* M_1 & \tau^* Y M_2^t & \tau^* S N_2^t \\ \tau^* M_1^t & -\tau^* I & 0 & 0 \\ \tau^* M_2 Y & 0 & -\tau^* I & 0 \\ \tau^* N_2 S & 0 & 0 & -\tau^* I \end{bmatrix} \quad < \quad 0$$

$$\mu_2 HH^t - I < 0, \quad \mu_4 E_d^t E_d - I < 0, \quad \mu_3 HH^t - I \quad < \quad 0 \qquad (5.68)$$

where

$$\Gamma(Y,\tau,\mu) = (A + A_d)Y + Y(A + A_d)^t + \mu_1 HH^t + SB^t + BS^t +$$
$$\mu_1^{-1}Y(E + E_d)^t(E + E_d)Y + S(R + \mu_1^{-1}E_b^t E_b)S +$$
$$S[\mu_1^{-1}E_b^t(E + E_d) + \tau N_1^t N_1]Y + Y[\mu_1^{-1}(E + E_d)^t E_b + \tau N_1^t N_1]S$$
$$(5.69)$$

Then, the state-feedback controller

$$u(t) = SY^{-1}x(t) \tag{5.70}$$

is a QGCC for system Σ_Δ with cost matrix $0 < P = P^t$.

Proof: Starting from (5.67) and substituting $Y = P^{-1}, K_s = SP$ and manipulating using **A.1**, one immediately obtains (5.68).

Remark 5.2: It is important to observe the difference between **Theorem 5.5** and **Theorem 5.6**. While the former gives closed-form solution for the feedback gain matrix after solving the ARI (5.60), the latter provides a numerical value of the gain matrix based on the solution of LMI (5.68).

Corollary 5.6: Consider the time-delay system

$$(\Sigma_{Do}): \quad \dot{x}(t) = A_\Delta(t)x(t) + A_d x(t-\tau) \tag{5.71}$$

with cost function (5.4) and matrix A is Hurwitz. Suppose that there exist matrix $0 < P = P^t$, scalars $r_1 > 0, r_2 > 0, \mu_1 > 0, \mu_2 > 0$ and $\tau^* > 0$ such that for any constant time-delay τ such that $0 \leq \tau \leq \tau^*$ satisfying the ARI

$$P(A + A_d) + (A + A_d)^t P + P[\mu_1 H H^t]P + [\mu_1^{-1} E^t E + \tau^* \hat{M}_2^t \hat{M}_2]$$
$$- \left\{ PB + \mu_1^{-1} E^t E_b + \tau^* \hat{N}_1^t \hat{N}_1 \right\} \left\{ R + \mu_1^{-1} E_b^t E_b + \tau^* \hat{N}_2^t \hat{N}_2 \right\}^{-1}$$
$$\left\{ B^t P + \mu_1^{-1} E_b^t E + \tau^* \hat{N}_1^t \hat{N}_1 \right\}^t < 0 \tag{5.72}$$

with

$$\begin{aligned}
\hat{M}_2^t \hat{M}_2 &= r_1 \mu_2^{-1} E^t E + r_1 A^t (I - \mu_2 H H^t)^{-1} A + r_2 A_d^t A_d \\
\hat{N}_1^t \hat{N}_1 &= r_1 \left(A^t (I - \mu_2 H H^t)^{-1} B + \mu_2^{-1} E^t E_b \right) \\
\hat{N}_2^t \hat{N}_2 &= r_1 \left(\mu_2^{-1} E_b^t E_b + B^t (I - \mu_2 H H^t)^{-1} B \right)
\end{aligned} \tag{5.73}$$

Then, the state-feedback controller

$$\begin{aligned}
u(t) &= K_s x(t) \\
K_s &= - \left\{ R + \mu_1^{-1} E_b^t E_b + \tau^* \hat{N}_2^t \hat{N}_2 \right\}^{-1} \\
&\quad \left\{ B^t P + \mu_1^{-1} E_b^t E + \tau^* \hat{N}_1^t \hat{N}_1 \right\}
\end{aligned} \tag{5.74}$$

is a QGCC for system Σ_Δ with cost matrix $0 < P = P^t$.

Corollary 5.7: *Consider the time-delay system*

$$(\Sigma_{Do}): \qquad \dot{x}(t) = A_\Delta(t)x(t) + A_d x(t - \tau) \qquad (5.75)$$

with cost function (5.4) and matrix A is Hurwitz. Suppose that there exist matrix $0 < P = P^t$, scalars $r_1 > 0, r_2 > 0, \mu_1 > 0, \mu_2 > 0$ and $\tau^* > 0$ such that for any constant time-delay τ such that $0 \leq \tau \leq \tau^*$ satisfying the LMIs

$$\begin{bmatrix} \hat{\Gamma}(Y,\tau^*,\mu) & \tau^* \hat{M}_1 & \tau^* Y \hat{M}_2^t & \tau^* S \hat{N}_2^t \\ \tau^* \hat{M}_1^t & -\tau^* I & 0 & 0 \\ \tau^* \hat{M}_2 Y & 0 & -\tau^* I & 0 \\ \tau^* \hat{N}_2 S & 0 & 0 & -\tau^* I \end{bmatrix} < 0$$

$$\mu_2 H H^t - I < 0 \qquad (5.76)$$

where

$$\begin{aligned} \hat{\Gamma}(Y,\tau,\mu) &= (A + A_d)Y + Y(A + A_d)^t + \mu_1 H H^t + \mu_1^{-1} Y E^t E Y \\ &+ SB^t + BS^t + S(R + \mu_1^{-1} E_b^t E_b)S \\ &+ S[\mu_1^{-1} E_b^t E + \tau N_1^t N_1]Y + Y[\mu_1^{-1} E^t E_b + \tau N_1^t N_1]S \quad (5.77) \end{aligned}$$

Then, the state-feedback controller

$$u(t) = SY^{-1}x(t) \qquad (5.78)$$

is a QGCC for system Σ_Δ with cost matrix $0 < P = P^t$.

5.2   Discrete-Time Systems

5.2.1   Problem Formulation

We consider a class of uncertain time-delay systems represented by:

$$\begin{aligned} (\Sigma_\Delta): \quad x(k+1) &= [A + \Delta A(k)]x(k) + [B + \Delta B(k)]u(k) + A_d x(k - \tau) \\ &= A_\Delta(k)x(k) + B_\Delta(k)u(k) + A_d x(k - \tau) \quad (5.79) \end{aligned}$$

where $x(t) \in \Re^n$ is the state, $u(t) \in \Re^m$ is the control input and the matrices $A \in \Re^{n \times n}$, $B \in \Re^{n \times m}$ and $A_d \in \Re^{n \times n}$ are real constant matrices representing the nominal plant. Here, τ is an unknown constant integer representing the number of delay units in the state. For all practical purposes, we consider $0 \leq \tau \leq \tau^*$ with τ^* being known. The matrices $\Delta A(k)$ and $\Delta B(k)$ and represent parameteric uncertainties which are of the form:

$$[\Delta A(k) \quad \Delta B(k)] \quad = \quad H \, \Delta(k) \, [E \quad E_b] \tag{5.80}$$

where $H \in \Re^{n \times \alpha}$, $E \in \Re^{\beta \times n}$ and $E_b \in \Re^{\beta \times m}$ are known constant matrices and $\Delta(k) \in \Re^{\alpha \times \beta}$ is an unknown matrix satisfying the norm-bound condition

$$\Delta^t(k) \, \Delta(k) \quad \leq \quad I \quad \forall k \tag{5.81}$$

The initial condition is specified as $\langle x(0), x(s) \rangle = \langle x_o, \phi(s) \rangle$, where $\phi(.) \in \ell_2[-\tau, 0]$.

Associated with the uncertain system (Σ_Δ) is the cost function:

$$J = \sum_{k=0}^{\infty} [\, x^t(k) Q x(k) + u^t(k) R u(k) \,] \tag{5.82}$$

where $0 < Q = Q^t \in \Re^{n \times n}$, $0 < R = R^t \in \Re^{m \times m}$ are given state and control weighting matrices.

Distinct from system (Σ_Δ) is the free-system

$$\begin{aligned}
(\Sigma_{DD}): \quad x(k+1) \quad &= \quad [A_o + \Delta A(k)] x(k) + D_o x(k - \tau) \\
&= \quad A_\Delta(k) x(k) + D_o x(k - \tau) \tag{5.83}
\end{aligned}$$

for which we associate the cost function:

$$J_o = \sum_{k=0}^{\infty} [\, x^t(k) Q x(k) \,] \tag{5.84}$$

5.2.2 Robust Performance Analysis III

Since the stability of system (Σ_{DD}) is crucial to the development of the guaranteed cost control for (Σ_Δ), we adopt the notion of robust stability independent of delay as discussed in Chapter 2. Reference is particularly made to Lemma 2.3 and based upon which we introduce the following:

Definition 5.6: *System* (Σ_{DD}) *with cost function (5.84) is said to be robustly stable (RS) with a quadratic cost matrix (QCM)* $0 < P = P^t \in \Re^{n \times n}$ *if there exists a matrix* $0 < W = W^t \in \Re^{n \times n}$ *such that* $W - A_d^t P A_d > 0$ *and*

$$A_\Delta^t(k) P A_\Delta(k) - P + A_\Delta^t(k) P A_d (W - A_d^t P A_d)^{-1} A_d^t P A_\Delta(k) + W < 0$$
$$\forall \Delta : \Delta^t(k)\, \Delta(k) \leq I \;\; \forall k$$

It is interesting to observe that **Definition 5.6** extends the result of **Corollary 2.3** by treating P as a quadratic cost matrix. Based thereon, we have the following results:

Theorem 5.7: *Consider system* (Σ_{DD}) *and cost function (5.84). If* $0 < P = P^t$ *is a QCM, then* (Σ_{DD}) *is RS and the cost function satisfies the bound*

$$J_o \leq x_o^t P x_o + \sum_{\alpha=-\tau}^{0} x^t(\alpha) W x(\alpha) \tag{5.85}$$

Conversely, if system (Σ_D) *is RS then there will be a QCM for this system and cost function (5.84).*

Proof:$(\Longrightarrow)$ Let $0 < P = P^t$ be a QCM for system (Σ_{DD}) and cost function (5.84). It follows from **Definition 5.6** that there exists a matrix $0 < W = W^t \in \Re^{n \times n}$ such that

$$X^t(k) \begin{bmatrix} A_\Delta^t(k) P A_\Delta(k) - P + W + Q & A_\Delta^t(k) P A_d \\ A_d^t P A_\Delta(k) & -(W - A_d^t P A_d) \end{bmatrix} X(k) < 0$$
$$\forall \Delta : \;\; \Delta^t\, \Delta \;\; \leq \;\; I \;\;, \;\; X(k) \neq [0 \;\; 0] \tag{5.86}$$

where

$$X(k) = [x^t(k) \;\; x^t(k - \tau)]^t \tag{5.87}$$

Therefore, system (Σ_{DD}) is RS.

Now by evaluating the first-forward difference $\Delta V_6(x_k)$ of the Lyapunov-Krasovskii functional (2.78), we obtain:

$$\Delta V_6(x_k) = V_6(x_{k+1}) - V_6(x_k) \leq -x^t(k) Q x(k) \tag{5.88}$$

from which we conclude that

$$x^t(k) Q x(k) \leq -\Delta V_6(x_k) \tag{5.89}$$

Summing up inequality (5.89) over the period $k = 0 \rightarrow \infty$ and using (5.84), we get:

$$J_o \leq V_6(x_o) - V_6(x_\infty) \tag{5.90}$$

By (5.86), system (Σ_{DD}) is RS leading to $V_6(x_k) \rightarrow 0$ as $k \rightarrow \infty$ and therefore (5.90) reduces to

$$J_o \leq x_o^t P x_o + \sum_{\alpha=-\tau}^0 x^t(\alpha) W x(\alpha)$$

($\Longleftarrow$) Let system (Σ_{DD}) be RS. It follows that there exist $0 < P = P^t$ and $0 < W = W^t$ such that $W - A_d^t P A_d > 0$ and

$$
\begin{aligned}
&A_\Delta^t(k) P A_\Delta(k) - P + W + \\
&A_\Delta^t(k) P A_d (W - A_d^t P A_d)^{-1} A_d^t P A_\Delta(k) < 0 \\
&\forall \Delta: \quad \Delta^t(k)\, \Delta(k) \leq I\ \forall k
\end{aligned}
\tag{5.91}
$$

Hence, one can find some $\rho > 0$ such that the following inequality holds:

$$
\begin{aligned}
&\rho^{-1}[A_\Delta^t(k) P A_\Delta(k) - P + W] + \\
&\rho^{-1}[A_\Delta^t(k) P D_o (W - D_o^t P D_o)^{-1} D_o^t P A_\Delta(k)] + Q = \\
&[A_\Delta^t(k) \check{P} A_\Delta(k) - \check{P} + \check{W}] + \\
&[A_\Delta^t(k) \check{P} D_o[\check{W} - D_o^t \check{P} D_o]^{-1} D_o^t \check{P} A_\Delta(k)] + Q < 0 \\
&\forall \Delta: \quad \Delta^t(k)\, \Delta(k) \leq I\ ,\quad \forall k
\end{aligned}
\tag{5.92}
$$

The above inequality implies that there exists a matrix $\check{W} = \rho^{-1}W$ such that the matrix $\check{P} = \rho^{-1}P$ is a QCM for system (Σ_{DD}).

Remark 5.3: In applying **Theorem 5.7** to a particular example we will need the bound τ^* on the time-delay factor τ to evaluate the upper bound on J_o.

Theorem 5.8: *A matrix $0 < P = P^t \in \Re^{n \times n}$ is a QCM for system (Σ_{DD}) and cost function (5.84) if and only if there exists a scalar $\mu > 0$ such that*

$$
\begin{aligned}
&P^{-1} - D_o W^{-1} D_o^t - \mu L L^t > 0\ , \\
&A_o^t \left\{ P^{-1} - D_o W^{-1} D_o^t - \mu L L^t \right\}^{-1} A_o - P + \\
&W + \mu^{-1} M_1^t M_1 + Q < 0
\end{aligned}
\tag{5.93}
$$

Proof: ($\Longrightarrow$) Suppose that $0 < P = P^t$ satisfies (5.86) for some $\mu > 0$. By **A.2**, we have:

$$A_\Delta^t(k)PA_\Delta(k) - P + W + Q +$$
$$A_\Delta^t(k)PD_o(W - D_o^tPD_o)^{-1}D_o^tPA_\Delta(k) =$$
$$A_\Delta^t(k)\left\{P^{-1} - D_oW^{-1}D_o^t\right\}^{-1}A_\Delta(k) - P + W + Q \quad (5.94)$$
$$\forall\Delta: \quad \Delta^t(k)\,\Delta(k) \leq I$$

Applying **B.1.5** and grouping like terms, we get:

$$A_\Delta^t(k)\left\{P^{-1} - D_oW^{-1}D_o^t\right\}^{-1}A_\Delta(k) - P + W + Q \leq$$
$$A_o^t\left\{P^{-1} - D_oW^{-1}D_o^t - \mu LL^t\right\}^{-1}A_o - P +$$
$$W + \mu^{-1}M_1^tM_1 + Q < 0 \quad (5.95)$$

Inequality (5.95) is implied from (5.93).

($\Longleftarrow$) Suppose that $0 < P = P^t$ is a QCM for system (Σ_{DD}) and cost function (5.89). It follows from **Corollary 2.2** that $\forall\,\Delta$ satisfying $\|\Delta\| \leq 1$

$$\begin{bmatrix} -P + W + Q & 0 \\ 0 & -W \end{bmatrix} + \begin{bmatrix} A_\Delta^t(k) \\ A_d^t \end{bmatrix} P\,[A_\Delta \quad A_d] < 0 \quad (5.96)$$

By **A.1**, inequality (5.96) can be expressed as

$$\begin{bmatrix} -P + W + Q & 0 & A_\Delta^t \\ 0 & -W & A_d^t \\ A_\Delta & A_d & -P^{-1} \end{bmatrix} < 0 \quad (5.97)$$

Using (5.80) and rearranging, we obtain:

$$\begin{bmatrix} -P + W + Q & 0 & A^t \\ 0 & -W & A_d^t \\ A & A_d & -P^{-1} \end{bmatrix} + \begin{bmatrix} 0 \\ 0 \\ H \end{bmatrix} \Delta(k)\,[E \quad 0 \quad 0] +$$
$$\begin{bmatrix} E^t \\ 0 \\ 0 \end{bmatrix} \Delta^t(k)\,[0 \quad 0 \quad H^t] < 0 \quad (5.98)$$

By **B.1.4**, (5.98) is equivalent to

$$\begin{bmatrix} -P+W+Q+\mu^{-1}E^tE & 0 & A^t \\ 0 & -W & A_d^t \\ A & A_d & -[P^{-1}-\mu HH^t] \end{bmatrix} < 0 \quad (5.99)$$

Application of **A.1** to (5.99) produces:

$$\begin{bmatrix} \begin{array}{l} -P+W+Q+\mu^{-1}E^tE \\ +A^t(P^{-1}-\mu HH^t)^{-1}A \end{array} & A^t(P^{-1}-\mu HH^t)^{-1}A_d \\ A_d^t(P^{-1}-\mu HH^t)^{-1}A & -\{W - A_d^t(P^{-1}-\mu HH^t)^{-1}A_d\} \end{bmatrix} < 0$$

which, in turn, corresponds to the algebraic matrix inequality

$$-P+W+Q+\mu^{-1}E^tE+A^t(P^{-1}-\mu HH^t)^{-1}A +$$
$$A^t(P^{-1}-\mu HH^t)^{-1}A_d\left\{W - A_d^t(P^{-1}-\mu HH^t)^{-1}A_d\right\}$$
$$A_d^t(P^{-1}-\mu HH^t)^{-1}A < 0 \quad (5.100)$$

Finally by **A.2**, inequality (5.100) is equivalent to (5.93).

At this stage, it is possible to cast **Theorem 5.8** into an H_∞−norm bound setting. Recall that **A.3.2** suggests that there exists a matrix $0 < P = P^t \in \Re^{n \times n}$ such that inequality (5.93) is satisfied if and only if the following conditions hold:

(a) A is Schur-stable
(b) The following H_∞−norm bound is satisfied

$$\left\| \begin{bmatrix} \mu^{-1/2}E \\ W^{1/2} \\ Q^{1/2} \end{bmatrix} [zI - A]^{-1} \begin{bmatrix} \mu^{1/2}H & A_dW^{-1/2} \end{bmatrix} \right\|_\infty < 1 \quad (5.101)$$

Much like the continuous-time case, we observe that the set $\Upsilon = \{(\mu, W) : \mu > 0, 0 < W = W^t\}$ over which inequality (5.93) has a solution $0 < P = P^t$ can be determined by finding those values of $\mu > 0$, $0 < W = W^t$ such that (5.101) is satisfied. More importantly, for any such μ, W the following ARE

$$A^t \left\{ \breve{P}^{-1} - A_dW^{-1}A_d^t - \mu HH^t \right\}^{-1} A - \breve{P} + W + \mu^{-1}E^tE + Q = 0 \quad (5.102)$$

has a stabilizing solution $\breve{P} \geq 0$.

5.2.3   Synthesis of Guaranteed Cost Control III

In this section, we focus attention on the problem of optimal guaranteed cost control based on state-feedback for the uncertain delay system (Σ_Δ) with uncertainties satisfying (5.81). Here the cost function is given by (5.82). To proceed further, we provide the following definition:

Definition 5.7: *A state-feedback controller* $u(k) = K_s x(k)$ *is said to define a quadratic guaranteed cost control (QGCC) associated with cost matrix* $0 < P = P^t \in \Re^{n \times n}$ *for system* (Σ_Δ) *and cost function (5.82) if there exists a matrix* $0 < W = W^t \in \Re^{n \times n}$ *such that* $W - A_d^t P A_d > 0$ *and* .

$$
\begin{aligned}
&[A_\Delta(k) + B_\Delta(k)K_s]^t P[A_\Delta(k) + B_\Delta(k)K_s] + K_s^t R K_s + \\
&[A_\Delta(k) + B_\Delta(k)K_s]^t P A_d (W - A_d^t P A_d)^{-1} A_d^t P[A_\Delta(k) + B_\Delta(k)K_s] \\
&-P + W + Q < 0 \\
&\forall \Delta : \Delta^t(k) \, \Delta(k) \leq I \; \forall k
\end{aligned}
\tag{5.103}
$$

or equivalently

$$
\begin{bmatrix}
A_{c\Delta}^t P A_{c\Delta} - P + W + Q & K_s^t & A_{c\Delta}^t P A_d \\
K_s & -R^{-1} & 0 \\
A_d^t P A_{c\Delta} & 0 & -(W - A_d^t P A_d)
\end{bmatrix} < 0
$$
$$
\forall \Delta : \quad \Delta^t \, \Delta \leq I \; \forall k
\tag{5.104}
$$

where

$$
\begin{aligned}
A_{c\Delta} &= A_\Delta + B_\Delta K_s \\
&= (A_o + B_o K_s) + H\Delta(k)(E + E_b K_s)
\end{aligned}
\tag{5.105}
$$

The following theorem establishes that the problem of determining a QGCC for system (Σ_Δ) and cost function (5.82) can be recast to an AMI feasibility problem.

Theorem 5.9: *Suppose that there exist a scalar* $\mu > 0$ *and a matrix* $0 < W = W^t$ *such that the* $\mu-$*dependent ARE*

$$
\begin{aligned}
&A^t P A - P + \mu^{-1} E^t E + Q + W \\
&-(A^t P \hat{B} + \mu^{-1} E^t E_b)(\hat{R} + \hat{B}^t P \hat{B})^{-1}(A^t P \hat{B} + \mu^{-1} E^t E_b)^t \\
&= 0,
\end{aligned}
\tag{5.106}
$$

where

$$\Lambda = (A_d W^{-1} A_d^t + \mu L L^t)^{1/2}$$
$$\hat{B} = [\Lambda \quad B]$$
$$\hat{R} = \begin{bmatrix} -I & 0 \\ 0 & R + \mu^{-1} E_b^t E_b \end{bmatrix} \qquad (5.107)$$

has a stabilizing solution $0 < P = P^t$. *In this case, the state-feedback controller*

$$u(k) = K_s x(k)$$
$$K_s = -[0 \quad I](\hat{R} + \hat{B}^t X \hat{B})^{-1}(A^t P \hat{B} + \mu^{-1} E^t E_b)^t \qquad (5.108)$$

is a QGCC for system Σ_Δ *with cost matrix* $\check{P}$ *which satisfies* $P < \check{P} < P + \rho I$ *for any* $\rho > 0$.

Conversely given any QGCC with cost matrix $0 < \check{P} = \check{P}^t$, *there exists a scalar* $\mu > 0$ *and a matrix* $0 < W = W^t$ *such that the ARE (5.106) has a stabilizing solution* $0 < P_* = P_*^t$ *where* $P_* < \check{P}$.

Proof: ($\Longrightarrow$) Define $X = \{\check{P}^{-1} - D_o W^{-1} D_o^t - \mu L L^t\}^{-1}$, then it is easy to see that (5.106) is equivalent to:

$$A^t P A + A^t P \Lambda (I - \Lambda^t \Lambda)^{-1} \Lambda^t P A - P + W + \mu^{-1} E^t E + Q$$
$$- \left\{ A^t P B + A^t P \Lambda (I - \Lambda^t \Lambda)^{-1} \Lambda^t P B + \mu^{-1} E^t E_b \right\}$$
$$\left\{ R + \mu^{-1} E_b^t E_b + B^t P B + B^t P \Lambda (I - \Lambda^t \Lambda)^{-1} \Lambda^t P B \right\}^{-1}$$
$$\left\{ B^t P A + B^t P \Lambda (I - \Lambda^t \Lambda)^{-1} \Lambda^t P A + \mu^{-1} E_b^t E \right\} = 0, \qquad (5.109)$$

Let u(k) be as defined in (5.108). Now, using (5.107)-(5.108) with repeated application of **A.2** and **B.1.5**, it can be shown that (5.109) reduces to:

$$(A + B K_s)^t \left\{ P^{-1} - A_d W^{-1} A_d^t - \mu H H^t \right\}^{-1} (A + A_d K_s) - P +$$
$$\mu^{-1}(E + E_b K_s)^t (E + E_b K_s) + K_s^t R K_s + W + Q = 0 \qquad (5.110)$$

By **A.3.2**, it follows that there exists a matrix $0 < \check{P} = \check{P}^t$ such that

$$(A + B K_s)^t \left\{ \check{P}^{-1} - A_d W^{-1} A_d^t - \mu H H^t \right\}^{-1} (A + B K_s) - \check{P} +$$
$$\mu^{-1}(E + E_b K_s)^t (E + E_b K_s) + K_s^t R K_s + W + Q < 0 \qquad (5.111)$$

which implies that there exists a matrix $\Phi > 0$ such that

$$(A + BK_s)^t \left\{ \breve{P}^{-1} - A_d W^{-1} A_d^t - \mu H H^t \right\}^{-1} (A + BK_s) - \breve{P} +$$
$$\mu^{-1}(E + E_b K_s)^t (E + E_b K_s) + K_s^t R K_s + W + Q + \Phi = 0 \quad (5.112)$$

Given $\sigma \in (0,1)$, it follows from (5.102) and the properties of the ARE that

$$(A + BK_s)^t \left\{ \hat{P}^{-1} - A_d W^{-1} A_d^t - \mu H H^t \right\}^{-1} (A + BK_s) - \hat{P} +$$
$$\mu^{-1}(E + E_b K_s)^t (E + E_b K_s) + K_s^t R K_s + W + Q + \sigma\Phi = 0 (5.113)$$

has a stabilizing solution $0 < \hat{P} = \hat{P}^t$. In addition, $\hat{P} > P$ and as $\sigma \to 0$, $\hat{P} \to P$. Therefore, given any $\rho > 0$, we can find a $\sigma > 0$ such that $P < \breve{P} = \hat{P} < P + \rho I$.

($\Longleftarrow$) Suppose that $u(k) = K_s x(k)$ is a QGCC with a cost matrix $\breve{P}$. By Theorem 5.8, it follows that there exist a scalar $\mu > 0$ and a matrix $0 < W = W^t$ such that

$$(A + BK_s)^t \left\{ \breve{P}^{-1} - A_d W^{-1} A_d^t - \mu H H^t \right\}^{-1} (A + BK_s) - \breve{P} +$$
$$\mu^{-1}(E + E_b K_s)^t (E + E_b K_s) + K_s^t R K_s + W + Q < 0 \quad (5.114)$$

Letting $\breve{X} = \{ \breve{P}^{-1} - A_d W^{-1} A_d^t - \mu H H^t \}^{-1}$ in (5.84) and completing the squares, we get:

$$A^t \breve{X} A - (\breve{X}^{-1} + \Lambda\Lambda^t)^{-1} + \mu^{-1} E^t E - \Omega\Gamma^{-1}\Omega^t +$$
$$(K_s + \Gamma^{-1}\Omega^t)^t \Gamma(K_s + \Gamma^{-1}\Omega^t) + W + Q < 0 \quad (5.115)$$

that is

$$A^t \breve{X} A - (\breve{X}^{-1} + \Lambda\Lambda^t)^{-1} - \Omega\Gamma^{-1}\Omega^t +$$
$$\mu^{-1} E^t E + W + Q < 0 \quad (5.116)$$

where

$$\Gamma = (R + \mu^{-1} E_b^t E_b + B^t \breve{X} B)$$
$$\Omega = (A^t \breve{X} B + \mu^{-1} E^t E_b) \quad (5.117)$$

Therefore, the ARE (5.106) has a stabilizing solution P_* and $P_* < \breve{P}$.

Corollary 5.8: *Consider the system*

$$(\Sigma_{o\Delta}): \quad x(k+1) = [A + \Delta A(k)]x(k) + Bu(k) + A_d x(k - \tau) \quad (5.118)$$

which is obtained from system (Σ_Δ) by setting $E_b = 0$; In this case, the ARE (5.106) and controller (5.108) reduce to:

$$A^t PA - P - A^t P\hat{B}(\hat{R} + \hat{B}^t P\hat{B})^{-1}\hat{B}^t PA +$$
$$\mu^{-1}E^t E + Q + W = 0 \qquad (5.119)$$
$$u(k) = -[0 \quad I](\hat{R} + \hat{B}^t P\hat{B})^{-1}\hat{B}^t AP \qquad (5.120)$$

where

$$\hat{R} = \begin{bmatrix} -I & 0 \\ 0 & R \end{bmatrix} \qquad (5.121)$$

Corollary 5.9: *Consider the system*

$$(\Sigma_{o\Delta}): \quad x(k+1) = Ax(k) + Bu(k) + A_d x(k - \tau) \qquad (5.122)$$

corresponding to the case of systems without uncertainties $(H = 0, E = 0)$. In such case, the ARE (5.89) and controller (5.90) reduce to:

$$A^t PA - P + Q + W - A^t P\tilde{B}(\tilde{R} + \tilde{B}^t P\tilde{B})^{-1}\tilde{B}^t PA = 0 \quad (5.123)$$
$$u(k) = -[0 \quad I](\tilde{R} + \tilde{B}^t P\tilde{B})^{-1}\tilde{B}^t AP \qquad (5.124)$$

where

$$\tilde{B} = [(A_d W^{-1} A_d^t)^{1/2} \quad B] \qquad (5.125)$$

which provides a guaranteed cost control for linear state-delay discrete-time systems without uncertainties.

5.3   Observer-Based Control

This section contains the design of robust observer-based controllers for a class of uncertain linear time-lag systems. Like most of our work, the uncertainties are assumed to be norm-bounded. We extend the previous results on quadratic-cost control (section 5.1) to the case of observer-based control for uncertain systems with unknown state and input time-lags.

5.3.1   Problem Description

The class of nominally linear systems with parametric uncertainties and time-lags under consideration is represented by:

$$
\begin{aligned}
(\Sigma_\Delta): \dot{x}(t) &= [A + \Delta A(t)]x(t) + [B + \Delta B(t)]u(t) \\
&\quad + [A_d + \Delta D(t)]x(t-\tau) + B_h u(t-\eta) \\
y(t) &= [C + \Delta C(t)]x(t) + [G + \Delta G(t)]u(t) \qquad (5.126)
\end{aligned}
$$

where $t \in \Re$ is the time, $x(t) \in \Re^n$ is the state, $u(t) \in \Re^m$ is the control input; $y(t) \in \Re^p$ is the measured output, and τ, η are constant scalars representing the amount of lags in the state and at the input of the system, respectively. There is no restriction placed on τ and η except that they are constants but otherwise unknown. The matrices $A \in \Re^{n \times n}$, $B \in \Re^{n \times m}$, $A_d \in \Re^{n \times n}$, $B_h \in \Re^{n \times m}$, $C \in \Re^{p \times n}$ and $G \in \Re^{p \times m}$ are real constant matrices representing the nominal plant with the pair (A, B) being controllable, and the pair (A, C) being observable. The uncertain matrices $\Delta A(t)$, $\Delta B(t)$, $\Delta C(t)$, $\Delta D(t)$ and $\Delta G(t)$ are assumed to be represented by:

$$
\begin{aligned}
\begin{bmatrix} \Delta A(t) & \Delta B(t) \\ \Delta C(t) & \Delta G(t) \end{bmatrix} &= \begin{bmatrix} H \\ H_c \end{bmatrix} \Delta(t) \begin{bmatrix} E & E_b \end{bmatrix} \\
\Delta D(t) &= H_d \Delta(t) E_d \qquad (5.127)
\end{aligned}
$$

where $\Delta(t) \in \Re^{\alpha_1 \times \alpha_2}$ is a matrix of uncertain parameters satisfying the bounds

$$
\Delta^t(t)\Delta(t) \leq I \qquad (5.128)
$$

and $H \in \Re^{n \times \alpha_1}$, $H_c \in \Re^{p \times \alpha_1}$, $H_d \in \Re^{n \times \alpha_1}$, $E \in \Re^{\alpha_2 \times n}$, $E_b \in \Re^{\alpha_2 \times m}$ and $E_d \in \Re^{\alpha_2 \times n}$ are known real constant matrices. The initial function of system (1)-(2) is specified as $\langle x_o, \Psi_1(\mu), \Psi_2(\mu) \rangle$ where $\Psi_1(\cdot) \in \mathcal{L}_2[-\tau, 0]$ and $\Psi_2(\cdot) \in \mathcal{L}_2[-\eta, 0]$.

In what follows, we extend section 5.1 to the case of uncertain systems with state and input delays. More importantly, we develop a linear dynamic output-feedback controller to yield a closed-loop system which is asymptotically stable for all admissible uncertainties and unknown time lags.

Associated with system (Σ_Δ) is the quadratic cost function:

$$
J = \int_0^\infty [x^t(t)Qx(t) + u^t(t)Ru(t)]dt \qquad (5.129)
$$

with $0 < Q = Q^t \in \Re^{n \times n}$ and $0 < R = R^t \in \Re^{m \times m}$. Our objective is to design an observer-based controller of the form:

$$(\Sigma_o): \quad \begin{aligned} \dot{\hat{x}}(t) &= A\hat{x}(t) + Bu(t) + K_o[y(t) - \hat{y}(t)] \\ \hat{y}(t) &= C\hat{x}(t) + Gu(t) \\ u(t) &= K_c\hat{x}(t) \end{aligned} \quad (5.130)$$

such that the closed-loop system (5.126) and (5.130) is robustly stable independent of delay and a prescribed level of performance is guaranteed independent of the amount of time-lags and for all admissible values of uncertainties satisfying (5.128). In (5.130), $K_o \in \Re^{n \times p}$ is the observer gain and $K_c \in \Re^{m \times n}$ is the feedback control gain.

5.3.2 Closed-Loop System

In terms of the error vector

$$e(t) = x(t) - \hat{x}(t) \quad (5.131)$$

and the extended state vector,

$$z(t) = \begin{bmatrix} x(t) \\ e(t) \end{bmatrix} \in \Re^{2n} \quad (5.132)$$

It should be noted that $u(t)$ in (5.130), can be written as,

$$u(t) = [K_c \quad -K_c]z(t) = K_t z(t) \quad (5.133)$$

We combine (5.126) and (5.130) to produce the closed-loop system:

$$(\Sigma_c): \quad \dot{z}(t) = [A_c + \Delta A_c(t)]z(t) + [D_c + \Delta D_c(t)]z(t-\tau) + E_c z(t-\eta) \quad (5.134)$$

where,

$$A_c = \begin{bmatrix} A + BK_c & -BK_c \\ 0 & A - K_oC \end{bmatrix} \quad (5.135)$$

$$D_c = \begin{bmatrix} A_d & 0 \\ A_d & 0 \end{bmatrix}, \quad E_c = \begin{bmatrix} B_hK_c & -B_hK_c \\ B_hK_c & -B_hK_c \end{bmatrix}$$

$$\Delta A_c = \begin{bmatrix} \Delta A + \Delta B K_c & -\Delta B K_c \\ \Delta A + \Delta B K_c - K_o \Delta C - K_o \Delta G K_c & -\Delta B K_c + K_o \Delta G K_c \end{bmatrix}$$

$$= \begin{bmatrix} H \\ H - K_o H_c \end{bmatrix} \Delta(t) \begin{bmatrix} E + E_b K_c & -E_b K_c \end{bmatrix}$$

$$= \widehat{H} \Delta(t) \widehat{E}$$

$$\Delta D_c = \begin{bmatrix} \Delta D(t) & 0 \\ \Delta D(t) & 0 \end{bmatrix}$$

$$= \begin{bmatrix} H_d \\ H_d \end{bmatrix} \Delta(t) \begin{bmatrix} E_d & 0 \end{bmatrix}$$

$$= \widehat{H}_d \Delta(t) \widehat{E}_d \qquad (5.136)$$

5.3.3   Robust Performance Analysis IV

Before embarking on the control design, we present hereafter some results in relation to the robust performance of system (Σ_c) when associated with the quadratic cost function

$$J_o = \int_o^\infty z^t(t) \Omega z(t) dt \quad ; \quad \Omega = diag(Q, 0) \qquad (5.137)$$

Theorem 5.10: *Given matrices* $0 < W = W^t$ *and* $0 < S = S^t$, *the uncertain time-lag system* (Σ_c) *is said to be robustly stable if there exists a matrix* $0 < \Pi = \Pi^t$ *satisfying the LMI:*

$$\Xi_\Delta = \begin{bmatrix} \begin{matrix} \Pi(A_c + \Delta A_c) + (A_c + \Delta A_c)^t \Pi \\ +W + S \end{matrix} & \Pi(D_c + \Delta D_c) & \Pi E_c \\ (D_c + \Delta D_c)^t \Pi & -W & 0 \\ E_c^t \Pi & 0 & -S \end{bmatrix} < 0$$

$$\forall \Delta : \Delta^t \Delta \le I \qquad (5.138)$$

Proof: We use a Lyapunov-Krasovskii functional of the form:

$$V(z_t) = z^t(t) \Pi z(t) + \int_{-\tau}^0 z^t(t+s) W z(t+s) ds + \int_{-\eta}^0 z^t(t+s) S z(t+s) ds$$

$$(5.139)$$

where the weighting matrices W, S are such that $0 < W = W^t \in \Re^{2n \times 2n}$ and $0 < S = S^t \in \Re^{2n \times 2n}$. Note that, $V(z_t) > 0$ for all $z(t) \ne 0$ and

$V(z_t) = 0$ for $z(t) = 0$. Let the extended state vector $Z(t)$ be such that $Z(t) = [z^t(t) \quad z^t(t - \tau) \quad z^t(t - \eta)]^t$. By differentiating (5.139) along the solutions of (5.134), it is easy to see that

$$\dot{V}(z) = Z^t(t)\Xi_\Delta Z(t) \tag{5.140}$$

If $\dot{V}(z) < 0$ then $z(t) \to 0$ as $t \to \infty$. This is implied by $\Xi_\Delta < 0$. By **A.1**, this yields (5.138).

Remark 5.4: It should be stressed that **Theorem 5.10** is independent of the time lags τ and η and extends the standard $RSID$ to the case of systems with input and state time-lags. By setting $\Delta A_c = 0$, $D_c = 0$, $\Delta D_c = 0$ and $E_c = 0$, one obtains the known definition of QS for norm-bounded uncertain systems (see **Lemma 2.1**).

Extending on **Definition 5.1**, we have the following:

Definition 5.8: A matrix $0 < \Pi = \Pi^t \in \Re^{2n \times 2n}$ is said to be a quadratic cost (QC) matrix for the system (Σ_c) and cost function (5.129) if it satisfies the LMI:

$$\Xi_\Delta = \begin{bmatrix} \Pi(A_c + \Delta A_c) + (A_c + \Delta A_c)^t\Pi & \Pi(D_c + \Delta D_c) & \Pi E_c \\ +W + S + \Omega & & \\ (D_c + \Delta D_c)^t\Pi & -W & 0 \\ E_c^t\Pi & 0 & -S \end{bmatrix} < 0 \tag{5.141}$$

for all admissible uncertainties satisfying (5.127).

Theorem 5.11 : Suppose that the matrix $0 < \Pi = \Pi^t$ is a QC matrix for system (Σ_c) and cost function (5.129). Then the closed-loop system (Σ_c) is RSID and an upper bound on the cost J_o is given by

$$J_o < z_o^t\Pi z_o + \int_{-\tau}^0 z^t(s)Wz(s)ds + \int_{-\eta}^0 z^t(s)Sz(s)ds \tag{5.142}$$

Conversely, if system (Σ_c) is RSID then there will exist a QC matrix for this system and the cost function (5.129).

Proof: ($\Rightarrow$) Suppose that the matrix $0 < \Pi = \Pi^t$ is a QC matrix for system (5.134) and cost function (5.129). It follows from **Definition 5.8**

that:

$$Z^t(t) \begin{bmatrix} \Pi(A_c + \Delta A_c) + (A_c + \Delta A_c)^t \Pi \\ +W + S + \Omega \\ (D_c + \Delta D_c)^t \Pi & -W & 0 \\ E_c^t \Pi & 0 & -S \end{bmatrix} Z(t) < 0$$

$$(5.143)$$

for all non-zero $Z(t) = [z^t(t) \quad z^t(t-\tau) \quad z^t(t-\eta)]^t$ and for all admissible uncertainty Δ satisfying (5.127). Hence system (5.134) is $RSID$. Using the Lyapunov functional (5.139), we get

$$\begin{aligned} \dot{V}(z) &= Z^t(t) \Xi_\Delta Z(t) \\ &< Z^t(t) \begin{bmatrix} -\Omega & 0 & 0 \\ 0 & 0 & 0 \\ 0 & 0 & 0 \end{bmatrix} Z(t) \\ &= -z^t(t)\Omega z(t) \end{aligned}$$

$$(5.144)$$

and therefore $z^t(t)\Omega z(t) < -\dot{V}(z_t)$. Thus,

$$\begin{aligned} J_o &= \int_0^\infty z^t(t)\Omega z(t)dt \\ &< V(z_o) - V(z_\infty) = V(z_o) \end{aligned}$$

$$(5.145)$$

since system (Σ_c) is $RSID$, it follows that $V(z_t) \to 0$ as $t \to \infty$ and therefore (5.145) reduces to (5.142).

($\Leftarrow$) Suppose that the closed-loop system (Σ_c) is $RSID$. By **Theorem 5.10**, there exist matrices $0 < \Pi = \Pi^t$, $0 < W = W^t$, $0 < S = S^t$ satisfying (5.138) for all admissible uncertainties. Thus, using **A.1** on (5.138), we get

$$\Pi(A_c + \Delta A_c) + (A_c + \Delta A_c)^t \Pi + W + S +$$
$$\Pi(D_c + \Delta D_c)W^{-1}(D_c + \Delta D_c)^t \Pi + \Pi E_c S^{-1} E_c^t \Pi < 0 \quad (5.146)$$

We can find a scalar $\mu > 0$ such that for $||\Delta|| \le 1$, the following inequality is satisfied:

$$\Omega + \mu^{-1}\{\Pi(A_c + \Delta A_c) + (A_c + \Delta A_c)^t \Pi + W + S +$$
$$\Pi(D_c + \Delta D_c)W^{-1}(D_c + \Delta D_c)^t \Pi + \Pi E_c S^{-1} E_c^t \Pi\} < 0 \quad (5.147)$$

It follows from inequality (5.147) that there exist matrices $\breve{W} = \mu^{-1}W$ and $\breve{S} = \mu^{-1}S$ such that the matrix $\breve{\Pi} = \mu^{-1}\Pi$ is a QC matrix for system (Σ_c).

Application of **B.1.2** yields for some $\epsilon > 0$

$$\Pi\Delta A_c + \Delta A_c^t\Pi \;=\; \Pi\widehat{H}_1\Delta\widehat{E}_1 + \widehat{E}_1^t\Delta^t\widehat{H}_1^t\Pi$$
$$\leq\; \epsilon\Pi\widehat{H}_1\widehat{H}_1^t\Pi + \epsilon^{-1}\widehat{E}_1^t\widehat{E}_1 \qquad (5.148)$$

In a similar way, application of **B.1.3** produces for some $\sigma > 0$

$$\Pi(D_c + \Delta D_c)W^{-1}(D_c + \Delta D_c)^t\Pi =$$
$$\Pi(D_c + \widehat{H}_3\Delta\widehat{E}_3)W^{-1}(D_c + \widehat{H}_3\Delta\widehat{E}_3)^t\Pi \leq$$
$$\Pi D_c(W - \sigma\widehat{E}_3^t\widehat{E}_3)^{-1}D_c^t\Pi + \sigma^{-1}\Pi\widehat{H}_3\widehat{H}_3^t\Pi \qquad (5.149)$$

The following theorem summarizes the main result by providing an LMI-based solution to the robust performance problem.

Theorem 5.12: *Given matrices* $0 < W = W^t$, $0 < S = S^t$, *a matrix* $0 < \Pi = \Pi^t$ *is a QC matrix for the closed-loop system* (Σ_c) *and cost function (18) if and only if there exist scalars* $\epsilon > 0$, $\sigma > 0$ *such that* $W > \sigma\widehat{E}_3^t\widehat{E}_3$ *and* Π *satisfies the LMI:*

$$\Xi_c = \begin{bmatrix} \Pi A_c + A_c^t\Pi + W + S & & & \\ +\Omega + \epsilon\Pi\widehat{H}_1\widehat{H}_1^t\Pi & \epsilon^{-1}\widehat{E}_1^t & \sigma^{-1}\Pi\widehat{H}_3 & \Pi E_c \\ +\Pi D_c(W - \sigma\widehat{E}_3^t\widehat{E}_3)^{-1}D_c^t\Pi & & & \\ \epsilon^{-1}\widehat{E}_1 & -I & 0 & 0 \\ \sigma^{-1}\widehat{H}_3^t\Pi & 0 & -I & 0 \\ E_c^t\Pi & 0 & 0 & -S \end{bmatrix} < 0$$
$$(5.150)$$

Proof: ($\Rightarrow$) Suppose that the matrix $0 < \Pi = \Pi^t$ satisfies (5.150) for some $\epsilon > 0$ and $\sigma > 0$ such that $W > \sigma\widehat{E}_3^t\widehat{E}_3$. By (5.148)-(5.149), inequality (5.150) is implied from (5.141) for all admissible uncertainties satisfying (5.127).

($\Leftarrow$) Suppose that the matrix $\Pi > 0$ is a QC matrix for system (Σ_c) and cost function (5.129). By **Definition 5.8**, inequality (5.141) is satisfied for all admissible uncertainties satisfying (5.127). By **B.1.3**, this implies that

$$z^t(t)[\Pi(A_c + \Delta A_c) + (A_c + \Delta A_c)^t\Pi + W + S + \Omega]z(t)$$
$$+2z^t(t)\Pi(D_c + \Delta D_c)z(t - \tau) + 2z^t(t)\Pi E_c z(t - \eta)$$
$$-z^t(t - \tau)Wz(t - \tau) - z^t(t - \eta)Sz(t - \eta) < 0 \qquad (5.151)$$

for all nonzero $z(t)$, $z(t-\tau)$ and $z(t-\eta)$ and for all admissible uncertainties satisfying (5.127). From the results of [175,176] and applying **B.1.1** to the term $(\Pi\Delta A_c + \Delta A_c^t\Pi)$ and **B.1.2** to the term $(2z^t(t)\Pi\Delta D_c z(t-\tau))$, inequality (5.151) implies that

$$z^t(t)[\Pi A_c + A_c^t\Pi + W + S + \Omega]z(t) + 2z^t(t)\Pi D_c z(t-\tau)$$
$$+2z^t(t)\Pi E_c z(t-\eta) - z^t(t-\tau)W z(t-\tau) - z^t(t-\eta)S z(t-\eta)$$
$$+2\|\widehat{H_1^t}\Pi z(t)\|\,\|\widehat{E_1} z(t)\| + 2\|\widehat{H_3^t}\Pi z(t-\tau)\|\,\|\widehat{E_3} z(t-\tau)\| < 0$$

$$(5.152)$$

for all nonzero $z(t)$, $z(t-\tau)$ and $z(t-\eta)$ and $\Delta : \Delta^t\Delta \le I$. Therefore

$$\Gamma_c = \begin{bmatrix} \Pi A_c + A_c^t\Pi + W + S + \Omega & \Pi D_c & \Pi E_c \\ D_c^t\Pi & -W & 0 \\ E_c^t\Pi & 0 & -S \end{bmatrix} < 0 \qquad (5.153)$$

and

$$\{Z^t(t)\Gamma_c Z\}^2 > 4\left(Z^t(t)\begin{bmatrix} \Pi\widehat{H}\widehat{H}^t\Pi & 0 & 0 \\ 0 & \Pi\widehat{H_d}\widehat{H_d}^t\Pi & 0 \\ 0 & 0 & 0 \end{bmatrix} Z(t)\right)\otimes$$
$$\left(Z^t(t)\begin{bmatrix} \widehat{E}^t\widehat{E} & 0 & 0 \\ 0 & \widehat{E_d}^t\widehat{E_d} & 0 \\ 0 & 0 & 0 \end{bmatrix} Z(t)\right)$$

$$(5.154)$$

for all nonzero $z(t)$, $z(t-\tau)$ and $z(t-\eta)$. From a known result of [222], inequalities (5.153) and (5.154) imply that there exist parameters $\epsilon > 0$ and $\sigma > 0$ such that

$$\Pi A_c + A_c^t\Pi + W + S + \Omega + \epsilon^{-1}\widehat{E}^t\widehat{E} + \Pi[\epsilon\widehat{H}\widehat{H}^t + \sigma^{-1}\widehat{H_d}\widehat{H_d}^t]\Pi$$
$$+\Pi[D_c(W - \sigma\widehat{E_d}^t\widehat{E_d})^{-1}D_c^t + E_c S^{-1} E_c^t]\Pi < 0$$

$$(5.155)$$

Finally, using **B.1.3** inequality (5.155) can be converted to the LMI (5.150).

5.3.4   Synthesis of Observer-Based Control

In this section, we consider the problem of guaranteed cost control via observer-based control for the uncertain time-lag system under consideration. By virtue of **Theorem 5.10** and **Definition 5.8**, we provide the following:

Definition 5.9: *An observer-based control law* $u(t) = K_t z(t)$ *is said to define quadratic guaranteed cost control (QGCC) with associated cost matrix* $\Pi > 0$ *for the system* (Σ_Δ) *and cost function (6) if it satisfies the following LMI:*

$$
\begin{bmatrix}
\begin{matrix} \Pi(A_c + \Delta A_c) + (A_c + \Delta A_c)^t \Pi \\ + W + S + \Omega_c \\ (D_c + \Delta D_c)^t \Pi \\ E_c^t \Pi \end{matrix} & \begin{matrix} \Pi(D_c + \Delta D_c) \\ -W \\ 0 \end{matrix} & \begin{matrix} \Pi E_c \\ 0 \\ -S \end{matrix}
\end{bmatrix} < 0 \quad (5.156)
$$

where

$$
\Omega_c = \begin{bmatrix} Q + K_c^t R K_c & -K_c^t R K_c \\ \\ -K_c^t R K_c & K_c^t R K_c \end{bmatrix} \quad (5.157)
$$

Before proceeding further, we define the weighting matrices:

$$
W = \begin{bmatrix} W_c & 0 \\ 0 & W_o \end{bmatrix}; \quad S = \begin{bmatrix} S_c & 0 \\ 0 & S_o \end{bmatrix} \quad (5.158)
$$

where $0 < W_c = W_c^t$, $0 < W_o = W_o^t$, $0 < S_c = S_c^t$, and $0 < S_o = S_o^t$.

The following theorem shows that the problem of determining an observer-based $QGCC$ for system (Σ_Δ) and cost function (5.129) can be solved via two matrix inequalities. One of these inequalities is a bilinear matrix inequality (BLMI) and the other is a linear matrix inequality (LMI). For simplicity in exposition, we introduce the following matrices:

$$
\begin{align}
\bar{A} &= A - \epsilon^{-1} B R_1^{-1} L_2^t L_1 & (5.159) \\
R_c R_c^t &= A_d (W_c - \sigma E_d^t E_d)^{-1} A_d^t + \epsilon H H^t + \sigma^{-1} H_d H_d^t & (5.160) \\
F_1 &= \epsilon^{-1}(E_b^t E + E_b^t E_d), \quad R_1 = R + \epsilon^{-1} E_b^t E_b & (5.161) \\
F_2 &= (\epsilon^{-1} I - \epsilon^{-2} E_b R_1^{-1} E_b^t)^{-1}, \quad S_t^{-1} = S_c^{-1} + S_o^{-1} & (5.162) \\
R_o R_o^t &= A_d (W_c - \sigma E_d^t E_d)^{-1} A_d^t + \epsilon H H^t \\
&+ \sigma^{-1} H_d H_d^t + B_h K_c S_t^{-1} K_c^t B_h^t & (5.163) \\
M_{12} &= H_c H^t, \quad \varphi = M_{12}^t (M_{12} M_{12}^t)^{-1}, \quad \theta = \epsilon^{-1} \varphi H_c H_c^t \varphi^t & (5.164) \\
\breve{A} &= \epsilon^{-1} R_o R_o^t (\varphi C - \theta X^{-1} L_3^t L_2 K_c) + \epsilon^{-1} X^{-1} E_d^t E_b K_c & (5.165)
\end{align}
$$

$$R_2 R_2^t = R_o R_o^t \theta R_o R_o^t - R_o R_o^t \tag{5.166}$$

$$\Upsilon = -\epsilon^{-2} K_c^t E_b^t E_d X^{-1} \varphi C - \epsilon^{-2} C^t \varphi^t X^{-1} E_d^t E_b K_c$$
$$+ \epsilon^{-2} K_c^t E_b^t E_d X^{-1} \theta X^{-1} E_d^t E_b K_c \tag{5.167}$$

The next theorem establishes the main result.

Theorem 5.13 : *Consider the uncertain time-delay system (5.125) satisfying (5.127). Suppose that there exist constants $\epsilon > 0$ and $\sigma > 0$ such that the matrices $0 < X = X^t$ and $0 < Y = Y^t$ satisfy the following BLMI:*

$$
\begin{bmatrix}
\begin{array}{c} X\bar{A} + \bar{A}^t X + S_c + \\ Q + W_c - X B_o R_1^{-1} B_o^t X \\ + X E_o R_1^{-1} (B_o^t X + F_1) S_t^{-1} \\ (B_o^t X + F_1)^t R_1^{-1} E_o^t X \end{array} & X R_c & L_1^t & L_3^t L_2 \\
R_c^t X & -I & 0 & 0 \\
L_1 & 0 & -F_2 & 0 \\
L_2^t L_3 & 0 & 0 & -\epsilon^2 R_1
\end{bmatrix} < 0 \tag{5.168}
$$

and the LMI:

$$
\begin{bmatrix}
\begin{array}{c} Y(A_o - \breve{A}) + (A_o^t - \breve{A}^t)Y \\ + W_o + S_o + \Upsilon \end{array} & Y R_2 & X B_o + F_1^t \\
R_2^t Y & -I & 0 \\
B_o^t X + F_1 & 0 & -R_1
\end{bmatrix} < 0 \tag{5.169}
$$

Then the closed-loop system is asymptotically stable with gains:

$$K_c = -R_1^{-1}(B_o^t X + F_1) \tag{5.170}$$

$$K_o = \epsilon^{-1}(R_o R_o^t + \epsilon^{-1} Y^{-1} K_c^t L_2^t L_3 X^{-1}) \varphi \tag{5.171}$$

Proof: Define Π such that

$$\Pi = \begin{bmatrix} X & 0 \\ 0 & Y \end{bmatrix} \tag{5.172}$$

Then expansion of (5.150), using **B.1.3**, yields:

$$\begin{bmatrix} \Phi_{11} & \Phi_{12} \\ \Phi_{12}^t & \Phi_{22} \end{bmatrix} < 0 \tag{5.173}$$

where

$$
\begin{aligned}
\Phi_{11} =\ & X(A + BK_c) + (A + BK_c)^t X + W_c + S_c \\
& + Q + K_c^t RK_c + XA_d(W_c - \sigma E_d^t E_d)^{-1} A_d^t X \\
& + XB_hK_c(S_o^{-1} + S_c^{-1})K_c^t B_h^t X + \epsilon XHH^t X + \sigma^{-1} XH_dH_d^t X \\
& + \epsilon^{-1}(E + E_bK_c)^t(E + E_bK_c) \qquad (5.174)
\end{aligned}
$$

$$
\begin{aligned}
\Phi_{12} =\ & -XBK_c - K_c^t RK_c + XA_d(W_c - \sigma L_3^t L_3)^{-1} A_d^t Y \\
& + XB_hK_c(S_o^{-1} + S_c^{-1})K_c^t B_h^t Y + \epsilon XHH^t Y - \epsilon XHH_c^t K_o^t Y \\
& - \epsilon^{-1} E^t E_b K_c - \epsilon^{-1} K_c^t E_b^t E_b K_c + \sigma^{-1} XH_dH_d^t Y \qquad (5.175)
\end{aligned}
$$

$$
\begin{aligned}
\Phi_{22} =\ & Y(A - K_oC) + (A - K_oC)^t Y + W_o + S_o \\
& + K_c^t RK_c + YA_d(W_c - \sigma E_d^t E_d)^{-1} A_d^t Y \\
& + YB_hK_c(S_o^{-1} + S_c^{-1})K_c^t B_h^t Y + \epsilon Y(H - K_oH_c)(H - K_oH_c)^t Y \\
& + \epsilon^{-1} K_c^t E_b^t E_b K_c + \sigma^{-1} YH_dH_d^t Y \qquad (5.176)
\end{aligned}
$$

A sufficient condition to satisfy (5.173) is that

$$
\Phi_{11} < 0, \quad \Phi_{22} < 0, \quad \Phi_{12} = 0 \qquad (5.177)
$$

Choosing the gain K_c as given by (5.170), and using (5.159)-(5.161), inequality $\Phi_{11} < 0$ reduces to the matrix inequality (5.168). From (5.175), and using (5.162)-(5.164), (5.170)-(5.171), it is easy to check that $\Phi_{12} = 0$, where it is assumed that $M_{12}M_{12}^t$ is nonsingular.

Finally considering the inequality $\Phi_{22} < 0$. Starting from (5.176) and using (5.163)-(5.171), after some algebraic manipulations we obtain inequality (5.169).

The following two corollaries represent some special cases of our results.

Corollary 5.10: *Consider the state-delay uncertain system*

$$
\begin{aligned}
\dot{x}(t) =\ & [A + \Delta A(t)]x(t) + [B + \Delta B(t)]u(t) \\
& + [A_d + \Delta D(t)]x(t - \tau) \qquad (5.178) \\
y(t) =\ & [C + \Delta C(t)]x(t) + [G + \Delta G(t)]u(t) \qquad (5.179) \\
x(t) =\ & \phi(t) \quad \forall t \in [-\tau, 0]
\end{aligned}
$$

combined with the observer-based controller (5.130). Let

$$
R_3 R_3^t = R_c R_c^t \theta R_c R_c^t - R_c R_c^t \qquad (5.180)
$$

$$\breve{A}' = \epsilon^{-1}R_cR_c^t(\varphi C - \theta X^{-1}E_d^tE_bK_c) + \epsilon^{-1}X^{-1}E_d^tE_bK_c \quad (5.181)$$

Then the resulting closed-loop controlled system is robustly stable if there exist constants $\epsilon > 0$ and $\sigma > 0$ such that the matrices $0 < X = X^t$ and $0 < Y = Y^t$ satisfy the following two LMIs:

$$\begin{bmatrix} X\bar{A} + \bar{A}^t X + S_c + Q & XR_c & E^t & E_d^t E_b \\ +W_c - XBR_1^{-1}B^tX & & & \\ R_c^t X & -I & 0 & 0 \\ E & 0 & -F_2 & 0 \\ E_b^t E_d & 0 & 0 & -\epsilon^2 R_1 \end{bmatrix} < 0 \quad (5.182)$$

and

$$\begin{bmatrix} Y(A - \breve{A}') + (A^t - \breve{A}')^t Y & YR_3 & XB + F_1^t \\ +W_o + S_o + \Upsilon & & \\ R_3^t Y & -I & 0 \\ B^t X + F_1 & 0 & -R_1 \end{bmatrix} < 0 \quad (5.183)$$

The gain matrices are such:

$$\begin{aligned} K_c &= -R_1^{-1}(B_o^t X + F_1) \\ K_o &= \epsilon^{-1}(R_o R_o^t + \epsilon^{-1}Y^{-1}K_c^t E_b^t E_d X^{-1})\varphi \end{aligned} \quad (5.184)$$

Proof: Set $B_h = 0$ in **Theorem 5.13**.

Corollary 5.11: Consider the uncertain system

$$\begin{aligned} \dot{x}(t) &= [A + \Delta A(t)]x(t) + [B + \Delta B(t)]u(t) \\ y(t) &= [C + \Delta C(t)]x(t) + [G + \Delta G(t)]u(t) \end{aligned} \quad (5.185)$$

combined with the observer-based controller (5.130). Let

$$R_4 R_4^t = \epsilon^2 HH^t\theta HH^t - \epsilon HH^t \quad (5.186)$$

Then the resulting closed-loop controlled system is robustly stable if there exist constants $\epsilon > 0$ and $\sigma > 0$ such that the matrices $0 < X = X^t$ and $0 < Y = Y^t$ satisfy the following two LMIs:

$$\begin{bmatrix} X\bar{A} + \bar{A}^t X + S_c & XH & E^t \\ +Q + W_c - XBR_1^{-1}B^tX & & \\ H^t X & -\epsilon^{-1}I & 0 \\ E & 0 & -F_2 \end{bmatrix} < 0 \quad (5.187)$$

and

$$\begin{bmatrix} Y(A - HH^t\varphi C) + (A - HH^t\varphi C)Y + \\ W_o + S_o & YR_4 & XB + \\ & & \epsilon^{-1}E_b^t E \\ R_4^t Y & -I & 0 \\ B^t X + & & \\ \epsilon^{-1}E^t E_b & 0 & -R_1 \end{bmatrix} < 0$$

(5.188)

The gain matrices are:

$$K_c = -R_1^{-1}(B^t X + \epsilon^{-1}E_b^t E) \qquad (5.189)$$
$$K_o = HH^t\varphi \qquad (5.190)$$

Proof: Set $A_d = 0$ and $B_h = 0$ in **Theorem 5.13.**

5.3.5 A Computational Algorithm

To implement the observer-based control design results, we apply the following computational algorithm:

(1) Read in the data A, B, C, A_d, B_h, G, H, H_c, H_d, E, E_b and E_d.
(2) Select scalars $\epsilon_k > 0$ and $\sigma_k > 0$ such that $F_2(\epsilon_k) > 0$ and set the iteration index $k = 1$.
(3) Solve matrix inequality (5.168) for X using the following iterative steps:
  (a) By using **A.1** and setting $P(\epsilon_k, \sigma_k) = X^{-1}$,
  $P_* = P^{-1}(\epsilon_k, \sigma_k)$, $W_{cc}W_{cc}^t = W_c + S_c + Q$,
  we rewrite inequality (5.168) as:

$$\begin{bmatrix} P\bar{A}^t(\epsilon_k) + \bar{A}(\epsilon_k)P \\ -B_o R_1^{-1}(\epsilon_k)B_o^t + R_c R_c^t(\epsilon_k, \sigma_k) \\ +E_o R_1^{-1}(\epsilon_k)(B_o^t P_* + F_1(\epsilon_k))S_t^{-1} & PW_{cc} & PE^t & PE_d^t E_b \\ (B_o^t P_* + F_1(\epsilon_k))^t R_1^{-1}(\epsilon_k)E_o^t \\ W_{cc}^t P & -I & 0 & 0 \\ EP & 0 & -F_2(\epsilon_k) & 0 \\ E_b^t E_d P & 0 & 0 & -\epsilon_k^2 R_1(\epsilon_k) \end{bmatrix} < 0$$

(5.191)

  (b) Select an initial value for P_* and solve inequality (5.191) for $P(\epsilon_k, \sigma_k)$ using the **LMI**-Control Toolbox.
  (c) Update P_* by $P_* \Leftarrow (1 - \zeta)P_* + \zeta P^{-1}(\epsilon_k, \sigma_k)$ with $0 < \zeta < 1$.

(d) Compute $d = ||P_* - P^{-1}(\epsilon_k, \sigma_k)||$.

(e) If $d \leq \delta$ where δ is a prespecified tolerance then set $\epsilon_k = \epsilon^*$, $\sigma_k = \sigma^*$ and go to step **4**.

(f) If $d > \delta$ then update ϵ and σ using

$$\epsilon_k \Leftarrow \epsilon_k + \Delta\epsilon_k$$

and $\sigma_k \Leftarrow \sigma_k + \Delta\sigma_k$

where $\Delta\epsilon_k$ and $\Delta\sigma_k$ are pre-chosen increments; go to step **3** with $k \Leftarrow k + 1$.

4) Compute the controller gain matrix K_c from (5.170).

5) Using A.1 on (5.169), setting $T = Y^{-1}$, $W_{oo}W_{oo}^t(\epsilon^*, \sigma^*)$ $= W_o + S_o + \Upsilon(\epsilon^*, \sigma^*)$ and premultiplying and postmultiplying the result by T yields:

$$\begin{bmatrix} \begin{array}{c} (A_o - \breve{A}(\epsilon^*, \sigma^*))T + \\ T(A_o^t - \breve{A}^t(\epsilon^*, \sigma^*)) + \\ R_2 R_2^t(\epsilon^*, \sigma^*) \end{array} & TW_{oo}(\epsilon^*, \sigma^*) & TK_c^t \\ W_{oo}^t(\epsilon^*, \sigma^*)T & -I & 0 \\ K_c T & 0 & -R_1^{-1}(\epsilon^*) \end{bmatrix} < 0 \qquad (5.192)$$

(6) Solve LMI (5.192) for T by using the **LMI**-Control Toolbox.

(7) Compute the gain matrix K_o from (5.171).

Remark 5.7: The proposed iterative steps (a)-(d) suggested to solve (5.191) may be heuristic, however they are simple to implement since (5.191) is convex in P. Computational experience with several examples has indicated that only few iterations are usually needed to reach a satisfactory solution with $0 < \epsilon < 1$, $0 < \sigma < 1$ and $d \leq 10^{-4}$. The LMI (5.192) is a standard linear matrix inequality. In this regard, the proposed computational algorithm is systematic and it would seem to be tangible for control system design using dynamic feedback controllers.

5.3.6 Example 5.1

Consider a second-order system modeled in the form (5.126) such that

$$A = \begin{bmatrix} -1 & 0 \\ 0.2 & -2 \end{bmatrix}; \quad B = \begin{bmatrix} -1 \\ 0.5 \end{bmatrix}; \quad C = [1 \ \ 0]$$

$$A_d = \begin{bmatrix} -0.2 & 0.5 \\ -1 & -1 \end{bmatrix}; \quad B_h = \begin{bmatrix} -0.1 \\ 0.2 \end{bmatrix}; \quad G = 1;$$

$$H = \begin{bmatrix} 0.1 & 0.05 \\ -0.05 & 0.08 \end{bmatrix}; \quad H_c = [0.06 \quad 0.1]; \quad H_d = \begin{bmatrix} -0.1 & 0.04 \\ -0.06 & 0.05 \end{bmatrix}$$

$$E = \begin{bmatrix} 0.05 & 0.08 \\ -0.08 & 0.1 \end{bmatrix}; \quad E_b = \begin{bmatrix} 0.1 \\ 0.06 \end{bmatrix}; \quad E_d = \begin{bmatrix} 0.1 & 0.1 \\ -0.2 & 0.1 \end{bmatrix}$$

$$\Delta(t) = \begin{bmatrix} 0.5\ sin(2t) & 0.3\ sin(t) \\ -0.4\ sin(t) & 0.6\ sin(3t) \end{bmatrix}$$

The gain matrices are chosen such that $Q = diag(1, 1)$, $W_c = diag(1, 1)$, $W_o = diag(0.1, 0.1)$, $S_c = diag(0.1, 0.1)$, $S_o = diag(0.1, 0.1)$ and $R = 1$. The scalars ϵ and σ are selected such that $\epsilon = 0.5$ and $\sigma = 1$.

By using the **LMI-Control Toolbox**, we solve the inequality (5.191) employing the computational algorithm to yield

$$X = \begin{bmatrix} 0.2983 & -0.0056 \\ -0.0056 & 0.0290 \end{bmatrix}$$

Using (5.170), the controller gain K_c is found to be:

$$K_c = \begin{bmatrix} 0.2966 & -0.078 \end{bmatrix}$$

Then, solving for Y using LMI (5.192), we obtain

$$Y = \begin{bmatrix} 13556 & -4031 \\ -4031 & 1199 \end{bmatrix}$$

Using (5.172), the observer gain K_o is found to be:

$$K_o = \begin{bmatrix} 197.08 \\ 664.86 \end{bmatrix}$$

5.4 Notes and References

For the case of nondelay systems, major results on guaranteed cost control are available in [156,224]. Extension to continuous time-delay systems is found in [165]. There are other related approaches [146,147] which would prove useful in the time-delay case. For classes of TDS, some promising results are found in [24,31,148,149,159,161]. Indeed more work is needed particularly in the direction of observer-based control.

Chapter 6

Passivity Analysis and Synthesis

6.1 Introduction

In this chapter, we provide results on the robust passivity analysis and synthesis problems for classes of time-delay systems (TDS) and uncertain time-delay systems (UTDS). This is equally true for continuous-time systems as well as discrete-time systems. For analytical tractability, we consider in the sequel the uncertainties to be time-varying norm-bounded and the delay factor an unknown constant within a prescribed interval. In systems theory, positive real (passivity) theory has played a major role in stability and systems theory [197, 200-202, 205-208]. A summary of the properties of positive real systems is given in **Appendix D**. The primary motivation for designing strict positive real controllers is for applications to positive real plants. It is well-known that when a strict positive real system is connected to a positive real plant in a negative-feedback configuration, the closed-loop is guaranteed to be stable for arbitrary plant variations as long as the plant remains to be positive real. Although a passivity problem can be converted into a small gain by the so-called Cayley transform [201], direct treatment is often desirable, especially when the system under consideration is subject to uncertainty. A natural problem is to design an internally stabilizing feedback controller such that a given closed-loop system is passive. In the context of linear systems, this problem is referred to as *a positive real control problem* [202] where a complete solution to the extended strict positive real (ESPR) is developed via two AREs or ARIs. In the context of nonlinear systems, a geometric characterization is introduced in [225] for systems that can be passified by a state-feedback.

From another angle, it is well-known that the notion of positive realness is closely related to the passivity of linear time-invariant systems [194]. Based thereon, two main problems will be addressed in this chapter: passivity analysis and passivity synthesis. For the passivity analysis problem, we will derive a sufficient condition for which the uncertain time-delay system is robustly stable and strictly passive for all admissible uncertainties. The condition is given in terms of an LMI. For the passivity synthesis problem, both state- and output-feedback designs will considered and it will be proved in both cases that the closed-loop uncertain time-delay system is stable and strictly passive for all admissible uncertainties. The controller gains can be determined from the solution of LMIs.

6.2   Continuous-Time Systems

6.2.1   A Class of Uncertain Systems

The class of linear time-invariant state-delay systems under consideration is represented by:

$$(\Sigma_\Delta): \quad \begin{aligned} \dot{x}(t) &= A_\Delta x(t) + Bw(t) + A_d x(t - \tau) \\ z(t) &= C_\Delta x(t) + Dw(t) \end{aligned} \tag{6.1}$$

where $x \in \Re^n$ is the state, $w \in \Re^p$ is the exogenous input, $z \in \Re^p$ is the output, τ is a time delay , $A_d \in \Re^{n \times n}$ is a constant delay matrix and $A_\Delta \in \Re^{n \times n}$, and $C_\Delta \in \Re^{p \times n}$ are uncertain matrices given by:

$$\begin{bmatrix} A_\Delta \\ C_\Delta \end{bmatrix} = \begin{bmatrix} A \\ C \end{bmatrix} + \begin{bmatrix} H \\ H_c \end{bmatrix} \Delta(t) \ E \tag{6.2}$$

where (A, C) are constant matrices representing the nominal matrices, and H, H_c, E are constant matrices with compatible dimensions with $\Delta(t) \in \Re^{\alpha \times \beta}$ being a matrix of time-varying uncertain parameters which satisfies the norm-bound form (2.46).

Distinct from system (6.1) are the free nominal system:

$$(\Sigma_f): \quad \begin{aligned} \dot{x}(t) &= Ax(t) + A_d x(t - \tau) \\ z(t) &= Cx(t) \end{aligned} \tag{6.3}$$

and the nominal system

$$(\Sigma_o): \quad \dot{x}(t) = Ax(t) + Bw(t) + A_d x(t - \tau)$$
$$z(t) = Cx(t) + Dw(t) \quad (6.4)$$

In the sequel, we examine the problem of passive analysis of (Σ_Δ) in relation to (Σ_f) and (Σ_o). In particular, it is required that system (Σ_f) to be asymptotically stable. With regards to the definition and properties of passive systems (Appendix D), we observe that the transfer function of (Σ_o)

$$T(s) = C(sI - A - A_d e^{-\tau s})^{-1}B + D \quad (6.5)$$

is analytic in $Re\,[s] > 0$ in view of the continuity and differentiability of $e^{\tau s}$. Also $T(s)$ is real for real positive s. This implies that $T(s)$ possesses the basic ingredients for positive realness. Since for linear time-invariant systems, positive realness corresponds to passivity [200, 201], we will use the terms, extended strictly passive (ESP), and extended strictly positive real (ESPR) interchangeably. Extending on these facts, we associate with system (Σ_o) the Hamiltonian:

$$H(x,t) = -\dot{V}(x) + 2z^t(t)w(t) \quad (6.6)$$

which depends on the input signal $u(t)$ and the output signal $z(t)$ with $V(x)$ being a Lyapunov functional for system (6.1). A final point to observe is that the passivity approach to system analysis is tightly linked with stability. Therefore, we have to state *a priori* the stability concept we are going to use.

6.2.2 Conditions of Passivity: Delay-Independent Stability

Initially, we adopt the notion of stability independent of delay and replace $V(x)$ in (6.6) by $V_1(x_t)$ given in (2.5), see Chapter 2. The first result is provided by the following theorem:

Theorem 6.1: *System (Σ_o) satisfying* **Assumption 2.1** *is asymptotically stable with extended strictly positive real (ESPR) independent of delay if the matrix $(D + D^t) > 0$ and there exist matrices $0 < P = P^t \in \Re^{n\times n}$ and $0 < Q = Q^t \in \Re^{n\times n}$ satisfying the LMI*

$$\begin{bmatrix} PA + A^t P + Q & PA_d & -(C^t - PB) \\ A_d^t P & -Q & 0 \\ -(C - B^t P) & 0 & -(D + D^t) \end{bmatrix} < 0 \quad (6.7)$$

or equivalently the matrix $(D + D^t) > 0$ and there exist matrices $0 < P = P^t \in \Re^{n \times n}$ and $0 < Q = Q^t \in \Re^{n \times n}$ satisfying the ARI

$$PA + A^t P + (C^t - PB)(D + D^t)^{-1}(C - B^t P) + Q + P A_d Q^{-1} A_d^t P < 0 \quad (6.8)$$

Proof: Differentiation of $V_1(x_t)$ along the solutions of (6.4) with some manipulations yields:

$$
\begin{aligned}
H(x,t) &= -x^t(t)[PA + A^t P + Q]x(t) + w^t(t)(D + D^t)w(t) \\
&\quad + x^t(t - \tau)Q x(t - \tau) - x^t(t - \tau)A_d^t P x(t) - x^t(t)P A_d x(t - \tau) \\
&\quad + w^t(t)(C - B^t P)x(t) + x^t(t)(C^t - PB)w(t) \\
&= -Z^t(t) \; \Omega(P) \; Z(t) \quad\quad\quad\quad\quad\quad\quad\quad\quad\quad\quad (6.9)
\end{aligned}
$$

where

$$
\begin{aligned}
Z(t) &= [x^t(t) \; x^t(t - \tau) \; w^t(t)]^t \\
\Omega(P) &= \begin{bmatrix} PA + A^t P + Q & PA_d & -(C^t - PB) \\ A_d^t P & -Q & 0 \\ -(C - B^t P) & 0 & -(D + D^t) \end{bmatrix} \quad (6.10)
\end{aligned}
$$

If $\Omega(P) < 0$, then $-\dot{V}_1(x_t) + 2z^t(t)w(t) > 0$, and from which it follows that

$$\int_{t_o}^{t_1} [z^t(t) \, w(t)] \, dt \; > \; 1/2 \, [V_1(x_1) \; - \; V_1(x_o)] \quad (6.11)$$

Since $V_1(x) > 0$ for $x \neq 0$ and $V_1(x) = 0$ for $x = 0$, it follows that as $t_1 \to \infty$ that system (Σ_o) is extended strictly positive real (passive).

Remark 6.1: Inequality (6.10) includes the effect of the delayed information on the positive realness condition through the matrix A_d. By setting $A_d = 0$, we recover the results of [202] for delayless systems.

Remark 6.2: Alternative forms of inequalities (6.7)-(6.8) are given by

$$\begin{bmatrix} AX + XA^t + A_d Q A_d^t & X & (XC^t - B) \\ X & -Q & 0 \\ (CX - B^t) & 0 & -(D + D^t) \end{bmatrix} < 0 \quad (6.12)$$

and

$$AX + XA^t + (XC^t - B)(D + D^t)^{-1}(CX - B^t) + XQ^{-1}X + A_d Q A_d^t < 0 \quad (6.13)$$

6.2.3   Conditions of Passivity: Delay-Dependent Stability

Next we consider the case of delay-dependent passivity, for which it would be more convenient to introduce the following Lyapunov-Krasovskii functional:

$$
\begin{aligned}
V_d(x_t) &= x^t(t)Px(t) \\
&+ \int_{-\tau}^{t}\Big\{ \int_{t+\theta}^{t} r_1[x^t(s)A^tAx(s)]ds + \int_{t-\tau+\theta}^{t} r_2[x^t(s)A_d^tA_dx(s)]ds \\
&+ \int_{t+\theta}^{t} r_3[w^t(s)B^tBw(s)]ds\Big\}d\theta
\end{aligned}
\tag{6.14}
$$

where $0 < P = P^t \in \Re^{n\times n}$ and $r_1 > 0$, $r_2 > 0$, $r_3 > 0$ are weighting factors. Now from (6.1) we get

$$
\begin{aligned}
x(t-\tau) &= x(t) - \int_{-\tau}^{0} \dot{x}(t+\theta)d\theta \\
&= x(t) - \int_{-\tau}^{0} Ax(t+\theta)d\theta - \int_{-\tau}^{0} A_dx(t-\tau+\theta)d\theta \\
&\quad - \int_{-\tau}^{0} Bw(t+\theta)d\theta
\end{aligned}
\tag{6.15}
$$

Hence, the state dynamics becomes:

$$
\begin{aligned}
(\Sigma_d):\ \dot{x}(t) &= [A+A_d]x(t) + Bw(t) - A_d\Big[\int_{-\tau}^{0} Ax(t+\theta)d\theta \\
&\quad + \int_{-\tau}^{0} A_dx(t-\tau+\theta)d\theta + \int_{-\tau}^{0} Bw(t+\theta)d\theta\Big]
\end{aligned}
\tag{6.16}
$$

The main result is summarized by the following theorem:

Theorem 6.2: *System (Σ_d) is asymptotically stable with ESPR for any τ satisfying $0 \le \tau \le \tau^*$ if the matrix $(D_o + D_o^t) > 0$ and there exist matrix $0 < X = X^t \in \Re^{n\times n}$ and scalars $\epsilon > 0 , \alpha > 0 , \sigma > 0$ satisfying the LMI:*

$$
\begin{bmatrix}
\begin{array}{c}(A+A_d)X + X(A+A_d)^t \\ +\tau^*(\epsilon+\alpha+\sigma)A_dA_d^t\end{array} & \tau^*XA_{do}^t & -(XC^t-B) \\
\tau^*A_{do}X & -J_{do} & 0 \\
-(XC-B^t) & 0 & \begin{array}{c}-D+D^t+ \\ \tau^*\sigma^{-1}B^tB\end{array}
\end{bmatrix} < 0
\tag{6.17}
$$

*or equivalently there exist matrices $0 < X = X^t \in \Re^{n \times n}$ and scalars $\epsilon >$
$0, \alpha > 0, \sigma > 0$ satisfying the ARI*

$$(A + A_d)X + X(A + A_d)^t + \tau^* X[\epsilon^{-1} A^t A + \alpha^{-1} A_d^t A_d]X +$$
$$\tau^*(\epsilon + \alpha + \sigma)A_d A_d^t +$$
$$(XC^t - B)(D + D^t - \tau^* \sigma^{-1} B^t B)^{-1}(XC - B^t) < 0 \qquad (6.18)$$

and such that $(D + D^t - \tau^* \sigma^{-1} B^t B) > 0$ where

$$A_{do} = [A_d^t \quad A^t]^t \ , \quad J_{do} = diag[\tau^* \alpha I \quad \tau^* \epsilon I] \qquad (6.19)$$

Proof: By differentiating $V_d(x_t)$ along the solutions of (6.16) and arranging terms, we obtain:

$$
\begin{aligned}
\dot{V}_d(x_t) \ = \ & x^t(t)[P(A + A_d) + (A + A_d)^t P]x(t) \\
+ \ & u^t(t)B^t Px(t) + x^t(t)PBu(t) - 2x^t(t)PA_d \int_{-\tau}^0 Bu(t + \theta)d\theta \\
- \ & 2x^t(t)PA_d \int_{-\tau}^0 Ax(t + \theta)d\theta - 2x^t(t)PA_d \int_{-\tau}^0 A_d x(t + \theta)d\theta \\
+ \ & \tau r_1 x^t(t)A^t Ax(t) + \tau r_2 x^t(t)A_d^t A_d x(t) + \tau r_3 u^t(t)B^t Bu(t) \\
- \ & \int_{-\tau}^0 r_1[x^t(t + \theta)A^t Ax(t + \theta)]d\theta \\
- \ & \int_{-\tau}^0 r_2[x^t(t - \tau + \theta)A_d^t A_d x(t - \tau + \theta)]d\theta \\
- \ & \int_{-\tau}^0 r_3[u^t(t + \theta)B^t Bu(t + \theta)]d\theta \qquad (6.20)
\end{aligned}
$$

By B.1.1, we have

$$
\begin{aligned}
-2x^t(t)PA_d \int_{-\tau}^0 Ax(t + \theta)d\theta \ & \leq \ r_1^{-1} \int_{-\tau}^0 [x^t(t)PA_d A_d^t Px(t)]d\theta \\
+r_1 \int_{-\tau}^0 [x^t(t + \theta)A^t Ax(t + \theta)]d\theta & \\
= \ \tau r_1^{-1} x^t(t)PA_d A_d^t Px(t) & \\
+ r_1 \int_{-\tau}^0 [x^t(t + \theta)A^t Ax(t + \theta)]d\theta & \qquad (6.21)
\end{aligned}
$$

$$
\begin{aligned}
-2x^t(t)PA_d \int_{-\tau}^0 A_d x(t - \tau + \theta)d\theta \ & \leq \ \tau r_2^{-1} x^t(t)PA_d A_d^t Px(t) \\
+ r_2 \int_{-\tau}^0 [x^t(t - \tau + \theta)A_d^t A_d x(t - \tau + \theta)]d\theta & \qquad (6.22)
\end{aligned}
$$

$$-2x^t(t)PA_d \int_{-\tau}^0 Bu(t+\theta)d\theta \leq \tau r_3^{-1} x^t(t)PA_d A_d^t Px(t)$$

$$+ r_3 \int_{-\tau}^0 [u^t(t+\theta)B^t Bu(t+\theta)]x d\theta \tag{6.23}$$

Using inequalities (6.21)-(6.23) into (6.20), there holds

$$\begin{aligned}
\dot{V}_d(x_t) &= x^t(t)[P(A+A_d)+(A+A_d)^t P+\tau r_1 A^t A+\tau r_2 A_d^t A_d \\
&+ \tau r_1^{-1} PA_d A_d^t P+\tau r_2^{-1} PA_d A_d^t P+\tau r_3^{-1} PA_d A_d^t P]x(t) \\
&+ u^t(t)B^t Px(t)+x^t(t)PBu(t)+\tau r_3 u^t(t)B^t Bu(t)
\end{aligned} \tag{6.24}$$

Finally, the Hamiltonian $H(x,t)$ can be written as:

$$\begin{aligned}
H(x,t) &= -x^t(t)[P(A+A_d)+(A+A_d)^t P+\tau r_1 A^t A \\
&+ \tau r_2 A_d^t A_d+\tau r_1^{-1} PA_d A_d^t P \\
&+ \tau r_2^{-1} PA_d A_d^t P+\tau r_3^{-1} PA_d A_d^t P]x(t) \\
&- \tau r_3 u^t(t)B^t Bu(t)+u^t(t)(D+D^t)u(t) \\
&+ x^t(t)[C^t-PB]u(t)+u^t(t)[C-B^t P]x(t) \\
&= -Y^t(t)\ \Pi(P)\ Y(t)
\end{aligned} \tag{6.25}$$

where

$$\begin{aligned}
Y(t) &= [x^t(t)\ u^t(t)]^t \\
\Pi(P) &= \begin{bmatrix} S(P) & (C^t-PB) \\ (C-B^t P) & -(D+D^t-\tau r_3 B^t B) \end{bmatrix} < 0 \\
S(P) &= P(A+A_d)+(A+A_d)^t P+\tau(r_1 A^t A+r_2 A_d^t A_d) \\
&+ \tau(r_1^{-1}+r_2^{-1}+r_3^{-1})PA_d A_d^t P
\end{aligned} \tag{6.26}$$

If $\Pi(P) < 0$, then $-\dot{V}_d + 2y^t(t)u(t) > 0$ and from which it follows that

$$\int_{t_o}^{t_1} [y^t(t)\ u(t)]\ dt\ >\ 1/2\ [V_d(x_1)\ -\ V_d(x_o)] \tag{6.27}$$

Since $V_d(x) > 0$ for $x \neq 0$ and $V_d(x) = 0$ for $x = 0$, it follows that as $t_1 \to \infty$ that system (Σ_d) is extended strictly passive. By **A.1** it is easy to verify that $\Pi(P) < 0$ is equivalent to:

$$\begin{aligned}
&P(A+A_d)+(A+A_d)^t P+\tau(r_1 A^t A+r_2 A_d^t A_d) \\
&+\tau(r_1^{-1}+r_2^{-1}+r_3^{-1})PA_d A_d^t P \\
&+(C^t-PB)(D+D^t-\tau r_3 B^t B)^{-1}(C-B^t P)\ <\ 0
\end{aligned} \tag{6.28}$$

Setting $r_1 = \varepsilon^{-1}, r_2 = \alpha^{-1}, r_3 = \sigma^{-1}$, premultiplying (6.28) by P^{-1}, post-multiplying the result by P^{-1} and letting $X = P^{-1}$, it shows that (6.18) implies (6.28) for any $0 \leq \tau \leq \tau^*$. By **A.1**, (6.17) is equivalent to (6.18).

Remark 6.3: It is important to note that the result of **Theorem 6.2** reduces to **Theorem 2.2** when only the stability of the system is concerned. This can be observed by setting $B = 0$ and $C = 0$ in (6.18).

6.2.4 μ-Parameterization

Now we proceed to examine the application of the passivity concept to system (Σ_Δ). First, motivated by the results of **Theorem 6.1** for stability independent of delay measure, we pose the following definition:

Definition 6.1: *System (Σ_Δ) is said to be strongly robustly stable with strict passivity (SP) if there exists a matrix $0 < P = P^t \in \Re^{n \times n}$ such that for all admissible uncertainties:*

$$\begin{bmatrix} PA_\Delta + A_\Delta^t P + Q & PA_d & (C_\Delta^t - PB) \\ A_d^t P & -Q & 0 \\ (C_\Delta - B^t P) & 0 & -(D + D^t) \end{bmatrix} < 0 \qquad (6.29)$$

Remark 6.4: It is readily evident from **Definition 6.1** that the concept of strong robust stability with SP implies both the robust stability and the SP for system (Σ_Δ). Note that the robust stability with SP is an extension of robust stability independent of delay (RSID) for uncertain time-delay system to deal with the extended strict passivity problem.

Now it is easy to realize that direct application of (6.29) would require tremendous efforts over all admissible uncertainties. To bypass this shortcoming, we introduce the following $\mu-$ parameterized linear time-invariant system:

$$(\Sigma_\mu): \quad \begin{aligned} \dot{x}(t) &= Ax(t) + B_\mu \breve{w}(t) + A_d x(t - \tau) \\ \breve{z}(t) &= C_\mu x(t) + D_\mu \breve{w}(t) \end{aligned} \qquad (6.30)$$

where

$$B_\mu = [B \quad 0 \quad - \mu H] \quad C_\mu = \begin{bmatrix} C \\ \mu^{-1} E \\ 0 \end{bmatrix} \qquad (6.31)$$

$$D_\mu = \begin{bmatrix} D & 0 & -\mu\,H_c \\ 0 & 1/2\,I & 0 \\ 0 & 0 & 1/2\,I \end{bmatrix} \tag{6.32}$$

The next theorem shows that the robust SP of system (Σ_Δ) can be ascertained from the strong stability with SP of (Σ_μ).

Theorem 6.3: *System (Σ_Δ) satisfying (6.2) is strongly robustly stable with SP if and only if there exists $\mu > 0$ such that (Σ_μ) is strongly stable with SP.*

Proof: By **Theorem 6.1**, system (Σ_μ) is strongly stable with SP if there exist matrices $0 < P = P^t \in \Re^{n\times n}$ and $0 < W = W^t \in \Re^{n\times n}$ such that

$$\begin{bmatrix} PA + A^tP + W & PA_d & (C_\mu^t - PB_\mu) \\ A_d^tP & -W & 0 \\ (C_\mu - B_\mu^tP) & 0 & -(D_\mu + D_\mu^t) \end{bmatrix} < 0 \tag{6.33}$$

Using (6.31)-(6.32), inequality (6.33) is equivalent to:

$$W = \begin{bmatrix} PA + A^tP + Q & PA_d & (C_o^t - PB_o) & \mu^{-1}E^t & \mu PH \\ A_d^tP & -Q & 0 & 0 & 0 \\ (C - B^tP) & 0 & -(D + D^t) & 0 & \mu H_c \\ \mu^{-1}E & 0 & 0 & -I & 0 \\ \mu H^tP & 0 & \mu H_c^t & 0 & -I \end{bmatrix} < 0 \tag{6.34}$$

By **A.1**, inequality (6.34) holds if and only if

$$\Omega_1 + \Omega_3^t\,\Omega_2^{-1}\,\Omega_3 < 0 \tag{6.35}$$

with

$$\Omega_1 = \begin{bmatrix} PA + A^tP + Q & PA_d & -(C^t - PB) \\ A_d^tP & -Q & 0 \\ -(C - B^tP) & 0 & -(D + D^t) \end{bmatrix}$$

$$\Omega_2 = \begin{bmatrix} -I & 0 \\ 0 & -I \end{bmatrix}, \quad \Omega_3^t = \begin{bmatrix} \mu^{-1}E^t & \mu PH \\ 0 & 0 \\ 0 & \mu H_c \end{bmatrix} \tag{6.36}$$

By **B.1.4**, inequality (6.35) holds if and only if

$$\Omega_1 + \begin{bmatrix} PH \\ 0 \\ H_c \end{bmatrix} \Delta(t)[E\ \ 0\ \ 0] + \begin{bmatrix} E^t \\ 0 \\ 0 \end{bmatrix} \Delta^t(t)[H^tP\ \ 0\ \ H_c^t] < 0$$

$$\forall\Delta : \Delta^t\,\Delta \leq I \tag{6.37}$$

Simple rearrangment of (6.37) using (6.2) produces

$$\begin{bmatrix} PA_\Delta + A_\Delta^t P + Q & PA_d & -(C_\Delta^t - PB) \\ A_d^t P & -Q & 0 \\ -(C_\Delta - B^t P) & 0 & -(D + D^t) \end{bmatrix} \quad < \quad 0 \qquad (6.38)$$

which in view of **Definition 6.1** implies that system (Σ_Δ) is strongly stable with SP.

Remark 6.5: **Theorem 6.3** establishes an uncertainty-independent procedure to evaluate if the uncertain time-delay system (Σ_Δ) is robustly stable with SP. Observe that inequality (6.34) with $\varepsilon = \mu^2$ is equivalent to:

$$W_t = \begin{bmatrix} G(P) & L(P) \\ L^t(P) & U(\varepsilon) \end{bmatrix} \quad < \quad 0$$

where

$$G(P) = \begin{bmatrix} PA + A^t P & PA_d & C^t - PB \\ A_d^t P & -Q & 0 \\ (C - PB^t) & 0 & -(D + D^t) \end{bmatrix}$$

$$L(P) = \begin{bmatrix} E^t & \varepsilon H \\ 0 & 0 \\ 0 & \varepsilon H_c \end{bmatrix}, \quad U(\varepsilon) = \begin{bmatrix} -\varepsilon I & 0 \\ 0 & -\varepsilon I \end{bmatrix}$$

Obviously, $W_t < 0$ is linear in P and ε which can be solved by employing the LMI Toolbox.

Had we adopted the delay-dependent stability measure, we would then follow the analysis in [9] and introduce the following definition:

Definition 6.2: *System (Σ_Δ) is said to be strongly robustly stable with SP for any τ satisfying $0 \le \tau \le \tau^*$ if there exist a matrix $0 < X = X^t \in \Re^{n \times n}$ and scalars $\varepsilon > 0, \alpha > 0, \sigma > 0$ satisfying the LMI*

$$\begin{bmatrix} \begin{matrix} (A_\Delta + A_d)X \\ +X(A_\Delta + A_d)^t \\ +\tau(\varepsilon + \alpha + \sigma)A_d A_d^t \end{matrix} & \tau^* X A_d^t & \tau^* X A_\Delta^t & (X C_\Delta^t - B) \\ \tau^* A_d X & -(\tau^* \alpha)I & 0 & 0 \\ \tau^* A_\Delta X & 0 & -(\tau^* \varepsilon)I & 0 \\ (C_\Delta X - B^t) & 0 & 0 & \begin{matrix} -(D + D^t) \\ +\tau^* \sigma^{-1} B^t B \end{matrix} \end{bmatrix} < 0$$

$$(6.39)$$

It should be observed that by setting $\Delta(t) \equiv 0$, Definition 6.2 reduces to (6.17). The next result is a delay-dependent counterpart of **Theorem 6.3**.

Theorem 6.4: *System* (Σ_Δ) *satisfying (6.2) is said to be strongly robustly stable with SP for any* τ *satisfying* $0 \le \tau \le \tau^*$ *if there exists a* $\mu > 0$ *satisfying the inequality:*

$$\begin{bmatrix} \Xi_1 & \Xi_3^t \\ \Xi_3 & \Xi_2 \end{bmatrix} \quad < \quad 0 \qquad\qquad (6.40)$$

where

$$\Xi_1 = \begin{bmatrix} \begin{matrix}(A+A_d)X \\ +X(A+A_d)^t \\ +\tau(\varepsilon+\alpha+\sigma)A_d A_d^t\end{matrix} & \tau^* X A_d^t & \tau^* X A^t & (XC^t - B) \\ \tau^* A_d X & -(\tau^*\alpha)I & 0 & 0 \\ \tau^* A X & 0 & -(\tau^*\varepsilon)I & 0 \\ (CX - B^t) & 0 & 0 & \begin{matrix}-(D+D^t)\\ +\tau^*\sigma^{-1}B^t B\end{matrix} \end{bmatrix}$$

$$\Xi_3^t = \begin{bmatrix} H & XE^t \\ 0 & 0 \\ 0 & 0 \\ H_c & 0 \\ \tau^* H & 0 \end{bmatrix}, \quad \Xi_2 = \begin{bmatrix} -I & 0 \\ 0 & -I \end{bmatrix} \qquad (6.41)$$

Proof: Note that (6.40) together with B.1.1 implies

$$\Xi_1 + \begin{bmatrix} H \\ 0 \\ \tau^* H \\ H_c \end{bmatrix} \Delta(t)[EX \ \ 0 \ \ 0 \ \ 0] + \begin{bmatrix} XE^t \\ 0 \\ 0 \\ 0 \end{bmatrix} \Delta^t(t)[H_1^t \ \ 0 \ \ \tau^* H_1 \ \ H_c^t] < 0$$

$$\forall \Delta : \Delta^t \Delta \le I$$

That is, (6.39) holds. By **Definition 6.2**, the system (Σ_Δ) is strongly robustly stable with SP.

Note that **Theorem 6.4** is basically an LMI feasibility result.

6.2.5   Observer-Based Control Synthesis

The analysis of robust stability with SP can be naturally extended to the corresponding synthesis problem. That is, we are concerned with the design of a feedback controller that not only internally stabilizes the uncertain time-delay system but also achieves SP for all admissible uncertainties and unknown delays. A controller which achieves the property of robust stability with SP is termed as a robust SP controller. To this end, we consider the class of uncertain systems of the form:

$$
\begin{aligned}
(\Sigma_{co}) \quad \dot{x}(t) &= A_\Delta x(t) + Bw(t) + B_{1\Delta} u(t) + A_d x(t - \tau) \\
z(t) &= C_\Delta x(t) + Dw(t) + D_{12\Delta} u(t) \\
y(t) &= C_{1\Delta} x(t) + D_{21} w(t) + D_{22\Delta} u(t)
\end{aligned}
\tag{6.42}
$$

where $y \in \Re^n$ is the measured output and $u \in \Re^n$ is the control input. The uncertain matrices are given by:

$$
\begin{bmatrix}
A_\Delta & B_{1\Delta} \\
C_\Delta & D_{12\Delta} \\
C_{1\Delta} & D_{22\Delta}
\end{bmatrix}
=
\begin{bmatrix}
A & B_1 \\
C & D_{12} \\
C_1 & D_{22}
\end{bmatrix}
+
\begin{bmatrix}
H \\
H_c \\
H_e
\end{bmatrix}
\Delta(t)
\begin{bmatrix}
E & E_e
\end{bmatrix}
\tag{6.43}
$$
$$
\Delta^t(t)\Delta(t) \leq I \quad \forall t
$$

In the sequel, we focus attention on the controller synthesis for system (Σ_{co}) by using an observer-based controller of the form

$$
\begin{aligned}
(\Sigma_{ob}): \quad \dot{\eta}(t) &= G_o\, \eta(t) + L_o\, [\, y(t) - C_1 \eta(t)] \\
u(t) &= K_o\, \eta(t)
\end{aligned}
\tag{6.44}
$$

where $(G_o,\ L_o,\ K_o)$ are constant matrices to be selected. Define the augmented state-vector by:

$$
\xi(t) = \begin{bmatrix} x(t) \\ \eta(t) \end{bmatrix}
\tag{6.45}
$$

Applying the observer-based controller (6.43) to system (Σ_{co}), we obtain the closed-loop system:

$$
\begin{aligned}
(\Sigma_{cc}): \quad \dot{\xi}(t) &= \widehat{A}_\Delta\, \xi(t) + \tilde{B}w(t) + \widehat{E}\, \xi(t - \tau) \\
z(t) &= \widehat{C}_\Delta\, \xi(t) + Dw(t)
\end{aligned}
\tag{6.46}
$$

where

$$\widehat{A}_\Delta = \begin{bmatrix} A_\Delta & B_{1\Delta}K_o \\ L_oC_{1\Delta} & G_o - L_oC_1 + L_oD_{22\Delta}K_o \end{bmatrix}$$

$$= \begin{bmatrix} A & B_1K_o \\ L_oC_1 & G_o - L_oC_1 + L_oD_{22}K_o \end{bmatrix} + \tilde{H}\,\Delta(t)\,\tilde{E}$$

$$= \widehat{A} + \tilde{H}\,\Delta(t)\,\tilde{E} \tag{6.47}$$

$$\widehat{B} = \begin{bmatrix} B \\ L_oD_{21} \end{bmatrix}, \quad \widehat{E} = \begin{bmatrix} A_d & 0 \\ 0 & 0 \end{bmatrix} \tag{6.48}$$

$$\widehat{C}_\Delta = [C_\Delta \quad D_{12\Delta}K_o] = \tilde{C} + H_2\Delta(t)\,\tilde{E} \tag{6.49}$$

and

$$\tilde{H} = \begin{bmatrix} H \\ L_oH_e \end{bmatrix}, \quad \tilde{E} = [E \quad E_eK_o], \quad \tilde{C} = [C_o \quad D_{12}K_o] \tag{6.50}$$

On the other hand, we introduce the following μ-parameterized linear time-invariant system:

$$(\Sigma_{\mu\mu}) \quad \dot{x}(t) = Ax(t) + B_\mu\breve{w}(t) + B_1u(t) + A_dx(t-\tau)$$
$$\breve{z}(t) = C_\mu x(t) + D_\mu\breve{w}(t) + D_{1\mu}u(t)$$
$$y(t) = C_1x(t) + D_{2\mu}\breve{w}(t) + D_{22}u(t) \tag{6.51}$$

where

$$D_{1\mu} = \begin{bmatrix} D_{12} \\ \mu^{-1}E_e \\ 0 \end{bmatrix}, \quad D_{2\mu} = [D_{21} \quad 0 \quad -\mu H_e] \tag{6.52}$$

and B_μ, C_μ, D_μ are given by (6.31)-(6.32). Now by combining systems $(\Sigma_{\mu\mu})$ and (Σ_{ob}), we obtain the closed-loop μ-parameterized system $(\Sigma_{c\mu})$:

$$(\Sigma_{c\mu}): \quad \dot{\xi}(t) = \widehat{A}\xi(t) + \widehat{B}\breve{w}(t) + \widehat{E}\,\xi(t-\tau)$$
$$z(t) = \widehat{C}\xi(t) + D_\mu\breve{w}(t) \tag{6.53}$$

where

$$\widehat{B} = \begin{bmatrix} B_\mu \\ L_oD_{2\mu} \end{bmatrix}, \quad \widehat{C} = [C_\mu \quad D_{1\mu}K_o] \tag{6.54}$$

The next theorem provides an interconnection between the observer-based passive real control of system (Σ_{cc}) and the passive control of system ($\Sigma_{c\mu}$).

Theorem 6.5: *Consider the uncertain time-delay system (Σ_Δ) satisfying (6.2). Then an observer-based controller of the form (6.44) achieves strong robust stability with SP for system (Σ_Δ) for all admissible uncertainties if and only if for some $\mu > 0$ this observer-based controller achieves strong stability with SP for system ($\Sigma_{c\mu}$).*

Proof: By **Theorem 6.1**, system ($\Sigma_{c\mu}$) is strongly stable with SP if there exist matrices $0 < X = X^t$ and $0 < \hat{Q} = \hat{Q}^t$ such that

$$
\begin{bmatrix} X\hat{A} + \hat{A}^t X + \hat{Q} & X\,\widehat{E} & (\hat{C}^t - X\hat{B}) \\ \widehat{E}^t X & -\hat{Q} & 0 \\ (\hat{C} - \hat{B}^t X) & 0 & -(D_\mu + D_\mu^t) \end{bmatrix} \quad < \quad 0 \qquad (6.55)
$$

Expansion of (6.56) using (6.31)-(6.32) and (6.55) yields:

$$
\begin{bmatrix} X\hat{A} + \hat{A}^t X + \hat{Q} & X\,\widehat{E} & (\tilde{C}^t - X\tilde{B}) & \mu^{-1}\tilde{E}^t & \mu X\tilde{H} \\ \widehat{E}^t X & -\hat{Q} & 0 & 0 & 0 \\ (\tilde{C} - \tilde{B}^t X) & 0 & -(D + D^t) & 0 & \mu H_c \\ \mu^{-1}\tilde{E} & 0 & 0 & -I & 0 \\ \mu \tilde{H}^t X & 0 & \mu H_c^t & 0 & -I \end{bmatrix} \quad < \quad 0 \ (6.56)
$$

Applying **A.1**, inequality (6.57) holds if and only if

$$
\Omega_1 + \Omega_3^t \Omega_2^{-1} \Omega_3 < 0 \qquad (6.57)
$$

with

$$
\Omega_1 \;=\; \begin{bmatrix} X\hat{A} + \hat{A}^t X + \hat{Q} & X\,\widehat{E} & (\tilde{C}^t - X\tilde{B}) \\ \widehat{E}^t X & -\hat{Q} & 0 \\ (\tilde{C} - \tilde{B}^t X) & 0 & -(D + D^t) \end{bmatrix}
$$

$$
\Omega_3^t \;=\; \begin{bmatrix} \mu^{-1}\tilde{E}^t & \mu X\tilde{H} \\ 0 & 0 \\ 0 & \mu H_c \end{bmatrix}, \; \Omega_2 = \begin{bmatrix} -I & 0 \\ 0 & -I \end{bmatrix} \qquad (6.58)
$$

Then by **B.1.1**, inequality (6.58) along with (6.59) holds if and only if

$$
\Omega_1 + \begin{bmatrix} X\tilde{H} \\ 0 \\ H_c \end{bmatrix} \Delta(t) [\tilde{E} \ \ 0 \ \ 0] + \begin{bmatrix} \tilde{E}^t \\ 0 \\ 0 \end{bmatrix} \Delta^t(t) \, [\tilde{H}^t X \ \ 0 \ \ H_c^t] < 0
$$

$$\forall \Delta(t) \leq I$$

or equivalently

$$
\begin{bmatrix}
X\widehat{A} + \widehat{A}^t X + \widehat{Q} + & X\widehat{E} & (\tilde{C}^t - X\tilde{B}) - \\
X\tilde{H}\Delta(t)\tilde{E} + \tilde{E}^t\Delta^t(t)\tilde{H}^t X & & +\tilde{E}^t\Delta^t(t)H_c^t \\
\widehat{E}^t X & -\widehat{Q} & 0 \\
(\tilde{C} - \tilde{B}^t X) + & 0 & -(D + D^t) \\
H_c\Delta(t)\tilde{E} & &
\end{bmatrix}
< 0 \quad (6.59)
$$

for all admissible uncertainties. The substitution of (6.48)-(6.50) into (6.61) produces:

$$
\begin{bmatrix}
X\widehat{A}_\Delta + \widehat{A}_\Delta^t X + \widehat{Q} & X\widehat{E} & (\widehat{C}_\Delta^t - X\tilde{B}) \\
\widehat{E}^t X & -\widehat{Q} & 0 \\
(\widehat{C}_\Delta - \tilde{B}^t X) & 0 & -(D + D^t)
\end{bmatrix}
< 0 \quad (6.60)
$$

which in view of **Theorem 6.3** implies that system (Σ_{cc}) is strongly robustly stable with SP.

Remark 6.6: In general, the gain matrices of the observer-based controller (6.43) can be determined by appropriately modifying the results of Theorem 4.1 in [202] to include the additional quadratic terms due to the state-delay.

6.3 Discrete-Time Systems

In this section, we continue our task of looking into the passivity analysis and control synthesis problems of time-delay systems by considering a class of discrete-time systems with unknown state-delay. The development here will be parallel to that of Section 2. As usual, we start by focusing on delay-independent stability and then move to delay-dependent stability.

6.3.1 A Class of Discrete-Delay Systems

The class of linear time-invariant state-delay systems under consideration is represented by:

$$
(\Sigma_\Delta): \quad
\begin{aligned}
x(k+1) &= A_\Delta x(k) + Bw(k) + A_d x(k - \tau) \\
z(k) &= C_\Delta x(k) + Dw(k)
\end{aligned}
\quad (6.61)
$$

where $x \in \Re^n$ is the state, $w \in \Re^p$ is the input signal, $z \in \Re^p$ is the output signal, τ is an unknown time delay within a known interval $0 \leq \tau^*$ and $B \in \Re^{n \times p}$, $D \in \Re^{p \times p}$ and $A_d \in \Re^{n \times n}$ are real constant matrices. The uncertain matrices $A_\Delta \in \Re^{n \times n}$ and $C_\Delta \in \Re^{p \times n}$ are given by:

$$\begin{bmatrix} A_\Delta \\ C_\Delta \end{bmatrix} = \begin{bmatrix} A \\ C \end{bmatrix} + \begin{bmatrix} H \\ H_c \end{bmatrix} \Delta(k) E \tag{6.62}$$

where $H \in \Re^{n \times \alpha}$, $H_c \in \Re^{p \times \alpha}$ and $M \in \Re^{\beta \times n}$ are known matrices and $\Delta(k) \in \Re^{\alpha \times \beta}$ is a matrix of uncertain parameters with

$$\Delta^t(k) \Delta(k) \leq I, \quad \forall k \tag{6.63}$$

Distinct from system (Σ_Δ) is the free nominal system:

$$(\Sigma_f): \quad \begin{aligned} x(k+1) &= Ax(k) + A_d x(k - \tau) \\ z(k) &= Cx(k) \end{aligned} \tag{6.64}$$

and the nominal system

$$(\Sigma_o): \quad \begin{aligned} x(k+1) &= Ax(k) + Bw(k) + A_d x(k - \tau) \\ z(k) &= Cx(k) + Dw(k) \end{aligned} \tag{6.65}$$

In the sequel, we examine the problem of passivity analysis of system (Σ_Δ) in relation to systems (Σ_f) and (Σ_o). In particular, it is required that system (Σ_f) to be asymptotically stable, that is, all the eigenvalues of matrix A are within the unit circle.

As a starting point, we recall from Chapter 2 that subject to **Assumption 2.3**, system (Σ_o) is asymptotically stable independent-of-delay if there exist matrices $0 < P = P^t \in \Re^{n \times n}$ and $0 < Q = Q^t \in \Re^{n \times n}$ satisfying the LMI

$$\begin{bmatrix} A^t PA - P + Q & A^t PA_d \\ A_d^t PA & -Q + A_d^t PA_d \end{bmatrix} < 0$$

or equivalently satisfying the ARI

$$A^t PA - P + A^t PA_d(Q - A_d^t PA_d)^{-1} A_d^t PA + Q < 0$$

Recall also that the transfer function of system (Σ_o) is given by:

$$T_o(z) = C[zI - A - z^{-\tau} A_d]^{-1} B + D$$

and note that $T_o(e^{i\theta}) + T_o^*(e^{i\theta}) > 0 \quad \forall \theta \in [0, 2\pi]$

6.3.2   Conditions of Passivity: Delay-Independent Stability

In line with the continuous-time systems, we provide the following definition:

Definition 6.3: *The dynamical system (Σ_o) is called passive if*

$$\sum_{j=0}^{\infty} w^t(j)z(j) \; > \; \beta \qquad \forall w \in \ell_2[0,\infty) \qquad (6.66)$$

where $w(k)$ is the input signal, $z(k)$ is the output signal and β is some constant depending on the initial condition of the system. It is said to be strictly passive (SP) if it is passive and $D + D^t > 0$.

Based on the foregoing facts, we now proceed to develop conditions under which systems with time-varying parameter uncertainty and unknown state-delay like (Σ_Δ), which is asymptotically stable independent of delay, can be guaranteed to be strictly passive (SP). Towards our goal, we associate with system (Σ_o) the Hamiltonian:

$$H(k) \; = \; - \,\Delta V(x_k) + 2z^t(k)w(k) \qquad (6.67)$$

which depends on the input signal $w(k)$ and the output signal $z(k)$ with $V(k)$ being the Lyapunov functional for the nominal system (Σ_o) identical to $V_3(x_k)$ as given by (2.23). The following result is now established.

Theorem 6.6: *System (Σ_o) satisfying Assumption 2.3 is asymptotically stable with SP independent-of-delay if there exist matrices $0 < P = P^t \in \Re^{n \times n}$ and $0 < Q = Q^t \in \Re^{n \times n}$ satisfying the LMI*

$$\begin{bmatrix} A^tPA - P + Q & A^tPA_d & (A^tPB - C^t) \\ A_d^tPA & -(Q - A_d^tPA_d) & A_d^tPB \\ (B^tPA - C) & B^tPA_d & -(D + D^t - B^tPB) \end{bmatrix} < 0 \quad (6.68)$$

or equivalently there exist matrices $0 < P = P^t \in \Re^{n \times n}$ and $0 < Q = Q^t \in \Re^{n \times n}$ satisfying the ARI

$$A^tPA - P + (A^tPB - C^t)\left\{D + D^t - B^tPB\right\}^{-1}(B^tPA - C) +$$
$$\left\{A^tPE_d + (A^tPB - C^t)[D + D^t - B^tPB]^{-1}B^tPA_d\right\}$$
$$\left\{Q - A_d^tPA_d - A_d^tPB[D + D^t - B^tPB]^{-1}B^tPA_d\right\}^{-1}$$
$$\left\{A_d^tPA + A_d^tPB[D + D^t - B^tPB]^{-1}(B^tPA - C)\right\} + Q < 0 \quad (6.69)$$

In this situation, system (Σ_o) is said to be *strongly stable with SP*. Obviously, *strong stability with SP* implies that system (Σ_o) is *asymptotically stable with SP*.

Proof: By evaluating $\Delta V_3(x_k)$ along the solutions of (6.65) and using (6.69), we get

$$
\begin{aligned}
H(k) \;=\; & -x^t(k+1)Px(k+1) - x^t(k)Px(k) + x^t(k)Qx(k) \\
& - x^t(k-\tau)Qx(k-\tau) + 2x^t(k)C^t w(k) + w^t(D^t + D)w(k) \\
\;=\; & -x^t(k)\{A^t PA - P + Q\}x(k) - w^t(k)\{B^t PB - (D^t + D)\}w(k) \\
& - x^t(k-\tau)\{Q - A_d^t PA_d\}x(k-\tau) - 2x^t(k)A^t PA_d x(k-\tau) \\
& - 2x^t(k)(A^t PB - C^t)w(k) - 2w^t(k)B^t PA_d x(k-\tau) \\
\;=\; & -\xi^t(k)\,\Omega(P)\,\xi(k) \qquad\qquad\qquad\qquad\qquad (6.70)
\end{aligned}
$$

$$
\xi(k) \;=\; [x^t(k) \;\; x^t(k-\tau) \;\; w^t(k)]^t
$$

$$
\Omega(k) \;=\; \begin{bmatrix}
A^t PA - P + Q & A^t PA_d & (A^t PB - C^t) \\
A_d^t PA & -(Q - A_d^t PA_d) & A_d^t PB \\
(B^t PA - C) & B^t PA_d & -(D + D^t - B^t PB)
\end{bmatrix} \tag{6.71}
$$

If $\Omega(P) < 0$, then $-\Delta V_3(x_k) + 2z^t(k)w(k) > 0$ and from which it follows that

$$
\begin{aligned}
\sum_{j=k_o}^{k_f} [z^t(j)w(j)] \;>\; & 1/2 \sum_{j=k_o}^{k_f} [\Delta V_3(x_k)] \\
\;=\; & 1/2\{V_1(x_o) - V_1(x_f)\} \tag{6.72}
\end{aligned}
$$

Since $V_1(k) > 0$ for $x \neq 0$ and $V_1(k) = 0$ for $x = 0$, it follows as $k_f \to \infty$ that system (Σ_o) is globally asymptotically stable with SP.

Remark 6.7: One can standardize the LMI (6.68) by applying **A.1** to yield:

$$
\begin{bmatrix}
-P + Q & 0 & -C^t & A^t P \\
0 & -Q & 0 & A_d^t P \\
-C & 0 & -(D + D^t) & B^t P \\
PA & PA_d & PB & -P
\end{bmatrix} \;<\; 0 \tag{6.73}
$$

which can be conveniently solved by the **LMI-Toolbox**.

Motivated by the foregoing result, we provide the following definition:

Definition 6.4: *The uncertain time-delay system* (Σ_Δ) *is said to be strongly robustly stable with SP if there exists a matrix* $0 < P = P^t \in \Re^{n \times n}$ *such that for all admissible uncertainties:*

$$
\left[
\begin{array}{ccc}
A_\Delta^t P A_\Delta - P + Q & A_\Delta^t P A_d & (A_\Delta^t P B - C_\Delta^t) \\
A_d^t P A_\Delta & -(Q - A_d^t P A_d) & A_d^t P B \\
(B^t P A_\Delta - C_\Delta) & B^t P A_d & -(D + D^t - B^t P B)
\end{array}
\right] < 0
$$

or equivalently there exist matrices $0 < P = P^t \in \Re^{n \times n}$ *and* $0 < Q = Q^t \in \Re^{n \times n}$ *satisfying the ARI*

$$
A_\Delta^t P A_\Delta - P + (A_\Delta^t P B - C_\Delta^t) \left\{ D + D^t - B^t P B \right\}^{-1} (B^t P A_\Delta - C_\Delta) +
$$
$$
\left\{ A_\Delta^t P A_d + (A_\Delta^t P B - C_\Delta^t)[D + D^t - B^t P B]^{-1} B^t P A_d \right\}
$$
$$
\left\{ Q - A_d^t P A_d - A_d^t P B[D + D^t - B^t P B]^{-1} B^t P A_d \right\}^{-1}
$$
$$
\left\{ A_d^t P A_\Delta + A_d^t P B[D + D^t - B^t P B]^{-1} (B^t P A_\Delta - C_\Delta) \right\} + Q < 0
$$

for all admissible uncertainties.

Remark 6.8: It is readily evident from **Definition 6.4** that the concept of strong robust stability with SP implies both the robust stability and the SP for system (Σ_Δ). Note that the robust stability with SP is an extension of robust stability (RS) for uncertain time-delay system to deal with the extended strict passivity problem. By setting $\Delta(t) \equiv 0$, **Definition 6.4** reduces to (6.68).

6.3.3   Conditions of Passivity: Delay-Dependent Stability

Here, we direct attention to the case of delay-dependent stability. By following parallel development to Section 6.2, we establish the following result.

Theorem 6.8: *Consider the time-delay system* (Σ_Δ) *satisfying* **Assumption 2.4.** *Then given a scalar* $\tau^* > 0$, *the system* (Σ_f) *is globally asymptotically stable for any constant time-delay* τ *satisfying* $0 \leq \tau \leq \tau^*$ *if one of the following two equivalent conditions holds:*

(1) There exist matrix $0 < P = P^t \in \Re^{n \times n}$ and scalars $\varepsilon_1 > 0$, $\varepsilon_2 > 0$, $\varepsilon_3 > 0$ $\varepsilon_4 > 0$ and $\varepsilon_5 > 0$, satisfying the LMI:

$$\Sigma = \begin{bmatrix} \Sigma_1 & \Sigma_2 \\ \Sigma_2^t & \Sigma_3 \end{bmatrix} < 0 \qquad (6.74)$$

where

$$\Sigma_1 = \begin{bmatrix} (A + A_d)^t P(A + A_d) - P & \tau^*(A + A_d)^t & \tau^*(A + A_d)^t \\ \tau^*(A + A_d) & -\tau^*(\varepsilon_1 P)^{-1} & 0 \\ \tau^*(A + A_d) & 0 & -\tau^*(\varepsilon_2 P)^{-1} \end{bmatrix}$$

$$\Sigma_2 = \begin{bmatrix} \tau^*(A^t A_d^t) & \tau^*(A^t A_d^t) & (A + A_d)^t P A_d - C^t \\ 0 & 0 & 0 \\ 0 & 0 & 0 \end{bmatrix}$$

$$\Sigma_3 = \begin{bmatrix} -\tau^*(\varepsilon_3 P)^{-1} & 0 & 0 \\ 0 & -\tau^*(\varepsilon_4 P)^{-1} & 0 \\ 0 & 0 & \begin{array}{c} -(D^t + D) + B^t P B \\ + \tau \varepsilon_5 B^t A^t P A_d B \end{array} \end{bmatrix} \qquad (6.75)$$

(2) There exist a matrix $0 < P = P^t \in \Re^{n \times n}$ and $\varepsilon_1 > 0$, $\varepsilon_2 > 0$, $\varepsilon_3 > 0$, $\varepsilon_4 > 0$ and $\varepsilon_5 > 0$, satisfying the ARI

$$[(A + A_d)^t P(A + A_d) - P] + \tau \varepsilon_1 (A + A_d)^t P(A + A_d)$$
$$+ \tau \varepsilon_3 A^t A_d^t P A_d A + \tau \varepsilon_2 (A + A_d)^t P(A + A_d)$$
$$+ \tau \varepsilon_4 A^t A_d^t P A_d A + \tau \varepsilon_3 A^t A_d^t P A_d A$$
$$+ \tau \varepsilon_4 A^t A_d^t P A_d A - [(A + A_d)^t P A_d - C^t]$$
$$\left\{ (D^t + D) - B^t P B + \tau \varepsilon_5 B^t A_d^t P A_d B \right\}^{-1} [A_d^t P(A + A_d) - C] < 0$$

$$(6.76)$$

Proof: (1) Introduce a discrete-type Lyapunov-Krasovskii functional $V_c(x_k)$ of the form:

$$V_c(x_k) = x^t(k) P x(k) + \sum_{\theta=-\tau}^{0} \{ \rho_1 \sum_{j=k+\theta-1}^{k-2} \Delta x^t(j) A^t A_d^t P A_d A \Delta x(j) \}$$

$$+ \{ \sum_{\theta=-\tau}^{0} \rho_2 \sum_{j=k+\theta-\tau-1}^{k-2} \Delta x^t(j) A_d^t A_d^t P A_d A_d \Delta x(j) \}$$

$$+ + \sum_{\theta=-\tau}^{0} \{ \rho_3 \sum_{j=k+\theta-1}^{k-2} \Delta w^t(j) B^t A_d^t P A_d B \Delta w(j) \} \qquad (6.77)$$

where $0 < P = P^t \in \Re^{n \times n}$ and $\rho_1 > 0$, $\rho_2 > 0$, $\rho_3 > 0$ are weighting factors to be selected. We start from (6.65) and get

$$
\begin{aligned}
x(k - \tau) &= x(k) - \sum_{\theta=-\tau}^{0} \Delta x(k + \theta) \\
&= x(k) - \sum_{\theta=-\tau}^{0} A\Delta x(k + \theta - 1) - \sum_{\theta=-\tau}^{0} A_d \Delta x(k + \theta - \tau - 1) \\
&\quad - \sum_{\theta=-\tau}^{0} B\Delta w(k + \theta - 1)
\end{aligned}
\tag{6.78}
$$

Substituting (6.78) back into (6.65) we obtain:

$$
\begin{aligned}
x(k + 1) &= (A + A_d)x(k) - A_d \sum_{\theta=-\tau}^{0} A\Delta x(k + \theta - 1) \\
&\quad + A_d \sum_{\theta=-\tau}^{0} A_d \Delta x(k + \theta - \tau - 1) \\
&\quad + A_d \sum_{\theta=-\tau}^{0} B\Delta w(k + \theta - 1) + Bw(k) \\
&= (A + A_d)x(k) + Bw(k) - A_d\xi_1(k) - A_d\xi_2(k) - A_d\xi_3(k)
\end{aligned}
\tag{6.79}
$$

Now by evaluating the first-forward difference $\Delta V_c(x_k)$ along the solutions of (6.79) and arranging terms, we obtain:

$$
\Delta V_c(k) = \Delta V_{c1}(k) + \Delta V_{c2}(k) + \Delta V_{c3}(k) + \Delta V_{c4}(k)
\tag{6.80}
$$

where

$$
\begin{aligned}
\Delta V_{c1}(k) &= x^t(k)[(A + A_d)^t P(A + A_d) - P]x(k) \\
&\quad + \xi_1^t(k)A_d^t P A_d \xi_1(k) + \xi_2^t(k)A_d^t P A_d \xi_2(k) \\
&\quad + w^t(k)[B^t P B]w(k) + \xi_3^t(k)A_d^t P A_d \xi_3(k) \tag{6.81} \\
\Delta V_{c2}(k) &= -2x^t(k)(A + A_d)^t P A_d \xi_1(k) - 2x^t(k)(A + A_d)^t P A_d \xi_2(k) \\
&\quad - 2x^t(k)(A + A_d)^t P A_d \xi_3(k) - 2\xi_1^t(k)A_d^t P A_d \xi_2(k) \\
&\quad - 2\xi_1^t(k)A_d^t P A_d \xi_3(k) - 2\xi_2^t(k)A_d^t P A_d \xi_3(k) \tag{6.82} \\
\Delta V_{c3}(k) &= \tau \rho_1 \Delta x^t(k - 1)[A^t A_d^t P A_d A]\Delta x(k - 1)
\end{aligned}
$$

$$
\begin{aligned}
&+\ \tau\rho_2\Delta x^t(k-1)[A_d^t A_d^t P A_d A_d]\Delta x(k-1)\\
&+\ \tau\rho_3\Delta w^t(k-1)[B^t A_d^t P A_d B]\Delta w(k-1)\\
&+\ 2w^t(k)B_o^t P(A+A_d)x(k) \hspace{4cm} (6.83)
\end{aligned}
$$

$$
\begin{aligned}
\Delta V_{c4}(k)\ =\ &-2w^t(k)B^t P A_d\xi_1(k)-2w^t(k)B^t P A_d\xi_2(k)\\
&-\ 2w^t(k)B^t P A_d\xi_3(k)\\[4pt]
&-\ \rho_1\sum_{\theta=-\tau}^{0}[\Delta x^t(k+\theta-1)A^t A_d^t P A_d A\Delta x(k+\theta-1)]\\
&-\ \rho_2\sum_{\theta=-\tau}^{0}[\Delta x^t(k+\theta-\tau-1)A_d^t A_d^t P A_d A_d\Delta x(k+\theta-\tau-1)]\\
&-\ \rho_3\sum_{\theta=-\tau}^{0}[\Delta w^t(k+\theta-1)B^t A_d^t P A_d B\Delta w(k+\theta-1)] \hspace{0.8cm} (6.84)
\end{aligned}
$$

With the help of B.1.1, we obtain the following set of inequalities:

$$
\begin{aligned}
&-2x^t(k)(A+A_d)^t P A_d\xi_1(k)\ =\\
&-2x^t(k)(A+A_d)^t P A_d\sum_{\theta=-\tau}^{0}A\Delta x(k+\theta-1)\\[4pt]
&\leq\ r_1^{-1}\sum_{\theta=-\tau}^{0}[x^t(k)(A+A_d)^t P(A+A_d)x(k)]\\
&+r_1\sum_{\theta=-\tau}^{0}[\Delta x^t(k+\theta-1)A^t A_d^t P A_d A\Delta x(k+\theta-1)]\\
&=\ \tau r_1^{-1}x^t(k)(A+A_d)^t P(A+A_d)x(k)\\
&+\ r_1\sum_{\theta=-\tau}^{0}[\Delta x^t(k+\theta-1)A^t A_d^t P A_d A\Delta x(t+\theta-1)] \hspace{0.8cm} (6.85)
\end{aligned}
$$

$$
\begin{aligned}
&-2x^t(k)(A+A_d)^t P A_d\xi_2(k)\ =\\
&-2x^t(k)(A+A_d)^t P A_d\sum_{\theta=-\tau}^{0}A_d\Delta x(k+\theta-\tau-1)\\
&\leq\ \tau r_2^{-1}x^t(k)(A+A_d)^t P(A+A_d)x(k)\\
&+\ r_2\sum_{\theta=-\tau}^{0}[\Delta x^t(k+\theta-\tau-1)A_d^t A_d^t P A_d A_d\Delta x(t+\theta-\tau-1)]
\end{aligned}
$$

$$
(6.86)
$$

$$-2\xi_1^t(k)A_d^tPA_d\xi_2(k) =$$

$$-2\sum_{\theta=-\tau}^{0} A\triangle x^t(k+\theta-1)E_o^tPE_o \sum_{\theta=-\tau}^{0} A\triangle x(k+\theta-\tau-1)$$

$$\leq r_3^{-1} \sum_{\theta=-\tau}^{0} \triangle x^t(k+\theta-1)A^tA_d^tPA_dA\triangle x(k+\theta-1)$$

$$+r_3 \sum_{\theta=-\tau}^{0} [\triangle x^t(k+\theta-\tau-1)A_d^tA_d^tPA_dA_d\triangle x(t+\theta-\tau-1)]$$

$$(6.87)$$

$$-2x^t(k)(A+A_d)^tPA_d\xi_3(k) =$$

$$-2x^t(k)(A+A_d)^tPA_d \sum_{\theta=-\tau}^{0} B\triangle w(k+\theta-\tau-1)$$

$$\leq \tau r_4^{-1}x^t(k)(A+A_d)^tP(A+A_d)x(k)$$

$$+r_4 \sum_{\theta=-\tau}^{0} [\triangle w^t(k+\theta-\tau-1)B^tA_d^tPA_dB\triangle w(t+\theta-\tau-1)]$$

$$(6.88)$$

$$-2\xi_1^t(k)A_d^tPA_d\xi_3(k) =$$

$$-2\sum_{\theta=-\tau}^{0} A\triangle x^t(k+\theta-1)A_d^tPA_d \sum_{\theta=-\tau}^{0} A\triangle w(k+\theta-1)$$

$$\leq r_5^{-1} \sum_{\theta=-\tau}^{0} \triangle x^t(k+\theta-1)A^tA_d^tPA_dA\triangle x(k+\theta-1)$$

$$+r_5 \sum_{\theta=-\tau}^{0} [\triangle w^t(k+\theta-1)B^tA_d^tPA_dB\triangle w(t+\theta-1)] \qquad (6.89)$$

$$-2\xi_2^t(k)A_d^tPA_d\xi_3(k) =$$

$$-2\sum_{\theta=-\tau}^{0} A\triangle x^t(k+\theta-\tau-1)A_d^tPA_d \sum_{\theta=-\tau}^{0} A\triangle w(k+\theta-1)$$

$$\leq r_6^{-1} \sum_{\theta=-\tau}^{0} \triangle x^t(k+\theta-\tau-1)A_d^tA_d^tPA_dA_d\triangle x(k+\theta-\tau-1)$$

$$+ r_6 \sum_{\theta=-\tau}^{0} [\Delta w^t(k + \theta - 1)B^t A_d^t P A_d B \Delta w(t + \theta - 1)] \qquad (6.90)$$

$$-2w^t(k)B^t P A_d \xi_1(k) \leq \tau r_7^{-1} w^t(k)B^t P B w(k)$$
$$+ r_7 \sum_{\theta=-\tau}^{0} [\Delta x^t(k + \theta - 1)A^t A_d^t P A_d A \Delta x(t + \theta - 1) \qquad (6.91)$$

$$-2w^t(k)B^t P A_d \xi_2(k) \leq \tau r_8^{-1} w^t(k)B^t P B w(k)$$
$$+ r_8 \sum_{\theta=-\tau}^{0} [\Delta x^t(k + \theta - \tau - 1)A_d^t A_d^t P A_d A_d \Delta x(t + \theta - \tau - 1)]$$
$$(6.92)$$

$$-2w^t(k)B^t P A_d \xi_3(k) \leq \tau r_9^{-1} w^t(k)B^t P B w(k)$$
$$+ r_9 \sum_{\theta=-\tau}^{0} [\Delta x^t(k + \theta - 1)A^t A_d^t P A_d A \Delta x(t + \theta - 1)] \qquad (6.93)$$

Hence, it follows by substituting (6.85)-(6.93) into (6.80)-(6.84) and letting $\rho_1 = 1 + r_1 + r_3^{-1} + r_5^{-1} + r_7 + r_9$, $\rho_2 = 1 + r_2 + r_3 + r_6^{-1} + r_8$, and $\rho_3 = 1 + r_4 + r_5 + r_6$ that

$$\begin{aligned}
\Delta V_c(k) =\ & x^t(k)[(A + A_d)^t P(A + A_d) - P]x(k) \\
& + w^t(k)B^t P B w(k) + \tau r_9^{-1} w^t(k)B^t P B w(k) \\
& + \tau r_1^{-1} x^t(k)(A + A_d)^t P(A + A_d)x(k) \\
& + \tau r_2^{-1} x^t(k)(A + A_d)^t P(A + A_d)x(k) \\
& + \tau r_7^{-1} w^t(k)B^t P B w(k) + \tau r_8^{-1} w^t(k)B^t P B w(k) \\
& + \tau \rho_1 \Delta x^t(k - 1)[A^t A_d^t P A_d A]\Delta x(k - 1) \\
& + \tau \rho_2 \Delta x^t(k - 1)[A_d^t A_d^t P A_d A_d]\Delta x(k - 1) \\
& + \tau \rho_3 \Delta w^t(k - 1)[B^t A_d^t P A_d B]\Delta w(k - 1) \\
& + 2w^t(k)B^t P(A + A_d)x(k) \qquad (6.94)
\end{aligned}$$

On using (6.65),(6.67) with $\Delta V(x_k) = \Delta V_c(x_k)$ and (6.94) with some arrangement, we express the Hamiltonian into the form:

$$H(k) = -\xi^t(k)\, \Upsilon(P)\, \xi(k)$$

$$\Upsilon(P) = \begin{bmatrix} R_1(P) & -R_2(P) & R_3(P) & 0 \\ -R_2(P) & R_2(P) & 0 & 0 \\ R_3^t(P) & 0 & -[(D^t + D) - B^t PB] & -R_4(P) \\ 0 & 0 & -R_4(P) & R_4(P) \end{bmatrix}$$

$$= \begin{bmatrix} \Upsilon_1(P) & \Upsilon_2(P) \\ \Upsilon_2^t(P) & \Upsilon_3(P) \end{bmatrix} \tag{6.95}$$

$$\xi(k) = [x^t(k) \quad x^t(k-1) \quad w^t(k) \quad w^t(k-1)]^t$$

where

$$\begin{aligned} R_1(P) &= [(A + A_d)^t P(A + A_d) - P] + \tau r_1^{-1}(A + A_d)^t P(A + A_d) \\ &+ \tau \rho_1[A^t A_d^t P A_d A] + \tau r_2^{-1}(A + A_d)^t P(A + A_d) + \tau \rho_2[A^t A_d^t P A_d A] \\ &+ \tau \rho_1 A^t A_d^t P A_d A + \tau \rho_2 A^t A_d^t P A_d A \end{aligned} \tag{6.96}$$

$$R_2(P) = \tau \rho_1 A^t A_d^t P A_d A + \tau \rho_2 A^t A_d^t P A_d A \tag{6.97}$$

$$R_3(P) = [(A + A_d)^t P A_d - C^t] \;, \quad R_4(P) = \tau \rho_3 B^t A_d^t P A_d B \tag{6.98}$$

By **A.1**, $\Upsilon(P) < 0$, if and only if

$$\Upsilon_1(P) < 0, \quad \Upsilon_1(P) - \Upsilon_2(P)\Upsilon_3^{-1}(P)\Upsilon_2^t(P) < 0 \tag{6.99}$$

Simple matrix manipulations of (6.95)-(6.98) shows that (6.99) is equivalent to

$$\begin{aligned} R_1(P) &< 0 \\ R_1(P) &- \\ R_3(P)&\left\{(D^t + D) - B^t PB + \tau \rho_3 B^t A_d^t P A_d B\right\}^{-1} R_3^t(P) < 0 \end{aligned} \tag{6.100}$$

Finally, substitution of (6.96) and (6.98) into (6.100) yields

$$\begin{aligned} &[(A + A_d)^t P(A + A_d) - P] + \tau r_1^{-1}(A + A_d)^t P(A + A_d) \\ &+ \tau \rho_1[A^t A_d^t P A_d A] + \tau r_2^{-1}(A + A_d)^t P(A + A_d) + \tau \rho_2[A^t A_d^t P A_d A] \\ &+ \tau \rho_1 A^t A_d^t P A_d A + \tau \rho_2 A^t A_d^t P A_d A - [(A + A_d)^t P A_d - C^t] \\ &\left\{(D^t + D) - B^t PB + \tau \rho_3 B^t A_d^t P A_d B\right\}^{-1} [(A + A_d)^t P A_d - C^t]^t < 0 \end{aligned} \tag{6.101}$$

By defining $\varepsilon_1 = r_1^{-1}$, $\varepsilon_2 = r_2^{-1}$, $\varepsilon_3 = \rho_1$, $\varepsilon_4 = \rho_2$ and $\varepsilon_5 = \rho_3$, it follows from (6.101) for any $\tau \in [0, \tau^*]$ that the stability requirement

$\Upsilon(P) < 0$, corresponds to (6.76) and this, via the Schur complements, yields the LMI (6.74)-(6.75).

Now since $\Upsilon(P) < 0$, then $-\Delta V_c(x_k) + 2z^t(k)w(k) > 0$ and from which it follows that

$$\sum_{j=k_o}^{k_f} [z^t(j)w(j)] > 1/2 \sum_{j=k_o}^{k_f} [\Delta V_2(k)]$$

$$= 1/2\{V_2(k_o) - V_2(k_f)\} \qquad (6.102)$$

In view of the fact that $\Delta V_c(x_k) < 0$ when $x \neq 0$, then $x(k) \to 0$ as $k \to \infty$ and the time-delay system (Σ_o) is globally asymptotically stable.

Remark 6.9: An alternative form of the LMI (6.74)-(6.75) is

$$\Pi(P) = \begin{bmatrix} \Pi_1(P) & \Pi_2(P) \\ \Pi_2^t(P) & \Pi_3(P) \end{bmatrix} < 0 \qquad (6.103)$$

where

$$\Pi_1(P) = \begin{bmatrix} (A + A_d)^t P(A + A_d) - P & \tau^* A_{doo} & \tau^*(A^t A_d^t) \\ \tau^* A_{doo}^t & -J_{oo} & 0 \\ \tau^*(A_d A) & 0 & -\tau^*(\varepsilon_3 P)^{-1} \end{bmatrix}$$

$$\Pi_2(P) = \begin{bmatrix} \tau^*(A^t A_d^t) & (A + A_d)^t P A_d - C^t & 0 & 0 \\ 0 & 0 & 0 & 0 \\ 0 & 0 & 0 & 0 \\ 0 & 0 & 0 & 0 \end{bmatrix}$$

$$\Pi_3(P) = \begin{bmatrix} -\tau^*(\varepsilon_4 P)^{-1} & 0 & 0 & 0 \\ 0 & -(D^t + D) & B_o^t & 0 \\ 0 & B & -P^{-1} & \tau^* B^t A_d \\ 0 & 0 & \tau^* A_d^t B & -\tau^*(\varepsilon_5 P)^{-1} \end{bmatrix}$$

$$A_{doo} = [(A + A_d)^t \quad (A + A_d)^t] \ , \quad J_{oo} = diag[\tau^*(\varepsilon_1 P)^{-1} \quad \tau^*(\varepsilon_2 P)^{-1}]$$

To standardize the LMI (6.74)-(6.75) or (6.103)-(6.104), we can employ the gridding procedure of **Remark 2.6** with $\sigma_j = \varepsilon_j(1 + \varepsilon_j)^{-1}$; $j = 1, ..., 5$. Alternatively, we can use the following iterative scheme:

(1) Add an (αI) term to the left-hand side of inequality (6.74) or (6.103), where α is a scalar variable.

(2) Guess initial value for $(\varepsilon_1, ..., \varepsilon_5)$.

(3) Search for P so that α is maximized. This is now an LMI problem.

(4) Fix P and find $(\varepsilon_1, ..., \varepsilon_5)$ to further maximize α which, again, is an LMI problem.

(5) If the maximum α is negative, repeat steps **(1)**-**(4)** until $max(\alpha) \geq 0$ (implying a feasible solution is found) or a prescribed number of iterations is reached in which the iterative scheme fails.

Remark 6.10: In principle, the gridding procedure, with a sufficiently small grid size, can guarantee a near-optimal solution. The foregoing iterative scheme is generally more efficient in numerical computations but a global feasible solution would require considerable effort.

To deal with system (Σ_Δ), we provide the following definition:

Definition 6.5: *The uncertain time-delay system (Σ_Δ) is said to be delay-dependent strongly robustly stable with SP if for any τ satisfying $0 \leq \tau \leq \tau^*$ there exists a matrix $0 < P = P^t \in \Re^{n \times n}$ such that for all admissible uncertainties:*

$$\begin{bmatrix} \Pi_1(P, \Delta) & \Pi_2(P, \Delta) \\ \Pi_2^t(P, \Delta) & \Pi_3(P, \Delta) \end{bmatrix} < 0 \qquad (6.104)$$

where

$$\Pi_1(P, \Delta) = \begin{bmatrix} (A_\Delta + A_d)^t P(A_\Delta + A_d) - P & \tau^* A_{\Delta oo} & \tau^*(A_\Delta^t A_d^t) \\ \tau^* A_{\Delta oo} & -J_{oo} & 0 \\ \tau^*(A_d A_\Delta) & 0 & -\tau^*(\varepsilon_3 P)^{-1} \end{bmatrix}$$

$$\Pi_2(P, \Delta) = \begin{bmatrix} \tau^*(A_\Delta^t E_o^t) & (A_\Delta + E_o)^t P E_o - C_\Delta^t & O & O \\ 0 & 0 & 0 & 0 \\ 0 & 0 & 0 & 0 \\ 0 & 0 & 0 & 0 \end{bmatrix}$$

$$\Pi_3(P) = \begin{bmatrix} -\tau^*(\varepsilon_4 P)^{-1} & 0 & 0 & 0 \\ 0 & -(D^t + D) & B^t & O \\ 0 & B & -P^{-1} & \tau^* B^t A_d \\ 0 & 0 & \tau^* A_d^t B & -\tau^*(\varepsilon_5 P)^{-1} \end{bmatrix}$$

$$A_{\Delta oo} = [(A_\Delta + A_d)^t \quad (A_\Delta + A_d)^t] \qquad (6.105)$$

Remark 6.11: It is readily evident from **Definition 6.5** that the concept of delay-dependent strong robust stability with SP implies both the robust stability and the delay-dependent SP for system (Σ_Δ). By setting $\Delta(t) \equiv 0$, Definition 6.5 reduces to (6.74).

6.3.4   Parameterization

Here, we restrict attention on delay-independent analysis. It is easy to realize that direct application of (6.106) would require tremendous efforts over all admissible uncertainties. To bypass this shortcoming, we introduce the following $\mu-$ parameterized linear time-invariant system:

$$(\Sigma_\mu): \quad \begin{aligned} x(k+1) &= Ax(k) + B_\mu \breve{w}(k) + A_d x(k - \tau) \\ \breve{z}(k) &= C_\mu x(k) + D_\mu \breve{w}(k) \end{aligned} \tag{6.106}$$

where

$$B_\mu = [B \quad 0 \quad \mu\,H] \quad C_\mu = \begin{bmatrix} C \\ -\mu^{-1}\,E \\ 0 \end{bmatrix} \tag{6.107}$$

$$D_\mu = \begin{bmatrix} D_o & 0 & -\mu\,L_2 \\ 0 & 1/2\,I & 0 \\ 0 & 0 & 1/2\,I \end{bmatrix} \tag{6.108}$$

The next theorem shows that the delay-independent robust SP of system (Σ_Δ) can be ascertained from the strong stability with SP of (Σ_μ).

Theorem 6.9: *System (Σ_Δ) satisfying (6.2) is strongly robustly stable with SP if and only if there exists $\mu > 0$ such that (Σ_μ) is strongly stable with SP.*

Proof: By Definition 6.4 system (Σ_Δ) is strongly robustly stable with SP if there exists a matrix $0 < P = P^t \in \Re^{n \times n}$ such that for all admissible uncertainties:

$$\begin{bmatrix} A_\Delta^t P A_\Delta - P + Q & A_\Delta^t P A_d & (A_\Delta^t PB - C_\Delta^t) \\ A_d^t P A_\Delta & -(Q - A_d^t P A_d) & A_d^t PB \\ (B^t P A_\Delta - C_\Delta) & B^t P A_d & -(D + D^t - B^t PB) \end{bmatrix} < 0 \tag{6.109}$$

Inequality (6.110) can be expressed conveniently as

$$\begin{bmatrix} -P + Q & 0 & -C_\Delta^t \\ 0 & -Q & 0 \\ -C_\Delta & 0 & -(D + D^t) \end{bmatrix} +$$

$$
\begin{bmatrix} A_\Delta^t \\ A_d^t \\ B^t \end{bmatrix} P[A_\Delta \ A_d \ B] \ < \ 0 \tag{6.110}
$$

Application of the Schur complements to (6.111) puts it into the form:

$$
\begin{bmatrix} -P+Q & 0 & -C_\Delta^t & A_\Delta^t \\ 0 & -Q & 0 & A_d^t \\ -C_\Delta & 0 & -(D+D^t) & B^t \\ A_\Delta & A_d & B & -P^{-1} \end{bmatrix} \ < \ 0 \tag{6.111}
$$

Substituting the uncertainty structure (6.62) into (6.112) and rearranging, we get:

$$
\begin{bmatrix} -P+Q & 0 & -C^t & A^t \\ 0 & -Q & 0 & A_d^t \\ -C_o & 0 & -(D+D^t) & B^t \\ A & A_d & B & -P^{-1} \end{bmatrix} + \begin{bmatrix} 0 \\ 0 \\ -H_c \\ H \end{bmatrix} \Delta(k)[E \ 0 \ 0 \ 0]
$$

$$
+ \begin{bmatrix} E^t \\ 0 \\ 0 \\ 0 \end{bmatrix} \Delta^t(k)[0 \ 0 \ -H_c^t \ H^t] \ < \ 0 \tag{6.112}
$$

By [229], inequality (6.113) holds if and only if for some $\mu > 0$

$$
\begin{bmatrix} -P+Q & 0 & -C^t & A^t \\ 0 & -Q & 0 & A_d^t \\ -C_o & 0 & -(D+D^t) & B^t \\ A & A_d & B & -P^{-1} \end{bmatrix} +
$$

$$
\begin{bmatrix} \mu^{-1}E^t & 0 \\ 0 & 0 \\ 0 & -\mu H_c \\ 0 & \mu H \end{bmatrix} \begin{bmatrix} \mu^{-1}E & 0 & 0 & 0 \\ 0 & 0 & -\mu H_c^t & \mu H^t \end{bmatrix} \ < \ 0
$$

$$
\tag{6.113}
$$

for all admissible uncertainties satisfying (6.62)-(6.63). On using the Schur complements in (6.114), it becomes:

$$
\begin{bmatrix}
-P+Q & 0 & -C^t & A^t & \mu^{-1}E^t & 0 \\
0 & -Q & 0 & A_d^t & 0 & 0 \\
-C_o & 0 & -(D+D^t) & B^t & 0 & -\mu H_c \\
A & A_d & B & -P^{-1} & 0 & \mu H \\
\mu^{-1}E & 0 & 0 & 0 & -I & 0 \\
0 & 0 & -\mu H_c^t & \mu H^t & 0 & -I
\end{bmatrix} < 0 \quad (6.114)
$$

Define

$$
S = \begin{bmatrix}
I & 0 & 0 & 0 & 0 & 0 \\
0 & I & 0 & 0 & 0 & 0 \\
0 & 0 & I & 0 & 0 & 0 \\
0 & 0 & 0 & 0 & 0 & I \\
0 & 0 & 0 & I & 0 & O \\
0 & 0 & 0 & 0 & I & 0
\end{bmatrix} \quad (6.115)
$$

Premultiplying (6.115) by S and postmultiplying the result by S^t , we get:

$$
\begin{bmatrix}
-P+Q & 0 & -C^t & \mu^{-1}E^t & 0 & A^t \\
0 & -Q & 0 & 0 & 0 & A_d^t \\
-C & 0 & -(D+D^t) & 0 & -\mu H_c & B^t \\
\mu^{-1}E & 0 & 0 & -I & 0 & 0 \\
0 & 0 & -\mu H_c^t & 0 & -I & \mu H^t \\
A & A_d & B & 0 & \mu H & -P^{-1}
\end{bmatrix} < 0 \quad (6.116)
$$

In terms of the μ−parameterized matrices of (6.108) and (6.109), inequality (6.119) can be written as:

$$
\begin{bmatrix}
-P+Q & 0 & -C_\mu^t & A^t \\
0 & -Q & 0 & A_d^t \\
-C_\mu & 0 & -(D_\mu+D_\mu^t) & B_\mu^t \\
A & A_d & B_\mu & -P^{-1}
\end{bmatrix} < 0 \quad (6.117)
$$

which is equivalent to

$$
\begin{bmatrix}
-P+Q & 0 & -C_\mu^t \\
0 & -Q & 0 \\
-C_\mu & 0 & -(D_\mu+D_\mu^t)
\end{bmatrix}
+
\begin{bmatrix}
A^t \\
A_d^t \\
B_\mu^t
\end{bmatrix}
P[A \ A_d \ B_\mu] < 0 \quad (6.118)
$$

if and only if

$$\begin{bmatrix} A^t P A - P + Q & A^t P A_d & (A^t P B_\mu - C_\mu^t) \\ A_d^t P A & -(Q - A_d^t P A_d) & A_d^t P B_\mu \\ (B_\mu^t P A - C_\mu) & B_\mu^t P A_d & -(D_\mu + D_\mu^t - B_\mu^t P B_\mu) \end{bmatrix} < 0$$

(6.119)

which eventually corresponds to the stability of system (Σ_μ) with SP.

Remark 6.12: Theorem 6.9 establishes an uncertainty-independent procedure to evaluate if the uncertain time-delay system (Σ_Δ) is robustly stable with SP. Observe that inequality (6.115) with $\varepsilon = \mu^2$ is equivalent to:

$$W_t = \begin{bmatrix} G(P) & L(P) \\ L^t(P) & U(\varepsilon) \end{bmatrix} < 0$$

$$G(P) = \begin{bmatrix} -P+Q & 0 & -C^t & A^t \\ 0 & -Q & 0 & A_d^t \\ -C_o & 0 & -(D+D^t) & B^t \\ A & A_d & B & -P^{-1} \end{bmatrix}$$

$$L(P) = \begin{bmatrix} E^t & 0 \\ 0 & 0 \\ 0 & -H_c^t \\ 0 & H^t \end{bmatrix} \qquad U(\varepsilon) = \begin{bmatrix} -\varepsilon I & 0 \\ 0 & -\varepsilon^{-1} I \end{bmatrix}$$

Define $T = diag[I, I, I, P, I, I]$, then it is obvious that

$$T W_t T = \begin{bmatrix} G_T(P) & L_T(P) \\ L_T^t(P) & U(\varepsilon) \end{bmatrix} < 0$$

where

$$G_T(P) = \begin{bmatrix} -P+Q & 0 & -C^t & A^t P \\ 0 & -Q & 0 & A_d^t P \\ -C_o & 0 & -(D+D^t) & B^t P \\ PA & PA_d & PB & -P \end{bmatrix}$$

$$L(P) = \begin{bmatrix} E^t & 0 \\ 0 & 0 \\ 0 & -H_c^t \\ 0 & PH^t \end{bmatrix}$$

and that $T W_t T$ is a standard LMI since it is jointly linear in P and ε. Hence, it can be effectively solved by employing the LMI Toolbox.

6.3.5 State-Feedback Control Synthesis

The analysis of robust stability with SP can be naturally extended to the corresponding synthesis problem. That is, we are concerned with the design of a feedback controller that not only internally stabilizes the uncertain time-delay system but also achieves SP for all admissible uncertainties and unknown delays. A controller which achieves the property of robust stability with SP is termed as a robust SP controller. To this end, we consider the class of uncertain systems of the form:

$$
\begin{aligned}
(\Sigma_{\Delta_s}) \quad x(k+1) &= A_\Delta x(k) + B_o w(k) + B_{1\Delta} u(k) \\
&\quad + E_o x(k-\tau) \\
z(k) &= C_\Delta x(k) + D_o w(k) + D_{12\Delta} u(k) \\
y(k) &= x(k)
\end{aligned}
\tag{6.120}
$$

in which all the states are assumed available for instantaneous measurements and the uncertain matrices are given by:

$$
\begin{bmatrix} A_\Delta & B_{1\Delta} \\ C_\Delta & D_{12\Delta} \end{bmatrix} = \begin{bmatrix} A & B_1 \\ C & D_{12} \end{bmatrix} + \begin{bmatrix} H \\ H_c \end{bmatrix} \Delta(k)[E \ E_b]
\tag{6.121}
$$
$$
\Delta^t(k)\Delta(k) \leq I \quad \forall k
$$

The objective of the robust SP control design problem is to determine a static feedback controller $u(k) = K_s x(k)$ that internally stabilizes system (Σ_{Δ_s}) for all admissible uncertainties and achieves the strict passivity property.

The next theorem establishes the main design result.

Theorem 6.10: *The state-feedback control* $u(k) = K_s x(k)$ *stabilizes system* (Σ_{Δ_s}) *if there exists a scalar* $\epsilon > 0$ *and matrices* $0 < Y = Y^t \in \Re^{n \times n}$ *and* $S \in \Re^{m \times n}$ *satisfying the LMI:*

$$
W_s = \begin{bmatrix} \Xi_1(Y,S) & \Xi_2(Y,S) \\ \Xi_2^t(Y,S) & U(\epsilon) \end{bmatrix} < 0
\tag{6.122}
$$

where

$$
\Xi_1(Y,S) = \begin{bmatrix} -Y+R & 0 & -YC^t+S^t D_{12}^t & YA^t+S^t B_1^t \\ 0 & -Q & 0 & A_d^t \\ -CY+D_{12}S & 0 & -(D+D^t) & B^t \\ AY+B_1 S & A_d & B & -Y \end{bmatrix}
$$

$$\Xi_2(Y,S) \;=\; \begin{bmatrix} YE^t + S^t E_b^t & 0 \\ 0 & 0 \\ 0 & -H_c \\ 0 & H \end{bmatrix} \tag{6.123}$$

Moreover, the gain matrix is $K_s = S Y^{-1}$.

Proof: System (6.123) under the action of the controller $u(k) = K_s x(k)$ becomes:

$$(\Sigma_{\Delta K}) \quad x(k+1) \;=\; [A_\Delta + B_{1\Delta} K_s] x(k) + B_\Delta w(k)$$
$$+ \; A_d x(k-\tau)$$
$$z(k) \;=\; [C_\Delta + D_{12\Delta} K_s] x(k) + D_\Delta w(k) \tag{6.124}$$

for all admissible uncertainties satisfying (6.124). Applying **Theorem 6.8**, it follows that system $(\Sigma_{\Delta K})$ is strongly robustly stable with SP if and only if there exists a $\mu > 0$ such that

$$\Omega = \begin{bmatrix} \Omega_1 & \Omega_2 \\ \Omega_2^t & \Omega_3 \end{bmatrix}$$

$$\Omega_1 = \begin{bmatrix} -P+Q & 0 \\ 0 & -Q \end{bmatrix}, \quad \Omega_2^t = \begin{bmatrix} -C + D_{12}K_s & 0 \\ A + B_1 K_s & A_d \\ \mu^{-1}(E + E_b K_s) & 0 \\ 0 & 0 \end{bmatrix}$$

$$\Omega_3 = \begin{bmatrix} -(D+D^t) & B^t & 0 & -\mu H_c \\ B & -P^{-1} & 0 & \mu H \\ 0 & 0 & -I & 0 \\ -\mu H_c^t & \mu H^t & 0 & -I \end{bmatrix} \tag{6.125}$$

Introducing

$$S_1 \;=\; diag[I, I, I, I, \mu I, \mu^{-1}I], \quad S_2 \;=\; diag[P^{-1}, I, I, I, I, I] \tag{6.126}$$

and evaluating the matrix $\Omega_s = (S_2 S_1 \Omega S_1^t S_2^t)$ with $Y = P^{-1}$, $K_s = SP$, $R = P^{-1}QP^{-1}$, $\varepsilon = \mu^2$, we obtain

$$\Omega_s \;=\; \begin{bmatrix} \Omega_{s1} & \Omega_{s2} \\ \Omega_{s2}^t & \Omega_{s3} \end{bmatrix}$$

$$\Omega_{s1} = \begin{bmatrix} -Y+R & 0 \\ 0 & -Q \end{bmatrix}, \quad \Omega_{s2}^t = \begin{bmatrix} -CY + D_{12}S & 0 \\ AY + B_1 S & A_d \\ EY + E_b S & 0 \\ 0 & 0 \end{bmatrix}$$

$$\Omega_3 = \begin{bmatrix} -(D+D^t) & B^t & 0 & -H_c \\ B & -Y & 0 & H \\ 0 & 0 & -\varepsilon I & 0 \\ -H_c^t & H^t & 0 & -\varepsilon^{-1}I \end{bmatrix} \tag{6.127}$$

In terms of $\Xi_1(Y,S)$ and $\Xi_2(Y,S)$ as given by (6.1254), it is easy to see that $\Omega_s \equiv W_s$ and $K_s = SY^{-1}$.

6.3.6   Output-Feedback Control Synthesis

Consider the class of uncertain time-delay systems

$$\begin{aligned} (\Sigma_{\Delta o}) \quad x(k+1) &= A_\Delta x(k) + Bw(k) + B_{1\Delta}u(k) + A_d x(k-\tau) \\ z(k) &= C_\Delta x(k) + Dw(k) + D_{12\Delta}u(k) \\ y(k) &= C_{1\Delta}x(k) + D_{21}w(k) + D_{22\Delta}u(k) \end{aligned} \tag{6.128}$$

where the uncertainties are given by (6.122) in addition to

$$[C_{1\Delta} \quad D_{22\Delta}] = H_d\Delta(k)[E \quad E_b] \tag{6.129}$$

Here we consider the problem of designing dynamic output-feedback controllers of order r and represented by

$$\begin{aligned} (\Sigma_{df}) \quad \xi(k+1) &= F_o\xi(k) + G_o y(k) \\ u(k) &= K_o\xi(k) \end{aligned} \tag{6.130}$$

where $F_o \in \Re^{r\times r}$, $G_o \in \Re^{r\times m}$, $K_o \in \Re^{m\times r}$ are the controller matrices. The objective is to ensure internal closed-loop stability for system $(\Sigma_{\Delta o})$ for all admissible uncertainties satisfying (6.122). By augmenting the controller (Σ_{df}) with system $(\Sigma_{\Delta o})$, we get the closed-loop system:

$$\begin{aligned} (\Sigma_{cc}) \quad \eta(k+1) &= \widehat{A}_\Delta \, \eta(k) + \widehat{B}\, w(k) + \widehat{E}\, \eta(k-\tau) \\ z(k) &= \widehat{C}_\Delta \, \eta(k) + Dw(k) \end{aligned} \tag{6.131}$$

where

$$\begin{aligned} \widehat{A}_\Delta &= \begin{bmatrix} A_\Delta & B_{1\Delta}K_o \\ G_oC_{1\Delta} & F_o + G_oD_{22\Delta}K_o \end{bmatrix} = \breve{A} + \breve{L}\,\Delta(k)\,\breve{M} \\ &= \begin{bmatrix} A & B_1K_o \\ G_oC_1 & F_o + G_oD_{22}K_o \end{bmatrix} \end{aligned}$$

$$+ \begin{bmatrix} H \\ G_oH_d \end{bmatrix} \Delta(k) \, [E \quad E_bK_o] \tag{6.132}$$

$$\widehat{B} = \begin{bmatrix} B \\ G_oD_{21} \end{bmatrix} , \quad \widehat{E} = \begin{bmatrix} A_d \\ 0 \end{bmatrix} \tag{6.133}$$

$$\widehat{C}_\Delta = [C_\Delta \quad D_{12\Delta}K_o] \tag{6.134}$$

$$= [C \quad D_{12}K_o] + L_2 \, \Delta(k) \, [E \quad E_bK_o] \tag{6.135}$$

Now we introduce the following μ–parametrized linear shift-invariant system

$$(\Sigma_{\mu d}) \quad x_\mu(k+1) = \breve{A} \, x_\mu(k) + \hat{B}_\mu \breve{w}(k) + \widehat{E} \, x_\mu(k-\tau)$$
$$\breve{z}(k) = \hat{C}_\mu x_\mu(k) + \hat{D}_\mu \breve{w}(k) \tag{6.136}$$

where

$$\hat{B}_\mu = \begin{bmatrix} B & 0 & \mu H \\ G_oD_{21} & 0 & \mu G_oH_d \end{bmatrix}$$

$$\hat{C}_\mu = \begin{bmatrix} C & D_{12}K_o \\ -\mu^{-1}E & \mu^{-1}E_bK_o \\ 0 & 0 \end{bmatrix} \tag{6.137}$$

$$\hat{D}_\mu = \begin{bmatrix} D_o & 0 & -\mu H_c \\ \mu^{-1}E_b & 1/2I & 0 \\ 0 & 0 & 1/2I \end{bmatrix} \tag{6.138}$$

The next theorem shows that the robust stability with SP of system (Σ_{cc}) can be ascertained from the strong stability of system $(\Sigma_{\mu d})$.

Theorem 6.11: System (Σ_{cc}) is strongly robustly stable with SP if and only if there exists a scalar parameter $\mu > 0$ such that system $(\Sigma_{\mu d})$ is strongly stable with SP.

Proof: By **Remark 6.7** and **Definition 6.4**, it follows that system (Σ_{cc}) is stongly robustly stable with SP if

$$\begin{bmatrix} -Y+W & 0 & -\widehat{C}_\Delta^t & \widehat{A}_\Delta^t \\ 0 & -Q & 0 & \widehat{E}^t \\ -\widehat{C}_\Delta & 0 & -(D_o+D_o^t) & \widehat{B}_o^t \\ \widehat{A}_\Delta & \widehat{E} & \widehat{B}_o & -Y^{-1} \end{bmatrix} < 0 \tag{6.139}$$

where

$$Y = \begin{bmatrix} Y_1 & Y_2 \\ Y_2^t & Y_3 \end{bmatrix} \in \Re^{(n+r)\times(n+r)}, \ W = \begin{bmatrix} Q & 0 \\ 0 & 0 \end{bmatrix},$$

$$X = Y^{-1} = \begin{bmatrix} X_1 & X_2 \\ X_2^t & X_3 \end{bmatrix} \tag{6.140}$$

The substitution of (6.133)-(6.135) into (6.139) with some manipulations yields:

$$\begin{bmatrix} -Y_1 + Q & -Y_2 & 0 & -C^t & A^t & C_1^t G_o^t \\ -Y_2^t & -Y_3 & 0 & -K_o^t D_{12}^t & K_o^t B_1^t & F_o^t + K_o^t D_{22}^t G_o^t \\ 0 & 0 & -Q & 0 & A_d^t & 0 \\ -C & -D_{12}K_o & 0 & -(D+D^t) & B^t & D_{21}^t G_o^t \\ A & B_1 K_o & A_d & B & -X_1 & -X_2 \\ G_o C_1 & F_o + G_o D_{22} K_o & 0 & G_o D_{21} & -X_2^t & -X_3 \end{bmatrix}$$

$$+ \begin{bmatrix} 0 \\ 0 \\ 0 \\ 0 \\ H \\ G_o H_d \end{bmatrix} \Delta(k)[E \ \ E_b K_o \ 0 \ 0 \ 0 \ 0]$$

$$+ \begin{bmatrix} E^t \\ K_o^t E_b^t \\ 0 \\ 0 \\ 0 \\ 0 \end{bmatrix} \Delta^t(k)[0 \ 0 \ 0 \ 0 \ H^t \ H_d^t G_o^t] < 0 \tag{6.141}$$

By [229], inequality (6.141) holds if and only if for some $\mu > 0$

$$\begin{bmatrix} -Y_1 + Q & -Y_2 & 0 & -C_o^t & A^t & C_1^t G_o^t \\ -Y_2^t & -Y_3 & 0 & -K_o^t D_{12}^t & K_o^t B_1^t & F_o^t + K_o^t D_{22}^t G_o^t \\ 0 & 0 & -Q & 0 & A_d^t & 0 \\ -C_o & -D_{12}K_o & 0 & -(D_o+D_o^t) & B_o^t & D_{21}^t G_o^t \\ A & B_1 K_o & A_d & B & -X_1 & -X_2 \\ G_o C_1 & F_o + G_o D_{22} K_o & 0 & G_o D_{21} & -X_2^t & -X_3 \end{bmatrix}$$

$$+ \begin{bmatrix} \mu^{-1}E^t & 0 \\ \mu^{-1}K_o^t E_b^t & 0 \\ 0 & 0 \\ 0 & 0 \\ 0 & \mu H \\ 0 & \mu G_o H_d \end{bmatrix} \begin{bmatrix} \mu^{-1}E & \mu^{-1}E_b K_o & 0 & 0 & 0 & 0 \\ 0 & 0 & 0 & 0 & \mu H^t & \mu H_d^t G_o^t \end{bmatrix}$$

$$< 0$$

$$(6.142)$$

for all admissible uncertainties satisfying (6.122) and (6.130). By similarity to **Theorem 6.8** and using (6.133)-(6.135) with some manipulations, inequality (6.142) is equivalent to:

$$\begin{bmatrix} -Y+W & 0 & -C_\mu^t & \breve{A}^t \\ 0 & -Q & 0 & \widehat{E}^t \\ -C_\mu & 0 & -(D_\mu + D_\mu^t) & \hat{B}_\mu^t \\ \breve{A} & \widehat{E} & \hat{B}_\mu & -Y^{-1} \end{bmatrix} < 0 \qquad (6.143)$$

which, in turn, is equivalent to

$$\begin{bmatrix} \breve{A}^t Y \breve{A} -Y+W & \breve{A}^t Y \widehat{E} & (\breve{A}^t Y \hat{B}_\mu - \hat{C}_\mu^t) \\ \widehat{E}^t Y \breve{A} & -(Q- \widehat{E}^t Y \widehat{E}) & \widehat{E}^t Y \hat{B}_\mu \\ (\hat{B}_\mu^t Y \breve{A} -\hat{C}_\mu) & \hat{B}_\mu^t Y \widehat{E} & -(D_\mu + D_\mu^t - \hat{B}_\mu^t Y \hat{B}_\mu) \end{bmatrix} < 0$$

$$(6.144)$$

This obviously corresponds to the stability of system $(\Sigma_{\mu d})$ with SP.

6.4 Notes and References

The results reported in Chapter 6 are essentially based on [203,226,228,232] which extend the results of [197,200-202,205-208] for delayless systems. Obviously, much more research work is needed.

Chapter 7

Interconnected Systems

7.1 Introduction

In control engineering research, problems of decentralized control and stabilization of interconnected systems are receiving considerable interest in
recent years [151-154, 183, 184, 187, 188] where most of the effort is focused
on dealing with the interaction patterns. Quadratic stabilization of classes
of interconnected systems have been presented in [189, 190, 233] where the
closed-loop feedback subsystems are cast into H_∞ control problems. Decentralized output tracking has been considered in [153, 154] for a class
of interconnected systems with matched uncertainties and in [234, 235] using norm-bounded uncertainties. When the interconnected system involves
delays, only few studies are available. In [154, 189], the focus has been on
delays in the interaction patterns with the subsystem dynamics being known
completely. The same problem within the scope of uncertain systems was
treated in [238]. In [236, 237], a class of uncertain systems is considered
where the delays occur within the subsystems.

In this chapter, we look into two broad classes of problems:
(1) Decentralized robust stabilization and robust H_∞ performance for a class
of interconnected systems with unknown delays and norm-bounded parametric uncertainties, and
(2) Decentralized control of interconnected systems with mismatched uncertainties.

In both problems, the delays are time-varying in the state of each sub-

system as well as in the interconnections among the subsystems. The main
objective is to design a decentralized feedback controller which renders the
closed-loop interconnected system asymptotically stable for all admissible
uncertainties and unknown-but-bounded delays. All the design effort is un-
dertaken at the subsystem level.

7.2   Problem Statement and Definitions

Consider a class of linear systems S composed of n_s coupled subsystems S_j
and modeled in state-space form by

$$
\begin{aligned}
S_j: \quad \dot{x}_j(t) &= A_j(t)x_j(t) + B_j(t)u_j(t) \\
&+ \ \Gamma_j w_j(t) + E_j(t)x_j(t - \tau_j) + \sum_{k=1}^{n_s} A_{jk}(t)x_k(t - \eta_{jk}) \\
y_j(t) &= C_j(t)x_j(t) + D_j(t)u_j(t) + \Pi_j w_j(t) \qquad (7.1) \\
z_j(t) &= G_j x_j(t) + F_j u_j(t)
\end{aligned}
$$

where $j \in \{1, ..., n_s\}$; $x_j(t) \in \Re^{n_j}$ is the state vector; $u_j(t) \in \Re^{m_j}$ is the
control input; $w_j(t) \in \Re^{q_j}$ is the disturbance input; $y_j(t) \in \Re^{p_j}$ is the measured output; $z_j(t) \in \Re^{r_j}$ is the controlled output, and η_{jk}, τ_j are unknown
time-delays satisfying $0 \leq \tau_j \leq \tau_j^o$, $0 \leq \eta_{jk} \leq \eta_{jk}^o$, $j,k \in \{1,..,n_s\}$ where
the bounds $\tau_j^o, \eta_{jk}^o; j, k \in \{1, ..., n_s\}$ are known constants in order to guarantee smooth growth of the state trajectories. In this work, the matri-
ces G_j, F_j, Γ_j, Π_j are known, real and constants but the system matrices
$A_j(t), A_{jk}(t), B_j(t), C_j(t), D_j(t), E_j(t)$ are uncertain and possibly fast time
varying. The inclusion of the term $E_j(t)x_j(t - \tau_j)$ as a separate term is
meant to emphasize the delay within the jth subsystem. Finally, the initial
condition for (1) is specified as $\langle x_j(0), x_j(r) \rangle = \langle x_{oj}, \phi_j(r) \rangle; j = 1, .., s$ where
$\phi_j(.) \in \mathcal{L}_2[-\tau_j^*, 0]; j = 1, .., s$

7.2.1   Uncertainty Structures

In the literature on large-scale state-space models containing parametric uncertainties, there has been different methods to characterize the uncertainty.
In one method, the uncertainty is assumed to satisfy the so-called matching
condition [233, 243]. Loosely speaking, this condition implies that the uncertainties cannot enter arbitrarily into the system dynamics but are rather

restricted to lie in the range space of the input matrix. By a second method, the uncertainty is represented by rank-1 decomposition. It is well-known that both methods are quite restrictive in practice. This limitation can be overcome by using the generalized matching conditions [239] through an iterative procedure of constructing stabilizing controllers. There is a fourth method in which the dynamic model is cast into the polytopic format [3] which implies that the systems of the associated state-space model depend on a single parameter vector.

In this section, we follow a linear fractional representation method in which the uncertainty is expressed in terms of a norm-bounded form as [215,218]:

$$\begin{bmatrix} A_j(t) & B_j(t) \\ C_j(t) & D_j(t) \end{bmatrix} = \begin{bmatrix} A_{oj} & B_{oj} \\ C_{oj} & D_{oj} \end{bmatrix} + \begin{bmatrix} S_j \\ L_j \end{bmatrix} H_j(t) \begin{bmatrix} M_j & N_j \end{bmatrix} \qquad (7.2)$$

$$E_j(t) = E_{oj}(t) + M_{ej}H_{ej}(t)N_{ej} \;\; ; \;\; A_{jk}(t) = A_{jko}(t) + M_{jk}H_{jk}(t)N_{jk}$$
$$H_j^t(t)H_j(t) \leq I \;\; ; \;\; H_{jk}(t)H_{jk}^t(t) \leq I \;\; ; \;\; H_{ej}(t)H_{ej}^t(t) \leq I \qquad (7.3)$$
$$\forall t, \forall j, \, k \in \{1, ..., n_s\}$$

where $H_j(t) \in \Re^{\phi_j \times \varphi_j}$, $H_{jk}(t) \in \Re^{\nu_j \times \epsilon_k}$, $H_{ej}(t) \in \Re^{\nu_j \times \varsigma_j}$ are unknown time-varying matrices whose elements are Lebsegue measurable; the matrices S_j, L_j, M_j, N_j, M_{jk}, N_{jk}, M_{ej}, N_{ej} are real, known and constant matrices with appropriate dimensions and A_{oj}, B_{oj}, C_{oj}, D_{oj}, E_{oj} are real constant and known matrices representing the nominal decoupled system (without uncertainties and interactions):

$$S_j^n : \quad \dot{x}_j(t) = A_{oj}x_j(t) + B_{oj}u_j(t) + \Gamma_j w_j(t)$$
$$y_j(t) = C_{oj}x_j(t) + D_{oj}u_j(t) + \Pi_j w_j(t) \qquad (7.4)$$
$$z_j(t) = G_j x_j(t) + F_j u_j(t)$$

In the sequel, we assume $\forall j \in \{l, ..., n_s\}$ that the n_s-pairs (A_{oj}, B_{oj}) and (A_{oj}, C_{oj}) are stabilizable and detectable, respectively.

The interest in the uncertainty characterization (7.4) is supported by the fact that quadratic stabilizability of feedback systems with norm-bounded uncertainties is equivalent to the standard H_∞ control problem [215-219].

For sake of completeness, we tackle the decentralized control problem when the system uncertainties are mismatched in a later section.

7.3   Decentralized Robust Stabilization I

The objective of the decentralized robust stabilization problem is to design decentralized dynamic feedback controllers for the interconnected system (7.1) such that the resulting closed-loop system is asymptotically stable for all admissible uncertainties. In this case, the interconnected system (7.1) is said to be *robustly stabilizable via decentralized dynamic feedback control.*

In the following, stabilization of system (7.1) in the absence of the external disturbance, $w_j(t) = 0$, is achieved through the use of a class of observer-based feedback controllers of the form:

$$S_j^c: \quad \dot{\hat{x}}_j(t) = [A_{oj} + \delta A_j]\hat{x}_j(t) + \Theta_j[y_j(t) - \hat{y}_j(t)] + B_{oj}u_j(t)$$
$$\hat{y}_j(t) = [C_{oj} + \delta C_j]\hat{x}_j(t) + D_{oj}u_j(t)$$
$$u_j(t) = -K_j\hat{x}_j(t) \tag{7.5}$$

where the matrices introduced in (7.5) will be specified shortly. We now look for a solution to the decentralized robust stabilization problem. For this purpose, consider that for $j = 1, ..., n_s$ there exist two matrices $0 < P_j = P_j^t \in \Re^{n_j \times n_j}$, $0 < Q_j = Q_j^t \in \Re^{n_j \times n_j}$, satisfying the two ARIs:

$$P_j A_{oj} + A_{oj}^t P_j - \rho_j^{-2} \tilde{B}_{oj}^t (N_j^t N_j)^{-1} \tilde{B}_{oj} + P_j X_j X_j^t P_j$$
$$+\rho_j^2 M_j^t M_j + (n_s + 1)I_j < 0 \tag{7.6}$$
$$Q_j(A_{oj} + X_j X_j^t P_j) + (A_{oj} + X_j X_j^t P_j)^t Q_j + Q_j X_j X_j^t Q_j -$$
$$\rho_j^2 J_j^t (L_j L_j^t)^{-1} J_j + \rho_j^{-2} \tilde{C}_{oj}^t (L_j L_j^t)^{-1} \tilde{C}_{oj} < 0 \tag{7.7}$$

along with

$$X_j X_j^t = \rho_j^{-2} S_j S_j^t + \Omega_{ej} \Omega_{ej}^t + \Omega_{aj} \Omega_{aj}^t \; ; \; \tilde{B}_{oj} = B_{oj}^t P_j + \rho_j^2 N_j^t M_j$$
$$\Omega_{ej} \Omega_{ej}^t = E_{oj}(I_j - \alpha_j^2 N_{ej}^t N_{ej})^{-1} E_{oj}^t + \alpha_j^{-2} M_{ej} M_{ej}^t$$
$$\Omega_{aj} \Omega_{aj}^t = \sum_{k=1}^{n_s} [A_{jko}(I_j - \beta_{jk}^2 N_{jk}^t N_{jk})^{-1} A_{jko}^t + \beta_{jk}^{-2} M_{jk} M_{jk}^t]$$
$$J_j = C_{oj} + \rho_j^{-2} L_j S_j^t [P_j + Q_j] \; ; \; \tilde{C}_{oj}^t = C_{oj} + \rho_j^2 L_j S_j^t P_j \tag{7.8}$$

where the scalars $\rho_j > 0$, $\alpha_j > 0$, $\beta_{jk} > 0$, and $I_j \in \Re^{n_j \times n_j}$ is the identity matrix. Introducing the error $e_j(t) = \hat{x}_j(t) - x_j(t)$, then it follows from (7.1) and (7.3)-(7.5) that it has the dynamics:

$$\dot{e}_j(t) = [A_{oj} - \Theta_j C_j + \delta A_j + \Theta_j \delta C_j + (S_j - \Theta_j L_j)H_j(t)N_j K_j]e_j$$

$$+[\delta A_j - \Theta_j \delta C_j - (S_j - \Theta_j L_j)H_j(t)(M_j - N_j K_j)]x_j$$

$$-E_j(t)x_j(t - \tau_j) - \sum_{k=1}^{n_s} A_{jk}(t)x_k(t - \eta_{jk}) \qquad (7.9)$$

In terms of $\xi_j^t = [x_j^t \quad e_j^t] \in \Re^{2n_j}$, the dynamics of the augmented system (7.1) and (7.9) take the form

$$\dot{\psi}_j(t) = \tilde{A}_j(t)\psi_j(t) + \tilde{E}_j(t)x_j(t - \tau_j) + \sum_{k=1}^{n_s} \tilde{A}_{jk}(t)x_k(t - \eta_{jk}) \qquad (7.10)$$

where

$$\tilde{A}_j = \breve{A}_j + \widehat{S}_j H_j(t)\widehat{M}_j \quad ; \quad \widehat{M}_j = [M_j - N_j K_j \quad - N_j K_j]$$

$$\breve{A}_j = \begin{bmatrix} A_{oj} - B_{oj}K_j & -B_{oj}K_j \\ \delta A_j - \Theta_j \delta C_j & A_{oj} - \Theta_j C_{oj} + \delta A_j - \Theta_j \delta C_j \end{bmatrix}$$

$$\widehat{S}_j = \begin{bmatrix} S_j \\ -S_j + \Theta_j L_j \end{bmatrix}$$

$$\tilde{E}_j = \begin{bmatrix} E_{oj} \\ -E_{oj} \end{bmatrix} + \begin{bmatrix} M_{ej} \\ -M_{ej} \end{bmatrix} H_{ej}(t)N_{ej}$$

$$= \widehat{E}_{oj} + \widehat{M}_{ej}H_{ej}(t)N_{ej}$$

$$\tilde{A}_{jk} = \begin{bmatrix} A_{jko} \\ -A_{jko} \end{bmatrix} + \begin{bmatrix} M_{jk} \\ -M_{jk} \end{bmatrix} H_{jk}(t)N_{jk}$$

$$= \widehat{A}_{jko} + \widehat{M}_{jk}H_{jk}(t)N_{jk} \qquad (7.11)$$

A preliminary result is established first:

Lemma 7.1: *Let the matrices δA_j, δC_j, Θ_j and K_j be defined as*

$$\delta A_j = X_j X_j^t P_j \quad , \quad \delta C_j = \rho_j^{-2} L_j S_j^t P_j \qquad (7.12)$$

$$\Theta_j = \rho_j^2 Q_j^{-1} J_j^t (L_j L_j^t)^{-1} \qquad (7.13)$$

$$K_j = \rho_j^{-2}(N_j^t N_j)^{-1}[B_{oj}^t P_j + \rho_j^2 N_j^t M_j] \qquad (7.14)$$

then, the following inequality holds

$$\tilde{A}_j^t W_j + W_j \tilde{A}_j + W_j(\tilde{E}_j \tilde{E}_j^t + \sum_{k=1}^{n_s} \tilde{A}_{jk}\tilde{A}_{jk}^t)W_j + U_s < 0 \qquad (7.15)$$

where P_j, Q_j satisfy inequalities (7.6) and (7.7) and

$$W_j := \begin{bmatrix} P_j & 0 \\ 0 & Q_j \end{bmatrix}, \quad U_s = \begin{bmatrix} (n_s+1)I_j & 0 \\ 0 & 0 \end{bmatrix} \tag{7.16}$$

Proof: Define the augmented matrices:

$$\Sigma_j = \begin{bmatrix} \Sigma_{1j} & \Sigma_{2j} \\ \Sigma_{3j} & \Sigma_{4j} \end{bmatrix}$$

$$= \breve{A}_j^t W_j + W_j \breve{A}_j + W_j \widehat{X}_j \widehat{X}_j^t W_j + \rho_j^2 \widehat{M}_j^t \widehat{M}_j + U_s \tag{7.17}$$

and the matrix expressions

$$\widehat{X}_j \widehat{X}_j^t = \rho_j^{-2} \widehat{S}_j \widehat{S}_j^t + \widehat{\Omega}_{ej} \widehat{\Omega}_{ej}^t + \widehat{\Omega}_{aj} \widehat{\Omega}_{aj}^t$$

$$\widehat{\Omega}_{ej} \widehat{\Omega}_{ej}^t = \widehat{E}_{oj}(I_j - \alpha_j^2 N_{ej}^t N_{ej})^{-1} \widehat{E}_{oj}^t + \alpha_j^{-2} \widehat{M}_{ej} \widehat{M}_{ej}^t$$

$$\widehat{\Omega}_{aj} \widehat{\Omega}_{aj}^t = \sum_{k=1}^{n_s} [\widehat{A}_{jko}(I_j - \beta_{jk}^2 N_{jk}^t N_{jk})^{-1} \widehat{A}_{jko}^t + \beta_{jk}^{-2} \widehat{M}_{jk} \widehat{M}_{jk}^t] \tag{7.18}$$

Considering (7.17) and after some algebraic manipulations using (7.11) and (7.18), it can be shown that

$$\begin{aligned}
\Sigma_{1j} =\ & (A_{oj}^t - K_j^t B_{oj}^t)P_j + P_j(A_{oj} - B_{oj}K_j) + \rho_j^{-2} P_j S_j S_j^t P_j \\
& + P_j E_{oj}(I_j - \alpha_j^2 N_{ej}^t N_{ej})^{-1} E_{oj}^t P_j + \alpha_j^{-2} P_j M_{ej} M_{ej}^t P_j \\
& + \sum_{k=1}^{n_s} P_j [A_{jko}(I_j - \beta_{jk}^2 N_{jk}^t N_{jk})^{-1} A_{jko}^t + \beta_{jk}^{-2} M_{jk} M_{jk}^t] P_j \\
& + (M_j^t - K_j^t N_j^t)(M_j - N_j K_j) + (n_s + 1)I_j \tag{7.19}
\end{aligned}$$

$$\begin{aligned}
\Sigma_{2j} =\ & \Sigma_{3j}^t = (\delta A_j^t - \delta C_j^t \Theta_j^t)Q_j - P_j B_{oj} K_j + \rho_j^{-2} P_j S_j(-S_j^t + L_j^t \Theta_j)Q_j \\
& - P_j E_{oj}(I_j - \alpha_j^2 N_{ej}^t N_{ej})^{-1} E_{oj}^t Q_j - \alpha_j^{-2} P_j M_{ej} M_{ej}^t Q_j \\
& - \sum_{k=1}^{n_s} P_j [A_{jko}(I_j - \beta_{jk}^2 N_{jk}^t N_{jk})^{-1} A_{jko}^t + \beta_{jk}^{-2} M_{jk} M_{jk}^t] Q_j \\
& - (M_j^t - K_j^t N_j^t)N_j K_j \tag{7.20}
\end{aligned}$$

$$\begin{aligned}
\Sigma_{4j} =\ & (A_{oj}^t - C_{oj}^t \Theta_j^t + \delta A_j^t - \delta C_j^t \Theta_j^t)Q_j \\
& + Q_j(A_{oj} - \Theta_j C_{oj} + \delta A_j - \Theta_j \delta C_j) \\
& + \rho_j^{-2} Q_j(-S_j + \Theta_j L_j)(-S_j^t + L_j^t \Theta_j)Q_j
\end{aligned}$$

$$+ \quad Q_j E_{oj}(I_j - \alpha_j^2 N_{ej}^t N_{ej})^{-1} E_{oj}^t Q_j$$

$$+ \quad \sum_{k=1}^{n_s} Q_j [A_{jko}(I_j - \beta_{jk}^2 N_{jk}^t N_{jk})^{-1} A_{jko}^t + \beta_{jk}^{-2} M_{jk} M_{jk}^t] Q_j$$

$$- \quad \alpha_j^{-2} Q_j M_{ej} M_{ej}^t Q_j + K_j^t N_j^t N_j K_j \qquad (7.21)$$

Using (7.12)-(7.14) into (7.19)-(7.21) with some lengthy but standard matrix manipulations, we obtain,

$$\Sigma_{1j} < 0, \quad \Sigma_{2j} = 0, \quad \Sigma_{3j} = 0, \quad \Sigma_{4j} < 0, \quad j = 1, ..., n_s \qquad (7.22)$$

This implies that

$$\breve{A}_j^t W_j + W_j \breve{A}_j + W_j \widehat{X}_j \widehat{X}_j^t W_j + \rho_j^2 \widehat{M}_j^t \widehat{M}_j + U_s < 0, \quad j = 1, ..., n_s \quad (7.23)$$

Then, by applying **Lemma 7.1** to (7.24) with the help of (7.6), (7.7), (7.12) and (7.23) we get

$$\tilde{A}_j^t W_j + W_j \tilde{A}_j + W_j (\tilde{E}_j \tilde{E}_j^t + \sum_{k=1}^{n_s} \tilde{A}_{jk} \tilde{A}_{jk}^t) W_j + U_s < 0 \qquad (7.24)$$

which corresponds to (7.15) as desired.

Remark 7.1: Note that the conditions that $N_j^t N_j$ and $L_j L_j^t$ being invertible are necessary to yield separable expressions for the matrices δA_j and δC_j that have been introduced to take care of the uncertainties. If these conditions do not hold, the design results can be obtained by applying the technique developed in [158]. In the absence of uncertainties, $\delta A_j \equiv 0$ and $\delta C_j \equiv 0$, we have to follow another route in the analysis.

Remark 7.2: Taking into account the quadratic nature of the term $W_j (\tilde{E}_j \tilde{E}_j^t + \sum_{k=1}^{n_s} \tilde{A}_{jk} \tilde{A}_{jk}^t) W_j$ it follows from that

$$\tilde{A}_j^t W_j + W_j \tilde{A}_j + W_j (\tilde{E}_j \tilde{E}_j^t + \sum_{k=1}^{r} \tilde{A}_{jk} \tilde{A}_{jk}^t) W_j + U_s < 0 \quad \forall r \leq n_s \quad (7.25)$$

Now, the main stability result is established by the following theorem:

Theorem 7.1: *System (7.1) is robustly stabilizable via the decentralized observer-based controller (7.5) if $\forall j, k \in \{l, ..., n_s\}$ there exist positive scalars ρ_j, α_j, β_{jk} such that the following conditions are met:*

(1) *the matrices $N_j^t N_j$, $L_j L_j^t$, $(I_j - \alpha_j^2 N_{ej}^t N_{ej})$ and $(I_j - \beta_{jk}^2 N_{jk}^t N_{jk})$ are invertible,*

(2) *there exist matrices $0 < P_j^t = P_j \in \Re^{n_j \times n_j}$, $0 < Q_j^t = Q_j \in \Re^{n_j \times n_j}$, satisfying (7.6) and (7.7), respectively.*

The matrices of the stabilizing decentralized controller (7.5) are given by (7.12)-(7.14).

Proof: Introduce the Lyapunov-Krasovskii functional

$$
\begin{aligned}
V(\xi_t) &= \sum_{j=1}^{n_s} V_j(\xi_{jt}) \ , \\
&= \sum_{j=1}^{n_s} \xi_j^t W_j \xi_j + \sum_{j=1}^{n_s} \int_{t-\tau_j}^{t} x_j^t(\alpha) x_j(\alpha) d\alpha \\
&\quad + \sum_{j=1}^{n_s} \sum_{k=1}^{n_s} \int_{t-\eta_{jk}}^{t} x_k^t(\beta) x_k(\beta) d\beta
\end{aligned}
\tag{7.26}
$$

which takes into account the present as well as the delayed states. Note that $V(\xi_t) > 0$ for $\xi \neq 0$. The total time derivative $dV(\xi_t)/dt$ of the function $V(\xi_t)$, which is computed with respect to (7.10), is obtained as:

$$
\begin{aligned}
dV(\xi_t)/dt &= \sum_{j=1}^{n_s} \{ \xi_j^t [W_j \tilde{A}_j^t + \tilde{A}_j W_j] \xi_j \\
&\quad + [\tilde{E}_j x_j(t - \tau_j)]^t W_j \xi_j + \xi_j^t W_j [\tilde{E}_j x_j(t - \tau_j)] \} \\
&\quad + \sum_{j=1}^{n_s} \{ x_j^t x_j - x_j^t(t - \tau_j) x_j(t - \tau_j) \} \\
&\quad + \sum_{j=1}^{n_s} \sum_{k=1}^{n_s} \{ x_k^t x_k - x_k^t(t - \eta_{jk}) x_k(t - \eta_{jk}) \} \\
&\quad + \sum_{j=1}^{n_s} \{ [\sum_{j=1}^{n_s} \tilde{A}_{jk} x_k(t - \eta_{jk})]^t W_j \xi_j \\
&\quad + \xi_j^t W_j [\sum_{j=1}^{n_s} \tilde{A}_{jk} x_k(t - \eta_{jk})] \}
\end{aligned}
\tag{7.27}
$$

Observing the fact

$$
\sum_{j=1}^{n_s} \sum_{k=1}^{n_s} x_k^t x_k = \sum_{j=1}^{n_s} \sum_{k=1}^{n_s} x_j^t x_j
\tag{7.28}
$$

and defining the extended state-vector

$$\zeta_j^t = [\xi_j^t,\ x_j^t(t - \tau_j),\ x_1^t(t - \eta_{j1}),\ ...,\ x_{n_s}^t(t - \eta_{jn_s})]$$ (7.29)

it then follows from (7.27) that

$$dV(\xi_t)/dt = \sum_{j=1}^{n_s} \zeta_j^t \Upsilon_j \zeta_j$$ (7.30)

where

$$\Upsilon_j = \begin{bmatrix} W_j\tilde{A}_j + \tilde{A}_j^t W_j + U_s & W_j\tilde{E}_j & W_j\tilde{A}_{j1} & ... & W_j\tilde{A}_{jn_s} \\ \tilde{E}_j^t W_j & -I & 0 & ... & 0 \\ \tilde{A}_{j1}^t W_j & 0 & -I & ... & 0 \\ . & 0 & 0 & . & . \\ . & . & . & . & . \\ \tilde{A}_{jn_s}^t W_j & 0 & 0 & . & -I \end{bmatrix}$$ (7.31)

In view of (7.25), it directly follows from (7.31) that $\Upsilon_j < 0$. Hence from (7.30), we get $dV(\xi_t)/dt < 0 \quad \forall \xi \neq 0$. Therefore, $V(\xi_t)$ of (7.26) is a Lyapunov function for the augmented system (7.10) and hence this system is asymptotically stable for all admissible uncertainties. It then follows that the interconnected system (7.1) is robustly stabilizable by the decentralized dynamic feedback controller (7.5).

Remark 7.3: It is readily evident from the preceding result that the closed-loop system stability is independent of delay. Although it is a conservative result, it is computationally simpler than those available in the literature; see [2,233] and the references cited therein. Had we followed another approach, we would have obtained delay-dependent stability result which requires initial data over the period $[-2\tau_j^*, 0], j = 1, .., s]$.

Remark 7.4: It is to be noted that the developed conditions of **Theorem 7.1** are only sufficient and therefore the results can be generally conservative. In order to reduce this conservativeness, some parameters $(\alpha_j, \beta_{jk}, \rho_j)$ are left to be adjusted by the designer.

Now to utilize **Theorem 7.1** in computer implementation, the following computational procedure is recommended:

(1)   Read the nominal matrices of subsystem j, $\forall j \in \{1, ..., n_s\}$ as given in model (5).

(2)   Identify the matrices of the uncertainty structure (7.2) and (7.3).

(3)   (a) Select initial values for the scalars $\alpha_j^2 = \alpha_o^2$, $\beta_{jk}^2 = \beta_o^2$
$\forall j, k \in \{1, ..., n_s\}$ such that $\alpha_o^2 \in [0.1, 0.2]$ and $\beta_o^2 \in [0.1, 0.2]$,
(b) Set counter $= 1$
(c) If $I_j - \alpha_j^2 N_{ej}^t N_{ej} > 0$ and $I_j - \beta_{jk}^2 N_{jk}^t N_{jk} > 0$ then go to (4). otherwise, update α_j^2 and β_{jk}^2 such that,
$\alpha_j^2 \Rightarrow \alpha_j^2 (1 + \Delta\alpha_j^2)$, $\beta_{jk}^2 \Rightarrow \beta_{jk}^2 (1 + \Delta\beta_{jk}^2)$ and
$\Delta\alpha_j^2$, $\Delta\beta_{jk}^2 \in [0.05, 0.1]$.
(d) Increase counter by 1.
(e) If counter exceeds 20 then go back to (2), else go to (3-c).

4)   Select the scalar ρ_j $\forall j \in \{1, ..., n_s\}$ and solve inequalities (7.6) and (7.7) for P_j, Q_j. Change ρ_j whenever necessary to ensure the positive-definiteness of the resulting matrices.
If no solution exists, update α_j, β_{jk} and go to (3).
If a solution exists, record the result and **STOP**.

Corollary 7.1: *Consider system (7.1) without time-delay; that is* $\tau_j = 0$, $\eta_{jk} = 0$ $\forall j, k \in \{1, ..., n_s\}$. *In this case we set* $E_j = 0$ $\Rightarrow$ $\Omega_{ej}\Omega_{ej}^t = 0$ $\forall j \in \{1, ..., n_s\}$. *It follows that the ARIs (7.6) and (7.7) reduce to:*

$$P_j A_{oj} + A_{oj}^t P_j - \rho_j^{-2} \tilde{B}_{oj}^t (N_j^t N_j)^{-1} \tilde{B}_{oj} + P_j \Omega_{sj}\Omega_{sj}^t P_j$$
$$+\rho_j^2 M_j^t M_j + (n_s + 1)I_j < 0 \tag{7.32}$$

$$Q_j (A_{oj} + \Omega_{sj}\Omega_{sj}^t P_j) + (A_{oj} + \Omega_{sj}\Omega_{sj}^t P_j)^t Q_j + Q_j \Omega_{sj}\Omega_{sj}^t Q_j$$
$$-\rho_j^2 J_j^t (L_j L_j^t)^{-1} J_j + \rho_j^{-2} \tilde{C}_{oj}^t (L_j L_j^t)^{-1} \tilde{C}_{oj} < 0 \tag{7.33}$$

$$\Omega_{sj}\Omega_{sj}^t = \Omega_{aj}\Omega_{aj}^t + \rho_j^{-2} S_j S_j^t \tag{7.34}$$

Remark 7.5: Note that the term $(n_s + 1)I$ in (7.32) implies that for an interconnected system composed of a large number of subsystems even though weakly coupled, it may not have a positive-definite solution. In this regard, this gives a measure of the conservativeness of the proposed design method.

Corollary 7.2: *Consider system (7.1) in the case that all state variables are measurable and available for feedback. It is readily seen that this system is robustly stabilizable via a decentralized state feedback controller structure*

$$u_j(t) \;\;=\;\; -K_j x_j(t)$$

$$= -\rho_j^{-2}(N_j^t N_j)^{-1}\tilde{B}_{oj}x_j(t) \ , \ \forall j \in \{1,...,n_s\} \tag{7.35}$$

if there exist $\rho_j > 0$ *and* $0 < P_j = P_j^t \in \Re^{n_j \times n_j}$ *satisfying*

$$P_j A_{oj} + A_{oj}^t P_j - \rho_j^{-2}\tilde{B}_{oj}^t(N_j^t N_j)^{-1}\tilde{B}_{oj} + P_j X_j X_j^t P_j +$$
$$\rho_j^2 M_j^t M_j + (n_s + 1)I_j < 0 \tag{7.36}$$

and the inverse $(N_j^t N_j)^{-1}$ *exists.*

Corollary 7.3: *The standard centralized solution of the* H_∞ *robust stabilization of dynamical systems without delay can be readily deduced from our results by dropping out the subscripts* j *and* k, *setting* $n_s = 1$, $A_{jk} = 0 \Rightarrow A_{jko} = 0$, $N_{jk} = 0$, $M_{jk} = 0 \Rightarrow \Omega_{jk} = 0$ *in addition to* $E_j = 0 \Rightarrow E_{oj} = 0$, $M_{ej} = 0$, $N_{ej} = 0$. *It then follows that*

$$PA_o + A_o^t P - \rho^{-2}[B_o^t P + \rho^2 N^t M]^t(N^t N)^{-1}[B_o^t P + \rho^2 N^t M] +$$
$$\rho^{-2}PSS^t P + \rho^2 M^t M + 2I < 0$$

which implies that

$$PA_o + A_o^t P - \rho^{-2}[B_o^t P + \rho^2 N^t M]^t(N^t N)^{-1}[B_o^t P + \rho^2 N^t M] +$$
$$\rho^{-2}PSS^t P + \rho^2 M^t M < 0 \tag{7.37}$$

$$Q(A_o + \rho^{-2}SS^t P) + (A_o + \rho^{-2}SS^t P)^t Q$$
$$+\rho^{-2}[C_o + \rho^2 LS^t P](LL^t)^{-1}[C_o + \rho^{-2}LS^t P]^t + \rho^{-2}QSS^t Q$$
$$-\rho^2[C_o + \rho^{-2}LS^t(P + Q)]^t(LL^t)^{-1}[C_o + \rho^{-2}LS^t(P + Q)] < 0$$
$$\tag{7.38}$$

7.4 Decentralized Robust H_∞ Performance

Now, we move to consider the stabilization of the interconnected system (7.1) by solving the problem of decentralized robust H_∞ performance. This problem is phrased as follows:

Given the n_s-positive scalars $\{\gamma_1, \gamma_2, ..., \gamma_{n_s}\}$, design decentralized dynamic feedback controllers for the interconnected system (7.1) such that the resulting closed-loop system is asymptotically stable and guarantees that under zero initial conditions

$$\sum_{j=1}^{n_s} ||z_j||_2 < \sum_{j=1}^{n_s} \gamma_j ||w_j||_2 \tag{7.39}$$

for all nonzero $w_j \in \mathcal{L}_2[0, \infty)$ and for all admissible uncertainties.

In this case, the interconnected system (7.1) is said to be *robustly stabilizable with disturbance attenuation* γ *via decentralized dynamic feedback control.*

Extending on (7.5), we use the observer-based feedback controller

$$S_j^o : \quad \begin{aligned} \dot{\hat{x}}_j(t) &= [A_{oj} + \delta A_j]\hat{x}_j(t) + \tilde{\Theta}_j[y_j(t) - \hat{y}_j(t)] + B_{oj}u_j(t) + \Gamma_j \hat{w}_j(t) \\ \hat{y}_j(t) &= [C_{oj} + \delta C_j]\hat{x}_j(t) + D_{oj}u_j(t) + \Pi_j \hat{w}_j(t) \\ u_j(t) &= -\tilde{K}_j \hat{x}_j(t) \end{aligned} \tag{7.40}$$

which has the same structure as (7.5) in addition to the auxiliary signal $\hat{w}_j(t) \in \mathcal{L}_2[0, \infty)$. This signal affects both the systems dynamics and the measured output and is introduced in order to cope with the external disturbance w_j. The different constant matrices in (7.40) are those of (7.5) in addition to

$$\delta A_j = X_j X_j^t P_j \ ; \quad \delta C_j = \rho_j^{-2} L_j S_j^t P_j \ ; \quad \hat{w}_j(t) = \gamma_j^{-2} \Gamma_j^t P_j \hat{x}_j(t) \tag{7.41}$$

and the gains $\tilde{\Theta}_j, \tilde{K}_j; \forall j \in \{1, ..., n_s\}$ will be specified shortly.

In terms of the error $e_j(t) = \hat{x}_j(t) - x_j(t)$, we obtain from (7.1), (7.2) and (7.40) the error model:

$$\begin{aligned} \dot{e}_j(t) = \ & [A_{oj} - \tilde{\Theta}_j C_{oj} + \delta A_j - \tilde{\Theta}_j \delta C_j \\ & + \ (S_j - \tilde{\Theta}_j L_j) H_j(t) N_j \tilde{K}_j + \gamma_j^{-2}(\Gamma_j - \tilde{\Theta}_j \Pi_j)\Gamma_j^t P_j] e_j \\ & + \ [(\delta A_j - \tilde{\Theta}_j \delta C_j) - (S_j - \tilde{\Theta}_j L_j) H_j(t)(M_j - N_j \tilde{K}_j) \\ & + \ \gamma_j^{-2}(\Gamma_j - \tilde{\Theta}_j \Pi_j)\Gamma_j^t P_j] x_j \\ & - \ [\Gamma_j - \tilde{\Theta}_j \Pi_j] w_j - E_j(t) x_j(t - \tau_j) - \sum_{k=1}^{n_s} A_{jk}(t) x_k(t - \eta_{jk}) \end{aligned} \tag{7.42}$$

In terms of $Z_j^t = [x_j^t \; e_j^t] \in \Re^{2n_j}$ the dynamics of the augmented system (7.1) and (7.42) can be put in the compact form:

$$\dot{Z}_j(t) = \hat{A}_j(t)Z_j(t) + \tilde{\Gamma}_j w_j + \tilde{E}x_j(t - \tau_j) + \sum_{k=1}^{n_s} \tilde{A}_{jk}(t)x_k(t - \eta_{jk}) \quad (7.43)$$

where

$$\hat{A}_j(t) = \widehat{A}_j + \tilde{S}H_j(t)\tilde{M}_j$$

$$\widehat{A}_j = \begin{bmatrix} A_{oj} - B_{oj}\tilde{K}_j & -B_{oj}\tilde{K}_j \\ \begin{matrix} \delta A_j - \tilde{\Theta}_j \delta C_j \\ +\gamma_j^{-2}(\Gamma_j - \tilde{\Theta}_j \Pi_j)\Gamma_j^t P_j \end{matrix} & \begin{matrix} A_{oj} - \tilde{\Theta}_j C_{oj} + \delta A_j \\ -\tilde{\Theta}_j \delta C_j \end{matrix} \end{bmatrix}$$

$$\tilde{\Gamma}_j = \begin{bmatrix} \Gamma_j \\ -(\Gamma_j - \tilde{\Theta}_j \Pi_j) \end{bmatrix} ; \quad \tilde{S}_j = \begin{bmatrix} S_j \\ -(S_j - \tilde{\Theta}_j L_j) \end{bmatrix} ;$$

$$\tilde{M}_j^t = \begin{bmatrix} (M_j - N_j \tilde{K}_j)^t \\ -(N_j \tilde{K}_j)^t \end{bmatrix} \quad (7.44)$$

and the remaining matrices are given by (7.12)-(7.14). From (7.1) and $e_j(t) = \hat{x}_j(t) - x_j(t)$, the controlled output has the form

$$z_j = [(G_j - F_j \tilde{K}_j) \quad -F_j \tilde{K}_j]\psi_j = \widehat{G}_j \Psi_j \quad (7.45)$$

Introduce a matrix $0 < W_j^t = W_j \in \Re^{2n_j \times 2n_j}$ such that $W_j = diag[P_j, Q_j]$ where the matrices $0 < P_j^t = P_j \in \Re^{n_j \times n_j}$, $0 < Q_j^t = Q_j \in \Re^{n_j \times n_j}$ satisfy the ARIs:

$$P_j A_{oj} + A_{oj}^t P_j + P_j Y_j Y_j^t P_j + G_j^t G_j + \rho_j^2 M_j^t M_j +$$
$$(n_s + 1)I_j - \hat{B}_{oj}^t \hat{N}_j^{-1} \hat{B}_{oj} < 0 \quad (7.46)$$

$$Q_j(A_{oj} + Y_j Y_j^t P_j) + (A_{oj} + Y_j Y_j^t P_j)^t Q_j + Q_j Y_j Y_j^t Q_j -$$
$$\hat{J}_j^t \hat{L}_j^{-1} \hat{J}_j + \hat{B}_{oj}^t \hat{N}_j^{-1} \hat{B}_{oj} < 0 \quad (7.47)$$

and

$$\hat{B}_{oj} = B_{oj}^t P_j + \rho_j^2 N_j^t M_j + F_j^t G_j \; , \quad \hat{N}_j^{-1} = (\rho_j^2 N_j^t N_j + F_j^t F_j)^{-1}$$
$$\hat{L}_j^{-1} = (\rho_j^{-2} L_j L_j^t + \gamma_j^{-2} \Pi_j \Pi_j^t)^{-1} \; , \quad Y_j Y_j^t = X_j X_j^t + \gamma_j^{-2} \Gamma_j \Gamma_j^t$$
$$\hat{J}_j = J_j + \gamma_j^{-2}(\Gamma_j \Pi_j^t Q_j + \Pi_j \Gamma_j^t P_j) \quad (7.48)$$

Theorem 7.2: *Given the desired levels of disturbance attenuation $\{\gamma_j > 0, j = 1, ..., n_s\}$, the augmented system (7.43) is robustly stabilizable with disturbance attenuation $\gamma = diag(\gamma_i)$ via the decentralized observer-based controller (41) if there exist positive scalars $\rho_j, \alpha_j, \beta_{jk}$, $\forall j = 1, ..., n_s$ such that the following conditions are met:*
   (1) *the matrices $(N_j^t N_j)$, $(L_j L_j^t)$, $(I_j - \alpha_j^2 N_{ej}^t N_{ej})$, $(I_j - \beta_{jk}^2 N_{jk}^t N_{jk})$, $(F_j^t F_j)$ and $(\Pi_j \Pi_j^t)$, are all invertible,*
   (2) *there exist $0 < P_j^t = P_j \in \Re^{n_j \times n_j}$, $0 < Q_j^t = Q_j \in \Re^{n_j \times n_j}$ satisfying inequalities (7.46) and (7.47), respectively. The feedback and observer gains are given by*

$$\tilde{\Theta}_j = Q_j^{-1}[J_j + \gamma_j^{-2}\Gamma_j\Pi_j^t Q_j]^t \tilde{L}_j^{-1} , \quad \tilde{K}_j = \hat{N}_j^{-1}\hat{B}_{oj} \qquad (7.49)$$

Proof: First, we have to establish the stability of the composite system (7.43)-(7.45). Given $0 < W_j^t = W_j \in \Re^{n_j \times n_j}$ such that $W_j = diag[P_j, Q_j]$ and $0 < P_j^t = P_j$, $0 < Q_j^t = Q_j$ satisfying inequalities (7.46) and (7.47), define the augmented matrix:

$$
\begin{aligned}
\Xi_j &= \begin{bmatrix} \Xi_{1j} & \Xi_{2j} \\ \Xi_{3j} & \Xi_{4j} \end{bmatrix} \\
&= \hat{A}_j^t W_j + W_j \hat{A}_j + W_j(\hat{X}_j\hat{X}_j^t + \gamma_j^{-2}\tilde{\Gamma}_j\tilde{\Gamma}_j^t)W_j \\
&\quad + \rho_j^2 \tilde{M}_j^t \tilde{M}_j + \hat{G}_j^t\hat{G}_j + U_s
\end{aligned}
\qquad (7.50)
$$

$\Xi_j \in \Re^{2n_j \times 2n_j}, \Xi_{kj} \in \Re^{n_j \times n_j}, k = 1, ..., 4.$

Algebraic manipulation of (7.50) using (7.12) and (7.19) in the manner of **Theorem 7.1** leads to:

$$\Xi_{1j} < 0, \quad \Xi_{2j} = 0, \quad \Xi_{3j} = 0, \quad \Xi_{4j} < 0, \quad j = 1, ..., n_s \qquad (7.51)$$

which in turn implies that

$$
\hat{A}_j^t W_j + W_j \hat{A}_j + W_j(\hat{X}_j\hat{X}_j^t + \gamma_j^{-2}\tilde{\Gamma}_j\tilde{\Gamma}_j^t)W_j + \rho_j^2 \tilde{M}_j^t \tilde{M}_j + \hat{G}_j^t\hat{G}_j + U_s < 0
\qquad (7.52)
$$

By applying **Lemma 7.1** to (7.50) with the help of (7.44), (7.46) and (7.47), we get

$$
\hat{A}_j^t W_j + W_j \hat{A}_j + \hat{G}_j^t\hat{G}_j + W_j(\tilde{E}_j\tilde{E}_j^t + \sum_{k=1}^{n_s} \tilde{A}_{jk}\tilde{A}_{jk}^t)W_j + \gamma_j^{-2}W_j\tilde{\Gamma}_j\tilde{\Gamma}_j^t W_j + U_s < 0
\qquad (7.53)
$$

Using a procedure similar to Theorem 7.1, the asymptotic stability of the closed-loop system (7.43) can be easily deduced.

Next, to establish that $||z_j||_2 < \gamma_j ||w_j||_2$ whenever $||w_j||_2 \neq 0$ which implies that the desired robust H_∞ performance is achieved, we introduce the performance measure

$$J_\infty = \sum_{j=1}^{n_s} \int_0^\infty [z_j^t(t)z_j(t) - \gamma_j^2 w_j^t(t)w_j(t)]dt \qquad (7.54)$$

which is bounded in view of the asymptotic stability of the closed-loop system (7.43) and since $w_j \in \mathcal{L}_2[0, \infty)$, $\forall j \in \{1, ..., n_s\}$. Consider the Lyapunov function candidate $V(\xi_t)$ of (7.27). The total time derivative $dV(\xi_t)/dt$, which is computed with respect to (7.43), is obtained as:

$$
\begin{aligned}
dV(\xi_t)/dt &= \sum_{j=1}^{n_s}\{\psi_j^t[W_j\hat{A}_j^t + \hat{A}_jW_j]\xi_j + [\tilde{E}_jx_j(t - \tau_j)]^tW_j\xi_j \\
&+ \xi_j^tW_j[\tilde{E}_jx_j(t - \tau_j)]\} \\
&+ \sum_{j=1}^{n_s}\{[\tilde{\Gamma}_jw_j(t)]^tW_j\xi_j + \xi_j^tW_j[\tilde{\Gamma}_jw_j(t)]\} \\
&+ \sum_{j=1}^{n_s}\{x_j^tx_j - x_j^t(t - \tau_j)x_j(t - \tau_j)\} \\
&+ \sum_{j=1}^{n_s}\sum_{k=1}^{n_s}\{x_k^tx_k - x_k^t(t - \eta_{jk})x_k(t - \eta_{jk})\} \\
&+ \sum_{j=1}^{n_s}\{[\sum_{j=1}^{n_s}\tilde{A}_{jk}x_k(t - \eta_{jk})]^tW_j\xi_j \\
&+ \xi_j^tW_j[\sum_{j=1}^{n_s}\tilde{A}_{jk}x_k(t - \eta_{jk})]\} \qquad (7.55)
\end{aligned}
$$

By augmenting (7.27), setting the initial conditions of system (7.1) to zero and using (7.43), we rewrite (7.54) in the form:

$$J_\infty^a = \sum_{j=1}^{n_s} \int_0^\infty [\xi_j^t(t)\widehat{G}_j^t\widehat{G}_j\xi_j(t) - \gamma_j^2 w_j^t(t)w_j(t) + \dot{V}_j(t)]dt - \sum_{j=1}^{n_s}V_j^\infty \quad (7.56)$$

where V_j^∞ is a non-negative and finite scalar representing the bound of $V_j(t)$ as $t \Rightarrow \infty$. Using (7.30) and the extended state-vector

$$\hat{\zeta}_j^t = [\xi_j^t,\ x_j^t(t-\tau_j),\ x_1^t(t-\eta_{j1}),\ ...,\ x_{n_s}^t(t-\eta_{jn_s}),\ w_j^t] \qquad (7.57)$$

we express (7.56) in the compact form:

$$J_\infty^a = \sum_{j=1}^{n_s} \int_0^\infty [\hat{\zeta}_j^t \hat{\Upsilon} \hat{\zeta}_j] dt - \sum_{j=1}^{n_s} V_j^\infty \qquad (7.58)$$

where

$$\hat{\Upsilon} = \begin{bmatrix}
\begin{array}{c} W_j \hat{A}_j + \hat{A}_j^t W_j \\ +\widehat{G_j^t G_j} + U_s \end{array} & W_j \tilde{E}_j & W_j \tilde{A}_{j1} & ... & W_j \tilde{A}_{jn_s} & W_j \tilde{\Gamma}_j \\
\tilde{E}_j^t W_j & -I & 0 & 0 & ... & 0 \\
\tilde{A}_{j1}^t W_j & 0 & -I & 0 & ... & 0 \\
0 & 0 & 0 & -I & . & 0 \\
. & . & . & 0 & 0 & . \\
\tilde{A}_{jn_s}^t W_j & . & . & . & . & . \\
\tilde{\Gamma}_j^t W_j & 0 & 0 & 0 & . & -\gamma_j^2 I
\end{array}
\end{bmatrix}$$

In view of the negative-definiteness of $\hat{\Upsilon}$, it is readily evident from (7.58) that $J_\infty^a < 0$. This in turn implies that condition **(2)** is satisfied for all admissible uncertainties and for all nonzero disturbances $w_j \in \mathcal{L}_2[0, \infty)$, $\forall j \in \{1, ..., n_s\}$.

Corollary 7.4: *Consider system (7.1) without time-delay; that is $\tau_j = 0$, $\eta_{jk} = 0$ $\forall j, k \in \{1, ..., n_s\}$. In this case we set $E_j = 0$ and thus $\Omega_{ej}\Omega_{ej}^t$ $\forall j \in \{1, ..., n_s\}$. It follows that the ARIs (7.46) and (7.47) reduce to:*

$$P_j A_{oj} + A_{oj}^t P_j + P_j(\Omega_{sj}\Omega_{sj}^t + \gamma_j^{-2}\Gamma_j\Gamma_j^t)P_j + G_j^t G_j +$$
$$\rho_j^2 M_j^t M_j + (n_s + 1)I_j - \hat{B}_{oj}^t \hat{N}_j^{-1} \hat{B}_{oj} < 0 \qquad (7.59)$$

$$Q_j(A_{oj} + [\Omega_{sj}\Omega_{sj}^t + \gamma_j^{-2}\Gamma_j\Gamma_j^t]P_j) + (A_{oj} + [\Omega_{sj}\Omega_{sj}^t + \gamma_j^{-2}\Gamma_j\Gamma_j^t]P_j)^t Q_j$$
$$+Q_j(\Omega_{sj}\Omega_{sj}^t + \gamma_j^{-2}\Gamma_j\Gamma_j^t)Q_j - \hat{J}_j^t \hat{L}_j^{-1} \hat{J}_j + \hat{B}_{oj}^t \hat{N}_j^{-1} \hat{B}_{oj} < 0 \qquad (7.60)$$

Corollary 7.5: *Consider system (7.1) in the case that all state variables are measurable and available for feedback. It is readily seen that this system* is robustly stabilizable via a decentralized state feedback controller structure

$$u_j(t) = -K_j x_j(t)$$
$$= -\rho_j^{-2}(N_j^t N_j)^{-1}\tilde{B}_{oj}x_j(t) \ , \ \forall j \in \{1,...,n_s\} \qquad (7.61)$$

if there exist $\rho_j > 0$ and $0 < P_j = P_j^t \in \Re^{n_j \times n_j}$ satisfying

$$P_j A_{oj} + A_{oj}^t P_j + P_j Y_j Y_j^t P_j + G_j^t G_j + \rho_j^2 M_j^t M_j$$
$$- \hat{B}_{oj}^t \hat{N}_j^{-1}\hat{B}_{oj} + (n_s + 1)I_j < 0 \qquad (7.62)$$

7.4.1 Example 7.1

To illustrate the design procedures developed in **Theorems 7.1 and 7.2**, we consider a representative water pollution model of three consecutive reaches of the River Nile. This linearized model forms an interconnected system of the type (7.1) for $n_s = 3$ along with the following information:

$$A_{o1} = \begin{bmatrix} 0.97 & -0.31 \\ 1.2 & 0 \end{bmatrix}; \ A_{o2} = \begin{bmatrix} 0 & 1 \\ 1.2 & -0.87 \end{bmatrix}; \ A_{o3} = \begin{bmatrix} 0 & 1 \\ 0.78 & 1.2 \end{bmatrix}$$

$$B_{o1} = \begin{bmatrix} 1 \\ 0 \end{bmatrix}; \ B_{o2} = \begin{bmatrix} 0 \\ 1 \end{bmatrix}; \ B_{o3} = \begin{bmatrix} 0 \\ 1 \end{bmatrix}$$

$$C_{o1} = \begin{bmatrix} 1 & -1 \end{bmatrix}; \ C_{o2} = \begin{bmatrix} 1 & 1 \end{bmatrix}; \ C_{o3} = \begin{bmatrix} -1 & -1 \end{bmatrix}$$

The delay and disturbance parameters are given by

$$\Gamma_1 = \begin{bmatrix} 0.1 \\ 0.1 \end{bmatrix}; \ \Gamma_2 = \begin{bmatrix} 0 \\ 0.3 \end{bmatrix}; \ \Gamma_3 = \begin{bmatrix} 0 \\ 0.2 \end{bmatrix}$$

$E_{o1} = 0.2$, $E_{o2} = 0.1$, $E_{o3} = 0.3$, $\Pi_1 = 0.08$, $\Pi_2 = 0.15$, $\Pi_3 = 0.06$, $D_{o1} = 0.02$, $D_{o2} = 0.03$, $D_{o3} = 0.01$, $G_1 = [0.1 \ 0.1 \ 0.2]$, $G_2 = [0.2 \ 0.1 \ 0.2]$, $G_3 = [0.2 \ 0.1 \ 0.1]$, $F_1 = 0.1$, $F_2 = 0.1$, $F_3 = 0.1$.
The coupling matrices are given by

$$A_{120} = \begin{bmatrix} -0.5 & 0 \\ 0 & 1 \end{bmatrix}; \ A_{130} = \begin{bmatrix} -0.3 & 0 \\ 0 & 0.4 \end{bmatrix}; \ A_{210} = \begin{bmatrix} -0.5 & 0 \\ 0 & 1 \end{bmatrix}$$

$$A_{230} = \begin{bmatrix} 0.5 & 0 \\ 0 & 1 \end{bmatrix}; \ A_{310} = \begin{bmatrix} -0.5 & 0 \\ 0 & 0.5 \end{bmatrix}; \ A_{320} = \begin{bmatrix} 0.3 & 0 \\ 0 & 0.4 \end{bmatrix}$$

$$M_{12} = \begin{bmatrix} 0.2 \\ 0.1 \end{bmatrix}; \quad M_{13} = \begin{bmatrix} -0.4 \\ 0.4 \end{bmatrix}; \quad M_{21} = \begin{bmatrix} 0.5 \\ -0.3 \end{bmatrix}; \quad M_{23} = \begin{bmatrix} 0.4 \\ 0.5 \end{bmatrix}$$

$$M_{31} = \begin{bmatrix} -0.6 \\ 0.5 \end{bmatrix}; \quad M_{32} = \begin{bmatrix} 0.2 \\ -0.1 \end{bmatrix}; \quad N_{12}^t = \begin{bmatrix} 0.3 \\ 0.1 \end{bmatrix}; \quad N_{13}^t = \begin{bmatrix} 0.3 \\ -0.3 \end{bmatrix}$$

$$N_{21}^t = \begin{bmatrix} -0.1 \\ 0.6 \end{bmatrix}; \quad N_{23}^t = \begin{bmatrix} 0.4 \\ 0.5 \end{bmatrix}; \quad N_{31}^t = \begin{bmatrix} 0.3 \\ -0.3 \end{bmatrix}; \quad N_{32}^t = \begin{bmatrix} 0.4 \\ -0.3 \end{bmatrix}$$

In terms of the uncertainty structure (7.4), the following data is made available:

$$S_1 = \begin{bmatrix} 0.1 \\ 0.2 \end{bmatrix}; \quad S_2 = \begin{bmatrix} 0.2 \\ 0.1 \end{bmatrix}; \quad S_3 = \begin{bmatrix} 0.2 \\ 0.2 \end{bmatrix}$$

$M_1 = [0.8 \ 0.3]$, $M_2 = [0.6 \ 0.4]$, $M_3 = [0.7 \ 0.2]$, $N_1 = 0.4$, $N_2 = -0.5$, $N_3 = -0.3$, $L_1 = 4$, $L_2 = 3$, $L_3 = 5$, $M_{e1} = 0.3$, $M_{e2} = 0.2$, $M_{e3} = 0.4$, $N_{e1} = 0.1$, $N_{e2} = 0.2$, $N_{e3} = 0.3$.

By selecting $\alpha_1 = 0.5$, $\alpha_2 = 0.5$, $\alpha_3 = 0.5$, $\beta_1 = 0.5$, $\beta_2 = 0.5$, $\beta_3 = 0.5$ as initial guess, it is found by applying the computational procedure set forth that $\rho_1 = 0.2$, $\rho_2 = 0.3$, $\rho_3 = 0.4$ is a successful (first) choice. Using the above plus the nominal data with the help of **MATLAB-Toolbox**, we obtain

$$P_1 = \begin{bmatrix} 0.1331 & 0.0364 \\ 0.0364 & 0.9109 \end{bmatrix}; \quad P_2 = \begin{bmatrix} 0.1445 & 0.0782 \\ 0.0782 & 0.7923 \end{bmatrix};$$

$$P_3 = \begin{bmatrix} 0.1557 & 0.1421 \\ 0.1421 & 0.1460 \end{bmatrix}; \quad Q_1 = \begin{bmatrix} 0.9070 & -0.5361 \\ -0.5361 & 2.3821 \end{bmatrix};$$

$$Q_2 = \begin{bmatrix} 2.2218 & 0.8047 \\ 0.8047 & 1.9533 \end{bmatrix}; \quad Q_3 = \begin{bmatrix} 2.6863 & 0.5944 \\ 0.5944 & 2.4828 \end{bmatrix}$$

as a candidate solution of inequalities (7.6) and (7.7). These give the following gain matrices

$$K_1 = [22.8166 \ \ 6.4302]; K_2 = [2.2767 \ \ 34.4125]; K_3 = [7.5354 \ \ 78.9143].$$

$$\Theta_1 = \begin{bmatrix} 0.0223 \\ 0.0235 \end{bmatrix}; \quad \Theta_2 = \begin{bmatrix} 0.0027 \\ 0.0202 \end{bmatrix}; \quad \Theta_3 = \begin{bmatrix} -0.0021 \\ 0.0187 \end{bmatrix}$$

Had we followed another route and took $\rho_1 = 0.02$, $\rho_2 = 0.03$, $\rho_3 = 0.04$ as a second choice, we then obtain

$$P_1 = \begin{bmatrix} 0.0138 & 0.0003 \\ 0.0003 & 0.1711 \end{bmatrix}; \quad P_2 = \begin{bmatrix} 0.0139 & 0.0013 \\ 0.0013 & 0.1693 \end{bmatrix};$$

$$P_3 = \begin{bmatrix} 0.0139 & 0.0021 \\ 0.0021 & 0.1876 \end{bmatrix}; \quad Q_1 = \begin{bmatrix} 0.0182 & -0.0181 \\ -0.0181 & 0.0191 \end{bmatrix};$$

$$Q_2 = \begin{bmatrix} 0.0275 & 0.0270 \\ 0.0270 & 0.0269 \end{bmatrix}; \quad Q_3 = \begin{bmatrix} 0.1610 & -0.0723 \\ -0.0723 & 0.1851 \end{bmatrix}$$

and the corresponding gain matrices

$$K_1 = [217.3081 \quad 5.6333]; \quad K_2 = [4.5533 \quad 751.6747]; \quad K_3 = [12.0 \quad 1302.2].$$

$$\Theta_1 = \begin{bmatrix} 8.1091 \\ 8.1291 \end{bmatrix}; \quad \Theta_2 = \begin{bmatrix} -17.5038 \\ 17.8486 \end{bmatrix}; \quad \Theta_3 = \begin{bmatrix} 0.0264 \\ 0.0510 \end{bmatrix}.$$

Note that the values of the first are 10 times that of the second choice. In both cases, the obtained results indicate the high-gain nature of the state-feedback controller. By reducing the weighting factor ρ_j, the norms $||P||$ and $||Q||$ decrease almost linearly. The rate of decrease of $||Q||$ is less than that of $||P||$. The reverse is true for the associated gain matrices where $||K||$ and $||\Theta||$ have increased with decreasing weighting factor ρ_j.

7.5 Decentralized Robust Stabilization II

In this section, we follow another route in dealing with the control problem of interconnected systems. The major difference stems from the way the uncertainties are modeled. Here we will consider that the uncertainties are mismatched and develop a decentralized stabilizing controller that renders the closed-loop controlled system exponentially stable to a calculable zone about the origin.

7.5.1 Problem Statement and Preliminaries

We knew from previous chapters that a wide class of uncertain systems with time-varying state delay can be modeled by:

$$(\Sigma): \quad \dot{x}(t) = Ax(t) + Bu(t) + A_d x(t - \eta(t))$$
$$+ \; e(x, u, t) + g(x, t) \tag{7.63}$$

where $A \in \Re^{n \times n}, B \in \Re^{n \times m}$ are known constant matrices describing the nominal system (without uncertainties), $x(t) \in \Re^n$ is the state vector, $u(t) \in \Re^m$ is the control input and the system uncertianties are lumped into three parts:

$A_d \in \Re^{n \times n}$, $e(.,.,.) : \Re^n \times \Re^m \times \Re_+ \to \in \Re^n$ and $g(.,.) : \Re^n \times \Re_+ \to \in \Re^n$. In (7.63), the information of the delay depends on the state variables and it is a time function

$$\eta(t) \in \aleph := \{\varphi(t) : 0 \le \varphi \le \eta^* < \infty; \dot\varphi \le \eta^+ < 1\}$$

where the bounds η^*, η^+ are known constants in order to guarantee smooth growth of the state trajectories.

In the absence of uncertainties, the resulting system becomes the familiar retarded type that we have studied in the previous chapters. Here, we build on this fact by treating system (Σ) as an interconnection of N_s subsystems and having the following description:

$$n = \sum_{j=1}^{N_s} n_j \;,\quad m = \sum_{j=1}^{N_s} m_j$$

$$A = block - diag[A_1 \; A_2 \; .... \; A_{N_s}] \;,\quad A_j \in \Re^{n_j \times n_j}$$

$$B = block - diag[B_1 \; B_2 \; .... \; B_{N_s}] \;,\quad B_j \in \Re^{n_j \times m_j}$$

$$A_d = block - diag[A_{d1} \; A_{d2} \; .... \; A_{dN_s}] \;,\quad A_{dj} \in \Re^{n_j \times n_j}$$

$$g(x,t) = block - diag[g_1^t(x,t) \; g_2^t(x,t) \; .... \; g_{N_s}^t(x,t)]^t \;,$$

$$e(x,u,t) = block - diag[e_1^t(x,u,t) \; e_2^t(x,u,t) \; .... \; g_{N_s}^t(x,u,t)]^t \;,$$

$$g_j(x,t) \in \Re_j^n \;,\quad e_j(x,u,t) \in \Re_j^n$$

$$x(t) = block - diag[x_1^t(t) \; x_2^t(t) \; .... \; x_{N_s}^t(t)]^t \;,$$

$$u(t) = block - diag[u_1^t(t) \; u_2^t(t) \; .... \; u_{N_s}^t(t)]^t \;,$$

$$x_j(t) \in \Re_j^n \;,\quad u_j(x,t) \in \Re_j^m$$

$$g_j(x,t) = \sum_{k=1,j \neq k}^{N_s} g_{jk}(x_k,t) \tag{7.64}$$

The dynamics of the uncertain subsystem (Σ_j) has the form:

$$(\Sigma_j): \quad \dot{x}_j(t) = A_j x_j(t) + B_j u_j(t) + A_{dj} x_j(t - \eta_j(t))$$

$$+ \ e_j(x_j, u_j, t) + \sum_{k=1, j\neq k}^{N_s} g_{jk}(x_k, t) \qquad (7.65)$$

From (7.65), it is clear that the uncertainties are due to local parameter changes within the subsystems (terms A_{dj} and $e_j(x_j, u_j, t)$) and variation in the subsystem couplings (term g_{jk}).

In the following, we focus attention on the robust decentralized stabilization problem using local state-feedback controllers

$$u_j(t) \ = \ - \ K_j \, x_j(t), \quad K_j \in \Re^{m_j \times n_j} \qquad (7.66)$$

such that the closed-loop uncertain system (7.63) under the action of a group of controllers of the type (7.67) is uniformly exponentially convergent about the origin. For this purpose, we recall the following definition and result [243]:

Definition 7.1: *The dynamical system*

$$\dot{x}(t) = f(x(t), t), \quad t \in \Re, \quad x(t) \in \Re^n$$

is (globally, uniformly) exponentially convergent to $\mathcal{G}(r)$ if and only if there exists a scalar $\alpha > 0$ with the property that

$$\forall t_o \in \Re \ , \ x(t) \in \Re^n$$

there exists $c(x_o) \geq 0$ such that, if $x(.) : [t_o, t_1] \rightarrow \Re^n$ is any solution of $\dot{x}(t) = f(x(t), t) \, , x(t_o) = x_o$ then

$$||x(t)|| \leq r + c(x_o) \, e^{-\alpha(t-t_o)} \ , \ t \geq t_o$$

Lemma 7.2: *Suppose there exists a continuously differentiable function $V : \Re^n \rightarrow \Re_+$ and scalars $\gamma_1 > 0, \gamma_2 > 0$ which satisfy*

$$\gamma_1 ||x||^2 \ \leq \ V(x) \ \leq \gamma_2 ||x||^2, \quad \forall x \in \Re^n$$

such that for some scalar $\alpha > 0, V_r \geq 0$

$$V(x) > V_r \implies \nabla V^t(x) f(x(t), t) \ \leq -2\alpha[V(x) - V_r]$$

Define $r := (\gamma_1^{-1} V_r) 1/2$. Then the system $\dot{x}(t) = f(x(t), t)$ is exponentially convergent to $\mathcal{G}(r)$ with rate α, where

$$c(x_o) = \begin{cases} 0 & \text{if } V(x_o) \leq V_r \\ [\gamma_1^{-1}(V(x_o) - V_r)]^{1/2} & \text{if } V(x_o) > V_r \end{cases}$$

To proceed further, the following structural assumptions are needed:

Assumption 7.1: *The N_s-pairs $\{A_j, B_j\}$ are stabilizable.*

Assumption 7.2: *For every $j, k \in \{1, ..., N_s\}$, the system uncertainties admit the following decompositions:*

$$
\begin{aligned}
A_{dj} &= B_j F_j + G_j \\
e(x_j, u_j, t) &= B_j d_j(x_j, u_j, t) + f_j(x_j, u_j, t) \\
g_{jk}(x_k, t) &= B_j h_{jk}(x_k, t) + r_{jk}(x_k, t)
\end{aligned}
\tag{7.67}
$$

and there exist known scalar constants $\delta_j, \vartheta_j, \xi_j, \phi_j, \kappa_j, \rho_j, \zeta_j, \alpha_{jk}, \omega_{jk}, \pi_{jk}, \sigma_{jk} \in [0, \infty), \beta_j \in (0, 1)$, *such that:*

$$
\begin{aligned}
\delta_j &= \lambda_M(G_j G_j^t) \ , \ \vartheta_j = \lambda_M(F_j F_j^t) \\
\|d_j(x_j, u_j, t)\| &\leq \phi_j + \xi_j \|x_j\| + \beta_j \|u_j\| \\
\|f(x_j, u_j, t)\| &\leq \kappa_j + \rho_j \|x_j\| + \zeta_j \|u_j\| \\
\|h_{jk}(x_k, t)\| &\leq \alpha_{jk} + \omega_{jk} \|x_k\| \\
\|r_{jk}(x_k, t)\| &\leq \sigma_{jk} + \pi_{jk} \|x_k\| \\
&\forall \ (x_j, u_j, \eta_j, t) \ \Re_j^n \times \Re_j^m \times [0, \eta^*] \times \Re
\end{aligned}
\tag{7.68}
$$

Remark 7.6: It should be noted that Assumption 7.1 is standard and pertains to the nominal part of the subsystems. Expressions (7.67) imply that the uncertainties (due to delay factors, parameters and coupling variables) do not satisfy the matching condition and therefore can be represented by a matched part and a mismatched part. Both representations are not unique and the functions involved are unknown-but-bounded with the corresponding bounds being known. Note that $F_j = 0, G_j = A_{dj}$ is an admissible decomposition.

7.6   Decentralized Stabilizing Controller

It is well-known that a key feature of reliable control of interconnected systems [184, 187, 188] is to base all the design effort on the subsystem level. For this purpose, we choose a Lyapunov-Krasovskii functional

$$
V(x_t) = \sum_{j=1}^{n_s} V_j(x_{jt}),
$$

$$= \sum_{j=1}^{n_s} x_j^t P_j x_j + \sum_{j=1}^{n_s} \mu_j \int_{t-\eta_j}^{t} x_j^t(\alpha) x_j(\alpha) d\alpha \qquad (7.69)$$

which takes into account the present as well as the delayed states at the subsystem level, where $\forall j \in [1, .., N_s]$; $\mu_j > 0$, $\varphi_j \geq 0$, $\Delta_j > 1$ are design parameters and $\forall j \in [1, .., N_s]$; $\tau_j \geq 0$; $0 < P_j = P_j^t \in \Re^{n_j \times n_j}$ such that

$$P_j(A_j + \tau_j I) + (A_j + \tau_j I)^t P_j - \varphi_j P_j B_j B_j^t P_j + \mu_j \Delta_j I \qquad (7.70)$$

Define

$$
\begin{aligned}
P &= block - diag[P_1 \ P_2 ..... \ P_{N_s}] \ , \ \Lambda = \lambda_m(P) \\
\tau_m &= min\{\tau_1, ..., \tau_{N_s}\}, \ \Omega = \lambda_M(P)
\end{aligned} \qquad (7.71)
$$

Note that $V(x_t) > 0$ for $x \neq 0$ and $V(x_t) = 0$ when $x = 0$.

Theorem 7.3: *Consider the uncertain system (7.65) subject to* **Assumption 7.1** *and* **Assumption 7.2** *with the decentralized control*

$$u_j = -\gamma_j B_j^t P_j x_j, \quad j = 1, ..., N_s \qquad (7.72)$$

Choose the local gain factors $\{\gamma_j\}$ to satisfy

$$\gamma_j > (1/2)(1 + \varphi_j)(1 - \beta_j)^{-1} \qquad (7.73)$$

If there exist scalar $\varphi_j, \mu_j \in (0, \infty), \Delta \in (1, \infty)$ such that

$$
\begin{aligned}
\mu_j (1 - \eta_j^+) &= 1 + \vartheta_j \\
(\Delta_j - 1) \mu_j &> a_j + b_j
\end{aligned} \qquad (7.74)
$$

where

$$a_j = \lambda_M(P_j) \left\{ 2\rho_j + \delta_j \lambda_M(P_j) + 2 \sum_{k=1, k \neq j}^{N_s} \pi_{kj} \right\} \qquad (7.75)$$

$$b_j = 2\|B_j^t P_j\|(1 - \beta_j)^{-1} \left\{ \lambda_M(P_j) \zeta_j(1 + \varphi_j) \right\} \qquad (7.76)$$

$$+ \ 2\|B_j^t P_j\| \left\{ \xi_j + \sum_{k=1, k \neq j}^{N_s} \omega_{kj} \right\}$$

$$s_j = 2\lambda_M(P_j) \left\{ \kappa_j + \sum_{k=1, k \neq j}^{N_s} \sigma_{kj} \right\} \qquad (7.77)$$

$$c_j = 2 \left\{ \phi_j + \sum_{k=1, k \neq j}^{N_s} \alpha_{kj} \right\} \qquad (7.78)$$

then the closed-loop uncertain system (7.63) is uniformly exponentially convergent to the ball $\mathcal{G}(\Gamma)$ of radius Γ at a rate τ_m where

$$\Gamma = \left\{1/2 \, \Theta \, \tau_m^{-1} \Lambda\right\}^{1/2}$$

$$\Theta = \sum_{j=1}^{N_s} \Theta_j,$$

$$\Theta_j = \left\{\frac{c_j^2}{4[2\gamma_j(1-\beta_j)-1-\varphi_j]}\right\} + \left\{\frac{s_j^2}{4[(\Delta_j-1)-a_j-b_j]}\right\} \quad (7.79)$$

Proof: The total time derivative $dV(x_t)/dt$ of the function $V(x_t)$, which is computed with respect to (7.65), is obtained as:

$$dV(x_t)/dt = \sum_{j=1}^{N_s} [x_j^t P_j \dot{x}_j + \dot{x}_j^t P_j x_j]$$

$$+ \sum_{j=1}^{N_s} \mu_j [x_j^t x_j - (1-\dot{\eta}) x_j^t (t - eta_j) x(t-\eta_j)] \quad (7.80)$$

A little algebra on (7.80) using (7.65) and (7.70) shows that:

$$dV(x_t)/dt = \sum_{j=1}^{N_s} -2\tau_j x_j^t P_j x_j - \sum_{j=1}^{N_s} (2\gamma_j - \varphi_j) x_j^t P_j B_j B_j^t P_j x_j$$

$$- \sum_{j=1}^{N_s} \mu_j [x_j^t x_j - (1-\dot{\eta}) x_j^t (t - eta_j) x(t-\eta_j)]$$

$$- \sum_{j=1}^{N_s} (\Delta_j - 1) \mu_j x_j^t x_j + \sum_{j=1}^{N_s} 2 x_j^t P_j A_{dj} x_j (t-\eta_j)$$

$$+ \sum_{j=1}^{N_s} 2 x_j^t P_j \left\{ e_j + \sum_{k=1, j \neq k}^{N_s} g_{jk} \right\} \quad (7.81)$$

By **B.1.1** and using (7.67)-(7.68), it follows that:

$$\sum_{j=1}^{N_s} 2 x_j^t P_j A_{dj} x_j (t-\eta_j) = \sum_{j=1}^{N_s} 2 x_j^t P_j F_j x_j (t-\eta_j) + \sum_{j=1}^{N_s} 2 x_j^t P_j G_j x_j (t-\eta_j)$$

$$\leq \sum_{j=1}^{N_s} x_j^t P_j B_j B_j^t P_j x_j$$

$$+ \sum_{j=1}^{N_s} x_j^t(t-\eta_j) F_j^t F_j x_j(t-\eta_j)$$

$$+ \sum_{j=1}^{N_s} x_j^t P_j G_j G_j^t P x_j + \sum_{j=1}^{N_s} x_j^t(t-\eta_j) x_j(t-\eta_j)$$

$$\leq \sum_{j=1}^{N_s} \|B_j^t P_j x_j\|^2 + \sum_{j=1}^{N_s} [1+\vartheta_j] x_j^t(t-\eta_j) x_j(t-\eta_j)$$

$$+ \sum_{j=1}^{N_s} \|P_j x_j\|^2 \delta_j \qquad (7.82)$$

$$\sum_{j=1}^{N_s} 2x_j^t P_j \left\{ e_j + \sum_{k=1, j\neq k}^{N_s} g_{jk} \right\} = \sum_{j=1}^{N_s} 2x_j^t P_j \{[B_j d_j + f_j]\}$$

$$+ \sum_{k=1, j\neq k}^{N_s} \{[B_j h_{jk} + r_{jk}]\}$$

$$\leq \sum_{j=1}^{N_s} 2\|B_j^t P_j x_j\|(\phi_j + \xi_j \|x_j\|)$$

$$+ \sum_{j=1}^{N_s} 2\|B_j^t P_j x_j\|(\beta_j \|\gamma_j B_j^t P_j x_j\|)$$

$$+ \sum_{j=1}^{N_s} 2\|P_j x_j\|(\kappa_j + \rho_j \|x_j\|)$$

$$+ \sum_{j=1}^{N_s} 2\|P_j x_j\|(\zeta_j \|\gamma_j B_j^t P_j x_j\|)$$

$$+ \sum_{j=1}^{N_s} 2\|B_j^t P_j x_j\|(\sum_{j=1, j\neq k}^{N_s} [\alpha_{jk} + \omega_{jk} \|x_{jk}\|])$$

$$+ \sum_{j=1}^{N_s} 2\|P_j x_j\|(\sum_{j=1, j\neq k}^{N_s} [\sigma_{jk} + \pi_{jk} \|x_{jk}\|])$$

$$(7.83)$$

The substitution of (7.82)-(7.83) into (7.81) using (7.74)-(7.78) yields:

$$
dV(x_t)/dt \ \leq \ \sum_{j=1}^{N_s} -2\tau_j x_j^t P_j x_j - \sum_{j=1}^{N_s}(2\gamma_j(1-\beta_j)-1-\varphi_j)\|B_j^t P_j x_j\|^2
$$

$$
+ \ \sum_{j=1}^{N_s} c_j\|B_j^t P_j x_j\| - \sum_{j=1}^{N_s}[(\Delta_j-1)\mu_j - a_j - b_j]\|x_j^t\|^2
$$

$$
+ \ \sum_{j=1}^{N_s}[1+\vartheta_j - \mu_j(1-\eta_j^+)]\|x_j(t-\eta_j)\|2 \tag{7.84}
$$

Using **B.2.2**, it follows that

$$
c_j\|B_j^t P_j x_j\| \ - \ [2\gamma_j(1-\beta_j)-1-\varphi_j]\|B_j^t P_j x_j\|^2 \leq
$$
$$
c_j^2\{4[2\gamma_j(1-\beta_j)-1-\varphi_j]\}^{-2} \tag{7.85}
$$
$$
s_j\|x_j\| \ - \ [(\Delta_j-1)\mu_j - a_j - b_j]\|x_j\|^2 \leq
$$
$$
s_j^2\{4[(\Delta_j-1)\mu_j - a_j - b_j]\}^{-2} \tag{7.86}
$$

Then from (7.74) and (7.85)-(7.86), we get:

$$
dV(x_t)/dt \ \leq \ \sum_{j=1}^{N_s} -2\tau_j x_j^t P_j x_j + \sum_{j=1}^{N_s}[\frac{c_j^2}{4[2\gamma_j(1-\beta_j)-1-\varphi_j]}]
$$

$$
+ \ \sum_{j=1}^{N_s}[\frac{s_j^2}{4[(\Delta_j-1)\mu_j - a_j - b_j]}]
$$

$$
\leq \ \sum_{j=1}^{N_s}[-2\tau_j V_j + \Theta_j]
$$

$$
\leq \ -\tau_m[V_t - 1/2\,\tau_m^{-1}\,\Theta]
$$

$$
\leq \ -\tau_m[V_t - V_f] \tag{7.87}
$$

Since from (7.69) we have $\Lambda\|x\|^2 \leq V(x) \leq \Omega\|x\|^2$, it follows from **Lemma 7.2** in the light of (7.71) and (7.79) that the closed-loop system is uniformly exponentially convergent about the origin at a rate of τ_m.

Corollary 7.6: *In the case of affinely cone-bounded uncertainties, it follows that ϕ_j, κ_j, σ_{jk}, $\alpha_{jk} = 0$, $\forall j, k = 1, ..., N_s$. Hence from (7.76)-(7.78) we get $c_j \to 0, s_j \to 0$ leading to $\Theta \to 0$ and $\Gamma \to 0$. This implies that the zone around the origin reduces to zero corresponding to asymptotic stability* with a prescribed rate of convergence.

Corollary 7.7: *In the special case $F_j = 0, G_j = A_{dj}$,$\forall j, k = 1, ..., N_s$, it follows that $\vartheta_j = 0 \, \forall j$ and hence (7.74) reduces to*

$$\mu_j (1 - \eta_j^+) = 1 \; , \; (\Delta_j - 1) \mu_j > a_j + b_j$$

7.6.1 Adjustment Procedure

Based on **Theorem 7.3**, the following procedure is used to decide upon the design parameters:

(1): Read the input data from subsystem j ; $j = 1, ..., N_s$ as given by (7.65)-(7.66).
(2): Compute the parameters $\delta_j, \vartheta_j, \xi_j, \phi_j, \kappa_j, \rho_j, \zeta_j, \alpha_{jk}, \omega_{jk}, \pi_{jk}, \sigma_{jk}, \beta_j$.
(3): Select μ_j to satisfy (7.74) and select initial values of $\varphi_j, \Delta_j, \tau_j$.
(4): Compute P_j from (7.70) and Λ, Ω, τ_m from (7.71).
(5): Check condition (7.74). If it is satisfied, then STOP and record the results.
Otherwise, increment $\Delta_j + \delta\Delta_j \Leftarrow \Delta_j$ where $\delta\Delta_j = (0.05 \rightarrow 0.15)\Delta_j$ and go back to **(4)**.

7.6.2 Example 7.2

Consider an interconnected system of the type (7.63) composed of three subsystens ($N_s = 3$). In terms of model (7.65), the nominal matrices are given by:

$$A_1 = \begin{bmatrix} -0.2 & 0 \\ 1 & -0.1 \end{bmatrix}, \; B_1 = \begin{bmatrix} 1 & 0 \\ 0 & 2 \end{bmatrix}$$

$$A_{d1} = \begin{bmatrix} 1.35 & 0.1 \\ -0.1 & 2.25 \end{bmatrix}, \; \eta_1 = 0.4 \; sin(2t) \;,$$

$$A_2 = \begin{bmatrix} -1 & 0 \\ 1 & -2 \end{bmatrix}, \; B_2 = \begin{bmatrix} 3 & 0 \\ 0 & 2 \end{bmatrix}$$

$$A_{d2} = \begin{bmatrix} 3.7 & -0.04 \\ 0.04 & 2.9 \end{bmatrix}, \; \eta_2 = 0.1 \; sin(5t),$$

$$A_3 = \begin{bmatrix} -0.6 & 0 \\ 1 & -0.8 \end{bmatrix}, \; B_3 = \begin{bmatrix} 1 & 0 \\ 0 & 1 \end{bmatrix}$$

$$A_{d3} = \begin{bmatrix} 1.3 & -0.05 \\ 0.05 & 1.4 \end{bmatrix}, \quad \eta_3 = 0.3\ sin(3t)$$

The uncertainties are described by the linear forms:

$$(\Sigma_1): \quad d_1 = \begin{bmatrix} 0.2 \\ 0.3 \end{bmatrix} + \begin{bmatrix} 0.02 & 0.01 \\ 0 & 0.05 \end{bmatrix} x_1 + \begin{bmatrix} 0.01 & 0 \\ 0 & 0.01 \end{bmatrix} u_1,$$

$$f_1 = \begin{bmatrix} 0 \\ 0.01 \end{bmatrix} + \begin{bmatrix} 0.01 & 0.02 \\ 0 & 0.04 \end{bmatrix} x_1 + \begin{bmatrix} 0.01 & 0 \\ 0 & 0.01 \end{bmatrix} u_1,$$

$$h_{12} = \begin{bmatrix} 0 \\ 0.1 \end{bmatrix} + \begin{bmatrix} 0.1 & 0 \\ 0 & 0.02 \end{bmatrix} x_2,$$

$$r_{12} = \begin{bmatrix} 0 \\ 0.01 \end{bmatrix} + \begin{bmatrix} 0.02 & 0 \\ 0 & 0.1 \end{bmatrix} x_2,$$

$$h_{13} = \begin{bmatrix} 0.2 \\ 0 \end{bmatrix} + \begin{bmatrix} 0.2 & 0 \\ 0 & 0.02 \end{bmatrix} x_3,$$

$$r_{13} = \begin{bmatrix} 0.03 \\ 0 \end{bmatrix} + \begin{bmatrix} 0.03 & 0 \\ 0 & 0.2 \end{bmatrix} x_3,$$

$$(\Sigma_2): \quad d_2 = \begin{bmatrix} 0.2 \\ 0 \end{bmatrix} + \begin{bmatrix} 0.03 & 0.02 \\ 0 & 0.04 \end{bmatrix} x_2 + \begin{bmatrix} 0.02 & 0 \\ 0 & 0.02 \end{bmatrix} u_2,$$

$$f_2 = \begin{bmatrix} 0.02 \\ 0 \end{bmatrix} + \begin{bmatrix} 0.02 & 0.03 \\ 0 & 0.01 \end{bmatrix} x_2 + \begin{bmatrix} 0.02 & 0 \\ 0 & 0.02 \end{bmatrix} u_2,$$

$$h_{21} = \begin{bmatrix} 0.1 \\ 0 \end{bmatrix} + \begin{bmatrix} 0.03 & 0 \\ 0 & 0.15 \end{bmatrix} x_1,$$

$$r_{21} = \begin{bmatrix} 0.02 \\ 0 \end{bmatrix} + \begin{bmatrix} 0.1 & 0 \\ 0 & 0.02 \end{bmatrix} x_1,$$

$$h_{23} = \begin{bmatrix} 0 \\ 0.3 \end{bmatrix} + \begin{bmatrix} 0.3 & 0 \\ 0 & 0.1 \end{bmatrix} x_3,$$

$$r_{23} = \begin{bmatrix} 0.02 \\ 0 \end{bmatrix} + \begin{bmatrix} 0.1 & 0 \\ 0 & 0.1 \end{bmatrix} x_3,$$

$$(\Sigma_3): \quad d_3 = \begin{bmatrix} 0.4 \\ 0.3 \end{bmatrix} + \begin{bmatrix} 0.03 & 0.02 \\ 0.01 & 0.04 \end{bmatrix} x_3 + \begin{bmatrix} 0.03 & 0 \\ 0 & 0.03 \end{bmatrix} u_1,$$

$$f_3 = \begin{bmatrix} 0.01 \\ 0.01 \end{bmatrix} + \begin{bmatrix} 0.02 & 0.02 \\ 0.01 & 0.04 \end{bmatrix} x_3 + \begin{bmatrix} 0.01 & 0 \\ 0 & 0.01 \end{bmatrix} u_3,$$

$$h_{31} = \begin{bmatrix} 0.2 \\ 0 \end{bmatrix} + \begin{bmatrix} 0.2 & 0 \\ 0 & 0.02 \end{bmatrix} x_1,$$

$$r_{31} = \begin{bmatrix} 0.01 \\ 0 \end{bmatrix} + \begin{bmatrix} 0.15 & 0 \\ 0 & 0.12 \end{bmatrix} x_1,$$

$$h_{32} = \begin{bmatrix} 0 \\ 0.2 \end{bmatrix} + \begin{bmatrix} 0.4 & 0 \\ 0 & 0.2 \end{bmatrix} x_2,$$

$$r_{32} = \begin{bmatrix} 0 \\ 0.02 \end{bmatrix} + \begin{bmatrix} 0.2 & 0 \\ 0 & 0.2 \end{bmatrix} x_2$$

Evaluation of the decomposition (7.68)-(7.69) gives

$$F_1 = \begin{bmatrix} 1.3 & 0 \\ 0 & 1.1 \end{bmatrix}, G_1 = \begin{bmatrix} 0.05 & 0.1 \\ -0.1 & 0.05 \end{bmatrix}$$

$$F_2 = \begin{bmatrix} 1.2 & 0 \\ 0 & 1.4 \end{bmatrix}, G_2 = \begin{bmatrix} 0.1 & -0.04 \\ 0.04 & 0.1 \end{bmatrix}$$

$$F_3 = \begin{bmatrix} 1.15 & 0 \\ 0 & 1.25 \end{bmatrix}, G_3 = \begin{bmatrix} 0.15 & -0.05 \\ 0.05 & 0.15 \end{bmatrix}$$

The norms of the delayed states are estimated by

Σ	ϑ_j	δ_j
Σ_1	1.69	0.013
Σ_2	1.96	0.012
Σ_3	1.563	0.025

The norms of the local uncertainties are estimated by

Σ	ϕ_j	ξ_j	β_j	κ_j	ρ_j	ζ_j
Σ_1	0.3611	0.051	0.01	0.01	0.045	0.01
Σ_2	0.2	0.048	0.02	0.02	0.037	0.02
Σ_3	0.5	0.051	0.03	0.014	0.051	0.01

The norms of the uncertainties in the coupling terms are estimated by

Σ	α_{jk}		ω_{jk}		σ_{jk}		π_{jk}	
Σ_1	0.1	0.2	0.1	0.2	0.01	0.03	0.1	0.2
Σ_2	0.1	0.3	0.15	0.3	0.02	0.02	0.1	0.1
Σ_3	0.2	0.2	0.2	0.4	0.01	0.02	0.15	0.2

By selecting the design parameters $\{\tau_j, \mu_j, \varphi_j\}$ to take on the values $\{1, 0.1, 6\}$, $\{1, 0.1, 6\}$, $\{1, 0.1, 7\}$ for the three subsystems, respectively, it has been found to satisfy condition (7.74) that Δ_j can take values of 27.9, 30.6, and 26.625 for the corresponding subsystems. Subsequently, the minimum values of the gain factors γ_j satisfying (7.73) are 13.535, 13.571, 14.124 for the three subsystems. This, in turn, yields $\Theta = 0.126$ and $\Gamma = 0.517$. To examine the sensitivity of the developed design procedure, we have considered the changes in the parameters $\{\tau_j, \mu_j, \varphi_j\}$. By considering τ_j only and increasing it up to 50 percent, the radius of convergence Γ decreases by 7.7 percent to 0.477. On the other hand, decreasing τ_j by 20 percent has caused Γ to increase by 12.57 percent to 0.582. The effect of changing μ_j leaves Γ almost intact. Finally, by decreasing φ_j to $\{4, 4, 5\}$, Γ decreases to 0.481 and Θ increases to 0.133. These results illuminate two points:

(1) The developed design procedure is robust and insensitive to tolerance in the design parameters, and

(2) The radius of convergence can be adjusted by properly selecting design parameters at the subsystem level.

7.7   Notes and References

We have presented two broad approaches to the decentralized stabilization of uncertain time-delay interconnected systems. There are a lot of approaches in the deterministic case [187] that need to be cast within the time-delay framework. It is our view that more work is required to lessen the computation load on the subsystem level and to produce more flexible design criteria. The topics of nonlinear interconnected UTDS and discrete-time interconnected UTDS are still at their infancy.

Bibliography

[1] Dugard, L. and E. I. Verriest (Editors), "**Stability and Control of Time-Delay Systems**," Springer-Verlag, New York, 1997.

[2] Mahmoud, M. S., "Output Feedback stabilization of Uncertain Systems with State Delay," vol. 63 of **Advances in Theory and Applications; C.T. Leondes (Ed.)**, 1994, pp. 197-257.

[3] Boyd, S., L. El-Ghaoui, E. Feron and V. Balakrishnan, "**Linear Matrix Inequalities in System and Control Theory**," vol. 15, SIAM Studies in Appl. Math., Philadelphia, 1994.

[4] Gahinet, P., A. Nemirovski, A. J. Laub and M. Chilali, "**LMI-Control Toolbox**," The MathWorks, Boston, Mass., 1995.

[5] Niculescu, S. I., E. I. Verriest, L. Dugard and J. M. Dion, "Stability and Robust Stability of Time-Delay Systems: A Guided Tour," Chapter 1 of **Stability and Control of Time-Delay Systems**, edited by Dugard and Verriest, Springer-Verlag, NewYork, 1997, pp. 1-71.

[6] Lou, J. S. and P. P. J. Van den Bosch, "Independent-of-Delay Stability Criteria for Uncertain Linear State Space Models," **Automatica**, vol. 33, 1997, pp. 171-179.

[7] Mahmoud, M. S., "Stabilizing Control of Systems with Uncertain Parameters and State-Delay," **Journal of the University of Kuwait (Science)**, vol. 21, 1994, pp. 185-202.

[8] Mahmoud, M. S., "Design of Stabilizing Controllers for Uncertain Discrete-Time Systems with State-Delay," **Journal of Systems Analysis and Modelling Simulation**, vol. 21, 1995, pp. 13-27.

[9] Trinh, T. and M. Aldeen, "Stabilization of Uncertain Dynamic Delay Systems by Memoryless State Feedback Controllers," **Int. J. Control**, vol. 56, 1994, pp. 1525-1542.

[10] Shen, J., B. Chen and F. Kung, "Memoryless Stabilization of Uncertain Dynamic Delay Systems: Riccati Equation Approach," **IEEE Trans. Automatic Control**, vol. 36, 1991, pp. 638-640.

[11] Niculescu, S. , C. E. de Souza, J. M. Dion and L. Dugard, "Robust Stability and Stabilization of Uncertain Linear Systems with State-Delay: Single Delay Case (I)," **Proc. IFAC Symposium on Robust Control Design**, Rio de Janeiro, Brazil, 1994, pp. 469-474.

[12] Moheimani, S. O. R., A. V. Savkin and I. R. Petersen, "Robust Control of Uncertain Time-Delay Systems: A Minimax Optimal Approach," **Proc. the 35th Conference on Decision and Control**, Kobe, JAPAN, 1995, pp. 1362-1367.

[13] Li, H., S. I. Niculescu, L. Dugard and J. M. Dion, "Robust $\mathcal{H}_\infty$ Control of Uncertain Linear Time-Delay Systems: A Linear Matrix Inequality Approach: Parts I, II," **Proc. the 35th Conference on Decision and Control**, Kobe, Japan, 1995, pp. 1370-1381.

[14] Cheres, E., S. Gutman and Z. J. Palmor, "Robust Stabilization of Uncertain Dynamic Systems Including State-Delay," **IEEE Trans. Automatic Control**, vol. 34, 1989, pp. 1199-1203.

[15] Phoojaruenchanachai, S., and K. Furuta, "Memoryless Stabilization of Uncertain Linear Systems Including Time-Varying State-Delay," **IEEE Trans. Automatic Control**, vol. 37, 1992, pp. 1022-1026.

[16] Xie, L. H. and C. E. de Souza, "Robust Stabilization and Disturbance Attenuation for Uncertain Delay Systems," **Proc. 2nd European Control Conference**, Groingen, The Netherlands, 1993, pp. 667-672.

[17] Niculescu, S. I., C. E. de Souza, J. M. Dion and L. Dugard, "Robust $\mathcal{H}_\infty$ Memoryless Control for Uncertain Linear Systems with Time-Varying Delay," **Proc. 3rd European Control Conference**, Rome, Italy, 1995, pp. 1814-1819.

[18] Bourles, H., "α-Stability of Systems Governed by a Functional Differential Equation: extension of results concerning linear delay systems," **Int. J. Control**, vol. 46, 1987, pp. 2233-2234.

[19] Niculescu, S. I., "$\mathcal{H}_\infty$ Memoryless Control with an α-Stability Constraint for Time-Delay Systems: An LMI Approach," **Proc. 34th Conference on Decision and Control,** New Orleans, 1995, pp. 1507-1512 (see also **IEEE Trans. Automatic Control,** vol. 43, 1998, pp. 739-743).

[20] Shyu, K. -K., and J. -J. Yan, "Robust Stability of Uncertain Time-Delay Systems and its Stabilization by Variable Structure Control,"**Int. J. Control,** vol. 57, 1993, pp. 237-246.

[21] Thowsen, A., "Uniform Ultimate Boundedness of the Solutions of Uncertain Dynamic Delay Systems with State-Dependenet and Memoryless Feedback Control," **Int. J. Control,** vol. 37, 1983, pp. 1143-1153.

[22] Lee, J. H., S. W. Kim and W. H. Kwon, "Memoryless $\mathcal{H}_\infty$ Controllers for State-Delayed Systems," **IEEE Trans. Automatic Control,** vol. 39, 1994, pp. 158-162.

[23] Ahlfors, L. V., "**Complex Analysis,**" McGraw-Hill, New York, 1953.

[24] Kubo, T., and E. Shimemura, "LQ Regulator of Systems with Time-Delay by Memoryless Feedback", **Proc. the 35th Conference on Decision and Control,** Kobe, Japan, 1996, pp. 1619-1620.

[25] Brierley, S. D., J. N. Chiasson and S. H. Zak, " On Stability Independent of Delay for Linear Systems," **IEEE Trans. Automatic Control,** vol. AC-27, 1982, pp. 252-254.

[26] Cooke, K. L. and J. M. Ferreira , " Stability Conditions for Linear Retarded Differential Difference Equations," **J. Math. Anal. Appli.,** vol. 96, 1983, pp. 480-504.

[27] Slemrod, M. and E. F. Infante, "Asymptotic Stability Criteria for Linear Systems of Differential Difference Equations of Neutral Type and their Discrete Analogues," **J. Math. Anal. Appl.,** vol 38, 1972, pp. 399-415.

[28] Feliachi, A. and A. Thowsen, "Memoryless Stabilization of Linear Delay-Differential Systems," **IEEE Trans. Automatic Control,** vol AC-26, 1981, pp. 586-587.

[29] Ge, J. H., P. M. Frank and C. F. Lin, "Robust $\mathcal{H}_\infty$ State Feedback Control for Linear Systems with State-Delay and Parameter Uncertainty," Automatica, vol. 32, 1996, pp. 1183-1185.

[30] Kamen, E. W., P. P. Khargonekar and A. Tannenbaum, "Stability of Time-Delay Systems Using Finite-Dimensional Controllers," IEEE Trans. Automatic Control, vol. AC-30, 1985, pp. 75-78.

[31] Kwon, W., and A. Pearson, "A Note on Feedback Stabilization of a Differential-Difference System," IEEE Trans. Automatic Control, vol. AC-22, 1977, pp. 468-470.

[32] Shen, J. C. , B. S. Chen and F. C. Kung, "Memoryless $\mathcal{H}_\infty$ Stabilization of Uncertain Dynamic Delay Systems: Riccati Equation Approach," J. IEEE Trans. Automatic Control, vol. 36, 1991, pp. 638-640.

[33] Lewis, R. M. and B. D. O. Anderson, "Necessary and Sufficient Conditions for Delay Independent Stability of Linear Autonomous Systems," IEEE Trans. Automatic Control, vol. AC-25, 1980, pp. 735-739.

[34] Xi, L. and C. E. de Souza, "Criteria for Robust Stability of Uncertain Linear Systems with Time Varying State-Delay," Proc. the 13th IFAC World Congress, San Francisco, 1996, pp. 137-142.

[35] Mahmoud, M. S., "Dynamic Control of Systems with Variable State-Delay," Int. J. Robust and Nonlinear Control, vol. 6, 1996, pp. 123-146.

[36] Mahmoud, M. S. and N.F. Al-Muthairi, "Design of Robust Controllers for Time-Delay Systems," IEEE Trans. Automatic Control, vol. 39, 1994, pp. 995-999.

[37] Mahmoud, M. S. and N.F. Al-Muthairi, "Quadratic Stabilization of Continuous-Time Systems with State-Delay and Norm-Bounded Time-Varying Uncertainties," IEEE Trans. Automatic Control, vol. 39, 1994, pp. 2135-2139.

[38] Mori, T., N. Fukuma and M. Kuwahara, "Delay Independent Stability Criteria for Discrete-Delays Systems," IEEE Trans. Automatic Control, vol. AC-27, 1982, pp. 964-966.

[39] Mori, T., N. Fukuma and M. Kuwahara, "On an Estimate of the Delay Rate for Stable Linear Delay Systems," Int. J. Control, vol. 36, 1982, pp. 95-97.

[40] Mori, T., "Criteria for Asymptotic Stability of Linear Time-Delay Systems," IEEE Trans. Automatic Control, vol. AC-30, 1985, pp. 158-161.

[41] Su, T. -J., I. -K. Fong, T. -S. Kuo and Y. -Y. Sun, "Robust Stability of Linear Time Delay Systems with Linear Parameter Perturbations," Int. J. Systems Science, vol. 19, 1988, pp. 2123-2129.

[42] Su, T. -J. and C. G. Haung, "Robust Stability of Delay Dependence of Linear Uncertain Systems," IEEE Trans. Automatic Control, vol. 37, 1992, pp. 1656-1659.

[43] Su, T. -J. and P. -L. Liu, "Robust Stability for Linear Uncertain Time-Delay Systems with Delay Dependence," Int. J. Systems Science, vol. 24, 1993, pp. 1067-1080.

[44] Wang, S. -S., B. -S. Chen and T. -P. Lin, "Robust Stability of Uncertain Time-Delay Systems," Int. J. Control, vol. 46, 1987, pp. 963-976.

[45] Zhou, K. and P. P. Khargonekar, "An Algebraic Riccati Equation Approach to $\mathcal{H}_\infty$- Optimization," Systems and Control Letters, vol. 11, 1988, pp. 85-91.

[46] Li, X. and C. E. de Souza, "Delay-Dependent Robust Stability and Stabilization of Uncertain Linear Dealy Systems: A Linear Matrix Inequality," IEEE Trans. Automatic Control, vol. 42, 1997, pp. 1144-1148.

[47] Mahmoud, M. S. and M. Zribi, "$\mathcal{H}_\infty$ Controllers for Time-Delay Systems Using Linear Matrix Inequalities," J. Optimization Thoery and Applications, vol. 100, 1999, pp. 89-122.

[48] Haddad, W. M., V. Kapila and C. T. Abdallah, "Stabilization of Linear and Nonlinear Systems with Time-Delay," Chapter 9 of Stability and Control of Time-Delay Systems, edited by Dugard and Verriest, 1997, pp. 205-217.

[49] Verriest, E. I., "Stabilization of Deterministic and Stochastic Systems with Uncertain Time-Delay," **Proc. the 33rd IEEE Conference on Decision and Control**, Florida, 1994, pp. 3829-3834.

[50] Hale, J. K., E. F. Infante and F. S. P. Tseng, "Stability in Linear Delay Systems," **J. Math. Anal. Appl.**, vol. 105, 1985, pp. 533-555.

[51] Lonemann, H. and S. Townley, "The Effect of Small Delays in the Feedback Loop on the Stability of Neutral Systems," **Systems and Control Letters**, vol. 27, 1996, pp. 267-274.

[52] Verriest, E. I., "Graphical Test for Robust Stability with Distributed Delayed Feedback," Chapter 5 of **Stability and Control of Time-Delay Systems** edited by Dugard and Verriest, Springer-Verlag, New York, 1997, pp. 117-139.

[53] Verriest, E. I., "Stability of Systems with Distributed Delays," **Proc. IFAC Conference on System Structure and Control**, Nantes, France, 1995, pp. 294-299.

[54] Toker, O. and H. Ozbay, "Complexity Issues in Robust Stability of Linear Delay Differential Systems," **Mathematics of Control, Signals and Systems**, vol. 9, 1990, pp. 386-400.

[55] Hmamed, A., "On the Stability of Time-Delay Systems: New Results," **Int. J. Control**, vol. 43, 1986, pp. 321-324.

[56] Hale, J. L. and W. Huang, "Global Geometry of the Stable Regions for Two-Delay Differential Equations," **J. Math. Anal. Appl.**, vol. 178, 1993, pp. 344-362.

[57] Verriest, E. I. and P. Florchinger, "Stability of Stochastic Systems with Uncertain Time Delays," **Systems and Control Letters**, vol. 24, 1994, pp. 41-47.

[58] Wu, H. and K. Mizukami, "Robust Stability Conditions based on Eigenstructure Assignment for Uncertain Time-Delay Systems: The Continuous-Time Case," **Proc. 13th IFAC World Congress**, San Francisco, 1996, pp. 131-136.

[59] Wu, H. and K. Mizukami, "Robust Stability Conditions based on Eigenstructure Assignment for Uncertain Time-Delay Systems: The

Discrete-Time Case," **Proc. 13th IFAC World Congress**, San Francisco, 1996, pp. 107-112.

[60] Chou, J. H., I. R. Horng and B. S, Chen, "Dynamical Feedback Compensator for Uncertain Time-Delay Systems containing Saturating Actuators," **Int. J. Control**, vol. 49, 1989, pp. 961-968.

[61] Lee, C. H., T. H. S. Li and F. C. Kung, "D-Stability Analysis for Discrete Systems with a Time-Delay," **Systems and Control Letters**, vol. 19, 1992, pp. 213-216.

[62] Su, T. J. and W. J. Shyr, "Robust D-Stability for Linear Uncertain Discrete Time-Delay Systems," **IEEE Trans. Automatic Control**, vol. 39, 1992, pp. 425-428.

[63] Wang, S. S., and T. P. Lin, "Robust Stabilization of Uncertain Time-Delay Systems with Sampled Feedback," **Int. J. Systems Science**, vol. 19, 1988, pp. 399-404.

[64] Wu, H. S., and K. Mizukami, "Robust Stabilization of Uncertain Linear Dynamical Systems with Time-Varying Delay," **J. Optimization Theory and Applications**, vol. 82, 1994, pp. 361-378.

[65] Wu, H. S., R. A. Willgoss and K. Mizukami, "Robust Stabilization for a Class of Uncertain Dynamical Systems with Time Delay," **J. Optimization Theory and Applications**, vol. 82, 1994, pp. 593-606.

[66] Wu, H., "Decentralized Output Feedback Control of Large-Scale Interconnected Time-Delay Systems," **Proc. 35th Conference on Decision and Control**, Kobe, Japan, 1996, pp. 1611-1616.

[67] Townley, S. and A. J. Pritcard, "On Problems of Robust Stability for Uncertain Systems with Time-Delay," **Proc. 1st European Control Conference**, Grenoble, France, 1991, pp. 2078-2083.

[68] Hmamed, A., "Further Results on the Delay-Independent Asymptotic Stability of Linear Systems," **Int. J. Systems Science**, vol. 22, 1991, pp. 1127-1132.

[69] Su, J. H., I. K. Fong and C. L. Tseng, "Stability Analysis of Linear Systems with Time-Delay," **IEEE Trans. Automatic Control**, vol. 39, 1994, pp. 1341-1344.

[70] Verriest, E. I. and A. F. Ivanov, "Robust Stability of Delay-Difference Equations," **Proc. 34th IEEE Conference on Decision and Control**, New Orleans, 1994, pp. 386-391.

[71] Verriest, E. I., "Stability and Stabilization of Stochastic Systems with Distributed Delays," **Proc. 34th IEEE Conference on Decision and Control**, New Orleans, 1995, pp. 2205-2210.

[72] Verriest, E. I. and M. K. H. Fan, "Robust Stability of Nonlinearly Perturbed Delay Systems," **Proc. 35th IEEE Conference on Decision and Control**, Kobe, Japan, 1996, pp. 2090-2091.

[73] Rozkhov, V. I. and A. M. Popov, "Inequalities for Solutions of Certain Systems of Differential Systems Equations with Large Time-Lag," **J. Diff. Eq.**, vol. 7, 1971, pp. 271-278.

[74] Verriest, E. I. and W. Aggoune, "Stability of Nonlinear Differential Delay Systems," **Mathematics and Computers in Simulation**, vol. 16, 1997, pp. 180-193.

[75] Walton, K. and J. E. Marshall, "Direct Method for TDS Stability Analysis," **Proc. IEE, Part D**, vol. 134, 1987, pp. 101-107.

[76] Wang, W. J. and R. J. Wang, "New Stability Criteria for Linear Time-Delay Systems," **Control Theory and Advanced Technology**, vol. 10, 1995, pp. 1213-1222.

[77] Wu, H. and K. Mizukami, "Quantitative Measures of Robustness for Uncertain Time-Delay Dynamical Systems," **Proc. 35th IEEE Conference on Decision and Control**, San Antonio, 1993, pp. 2004-2005.

[78] Xi, L. and C. E. de Souza, "LMI Approach to Delay-Dependent Robust Stability and Stabilization of Uncertain Linear Delay Systems," **Proc. 35th IEEE Conference on Decision and Control**, New Orleans, 1995, pp. 3614-3619.

[79] Slamon, D., "Structure and Stability of Finite Dimensional Approximations for Functional Differential Equations," **SIAM J. Control and Optimization**, vol. 23, 1985, pp. 928-951.

[80] Tsypkin, Y. Z. and M. Fu, "Robust Stability of Time-Delay Systems with an Uncertain Time-Delay Constant," **Int. J. Control**, vol. 57, 1993, pp. 865-879.

[81] Suh, I. H. and Z. Bien, "Root-Locus Technique for Linear Systems with Delay," **IEEE Trans. Automatic Control**, vol. AC-27, 1982, pp. 205-208.

[82] Fu, M., "Robust Stability and Stabilization of Time-Delay Systems via Integral Quadratic Constraint Approach," Chapter 4 in **Stability and Control of Time-Delay Systems**, edited by Dugard and Verriest, Springer-Verlag, 1997, pp. 101-116.

[83] Yu, W., K. M. Sobel and E. Y. Shapiro, "A Time-Domain Approach to the Robustness of for Time-Delay Systems," **Proc. 31st IEEE Conference on Decision and Control**, Tucson, 1992, pp. 3726-3727.

[84] Cooke, K. L. and J. M. Ferreira, "Stability Conditions for Linear Retarded Differential Difference Equations," **J. Math. Anal. Appli.**, vol. 96, 1983, pp. 480-504.

[85] O'Conner, D. and T. J. Tarn, "On the Stabilization by State Feedback for Neutral Differential Difference Equations," **IEEE Trans. Automatic Control**, vol. AC-28, 1983, pp. 615-618.

[86] Zhang, D. N. , M. Saeki and K. Ando, "Stability Margin Calculation of Systems with Structured Time-Delay Uncertainties," **IEEE Trans. Automatic Control**, vol. 37, 1992, pp. 865-868.

[87] Singh, T. and S. R. Vadali, "Robust Time-Delay Control," **J. Dynamic Systems, Measurements and Control**, vol. 115, 1993, pp. 303-306.

[88] Fiala, J. and R. Lumia, "The Effect of Time-Delay and Discrete Control on the Contact Stability of Simple Position controllers," **IEEE Trans. Automatic Control**, vol. 39, 1994, pp. 870-873.

[89] Furumochi, T., "Stability and Boundedness in Functional Differential Equations," **J. Math. Anal. Appl.**, vol. 113, 1986, pp. 473-489.

[90] Su, J. -H., "Further Results on the Robust Stability of Linear Systems with a Single Time Delay," **Systems and Control Letters**, vol. 23, 1944, pp. 123-146.

[91] Kojima, A. and S. Ishijima, "Robust Controller Design for Delay Systems in the Gap Metric," **IEEE Trans. Automatic Control**, vol. 40, 1995, pp. 370-374.

[92] Zheng, F. M. Cheng and W. B. Gao, "Feedback Stabilization of Linear Systems with Distributed Delays in State and Control Variable," **IEEE Trans. Automatic Control**, vol. 39, 1994, pp. 1714-1718.

[93] Glader, C., G. Hognas, P. Makila and H. T. Toivonen, "Approximation of Delay Systems: A Case Study," **Int. J. Control**, vol. 53, 1991, pp. 369-390.

[94] Gu, G. and E. B. Lee, "Stability Testing of Time-Delay Systems," **Automatica**, vol. 25, 1989, pp. 777-780.

[95] Gu, G., P. P. Khargonekar , E. B. Lee and P. Misrahou, "Finite Dimensional Approximations of Unstable Infinite-Dimensional Systems," **SIAM J. Control and Optimization**, vol. 30, 1992, pp. 704-716.

[96] Schoen, G. M. and H. P. Geering, "Stability Condition for a Delay Differential System," **Int. J. Contro**, vol. 58, 1993, pp. 432-439.

[97] Kamen, E. W., P. P. Khargonekar and A. Tannenbaum, "Stability of Time-Delay Systems Using Finite-Dimensional Controllers," **IEEE Trans. on Automatic Control**, vol. 30, 1985, pp. 75-78.

[98] Mori, T. and H. Kokame, "Stability of $\dot{x}(t) = Ax(t) + Bx(t - \tau)$," **IEEE Trans. Automatic Control**, vol. 34, 1989, pp. 460-462.

[99] Feliachi, A., and A. Thowsen, "Memoryless Stabilization of Linear Delay-Differential Systems," **IEEE Trans. Automatic Control**, vol. AC-26, 1981, pp. 586-587.

[100] Su, T. J. and C. G. Haung, "Robust Stability of Delay Dependence of Linear Uncertain Systems," **IEEE Trans. Automatic Control**, vol. 37, 1992, pp. 1656-1659.

[101] Chen, J., "On Computing the Maximal Delay Intervals for Stability of Linear Delay Intervals," **IEEE Trans. Automatic Control**, vol. 40, 1995, pp. 1087-1093.

[102] Chen, J. and H. A. Latchman, "A Frequency Sweeping Test for Stability Independent of Delay," **IEEE Trans. Automatic Control**, vol. 40, 1995, pp. 1640-1645.

[103] Chen, T., G. Gu and C. N. Nett, "A New Method for Computing Delay Margins for Stability of Linear Delay Systems," **Proc. the 33rd Conference on Decision and Control**, Lake Buena Vista, 1994, pp. 433-437.

[104] Chen, T., D. Xu and B. Shafai, "On Sufficient Conditions for Stability Independent of Delay," **Proc. American Control Conference**, Baltimore, 1994, pp. 1929-19336.

[105] Chiasson, J., "A Method for Computing the Interval of Delay Values for which a Differential-Delay System is Stable," **IEEE Trans. Automatic Control**, vol. 33, 1988, pp. 1176-1178.

[106] Verriest, E. I., "Robust Stability of Time-Varying Systems with Unknown Bounded Delays," **Proc. 33rd Conference on Decision and Control**, Lake Buena Vista, 1994, pp. 417-422.

[107] Verriest, E. I., M. K. Fan and J. Kullstam, "Frequency Domain Robust Stability of Linear Delay Systems," **Proc. the 32nd Conference on Decision and Control**, San Antonio, 1993, pp. 3473-3477.

[108] Verriest, E. I. and A. F. Ivanov, "Robust Stability of Systems with Delayed Feedback," **Circuits, Systems and Signal Processing**, vol. 13, 1994, pp. 213-222.

[109] Zheng, F., M. Cheng and W. Gao, "Feedback Stabilization of Linear Systems with Point Delays in State and Control Variables," **Proc. 12th IFAC World Congress**, Sydney, Australia, 1993, pp. 375-378.

[110] Luo, J. S., A. Johnson and P. P. J. van den Bosch, "Delay-Independent Robust Stability of Uncertain Linear Systems," **Systems and Control Letters**, vol. 24, 1995, pp. 33-39.

[111] Wu, H. and Mizukami, "Exponential Stabilization of a Class of Uncertain Dynamical Systems with Time-Delay," **Control Theory and Advanced Technology**, vol. 10, 1995, pp. 1147-1157.

[112] Lee, C. J., T. H. S. Li and F. C. Kung, "New Stability Criteria for Discrete Time-Delay Systems with Uncertainties," **Control Theory and Advanced Technology**, vol. 10, 1995, pp. 1159-1168.

[113] Hsiao, F. H. "Stability Analysis of Nonlinear Multiple Time-Delay Systems with a Periodic Input," **Control Theory and Advanced Technology**, vol. 10, 1995, pp. 1223-1233.

[114] Shyu, W. J. and Y. T. Juang, "Stability of Time-Delay Systems," **Control Theory and Advanced Technology**, vol. 10, 1995, pp. 2099-2107.

[115] Wang, R. J., and W. J. Wang, "Disk Stability and Robustness for Discrete-Time Systems with Multiple Time-Delays," **Control Theory and Advanced Technology**, vol. 10, 1995, pp. 1505-1513.

[116] Freedman, H. I., J. W. H. So and P. Waltman, "Chemostat Competition with Time Delays," **IMACS Ann. Comput. and Appl. Math.**, vol. 5, 1989, pp. 171-173.

[117] Lee, R. C. and W. J. Wang, "Robust Variable Structure Control Synthesis in Discrete-Time Uncertain Systems," **Control Theory and Advanced Technology**, vol. 10, 1995, pp. 1785-1796.

[118] Shyu, K. K. and J. J. Yan, "Variable Structure Model Following Adaptive Control for Systems with Time-Varying Delay," **Control Theory and Advanced Technology**, vol. 10, 1995, pp. 513-521.

[119] Xu, B., "Comments on (Robust Stability of Delay Dependence of Linear Uncertain Systems)," **IEEE Trans. Automatic Control**, vol. 39, 1994, p. 2356.

[120] Jeung, E. T., D. C. Oh, J. H. Kim and H. B. Park, "Robust Controller Design for Uncertain Systems with Time Delays: LMI Approach," **Automatica**, vol. 32, 1996, pp. 1229-1231.

[121] Choi, H. H. and M. J. Chung, "Memoryless Stabilization of Uncertain Dynamic Systems with Time-Varying Delayed States and Controls," **Automatica**, vol. 31, 1995, pp. 1349-1351.

[122] Verriest, E. J. and S. I. Niculoscu, "Delay-Independent Stability of Linear Neutral Systems: A Riccati Equation Approach," Chapter 3 in

Stability and Control of Time-Delay Systems, edited by Dugard, L. and E. I. Verriest, Springer-Verlag, New York, 1997, pp. 93-100.

[123] Agathoklis, P. and S. G. Foda, "Stability and Matrix Lyapunov Equation for Delay Differential Systems," Int. J. Control, vol. 49, 1989, pp. 417-432.

[124] Amemyia, T., "Delay-Independent Stability of Higher-Order Systems," Int. J. Control, vol. 50, 1989, pp. 139-149.

[125] Barmish, B. R. and Z. Shi, "Robust Stability of Perturbed Systems with Time-Delays," Automatica, vol. 25, 1989, pp. 371-381.

[126] Boese, F. G., "Stability Conditions for the General Linear Difference-Differential Equations with Constant Coefficients and One Constant Delay," J. Math. Anal. Appl. vol. 140, 1989, pp. 136-176.

[127] Boese, F. G., "Stability in a Special Class of Retarded Difference-Differential Equations with Interval-Valued Parameters," J. Math. Anal. Appl. vol. 181, 1994, pp. 227-247.

[128] Boese, F. G., "Stability Criteria for a Second-Order Dynamical Systems Involving Several Time-Delays," SIAM J. Math. Anal. vol. 5, 1995, pp. 1306-1330.

[129] Brierley, S. D., J. N. Chiasson, E. B. Lee and S. H. Zak, "On Stability Independent of Delay for Linear Systems," IEEE Trans. Automatic Control, vol. AC-27, 1982, pp. 252-254.

[130] Buslowicz, M., "Sufficient Conditions for Instability of Delay Differential Systems," Int. J. Control, vol. 37, 1983, pp. 1311-1321.

[131] Hertz, D., E. I. Jury and E. Zeheb, "Stability Independent and Dependent of Delay for Delay Differential Systems," J. Franklin Institute, vol. 318, 1984, pp. 143-150.

[132] Infante, E. F. and W. B. Castelan, "A Lyapunov Functional for a Matrix Difference-Differential Equation," J. Diff. Eq., vol. 29, 1978, pp. 439-451.

[133] Kharitonov, V. L. and A. P. Zhabko, "Robust Stability of Time-Delay Systems," IEEE Trans. Automatic Control, vol. 39, 1994, pp. 2388-2397.

[134] Lee, E. B., W. S. Lu and N. E. Wu, "A Lyapunov Theory for Linear Time-Delay Systems", **IEEE Trans. Automatic Control**, vol. AC-31, 1986, pp. 259-261.

[135] Goubet, A. M. Dambrine and J. P. Richard, "An Extension of Stability Criteria for Linear and Nonlinear Time-Delay Systems," **Proc. IFAC Conference on System Structure in Control**, Nante, France, 1995, pp. 278-283.

[136] Niculescu, S. I., C. E. de Souza, L. Dugard and J. M. Dion, "Robust Exponential Stability of Uncerttain Linear Systems with Time-Varying Delays," **Proc. 3rd Eurpean Control Conference**, Rome, Italy, 1995, pp. 1802-1808 (see also **IEEE Trans. Automatic Control**, vol. 43, 1998, pp. 743-748).

[137] Niculescu, S. I., A. Trofino-Neto, J. M. Dion and L. Dugard, "Delay-Dependent Stability of Linear Systems with Delayed State: An LMI Approach," **Proc. 34th Decision and Control Conference**, New Orleans, 1995, pp. 1495-1497.

[138] Ikeda, M. and T. Ashida, "Stabilization of Linear Systems with Time-Varying Delay," **IEEE Trans. Automatic Control**, vol. AC-24, 1979, pp. 369-370.

[139] Johnson, R. A., "Functional Equations, Approximations and Dynamic Response of Systems with Variable Time-Delay," **IEEE Trans. Automatic Control**, vol. AC-17, 1972, pp. 398-401.

[140] Lehman, B. and K. Shujaee, "Delay Independent Stability Conditions and Delay Estimates for Time-Varying Functional Differential Equations," **IEEE Trans. Automatic Control**, vol. 39, 1994, pp. 1673-1676.

[141] Lee, C. H., "D-Stability of Continuous Time-Delay Systems Subjected to a Class of Highly Structured Pertubations," **IEEE Trans. Automatic Control**, vol. 40, 1995, pp. 1803-1807.

[142] Sun, Y. J., C. H. Lien and J. G. Hsieh, "Commnets on (D-Stability of Continuous Time-Delay Systems subjected to a Class of Highly Structured Pertubations)," **IEEE Trans. Automatic Control**, vol. 43, 1998, pp. 689.

[143] Bourles, H., "α-Stability of Systems Governed by a Functinal Differential Equation: Exytension of Results Concerning Linear Delay Systems," Int. J. Control, vol. 45, 1987, pp. 2233-2234.

[144] Xu, B. and Y. Liu, "An Improved Razumkhin-Type Theorem and Its Applications," IEEE Trans. Automatic Control, vol. 39, 1994, pp. 839-841.

[145] Mao, K., "Comments on (An Improved Razumkhin-Type Theorem and Its Applications)," IEEE Trans. Automatic Control, vol. 40, 1995, pp. 639-640.

[146] Doyle, J. C., "Guaranteed Margins for LQG Regulators," IEEE Trans. Automatic Control, vol. AC-26, 1978, pp. 756-757.

[147] Doyle, J. C. and G. Stein, "Multivariable Feedback Design: Concepts for a Classical/Modern Synthesis," IEEE Trans. Automatic Control, Vol. 26,1981, pp. 4-16.

[148] Flagbedzi, Y. A. and A. E. Pearson, "Output Feedback Stabilization of Delay Systems via Generalization of the Transformation Method," Int. J. of Control, vol. 51, No. 4, 1990, pp. 801-822.

[149] Douglas, J. and M. Athans, "Robust Linear Quadratic Designs with Real Parameter Uncertainty," IEEE Trans. Automatic Control, vol. 39, 1994, pp. 107-111.

[150] Mahmoud, M. S., and S. Bingulac, "Robust Design of Stabilizing Controllers for Interconnected Time-Delay Systems," Automatica, vol. 34, 1998, pp. 795-800.

[151] Ikeda, M. and D. D. Silijak, "Decentralized Stabilization of Large-Scale Systems with Time-Delay," Large-Scale Systems, vol. 1, 1980, pp. 273-279.

[152] Bakula, L., "Decentralized Stabilization of Uncertain Delayed Interconnected Systems," Proc. IFAC Conference on Large-Scale Systems, London, 1995, pp. 575-580.

[153] Hu, Z. "Decentralized Stabilization of Large-Scale Interconnected Systems with Delays," IEEE Trans. Automatic Control, vol. 39, 1994, pp. 180-182.

[154] Lee, T. N. and U. L. Radovic, "Decentralized Stabilization of Linear Continuous and Discrete-Time Systems with Delays in Interconnection," IEEE Trans. Automatic Control, vol. 33, 1988, pp. 757-760.

[155] Garcia, G. J. Bernussou and D. Arzelier, "Robust Stabilization of Discrete-Time Linear Systems with Norm-Bounded Time-Varying Uncertainties," Systems and Control Letters, vol. 22, 1994, pp. 327-339.

[156] Xie, L. and Y. C. Soh, "Guaranteed Cost Control of Uncertain Discrete-Time Systems," Control-Theory and Advanced Technology", vol. 10, 1995, pp. 1235-1251.

[157] de Souza, C. E., M. Fu and Xie, L., "H_∞ Analysis and Synthesis of Discrete-Time Systems with Time Varying Uncertainty," IEEE Trans. Automatic Control", vol. 38, 1993, pp. 459-462.

[158] Khargonekar, P. P., I. R. Petersen and K. Zhou, "Robust Stabilization of Uncertain Linear Systems: Quadratic Stability and H_∞-Control Theory," IEEE Trans. Automatic Control, vol. 35, 1990, 356-361.

[159] Alekal, Y., P. Brunovsky, D. H. Chyung and E. B. Lee, "The Quadratic Problem for Systems with Time-Delays," IEEE Trans. Automatic Control, vol. AC-16, 1971, pp. 673-687.

[160] Delfour, M. C., C. McCalla and S. K. Mitter, "Stability and the Infinite-Time Quadratic Cost Problem for Linear-Hereditary Differential Systems," SIAM J. Control, vol. 13, 1975, pp. 48-88.

[161] Ichikawa, A., "Quadratic Control of Evolution Equations with Delays in Control," SIAM J. Control and Optimization, vol. 20, 1982, pp. 645-668.

[162] Lee, E. B., "Generalized Quadratic Optimal Control for Linear Hereditary Systems," IEEE Trans. Automatic Control, vol. AC-25, 1980, pp. 528-530.

[163] Uchida, K., E. Shimemura, T. Kubo and N. Abe, "The Linear-Quadratic Optimal Control Approach to Feedback Design for Systems with Delay," Automatica, vol. 24, 1988, pp. 773-780.

[164] Uchida, K. and E. Shimemura, "Closed-Loop Properties of the Infinite-Time Linear Quadratic Optimal Regulators for Systems with Delays," Int. J. Control, vol. 43, 1986, pp. 773-7792.

[165] Moheimani, S. O. R., and I. R. Petersen, "Optimal Quadratic Guaranteed Cost Control of a Class of Uncertain Time-Delay Systems," Proc. IEE, Part D, vol. 144, 1997, pp. 183-188.

[166] Galimidi, A. R. and B. R. Barmish, "The Constrained Lyapunov Problem and its Application to Robust Output Feedback Stabilization," IEEE Trans. Automatic Control, vol. AC-31, 1986, pp. 411-419.

[167] Narendra, K. S. and A. M. Annaswamy, "Stable Adaptive Systems," Prentice-Hall, New York, 1989.

[168] Apkarian, P. , P. Gahinet and G. Becker, "Self-Scheduled H_∞ Control of Linear Parameter-Varying Systems: A Design Example," Automatica, vol. 31, 1995, pp. 1251-1261.

[169] Becker, G. and A. Packard, "Robust Performance of Linear Parametrically-Varying Systems using Parameterically-Dependent Linear Feedback," Systems and Control Letters, vol. 23, 1994, pp. 205-215.

[170] Apkarian, P. and P. Gahinet, "A Convex Characterization of Gain Scheduled H_∞ Controllers," IEEE Trans. Automatic Control, vol. 40, 1995, pp. 853-864.

[171] Gahinet, P. , P. Apkarian and M. Chilali, "Affine Parameter-Dependent Lyapunov Fuctions for Real Parametric Uncertainty," IEEE Trans. Automatic Control, vol. 41, 1996, pp. 436-442.

[172] Iwasaki, T. and R. E. Skelton, "All Controllers for the General $\mathcal{H}_\infty$ Control Design Problem: LMI Existence Conditions and State-Space Formulas," Automatica, vol. 30, 1994, pp. 1307-1317.

[173] Bhat, K. and H. Koivo, "Modal Characterization of Controllability and Observability in Time-Delay Systems," IEEE Trans. Automatic Control, vol. AC-21, 1976, pp. 292-293.

[174] Mahmoud, M. S and L. Xie, "Guaranteed Cost Control of Uncertain Discrete Systems with Delays," Int. J. Contro, to appear.

[175] Petersen, I. R., "A Stabilization Algorithm for a Class of Uncertain Linear Systems," **Systems and Control Letters**, vol. 8, 1987, pp. 351-357.

[176] Petersen, I. R., "Disturbance Attenuation and H_∞ Optimization: A Design Method Based on the Algebraic Riccati Equation," **IEEE Trans. Automatic Control**, vol. 32, 1987, pp. 427-429.

[177] Verriest, E. I. and A. F. Ivanov, "Robust Stabilization of Systems with Delayed Feedback," **Proc. 2nd Int. Symposium on Implicit and Robust Systems**, Warxzawa, Poland, 1991, pp. 190-193.

[178] Sampei, M. , T. Mita and M. Nakamichi, "An Algebraic Approach to $\mathcal{H}_\infty$ Output-Feedback Control Problems," **Systems and Control Letters**, vol. 14, 1990, pp. 13-24.

[179] de Souza, C. E. and L. Xie, "On the Discrete-Time Bounded Real Lemma with Application in the Characterization of Static State Feedback H_∞ Controllers," **Systems and Control Letters**, vol. 18, 1992, pp. 61-71.

[180] Yu, L., U. Holmberg and D. Bonvin, "Decentralized Robust Stabilization of a Class of Interconnected Uncertain Delay Systems," **Control Theory and Advanced Technology**, vol. 10, 1995, pp. 1475-1483.

[181] Sinha, A. S. C., "Stabilization of Large-Scale Nonlinear Infinite Delay Systems," **Int. J. Systems Science**, vol. 21, 1990, pp. 2679-2684.

[182] Yaesh, I. and U. Shaked, "A Transfer Function Approach to the Problems of Discrete-Time Systems: $H_\infty-$ Optimal Linear Control and Filtering," **IEEE Trans. Automatic Control**, vol. 36, 1991, pp. 1264-127.

[183] Mahmoud, M. S. and M. G. Singh, "**Discrete Systems: Analysis, Optimization and Control**," Springer-Verlag, Berlin, 1984.

[184] Mahmoud, M. S., "**Computer-Operated Systems Control**," Marcel Dekker Inc., New York, 1991.

[185] Xie, L., "Output Feedback H_∞ Control of Systems with Parameter Uncertainty," **Int. J. Control**, vol. 63, 1996, pp. 741-750.

[186] Petersen, I. R., B. D. O. Anderson and E. A. Jonckheere, "A First Principle Solution to The Non-Singular H_∞ Control Problem," **Int. J. Robust and Nonlinear Control,** vol. 1, 1991, pp. 171-185.

[187] Silijak, D. D., "Decentralized Control of Complex Systems," Academic Press, San Diego, 1991.

[188] Mahmoud, M. S., M. F. Hassan and M. G. Darwish, "**Large-Scale Control Systems: Theories and Techniques,**" Marcel Dekker Inc., New York, 1985.

[189] Mahmoud, M. S., "Stabilizing Controllers for a Class of Uncertain Interconncected Systems," **IEEE Trans. Automatic Control,** vol. 39, 1994, pp. 2484-2488.

[190] Shi, Z. C. and W. B. Gao, "Decentralized Stabilization of Time-Varying Large-Scale Interconnected Systems," **Int. J. Systems Science,** vol. 18, 1987, pp. 1523-1535.

[191] Mahmoud, M. S. and M. Zribi, "Robust and $\mathcal{H}_\infty$ Stabilization of Interconnected Systems with Delays," **Proc. IEE Control Theory and Applications,** vol. 145, 1998, pp. 559-567.

[192] Nagpal, K. M., P. P. Khargonekar and K. R. Poola, "$\mathcal{H}_\infty$ -Control with Transients," **SIAM J. Control and Optimization,** vol. 29, 1991, pp. 1373-1393.

[193] Haddad, W. M. and D. S. Berstein, "Explicit Construction of Quadratic Lyapunov Functions for the Small Gain, Positivity, Circle and Popov Theorem and their Applications to Robust Analysis, Part I: Continuous-Time Theory," **Int. J. Robust and Nonlinear Control,** vol. 4, 1994, pp. 233-248.

[194] Willems, J. C., "The Generation of Lyapunov Functions for Input-Output Stable Systems," **SIAM J. Control,** vol. 9, 1971, pp. 105-133.

[195] Shimizu, K. and T. Tamura, "A Derivation of the Riccati Equation for $\mathcal{H}_\infty$ Control in Time Domain," **Proc. the 12th IFAC Congress,** Sydney, Australia, vol. 2, 1993, pp. 33-37.

[196] Haddad, W. M. and D. S. Berstein, "Explicit Construction of Quadratic Lyapunov Functions for the Small Gain, Positivity, Circle and Popov Theorem and their Applications to Robust Analysis, Part II: Discrete-Time Theory," **Int. J. Robust and Nonlinear Control**, vol. 4, 1994, pp. 249-265.

[197] Safonov, M. G., E. A. Jonkheere, M. Verma and D. J. N. Limbeer, "Synthesis of Positive Real Multivariable Feedback Systems," **Int. J. Control**, vol. 45, 1987, pp. 817-842.

[198] Mahmoud, M. S., "Adaptive Control of a Class of Systems with Uncertain Parameters and State-Delay," **Int. J. Control**, vol. 63, 1996, pp. 937-950.

[199] Foda, S. G., and M. S. Mahmoud, "Adaptive Stabilization of Delay Differential Systems with Unknown Uncertainty Bounds," **Int. J. Control**, vol. 68, 1998, pp. 259-275.

[200] Anderson, B. D. O. and S. Vongpanitherd, "**Network Analysis and Synthesis: A Modern Systems Theory Approach**," Prentice-Hall, New Jersey, 1973.

[201] Anderson, B. D. O., "A System Theory Criterion for Positive Real Matrices," **SIAM J. Control and Optimization**, vol. 5, 1967, pp. 171-182.

[202] Sun, W., P. P. Khargonekar and D. Shim, "Solution to the Positive Real Control Problem for Linear Time-Invariant Systems," **IEEE Trans. Automatic Control**, vol. 39, 1994, pp. 2034-2046.

[203] Mahmoud, M. S. and L. Xie, "Stability and Positive Realness of Time-Delay Systems," **J. Mathematical Analysis and Applications**, to appear.

[204] Wang, W. J., R. J. Wang and C. S. Chen, "Stabilization, Estimation and Robustness for Discrete Large-Scale Systems with Delays," **Control Theory and Advanced Technology**, vol. 10, 1995, pp. 1717-1735.

[205] Desoer, C. A. and M. Vidyasagar, "**Feedback Systems: Input-Output Properties**," Academic Press, New York, 1975.

[206] Vidyasagar, M., "Nonlinear Systems," Prentice-Hall, New York, 1989.

[207] Lozano-Leal, R. and S. M. Joshi, "Strictly Positive Transfer Function Revisited," IEEE Trans. Automatic Control, vol. 35, 1990, pp. 1243-1245.

[208] Haddad, W. M. and D. S. Berstein, "Robust Stabilization of Positive Real Uncertainty: Beyond the Small Gain Theorem," Systems and Control Letters, vol. 17, 1991, pp. 191-208.

[209] Zhou, K. and P. P. Khargonekar, "An Algebraic Riccati Equation Approach to $\mathcal{H}_\infty$-Optimization," Systems and Control Letters, vol 111, 1988, pp. 85-91.

[210] Anderson, B. D. O. and J. B. Moore, "Optimal Control: Linear Quadratic Methods," Prentice-Hall, New York, 1990.

[211] Lakshmikantham, V., and S. Leela, "Differential and Integral Inequalities Theory and Applications: Vols I and II," Academic Press, New York, 1969.

[212] Halanay, A. "Differential Equations: Stability, Oscillations, Time Lags," Academic Press, New York, 1966.

[213] Dorato, P. and R. K. Yedavalli, "Recent Advances in Robust Control," IEEE Press, 1990, New York.

[214] Basar, T. and P. Bernard, "$\mathcal{H}_\infty$-Optmal Control and Related Minimax Design Problems: A Dynamic Game Approach," Birkhauser, Boston, 1991.

[215] Doyle, J. C., K. Glover, P. P. Khargonekar and B. A. Francis, "State-Space Solutions to Standard H_2 and H_∞ Control Problems," IEEE Trans. Automatic Control, vol. 34, 1989, pp. 831-847.

[216] Francis, B. A., "A Course in $\mathcal{H}_\infty$ Control Theory," Springer Verlag, New York, 1987.

[217] Kwakernaak, H., "Robust Control and $\mathcal{H}_\infty$-Optimization: Tutorial Paper," Automatica, vol. 29, 1993, pp. 255-273.

[218] Zhou, K., "Essentials of Robust Control," Prentice-Hall, New York, 1998.

[219] Stoorvogel, A., "The $\mathcal{H}_\infty$ Control Problem," Prentice-Hall, New York, 1992.

[220] Gahinet, P. and P. Apkarian, "A Linear Matrix Inequality Approach to $\mathcal{H}_\infty$ Control," Int. J. Robust and Nonlinear Control, vol. 4, 1994, pp. 421-448.

[221] Lee, C. S. and G. Leitmann, "Continuous Feedback Guaranteeing Uniform Ultimate Boundedness for Uncertain Linear Delay Systems: An Application to River Pollution Control," Computer and Mathematical Modeling, vol. 16, 1988, pp. 929-938.

[222] Mahmoud, M. S., "Robust Stability and Stabilization of a Class of Uncertain Nonlinear Systems with Delays," J. Mathematical Problems in Engineering, vol. 3, 1997, pp. 1-22.

[223] Petersen, I. R. and C. V. Hollot, "A Riccati Equation to the Stabilization of Uncertain Linear Systems," Automatica, Vol. 22, 1986, pp 397-411.

[224] Petersen, I. R., and D. C. McFarlane, "Optimal Guaranteed Cost Control and Filtering for Uncertain Linear Systems," IEEE Trans. Automatic Control, vol. 39, 1994, 1971-1977.

[225] Wang, Y., C. E. de Souza and L. Xie, "Decentralized Output Feedback Control of Interconnected Uncertain Delay Systems," Proceedings of the 12th IFAC Congress, 1993, Sidney, Australia, vol. 6, pp. 38-42.

[226] Lozano, R., B. Brogliato and I. Landau, "Passivity and Global Stabilization of Cascaded Nonlinear Systems," IEEE Trans. Automatic Control, Vol. 37, 1992, pp. 1386-1388.

[227] Mahmoud, M. S., "Passive Control Synthesis for Uncertain Time-Delay Systems," Proc. the 37th IEEE Conference on Decision and Control, Tampa, Florida, 1998, pp.

[228] Mahmoud, M. S. and L. Xie, "Positive Real Analysis and Synthesis for Uncertain Discrete-Time Systems," Technical Report NTU-EEE4-97002, 1997.

[229] Mahmoud, M. S., "Passivity Analysis and Synthesis for Uncertain Discrete-Time Delay Systems," Technical Report NTU-EEE4-97003, 1997.

[230] Mahmoud, M. S. and L. Xie, "Passivity Analysis and Synthesis for Uncertain Time-Delay Systems," **Technical Report NTU-EEE4-97004, 1997.**

[231] Mahmoud, M. S., "Robust Performance Results for Discrete-Time Systems," **J. Mathematical Problems in Engineering,** vol. 4, 1997, pp. 17-38.

[232] Mahmoud, "Control of Uncertain State-Delay Systems: Guaranteed Cost Control," Technical Report NTU-EEE4-98004, 1997.

[233] Bahnasawi, A. A. and M. S. Mahmoud, "**Control of Partially-Known Dynamical Systems,**" Springer-Verlag, 1989, Berlin.

[234] Yasuda, K. , "Decentralized Quadratic Stabilization of Interconnected Systems," **Proc. the 12th IFAC Congress,** Sydney, Australia, 1993, pp. 95-98.

[235] Zhai, G., K. Yasuda and M. Ikeda, "Decentralized Quadratic Stabilization of Large Scale Systems," **Proc. 33rd IEEE Conference on Decision and Control,** 1994, Tampa, Florida, pp. 2337-2339.

[236] Xie, L. Y. Wang and C. E. de Souza, "Decentralized Output Feedback Control of Discrete-Time Interconnected Uncertain Systems," **Proc. 32nd IEEE Conference on Decision and Control,** San Antonio, 1993, pp. 3762-3767.

[237] Wang, Y., C. E. de Souza and L. Xie, "Decentralized Output Feedback Control of Interconnected Uncertain Delay Systems," **Proc. the 12th IFAC Congress,** Sydney, Australia, 1993, vol. 6, pp. 38-42.

[238] Mahmoud, M. S., "Guaranteed Stabilization of Interconnected Discrete-Time Uncertain Systems," **Int. J. Systems Science,** vol. 26, 1994, pp. 337-358.

[239] Rotea, M. A. and P. P. Khargonekar, "Stabilization of Uncertain Systems with Norm-Bounded Uncertainty: A Control Lyapunov Function Approach," **SIAM J. Control & Optimization,** vol. 27, 1989, pp. 1462-1476.

[240] Anderson, B. D. O. and E. I. Jury, "On Robust Hurwitz Polynomials," **IEEE Trans. Automatic Control,** vol. AC-32, 1987, pp. 1001-1008.

[241] Barmish, B. R. and Z. Shi, "Robust Stability of Perturbed Systems with Time-Delays," **Automatica**, vol. 25, 1989, pp. 371-387.

[242] Fu, M., A. W. Olbrot and M. Polis, "Robust Stability of Time-Delayed Systems: The Edge Theorem and Graphical Tests," **IEEE Trans. Automatic Control**, vol. 34, 1989, pp. 813-821.

[243] Corless, M., "Control of Uncertain Systems," **J. Dynamic Systems, Measurement and Control**, vol. 115, 1993, pp. 362-372.

[244] Shaked, U., I. Yaesh and C. E. de Souza, "Bounded Real Criteria for Time-Delay Systems," **IEEE Trans. Automatic Control**, vol. 43, 1998, pp. 1022-1027.

[245] Choi, Y. H., M. J. Chung, "Observer-Based $\mathcal{H}_\infty$ Controller Design for State Delayed Linear Systems," **Automatica**, vol. 32, 1996, pp. 1073-1075.

[246] Jeung, E. T., J. H. Kim and H. B. Park, "$\mathcal{H}_\infty$-Output Feedback Controller Design for Linear Systems with Time-Varying Delayed State," **IEEE Trans. Automatic Control**, vol. 43, 1998, pp. 971-975.

[247] He, J. B., Q. G. Wang and T. H. Lee, "$\mathcal{H}_\infty$-Disturbance Attenuation for State Delay Systems," **Systems and Control Letters**, vol. 33, 1998, pp. 105-114.

[248] Choi, Y. H., and M. J. Chung, "Robust Observer-Based $\mathcal{H}_\infty$ Controller Design for Linear Uncertain Time-Delay Systems," **Automatica**, vol. 33, 1997, pp. 1749-1752.

[249] Kim, J. H., E. T. Jeung and H. B. Park, "Robust Control for Parameter Uncertain Delay Systems in State and Control Input," **Automatica**, vol. 32, 1996, pp. 1337-1339.

[250] Fridman, E. and U. Shaked, "$\mathcal{H}_\infty$-State Feedback Control of Linear Systems with Small State Delay," **Systems and Control Letters**, vol. 33, 1998, pp. 141-150.

[251] Amemiya, T. and G. Leitmann, "A Method for Designing a Stabilizing Control for a Class of Uncertain Linear Dynamic Systems," **Dynamics and Control**, vol. 4, 1994, pp. 147-167.

Part II

ROBUST FILTERING

Chapter 8

Robust Kalman Filtering

8.1 Introduction

State-estimation forms an integral part of control systems theory. Estimating the state-variables of a dynamic model is important to help in improving our knowledge about different systems for the purpose of analysis and control design. The celebrated Kalman filtering algorithm [3,4] is the optimal estimator over all possible linear ones and gives unbiased estimates of the unknown state vectors under the conditions that the system and measurement noise processes are mutually-independent Gaussian distributions. Robust state-estimation arises out of the desire to estimate unmeasurable state variables when the plant model has uncertain parameters. In [17], a Kalman filtering approach has been studied with an H_∞-norm constraint. For systems with bounded parameter uncertainty, the robust estimation problem has been addressed in [11,12,32]. Despite the frequent occurence of uncertain systems with state-delay in engineering applications, the problem of estimating the state of an uncertain system with state-delay has been overlooked. The purpose of this chapter is to consider the state-estimation problem for linear systems with norm-bounded parameter uncertainties and unknown state-delay. Specifically, we address the state-estimator design problem such that the estimation error covariance has a guaranteed bound for all admissible uncertainties and state-delay. We will divide efforts into two parts: one part for contiouous-time systems and the other part for discrete-time systems. Both time-varying and steady-state robust Kalman filtering are considered. The main tool for solving the foregoing problem is the Riccati equation approach and the end result is an extended robust Kalman filter the solution of which is expressed in terms of two Riccati equations involving scaling parameters.

8.2 Continuous-Time Systems

8.2.1 System Description

We consider a class of uncertain time-delay systems represented by:

$$
\begin{aligned}
\dot{x}(t) &= [A(t) + \Delta A(t)]x(t) + A_d(t)x(t - \tau) + w(t) \\
&= A_\Delta(t)x(t) + A_d(t)x(t - \tau) + w(t) \quad\quad (8.1) \\
y(t) &= [C(t) + \Delta C(t)]x(t) + v(t) \\
&= C_\Delta(t)x(t) + v(t) \quad\quad (8.2)
\end{aligned}
$$

where $x(t) \in \Re^n$ is the state, $y(t) \in \Re^m$ is the measured output and $w(t) \in \Re^n$ and $v(t) \in \Re^m$ are, respectively, the process and measurement noises. In (8.1)-(8.2), $A(t) \in \Re^{n \times n}$, $A_d(t) \in \Re^{n \times n}$ and $C(t) \in \Re^{m \times n}$ are piecewise-continuous matrix functions. Here, τ is a constant scalar representing the amount of delay in the state. The matrices $\Delta A(t)$ and $\Delta C(t)$ represent time-varying parametric uncertainties which are of the form:

$$
\begin{bmatrix} \Delta A(t) \\ \Delta C(t) \end{bmatrix} = \begin{bmatrix} H(t) \\ H_c(t) \end{bmatrix} \Delta(t)\, E(t) \quad\quad (8.3)
$$

where $H(t) \in \Re^{n \times \alpha}$, $H_c(t) \in \Re^{m \times \alpha}$ and $E(t) \in \Re^{\beta \times n}$ are known piecewise-continuous matrix functions and $\Delta(t) \in \Re^{\alpha \times \beta}$ is an unknown matrix with Lebsegue measurable elements satisfying

$$
\Delta^t(t)\Delta(t) \;\leq\; I \quad \forall\, t \quad\quad (8.4)
$$

The initial condition is specified as $\langle x(0), x(s) \rangle = \langle x_o, \phi(s) \rangle$, where $\phi(\cdot) \in \mathcal{L}_2[-\tau, 0]$ which is assumed to be a zero-mean Gaussian random vector. The following standard assumptions on noise statistics are recalled:

Assumption 8.1: $\forall t,\ s \geq 0$

$$
\begin{aligned}
&(a)\mathcal{E}[w(t)] = 0; \quad \mathcal{E}[w(t)w^t(s)] = W(t)\delta(t - s); \quad W(t) > 0 \quad (8.5) \\
&(b)\mathcal{E}[v(t)] = 0; \quad \mathcal{E}[v(t)v^t(s)] = V(t)\delta(t - s); \quad V(t) > 0 \quad (8.6) \\
&(c)\mathcal{E}[x(0)w^t(t)] = 0; \quad \mathcal{E}[x(0)v^t(t)] = 0 \quad\quad (8.7) \\
&(d)\mathcal{E}[w(t)v^t(s)] = 0; \quad \mathcal{E}[x(0)x^t(0)] = R_o \quad\quad (8.8)
\end{aligned}
$$

where $\mathcal{E}[\cdot]$ stands for the mathematical expectation and $\delta(\cdot)$ is the Dirac function.

8.2.2 Robust Filter Design

Our objective is to design a stable state estimator of the form:

$$\dot{\hat{x}}(t) = G(t)\,\hat{x}(t) + K(t)\,y(t), \quad \hat{x}(0) = 0 \tag{8.9}$$

where $G(t) \in \Re^{n \times n}$ and $K(t) \in \Re^{n \times m}$ are piecewise-continuous matrices to be determined such that there exists a matrix $\Psi \geq 0$ satisfying

$$\mathcal{E}[(x - \hat{x})(x - \hat{x})^t] \leq \Psi, \quad \forall \Delta : \Delta^t(t)\Delta(t) \leq I \tag{8.10}$$

Note that (8.10) implies

$$\mathcal{E}[(x - \hat{x})^t(x - \hat{x})] \leq tr(\Psi), \quad \forall \Delta : \Delta^t(t)\Delta(t) \leq I \tag{8.11}$$

In this case, the estimator (8.9) is said to provide a guaranteed cost (GC) matrix Ψ.

Examination of the proposed estimator proceeds by analyzing the estimation error

$$e(t) = x(t) - \hat{x}(t) \tag{8.12}$$

Substituting (8.1) and (8.9) into (8.12), we express the dynamics of the error in the form:

$$
\begin{aligned}
\dot{e}(t) = {} & G(t)\,e(t) + [A(t) - G(t) - K(t)C(t)]x(t) \\
& + [\Delta A(t) - K(t)\Delta C(t)]x(t) \\
& + A_d(t)x(t - \tau) + [w(t) - K(t)v(t)]
\end{aligned} \tag{8.13}
$$

By introducing the extended state-vector

$$\xi(t) = \begin{bmatrix} x(t) \\ e(t) \end{bmatrix} \in \Re^{2n} \tag{8.14}$$

it follows from (8.1)-(8.2) and (8.13) that

$$
\begin{aligned}
\dot{\xi}(t) &= [\widehat{A}(t) + \widehat{H}(t)F(t)\,\widehat{E}(t)]\xi(t) + \widehat{D}(t)\xi(t - \tau) + \widehat{B}(t)\eta(t) \\
&= \widehat{A}_\Delta(t)\xi(t) + \widehat{D}(t)\xi(t - \tau) + \widehat{B}(t)\eta(t)
\end{aligned} \tag{8.15}
$$

where $\eta(t)$ is a stationary zero-mean noise signal with identity covariance matrix and

$$\widehat{A}(t) = \begin{bmatrix} A(t) & 0 \\ A(t) - G(t) - K(t)C(t) & G(t) \end{bmatrix}, \qquad (8.16)$$

$$\widehat{H}(t) = \begin{bmatrix} H(t) \\ H(t) - K(t)H_c(t) \end{bmatrix}, \quad \widehat{E}(t) = [E(t) \quad 0] \qquad (8.17)$$

$$\widehat{BB}^t(t) = \begin{bmatrix} W(t) & W(t) \\ W(t) & W(t) + K(t)V(t)K^t(t) \end{bmatrix}, \qquad (8.18)$$

$$\widehat{D}(t) = \begin{bmatrix} A_d(t) & 0 \\ A_d(t) & 0 \end{bmatrix}, \quad \eta = \begin{bmatrix} w(t) \\ v(t) \end{bmatrix} \qquad (8.19)$$

Definition 8.1: *Estimator (8.9) is said to be a quadratic estimator (QE) associated with a matrix $\Omega(t) > 0$ for system (8.1) if there exists a scalar $\lambda(t) > 0$ and a matrix*

$$0 < \Omega(t) = \begin{bmatrix} \Omega_1(t) & \Omega_3(t) \\ \Omega_3^t(t) & \Omega_2(t) \end{bmatrix} \qquad (8.20)$$

satisfying the algebraic inequality:

$$-\dot{\Omega}(t) + \widehat{A}_\Delta(t)\,\Omega(t) + \Omega(t)\,\widehat{A}_\Delta^t(t) + \lambda(t)\,\Omega(t-\tau)$$
$$+\lambda^{-1}(t)\,\widehat{D}(t)\,\Omega(t-\tau)\,\widehat{D}^t(t) + \widehat{B}(t)\,\widehat{B}^t(t) \leq 0 \qquad (8.21)$$

The next result shows that if (8.9) is QE for system (8.1)-(8.2) with cost matrix $\Omega(t)$, then $\Omega(t)$ defines an upper bound for the filtering error covariance, that is,

$$\mathcal{E}[e(t)e^t(t)] \leq \Omega_2(t) \quad \forall\ t$$

for all admissible uncertainties satisfying (8.3)-(8.4).

Theorem 8.1: *Consider the time-delay (8.1)-(8.2) satisfying (8.3)-(8.4) and with known initial state. Suppose there exists a solution $\Omega(t) \geq 0$ to inequality (8.21) for some $\lambda(t) > 0$ and for all admissible uncertainties. Then the estimator (8.9) provides an upper bound for the filtering error covariance, that is,*

$$\mathcal{E}[e(t)e^t(t)] \leq \Omega_2(t)$$

Proof: Supose that the estimator (8.9) is QE with cost matrix $\Omega(t)$. By evaluating the derivative of the covariance matrix $\Sigma(t) = \mathcal{E}[\xi(t)\,\xi^t(t)]$ we get:

$$\dot{\Sigma}(t) = \widehat{A}_\Delta(t)\,\Sigma(t) + \Sigma(t)\,\widehat{A}_\Delta^t(t) + \widehat{D}(t)\mathcal{E}[\xi(t-\tau)\xi^t(t)]$$
$$+ \ \mathcal{E}[\xi(t)\,\xi^t(t-\tau)]\,\widehat{D}^t(t) + \mathcal{E}[\eta(t)\,\xi^t(t)] + \mathcal{E}[\xi(t)\,\eta^t(t)] \quad (8.22)$$

Using B.1.1, we get the inequality:

$$\widehat{D}(t)\mathcal{E}[\xi(t-\tau)\,\xi^t(t)] + \mathcal{E}[\xi(t)\,\xi^t(t-\tau)]\,\widehat{D}^t(t) =$$
$$\widehat{D}(t)\Sigma(t-\tau) + \Sigma(t-\tau)\,\widehat{D}^t(t) \leq$$
$$\lambda(t)\,\Sigma(t-\tau) + \lambda^{-1}(t)\,\widehat{D}(t)\Sigma(t-\tau)\,\widehat{D}^t(t) \quad (8.23)$$

Using (8.23) into (8.22) and arranging terms, we obtain:

$$\dot{\Sigma}(t) \leq \widehat{A}_\Delta(t)\,\Sigma(t) + \Sigma(t)\,\widehat{A}_\Delta^t(t) + \lambda(t)\,\Sigma(t-\tau)$$
$$+ \ \lambda^{-1}(t)\,\widehat{D}(t)\,\Sigma(t-\tau)\,\widehat{D}^t(t) + \widehat{B}(t)\,\widehat{B}^t(t) \quad (8.24)$$

Combining (8.21) and (8.24) and letting $\Xi(t) = \Sigma(t) - \Omega(t)$, we obtain:

$$\dot{\Xi}(t) \leq \widehat{A}_\Delta(t)\,\Xi(t) + \Xi(t)\,\widehat{A}_\Delta^t(t) + \lambda(t)\,\Xi(t-\tau)$$
$$+ \ \lambda^{-1}(t)\,\widehat{D}(t)\,\Xi(t-\tau)\,\widehat{D}^t(t) \quad (8.25)$$

On considering that the state is known over the period $[-\tau, 0]$ justifies letting $\Sigma(0) = 0$. Hence, inequality (8.25) implies that $\Xi(t) \leq 0 \ \forall\, t > 0$, that is $\Sigma(t) \leq \Omega(t) \ \forall\, t > 0$. Finally, it is obvious that

$$\mathcal{E}[e(t)e^t(t)] = [0 \ I]\Sigma(t)\begin{bmatrix} 0 \\ I \end{bmatrix} \leq \Omega_2(t)$$

8.2.3 A Riccati Equation Approach

We employ hereafter a Riccati equation approach to solve the robust Kalman filtering for time-delay systems. To this end, we define piecewise matrices $P(t) = P^t(t) \in \Re^{n \times n}$; $L(t) = L^t(t) \in \Re^{n \times n}$ as the solutions of the Riccati differential equations (RDEs):

$$\dot{P}(t) = A(t)P(t) + P(t)A^t(t) + \lambda(t)P(t-\tau) + \hat{W}(t)$$

$$+ \quad \lambda^{-1}(t)A_d(t)P(t-\tau)A_d^t(t) + \mu(t)P(t)E^t(t)E(t)P(t) \, ;$$

$$P(t-\tau) \quad = \quad 0 \quad \forall \, t \in [0,\tau] \tag{8.26}$$

$$\dot{L}(t) \quad = \quad A(t)L(t) + L(t)A^t(t) + \lambda(t)L(t-\tau) + \hat{W}(t)$$

$$+ \quad \lambda^{-1}(t)A_d(t)P(t-\tau)A_d^t(t) + \mu(t)L(t)E^t(t)E(t)L(t)$$

$$- \quad [L(t)C^t(t) + \mu^{-1}(t)H(t)H_c^t(t)]\hat{V}^{-1}(t)$$

$$[C(t)L(t) + \mu^{-1}(t)H_c(t)H^t(t)] \, ;$$

$$L(t-\tau) \quad = \quad 0 \quad \forall \, t \in [0,\tau] \tag{8.27}$$

where $\lambda(t) > 0$, $\mu(t) > 0 \; \forall \, t$ are scaling parameters and the matrices $\hat{A}(t), \hat{V}(t)$ and $\hat{W}(t)$ are given by:

$$\hat{W}(t) \quad = \quad W(t) + \mu^{-1}(t)H(t)H^t(t) \tag{8.28}$$

$$\hat{V}(t) \quad = \quad V(t) + \mu^{-1}(t)H_c(t)H_c^t(t) \tag{8.29}$$

$$\hat{A}(t) \quad = \quad A(t) + \delta A(t)$$

$$= \quad A(t) + \mu^{-1}(t)L^t(t)E^t(t)E(t) \tag{8.30}$$

Let the (λ, μ)–parameterized estimator be expressed as:

$$\dot{\hat{x}}(t) \quad = \quad \left\{ A(t) + \mu^{-1}(t)L^t(t)E^t(t)E(t) \right\} \hat{x}(t)$$

$$+ \quad K(t)\{y(t) - C(t)\hat{x}(t)\} \tag{8.31}$$

where the gain matrix $K(t) \in \Re^{n \times m}$ is to be determined. The following theorem summarizes the main result:

Theorem 8.2: *Consider system (8.1)-(8.2) satisfying the uncertainty structure (8.3)-(8.4) with zero initial condition. Suppose the process and measurement noises satisfy* **Assumption 8.1.** *For some $\mu(t) > 0$, $\lambda(t) > 0$, let $P(t) = P^t(t)$ and $L(t) = L^t(t)$ be the solutions of RDEs (8.26) and (8.27), respectively. Then the (λ, μ)–parameterized estimator (8.31) is QE estimator such that*

$$\mathcal{E}[\{x(t) - \hat{x}(t)\}^t \, \{x(t) - \hat{x}(t)\}] \quad \leq \quad tr[L(t)] \tag{8.32}$$

Moreover, the gain matrix $K(t)$ is given by

$$K(t) = \left\{ L(t)C^t(t) + \mu^{-1}(t)H(t)H_c^t(t) \right\} \hat{V}^{-1}(t) \tag{8.33}$$

Proof: Let

$$X(t) \quad = \quad \begin{bmatrix} P(t) & L(t) \\ L(t) & L(t) \end{bmatrix} \tag{8.34}$$

where $P(t)$ and $L(t)$ are the positive-definite solutions to (8.26)and (8.27), respectively. By combining (8.26)-(8.30) with some standard matrix manipulations, it is easy to see that

$$-\dot{X}(t)+ \widehat{A}\,(t)X(t) + X(t)\,\widehat{A}^{t}\,(t) \,+\, \lambda(t)X(t-\tau)+$$
$$\mu^{-1}(t)\,\widehat{H}\,(t)\,\widehat{H}^{t}\,(t) \,+\, \mu(t)X(t)\,\widehat{E}^{t}\,(t)\,\widehat{E}\,(t)X(t)+$$
$$\lambda^{-1}(t)\,\widehat{D}\,(t)\,X(t-\tau)\,\widehat{D}^{t}\,(t) + \,\widehat{B}\,(t)\,\widehat{B}^{t}\,(t) \,= 0 \qquad (8.35)$$

where $\widehat{A}\,(t), \widehat{B}\,(t), \widehat{H}\,(t), \widehat{D}\,(t)$ are given by (8.16)-(8.19). A simple comparison of (8.9) and (8.31) taking into consideration (8.28)-(8.31) and (8.33) shows that $G(t) = \hat{A}(t) - K(t)\,C(t)$. By making use of a version of **B.1.1** that for some $\mu(t) > 0$ we have

$$\widehat{H}\,(t)F(t)\,\widehat{E}\,(t)X(t) \,+\, X(t)\,\widehat{E}^{t}\,(t)F^{t}(t)\,\widehat{H}^{t}\,(t) \,\le$$
$$\mu(t)\,X(t)\,\widehat{E}^{t}\,(t)\,\widehat{E}\,(t)X(t) \,+\, \mu^{-1}(t)\,\widehat{H}\,(t)\,\widehat{H}^{t}\,(t)\,; \qquad (8.36)$$

Using (8.36), it is now a simple task to verify that (8.35) becomes:

$$-\dot{X}(t)+ \widehat{A}_{\Delta}\,(t)X(t) + X(t)\,\widehat{A}_{\Delta}^{t}\,(t) + \lambda(t)X(t-\tau)$$
$$+\lambda^{-1}(t)\,\widehat{D}\,(t)X(t-\tau)\,\widehat{D}^{t}\,(t)+ \widehat{B}\,(t)\,\widehat{B}^{t}\,(t) \,\le\, 0 \qquad (8.37)$$

$$\forall \Delta : \Delta^{t}(t)\,\Delta(t) \le I \;\; \forall t$$

By **Theorem 8.1**, it follows that for some $\mu(t) > 0$, $\lambda(t) > 0$, that (8.31) is a quadratic estimator and $\mathcal{E}[e(t)e^{t}(t)] \le L(t)$. This implies that $\mathcal{E}[e^{t}(t)e(t)] \le tr[L(t)]$

Remark 8.1: It is known that the uncertainty representation (8.3)-(8.4) is not unique. We note that $H(t)$, $H_c(t)$ may be postmultiplied and $E(t)$ may be premultiplied by any unitary matrix since eventually this unitary matrix may be absorbed in $\Delta(t)$. It is significant to observe that such unitary multiplication does not affect the solution developed in this section.

Remark 8.2: Had we defined

$$X(t) \;=\; \begin{bmatrix} P^{-1}(t) & 0 \\ 0 & L(t) \end{bmatrix} \qquad (8.38)$$

we would have obtained:

$$
\begin{aligned}
\dot{P}(t) &= P(t)A(t) + A^t(t)P(t) + \lambda(t)P(t-\tau) + P(t)\hat{W}(t)P(t) \\
&\quad + \lambda^{-1}(t)P(t-\tau)A_d(t)P^{-1}(t-\tau)A_d^t(t)P(t-\tau) \\
&\quad + \mu(t)E^t(t)E(t) \\
P(t-\tau) &= 0 \quad \forall\, t \in [0,\tau] \\
\dot{L}(t) &= A(t)L(t) + L(t)A^t(t) + \lambda(t)L(t-\tau) + \hat{W}(t) \\
&\quad + \lambda^{-1}(t)A_d(t)P(t-\tau)A_d^t(t) + \mu(t)L(t)E^t(t)E(t)L(t) \\
&\quad - [L(t)C^t(t) + \mu^{-1}(t)H(t)H_c^t(t)]\hat{V}^{-1}(t) \\
&\quad [C(t)L(t) + \mu^{-1}(t)H_c(t)H^t(t)] \;; \\
L(t-\tau) &= 0 \quad \forall\, t \in [0,\tau]
\end{aligned}
$$
$$(8.39)$$
$$(8.40)$$

We note that (8.39) is of non-standard form although X(t) in (8.39) is frequently used in similar situations for delayless systems [12,32]. Indeed, the difficulty comes from the delay-term $\lambda^{-1}(t)P(t-\tau)A_d(t)P^{-1}(t-\tau)A_d^t(t)P(t-\tau)$. This point emphasizes the fact that not every result of delayless systems are straightforwardly transformable to time-delay systems.

Remark 8.3: It is interesting to observe that the estimator (8.31) is independent of the delay factor τ and it reduces to the standard Kalman filtering algorithm in the case of systems without uncertainties and delay factor $H(t) \equiv 0$, $H_c(t) \equiv 0$, $E(t) \equiv 0$, $A_d(t) \equiv 0$, $\lambda(t) \equiv 0$.

Remark 8.4: In the delayfree case $(A_d(t) \equiv 0$, $\lambda(t) \equiv 0)$, we observe that (8.33) reduces to the Kalman filter for the system

$$
\begin{aligned}
\dot{x}(t) &= \hat{A}(t)\, x(t) + \hat{w}(t) \\
y(t) &= C(t)\, x(t) + \hat{v}(t)
\end{aligned}
$$
$$(8.41)$$
$$(8.42)$$

where $\hat{w}(t)$ and $\hat{v}(t)$ are zero-mean white noise sequences with covariance matrices $\hat{W}(t)$ and $\hat{V}(t)$, respectively, and having cross-covariance matrix $[\mu^{-1}(t)H(t)H_c^t(t)]$. Looked at in this light, our approach to robust filtering in **Theorem 8.2** corresponds to designing a standard Kalman filter for a related continuous-time system which captures all admissible uncertainties and time-delay, but does not involve parameter uncertainties. Indeed, the robust filter (8.31) using (8.28)-(8.30) can be rewritten as

$$
\dot{\hat{x}}(t) = [A(t) + \delta A(t)]\, \hat{x}(t) + K(t) \{y(t) - C(t)\, \hat{x}(t)\}
$$

where $\delta A(t)$ is defined in (8.30) and it reflects the effect of uncertainties $\{\Delta A(t), \Delta C(t)\}$ and time delay factor $A_d(t)$ on the structure of the filter.

8.2.4 Steady-State Filter

Now, we investigate the asymptotic properties of the Kalman filter developed in Section 8.2.3. For this purpose, we consider the uncertain time-delay system

$$
\begin{aligned}
\dot{x}(t) &= [A + H\Delta(t)E]x(t) + A_d x(t - \tau) + w(t) \\
&= A_\Delta x(t) + A_d x(t - \tau) + w(t) \quad (8.43) \\
y(t) &= [C + H_c\Delta(t)E]x(t) + v(t) \\
&= C_\Delta x(t) + v(t) \quad (8.44)
\end{aligned}
$$

where $\Delta(t)$ satisfies (8.4). The matrices $A \in \Re^{n\times n}$, $C \in \Re^{m\times n}$ are real constant matrices representing the nominal plant. It is assumed that A is Hurwitz. The objective is to design a time-invariant *a priori* estimator of the form:

$$
\dot{\hat{x}}(t) = \hat{A}\,\hat{x}(t) + K\,[y(t) - C\hat{x}(t)] \quad \hat{x}(t_o) = 0 \quad (8.45)
$$

that achieves the following asymptotic performance bound

$$
\lim_{t\to\infty} \mathcal{E}\left\{[\hat{x}(t) - x(t)][\hat{x}(t) - x(t)]^t\right\} \leq L \quad (8.46)
$$

Theorem 8.3: *Consider the uncertain time-delay system (8.44)-(8.45) with A being Hurwitz. If for some scalars $\mu > 0$, $\lambda > 0$, there exist stabilizing solutions for the AREs*

$$
AP + PA^t + \lambda P + \hat{W} + \lambda^{-1}A_d P A_d^t + \mu PE^t EP = 0 \quad (8.47)
$$

$$
\begin{aligned}
&AL + LA^t + \lambda L + \hat{W} + \lambda^{-1}A_d P A_d^t + \mu LE^t EL - \\
&[LC^t + \mu^{-1}HH_c^t]\hat{V}^{-1}[CL + \mu^{-1}H_c H^t] = 0 \quad (8.48)
\end{aligned}
$$

Then the estimator (8.45) is a stable quadratic (SQ) and achieves (8.46) with

$$
\begin{aligned}
\hat{W} &= W + \mu^{-1}HH^t, \quad \hat{V} = V + \mu^{-1}H_c H_c^t \quad (8.49) \\
\hat{A} &= A + \delta A \\
&= A + \mu^{-1}L^t E^t E \quad (8.50)
\end{aligned}
$$

$$
K = \left\{LC^t + \mu^{-1}HH_c^t\right\}\hat{V}^{-1} \quad (8.51)
$$

for some L ≥ 0.

Proof: To examine the stability of the closed-loop system, we augment (8.43)-(8.45) with $(w(t) = 0, v(t) = 0)$, to obtain

$$
\begin{aligned}
\dot{\xi}(t) &= \widehat{A}_\Delta\, \xi(t) + \widehat{D}\, \xi(t - \tau) \\
&= \begin{bmatrix} A_\Delta & 0 \\ A_\Delta - G - KC_\Delta & G \end{bmatrix} \xi(t) + \begin{bmatrix} A_d & 0 \\ A_d & 0 \end{bmatrix} \xi(t - \tau) \quad (8.52)
\end{aligned}
$$

By a similar argument as in the proof of **Theorem 8.2**, it is easy to see that

$$
X\,\widehat{A}_\Delta + \widehat{A}_\Delta^t\, X - \lambda X + \lambda^{-1}\,\widehat{D}\, X\, \widehat{D}^t < 0 \qquad (8.53)
$$

where

$$
X = \begin{bmatrix} P & L \\ L & L \end{bmatrix} \qquad (8.54)
$$

Introducing a Lyapunov-Krasovskii functional

$$
V(x_t) = \xi^t(t) X \xi(t) + \int_{t-\tau}^{t} \xi^t(\alpha)(\lambda^{-1})\,\widehat{D}\, X\, \widehat{D}^t\, \xi(\alpha)\, d\alpha \qquad (8.55)
$$

and observe that $V(x_t) > 0$, for $\xi(t) \neq 0$ for some $\lambda > 0$ and $V(x_t) = 0$ when $\xi = 0$. By differentiating the Lyapunov-Krasovskii functional (8.55) along the trajectories of system (8.52), we get:

$$
\begin{aligned}
\dot{V}(x_t) &= \xi^t(t)[X\,\widehat{A}_\Delta + \widehat{A}_\Delta^t\, X + \lambda^{-1}\,\widehat{D}\, X\, \widehat{D}^t]\xi(t) \\
&+ \xi^t(t)X\,\widehat{D}\, \xi(t - \tau) + \xi^t(t - \tau)\,\widehat{D}^t\, X\xi(t) \\
&- \lambda^{-1}\xi^t(t - \tau)\,\widehat{D}\, X\, \widehat{D}^t\, \xi(t - \tau) \\
&\leq \xi^t(t)[X\,\widehat{A}_\Delta + \widehat{A}_\Delta^t\, X + \lambda X + \lambda^{-1}\,\widehat{D}\, X\, \widehat{D}^t]\xi(t) \\
&- \lambda^{-1}\xi^t(t - \tau)\,\widehat{D}\, X\, \widehat{D}^t\, \xi(t - \tau) \\
&= \xi^t(t)[X\,\widehat{A}_\Delta + \widehat{A}_\Delta^t\, X + \lambda X + \lambda^{-1}\,\widehat{D}\, X\, \widehat{D}^t]\xi(t) \\
&< 0, \qquad \xi(t) \neq 0 \qquad (8.56)
\end{aligned}
$$

which means that the augmented system (8.52) is asymptotically stable. in turn, this implies that (8.45) is SQ. The guaranteed performance

$$
\mathcal{E}[e(t)e^t(t)] \leq L
$$

follows from similar lines of argument as in the proof of **Theorem 8.2.**

The next theorem provides an LMI-based solution to the steady-state robust Kalman.

Theorem 8.4: *Consider the uncertain time-delay system (8.44)-(8.45) with A being Hurwitz. The estimator*

$$\dot{\hat{x}}(t) = [A + \mu^{-1}L^t E^t E]\hat{x}(t)$$
$$+ [LC^t + \mu^{-1}HH_c^t]\hat{V}^{-1}[y(t) - C\hat{x}(t)] \qquad (8.57)$$

where

$$\hat{V} = V + \mu^{-1}H_c H_c^t \qquad (8.58)$$

is a stable quadratic and achieves (8.46) for some $L \geq 0$ if for some scalars $\mu > 0$, $\lambda > 0$, there exist matrices $0 < Y = Y^t$ and $0 < X = X^t$ satisfying the LMIs

$$\begin{bmatrix} AY + YA^t + Q_y(Y,\lambda) & A_d Y & YE^t \\ YA_d^t & -\lambda I & 0 \\ EY & 0 & -\mu^{-1}I \end{bmatrix} < 0 \qquad (8.59)$$

$$\begin{bmatrix} AX + XA^t + Q_x(X,\lambda) & A_d Y & XE^t \\ YA_d^t & -\lambda I & 0 \\ EX & 0 & -\mu^{-1}I \end{bmatrix} < 0 \qquad (8.60)$$

where

$$Q_y(Y,\lambda) = \lambda Y + W + \mu^{-1}HH^t,$$
$$Q_x(X,\lambda) = \lambda X + W + \mu^{-1}HH^t$$
$$- [XC^t + \mu^{-1}HH_c^t]\hat{V}^{-1}[CX + \mu^{-1}H_c H^t] \qquad (8.61)$$

Proof: By **A.3.1** and (8.47)-(8.48), it follows that there exist matrices $0 < Y = Y^t$ and $0 < X = X^t$ satisfying the ARIs:

$$AY + YA^t + \lambda Y + \hat{W} + \lambda^{-1}A_d Y A_d^t + \mu Y E^t EY \leq 0 \qquad (8.62)$$
$$AX + XA^t + \lambda X + \hat{W} + \lambda^{-1}A_d Y A_d^t + \mu X E^t EX -$$
$$[XC^t + \mu^{-1}HH_c^t]\hat{V}^{-1}[CX + \mu^{-1}H_c H^t] \leq 0 \qquad (8.63)$$

such that $Y > P, X > L$. Application of **A.3.1** to the ARIs (8.62)-(8.63) yields the LMIs (8.59)-(8.60).

Remark 8.6: It should be emphasized the AREs (8.47)-(8.48) do not have clear-cut monotonicity properties enjoyed by standard AREs. The main reason for this is the presense of the term $A_d P A_d^t$.

8.2.5   Example 8.1

For the purpose of illustrating the developed theory, we focus on the steady-state Kalman filtering and proceed to determine the estimator gains. Essentially, we seek to solve (8.47)-(8.50) when $\lambda \in [\lambda_1 \rightarrow \lambda_2]$, $\mu \in [\mu_1 \rightarrow \mu_2]$, where $\lambda_1, \lambda_2, \mu_1, \mu_2$ are given constants. Initially, we observe that (8.47) depends on P only and it is not of the standard forms of AREs. On the other hand, (8.48) depends on both L and P and it can be put into the standard ARE form. For numerical simulation, we employ a Kronecker Product-like technique to reduce (8.47) into a system of nonlinear algebraic equations of the form

$$f(\alpha) \; = \; G\,\alpha \; + \; h(\alpha) \; + \; q \tag{8.64}$$

where $\alpha \in \Re^{n(n+1)/2}$ is a vector of the unknown elements of the P-matrix. The algebraic equation (8.64) can then be solved using an iterative Newton Raphson technique according to the rule:

$$\alpha_{(i+1)} \; = \; \alpha_{(i)} \; - \; \gamma_{(i)}[G + \nabla_\alpha h(\alpha_{(i)}]^{-1} f(\alpha_{(i)}) \tag{8.65}$$

where i is the iteration index, $\alpha_{(o)} = 0$, $\nabla_\alpha h(\alpha)$ is the Jacobian of $h(\alpha)$ and the step-size $\gamma_{(i)}$ is given by $\gamma_{(i)} = 1/[\||f(\alpha(i))\|| + 1]$.

Given the solution of (8.47), we proceed to solve (8.48) using a standard Hamiltonian/Eigenvector method. All the computations are carried out using the **MATLAB**-Software. As a typical case, consider a time-delay system of the type (8.43)-(8.44) with

$$A \; = \; \begin{bmatrix} -2 & 0.5 \\ 1 & -3 \end{bmatrix}, \; A_d \; = \; \begin{bmatrix} -0.2 & -0.1 \\ 0.1 & 0.4 \end{bmatrix}, \; W \; = \; \begin{bmatrix} 1 & 0 \\ 0 & 1 \end{bmatrix}$$

$$C \; = \; [1 \;\; -3], \; E \; = \; [0.5 \;\; 1], \; H_c \; = \; 2, \; V \; = \; 1, \; H \; = \; \begin{bmatrix} 0.5 \\ 0.5 \end{bmatrix}$$

A summary of the computational results is presented in Tables 8.1-8.2 and from which we observe the following:
(1) For a given $\lambda \in [0.1 - 0.9]$, increasing μ by 50% results in 0.3% increase in $\||K\||$ (for small λ) and about 1.12% increase in $\||K\||$ when λ is relatively

large.

(2) For a given μ, increasing λ from 0.1 to 0.9 causes $\|K\|$ to increase by about 5.35%.

(3) For $\mu < 0.6$, $\lambda in [0.1, 0.9]$, the estimator is unstable.

(4) Increasing (λ, μ) beyond $(1, 1)$ yields unstable estimator.

Therefore we conclude that: (1) The stable-estimator gains are practically insensitive to the (λ, μ)-parameters, and (2) There is a finite range for (λ, μ) that guarantees stable performance of the developed Kalman filter.

8.3 Discrete-Time Systems

It is well-known that the celebrated Kalman filtering provides an optimal solution to the filtering problem of dynamical systems subject to stationary Gaussian input and measurement noise processes [3]. Its original derivation was in discrete-time. As we steered through the previous chapters, we noted that most of the research efforts on UTDS have been concentrated on robust stability and stabilization and the problem of estimating the state of uncertain systems with state-delay has been overlooked despite its importance for control and signal processing. This is particularly true for discrete-time systems. Therefore, the purpose of this section is to consider the state estimation problem for linear discrete-time systems with norm-bounded parameter uncertainties and unknown state-delay. Specifically, we address the state estimator design problem such that the estimation error covariance has a guaranteed bound for all admissible uncertainties and state-delay. Looked at in this light, the developed results are the discrete-counterpart of the previous section. Although for convenience purposes we will follow parallel lines to the continuous case, we caution the reader that the discrete-time results cannot be derived from the continuous-time results and vice-versa.

8.3.1 Uncertain Discrete-Delay Systems

We consider a class of uncertain time-delay systems represented by:

$$
\begin{aligned}
x_{k+1} &= [A_k + \Delta A_k] x_k + A_{dk} x_{k-\tau} + w_k \\
&= A_{k,\Delta} x_k + A_{dk} x_{k-\tau} + w_k & (8.66) \\
y_k &= [C_k + \Delta C_k] x_k + v_k \\
&= C_{k,\Delta} x_k + v_k & (8.67) \\
z_k &= L_k x_k & (8.68)
\end{aligned}
$$

where $x_k \in \Re^n$ is the state, $y_k \in \Re^m$ is the measured output, $z_k \in \Re^p$ is a linear combination of the state variables to be estimated and $w_k \in \Re^r$ and $v_k \in \Re^m$ are, respectively, the process and measurement noise sequences. The matrices $A_k \in \Re^{n \times n}$, $A_{dk} \in \Re^{n \times n}$ and $C_k \in \Re^{m \times n}$ are real-valued matrices representing the nominal plant. Here, τ is a constant scalar representing the amount of delay in the state. The matrices ΔA_k and ΔC_k represent time-varying parametric uncertainties given by:

$$\begin{bmatrix} \Delta A_k \\ \Delta C_k \end{bmatrix} = \begin{bmatrix} H_k \\ H_{ck} \end{bmatrix} \Delta_k \, E_k \qquad (8.69)$$

where $H_k \in \Re^{n \times \alpha}$, $H_{ck} \in \Re^{m \times \alpha}$ and $E_k \in \Re^{\beta \times n}$ are known matrices and $\Delta_k \in \Re^{\alpha \times \beta}$ is an unknown matrix satisfying

$$\Delta_k^t \, \Delta_k \quad \leq \quad I \quad k = 0, 1, 2.... \qquad (8.70)$$

The initial condition is specified as $\langle x_o, \phi(s) \rangle$, where $\phi(.) \in \ell_2[-\tau, 0]$. The vector x_o is assumed to be a zero-mean Gaussian random vector. The following standard assumptions on x_o and the noise sequences $\{w_k\}$ and $\{v_k\}$, are assumed:

$$(a) \mathcal{E}[w_k] = 0; \quad \mathcal{E}[w_k w_j^t] = W_k \, \delta(k-j); W_k > 0; \, \forall k, j \qquad (8.71)$$
$$(b) \mathcal{E}[v_k] = 0; \quad \mathcal{E}[v_k v_j^t] = V_k \, \delta(k-j); V_k > 0; \, \forall k, j \qquad (8.72)$$
$$(c) \mathcal{E}[w_k v_j^t] = 0; \quad \mathcal{E}[x_o w_k^t] = 0, \, \forall k, j \qquad (8.73)$$
$$(d) \mathcal{E}[x_o x_o^t] = R_o \qquad (8.74)$$

where $\mathcal{E}[.]$ stands for the mathematical expectation and $\delta(.)$ is the Dirac function.

8.3.2   Robust Filter Design

Our objective is to design a stable state-estimator of the form:

$$\hat{x}_{k+1} = G_{o,k} \, \hat{x}_k + K_{o,k} \, y_k \qquad \hat{x}_o = 0 \qquad (8.75)$$

where $G_{o,k} \in \Re^{n \times n}$ and $K_{o,k} \in \Re^{n \times m}$ are real matrices to be determined such that there exists a matrix $\Psi \geq 0$ satisfying

$$E[\{x_k - \hat{x}_k\}\{x_k - \hat{x}_k\}^t] \quad \leq \quad \Psi$$
$$\forall \Delta : \Delta_k^t \, \Delta_k \quad \leq \quad I \qquad (8.76)$$

Note that (8.74) implies

$$E[\{x_k - \hat{x}_k\}^t\{x_k - \hat{x}_k\}] \leq tr(\Psi)$$
$$\forall \Delta : \Delta_k^t \Delta_k \leq I \tag{8.77}$$

In this case, the estimator (8.73) is said to provide a guaranteed cost (GC) matrix Ψ.

The proposed estimator is now analyzed by defining

$$G_{o,k} = (A_k + \delta A_k) - K_{o,k}C_k \tag{8.78}$$

where δA_k and $K_{o,k}$ are unknown matrices to be determined later on. Using (8.64)-(8.65) and (8.76) to express the dynamics of the state-estimator in the form:

$$\hat{x}_{k+1} = [(A_k + \delta A_k) - K_{o,k}C_k]\hat{x}_k$$
$$+ K_{o,k}[(C_k + \Delta C_k)x_k + v_k] \tag{8.79}$$

Introduce the augmented state vector

$$\xi_k = \begin{bmatrix} x_k \\ x_k - \hat{x}_k \end{bmatrix} = \begin{bmatrix} x_k \\ e_k \end{bmatrix} \in \Re^{2n} \tag{8.80}$$

It follows from (8.64) and (8.77) that:

$$\xi_{k+1} = [\widehat{A}_k + \widehat{H}_k \Delta_k \breve{E}_k] \xi_k + \widehat{D}_k \xi_{k-\tau} + \widehat{B}_k \eta_k$$
$$= \widehat{A}_{k,\Delta} \xi_k + \widehat{D}_k \xi_{k-\tau} + \widehat{B}_k \eta_k \tag{8.81}$$

where η_k is a stationary zero-mean noise signal with identity covariance matrix and

$$\widehat{A}_k = \begin{bmatrix} A_k & 0 \\ \delta A_k & (A_k + \delta A_k) - K_{o,k}C_k \end{bmatrix}, \widehat{D}_k = \begin{bmatrix} D_k & 0 \\ 0 & 0 \end{bmatrix} \tag{8.82}$$

$$\widehat{H}_k = \begin{bmatrix} H_{1,k} \\ H_{1,k} - K_{o,k}H_{2,k} \end{bmatrix}, \breve{E}_k = [E_k \quad 0] \tag{8.83}$$

$$\widehat{B}_k\widehat{B}_k^t = \begin{bmatrix} W_k & 0 \\ 0 & W_k + K_{o,k}V_kK_{o,k}^t \end{bmatrix}, \eta_k = \begin{bmatrix} w_k \\ v_k \end{bmatrix} \tag{8.84}$$

Definition 8.2: *Estimator (8.73) is said to be a quadratic estimator (QE) associated with a sequence of matrices* $\{\Omega_k\} > 0$ *for system (8.64)-(8.65) if there exist a sequence of scalars* $\{\lambda_k\} > 0$ *and a sequence of matrices* $\{\Omega_k\}$ *such that*

$$0 < \Omega_k = \begin{bmatrix} \Omega_{1,k} & \Omega_{3,k} \\ \Omega_{3,k}^t & \Omega_{2,k} \end{bmatrix} \tag{8.85}$$

satisfying the algebraic matrix inequality

$$(1+\lambda_k)\,\widehat{A}_{k,\Delta}\;\Omega_k\;\widehat{A}_{k,\Delta}^t \;-\; \Omega_{k+1} \;+\; (1+\lambda_k^{-1})\,\widehat{D}_k\;\Omega_{k-\tau}\;\widehat{D}_k^t \;+\; \widehat{B}_k\widehat{B}_k^t$$
$$\leq\; 0, \qquad k \geq 0 \tag{8.86}$$

for all admissible uncertainties satisfying (8.67)-(8.68).

Our next result shows that if (8.73) is QE for system (8.64)-(8.65) with cost matrix Ω_k, then Ω_k defines an upper bound for the filtering error covariance, that is, $E[e_k\, e_k^t] \leq \Omega_{2,k}, \; \forall\, k \geq 0$.

Theorem 8.5: *Consider the time-delay system (8.64)-(8.65) satisfying (8.67)-(8.68) and with known initial state. Suppose there exists a solution* $\Omega_k \geq 0$ *to inequality (8.84) for some* $\lambda_k > 0$ *and for all admissible uncertainties. Then the estimator (8.73) provides an upper bound for the filtering error covariance, that is,*

$$\mathcal{E}\,[e_k\, e_k^t] \;\leq\; [0 \;\; I]\Omega_k[0 \;\; I]^t \quad \forall\, k \geq 0 \tag{8.87}$$

Proof: Suppose that estimator (8.73) is QE with cost matrix Ω_k. By evaluating the one-step ahead covariance matrix $\Sigma_{\xi,k+1} = E[\xi_{k+1}\,\xi_{k+1}^t]$, we get

$$\Sigma_{\xi,k+1} = \mathcal{E}[\widehat{A}_{k,\Delta}\;\xi_k + \widehat{D}_k\;\xi_{k-\tau} + \widehat{B}_k\;\eta_k][\widehat{A}_{k,\Delta}\;\xi_k + \widehat{D}_k\;\xi_{k-\tau} + \widehat{B}_k\;\eta_k]^t$$
$$= \mathcal{E}[\widehat{A}_{k,\Delta}\;\xi_k\xi_k^t\;\widehat{A}_{k,\Delta}^t] + \mathcal{E}[\widehat{A}_{k,\Delta}\;\xi_k\xi_{k-\tau}^t\;\widehat{D}_k^t] + \mathcal{E}[\widehat{D}_k\;\xi_{k-\tau}\xi_k^t\;\widehat{A}_{k,\Delta}^t]$$
$$+\; \mathcal{E}[\widehat{D}_k\;\xi_{k-\tau}\xi_{k-\tau}^t\;\widehat{D}_k^t] + \mathcal{E}[\widehat{B}_k\;\eta_k\eta_k^t\;\widehat{B}_k^t] \tag{8.88}$$

Note that

$$\widehat{D}_k\;\mathcal{E}[\xi_{k-\tau}\xi_k^t]\;\widehat{A}_{k,\Delta}^t + \widehat{A}_{k,\Delta}\;\mathcal{E}[\xi_k\xi_{k-\tau}^t]\;\widehat{D}_k^t \leq$$
$$\lambda_k\;\widehat{A}_{k,\Delta}\;\mathcal{E}[\xi_k\,\xi_k^t]\;\widehat{A}_{k,\Delta}^t + \lambda_k^{-1}\;\widehat{D}_k\;\mathcal{E}[\xi_{k-\tau}\,\xi_{k-\tau}^t]\;\widehat{D}_k^t \tag{8.89}$$

Using (8.89) into (8.88) and arranging terms, we get:

$$
\begin{aligned}
\Sigma_{\xi,k+1} &\leq (1+\lambda_k)\,\widehat{A}_{k,\Delta}\,\Sigma_{\xi,k}\,\widehat{A}_{k,\Delta}^t + (1+\lambda_k^{-1})\,\widehat{D}_k\,\Sigma_{\xi,k-\tau}\,\widehat{D}_k^t \\
&\quad + \widehat{B}_k\widehat{B}_k^t
\end{aligned}
\tag{8.90}
$$

Letting $\Xi_k = \Sigma_{\xi,k} - \Omega_k$ with $e_k = x_k - \hat{x}_k$ and considering inequalities (8.86) and (8.90), we get:

$$
\Xi_{k+1} \leq (1+\lambda_k)\,\widehat{A}_{k,\Delta}\,\Xi_k\,\widehat{A}_{k,\Delta}^t + (1+\lambda_k^{-1})\,\widehat{D}_k\,\Xi_{k-\tau}\,\widehat{D}_k^t
\tag{8.91}
$$

By considering that the state is known over the period $[-\tau,0]$, it justifies letting $\Sigma_{\xi,k} = 0\ \forall k \in [-\tau,0]$. Then it follows from (8.91) that $\Xi_k \leq 0$ for $k > 0$; that is, $\Sigma_{\xi,k} \leq \Omega_k$ for $k > 0$. Hence, $\mathcal{E}[e_k e_k^t] \leq [0\ \ I]\Omega_k[0\ \ I]^t\ \forall k \geq 0$.

8.3.3 A Riccati Equation Approach

Motivated by the recent results of robust control theory [1,2,6-8], we employ hereafter a Riccati equation approach to solve the robust Kalman filtering for time-delay systems. To this end, we assume that A_k is invertible for any $k \geq 0$, and define matrices $P_k = P_k^t \in \Re^{n \times n}$; $S_k = S_k^t \in \Re^{n \times n}$ as the solutions of the Riccati difference equations (RDEs):

$$
\begin{aligned}
P_{k+1} &= (1+\lambda_k)\{A_k\,(I + \mu_k P_k Y_k)\,P_k A_k^t\} + (1+\lambda_k^{-1})D_k P_{k-\tau} D_k^t + \widehat{W}_k; \\
P_{k-\tau} &= 0 \quad \forall\,k \in [0,\tau] \tag{8.92} \\
S_{k+1} &= (1+\lambda_k)\hat{A}_k\,(I + \mu_k S_k Y_k)\,S_k \hat{A}_k^t \\
&\quad + (1+\lambda_k)\delta A_k\,(I + \mu_k P_k Y_k)\,P_k \delta A_k^t \\
&\quad + (1+\lambda_k)\mu_k \hat{A}_k P_k Y_k S_k \hat{A}_k^t + (1+\lambda_k)\hat{A}_k \mu_k S_k Y_k P_k \delta A_k^t \\
&\quad - \hat{M}_k^t \left(\hat{\Gamma}_k + \hat{V}_k\right)^{-1} \hat{M}_k \\
&\quad + (1+\lambda_k^{-1})D_k S_{k-\tau} D_k^t + \hat{W}_k \\
S_{k-\tau} &= 0 \quad \forall\,k \in [0,\tau] \tag{8.93}
\end{aligned}
$$

where $\mu_k > 0$, $\lambda_k > 0$ are scaling parameters such that $P_k^{-1} - \mu_k^{-1}E_k^t E_k > 0$ and the matrices $\hat{A}_k, \delta A_k, C_k, \hat{W}_k, \hat{\Gamma}_k$ and $\hat{M}_k$ are given by:

$$
Y_k = E_k^t \left(I - \mu_k E_k P_k E_k^t\right)^{-1} E_k
\tag{8.94}
$$

$$
\hat{W}_k = W_k + (1+\lambda_k)\mu_k^{-1}H_{1,k}H_{1,k}^t
\tag{8.95}
$$

$$\hat{V}_k = V_k + (1 + \lambda_k)\mu_k^{-1} H_{2,k} H_{2,k}^t \tag{8.96}$$

$$T_k = (1 + \lambda_k)(P_k - S_k)(I + \mu_k Y_k P_k) A_k^t$$

$$Z_k = (1 + \lambda_k)\hat{M}_k^t \left(\hat{\Gamma}_k + \hat{V}_k\right)^{-1} \left(C_k S_k (I + \mu_k Y_k P_k) A_k^t\right)$$
$$+ \mu_k^{-1} H_{2,k} H_{1,k}^t \tag{8.97}$$

$$X_k = (1 + \lambda_k)\mu_k A_k S_k Y_k P_k A_k^t + (1 + \lambda_k)\mu_k^{-1} H_{1,k} H_{1,k}^t$$
$$+ (1 + \lambda_k^{-1}) D_k S_{k-\tau} D_k^t \tag{8.98}$$

$$\hat{\Gamma}_k = (1 + \lambda_k) C_k S_k C_k^t \tag{8.99}$$

$$\hat{A}_k = A_k + \delta A_k \ ; \ \delta A_k = T_k^{-1}(X_k + Z_k) \tag{8.100}$$

$$\hat{M}_k = (1 + \lambda_k)[C_k S_k A_k^t + \mu_k S_k Y_k P_k \delta A_k^t + \mu_k H_{2,k} H_{1,k}^t] \tag{8.101}$$

Note that the assumption that A_k being invertible for all k is needed for the existence of T_k and δA_k. Let the (λ, μ)–parametrized estimator be expressed as:

$$\hat{x}_{k+1} = \left(A_k + T_k^{-1}(X_k + Z_k)\right)\hat{x}_k + K_{o,k}[y_k - C_k\hat{x}_k] \tag{8.102}$$

where the Kalman gain matrix $K_{o,k} \in \Re^{n \times m}$ is to be determined. The following theorem summarizes the main result:

Theorem 8.6: *Consider system (8.64)-(8.65) satisfying the uncertainty structure (8.67)-(8.68) with zero initial condition. Suppose the process and measurement noises satisfy (8.69)-(8.71). For some $\mu_k > 0$, $\lambda_k > 0$, let $0 < P_k = P_k^t$ and $0 < S_k = S_k^t$ be the solutions of RDEs (8.92) and (8.93), respectively. Then the (λ, μ)–parametrized estimator (8.102) is a QE estimator*

$$\mathcal{E}[\{\hat{x}_k - x_k\}^t\{\hat{x}_k - x_k\}] \leq \text{tr}(P_k - S_k) \tag{8.103}$$

Moreover, the gain matrix K is given by

$$K_{o,k} = \hat{M}_k^t \left\{\hat{\Gamma}_k + \hat{V}_k\right\}^{-1} \tag{8.104}$$

Proof: Let

$$X_k = \begin{bmatrix} P_k & S_k \\ S_k & S_k \end{bmatrix} \tag{8.105}$$

where P_k and S_k are the positive-definite solutions to (8.92) and (8.93), respectively. By using B.1.2, B.1.3 and combining (8.92)-(8.101), it is a simple task to verify that

$$(1+\lambda_k)\left\{\widehat{A}_k\, X_k\, \widehat{A}_k^t + \mu_k\, \widehat{A}_k\, X_k\, \breve{E}_k^t\, [I - \mu_k\, \breve{E}_k\, X_k\, \breve{E}_k^t]^{-1}\, \breve{E}_k\, X_k\, \widehat{A}_k^t\right\}$$

$$-X_{k+1} + (1+\lambda_k)\mu_k^{-1}\, \widehat{H}_k\widehat{H}_k^t + \widehat{B}_k\widehat{B}_k^t + (1+\lambda_k^{-1})\, \widehat{D}_k\, X_{k-\tau}\, \widehat{D}_k^t = 0 \quad (8.106)$$

where $\widehat{A}_k, \widehat{B}_k, \widehat{H}_k, \widehat{D}_k$ are given by (8.80)-(8.82).

Using [39], it is easy to see on using **B.1.3** with some algebraic manipulations that (8.106) implies that:

$$(1+\lambda_k)[\widehat{A}_k + \widehat{H}_k\, \Delta_k\, \breve{E}_k]X_k[\widehat{A}_k + \widehat{H}_k\, \Delta_k\, \breve{E}_k]^t - X_{k+1} +$$
$$(1+\lambda_k^{-1})\, \widehat{D}_k\, X_{k-\tau}\, \widehat{D}_k^t + \widehat{B}_k\widehat{B}_k^t =$$
$$(1+\lambda_k)\, \widehat{A}_{\Delta,k}\, X_k\, \widehat{A}_{\Delta,k}^t - X_{k+1} +$$
$$(1+\lambda_k^{-1})\, \widehat{D}_k\, X_{k-\tau}\, \widehat{D}_k^t + \widehat{B}_k\widehat{B}_k^t \leq 0 \quad (8.107)$$
$$\forall \Delta : \Delta_k^t\, \Delta_k \leq I \; \forall k$$

It follows from **Theorem 8.5** that (8.102) is a quadratic estimator and

$$\mathcal{E}[e_k e_k^t] = \mathcal{E}[0 \; I]X_k[0 \; I]^t \leq S_k$$

which implies that $\mathcal{E}[e_k^t\, e_k] \leq tr\, (S_k)$.

Remark 8.7: In the case of systems without uncertainties and delay factors, that is $H_{1,k} = 0$, $H_{2,k} = 0$, $E_k = 0$, $D_k = 0$, it can be easily shown that

$$Y_k = 0 \; ; \; \mathcal{X}_k = 0 \; ; \; \dot{W}_k = W_k \; ; \; T_k = (1+\lambda_k)(P_k - S_k)A_k^t$$
$$\mathcal{Z}_k = (1+\lambda_k)^2 A_k S_k C_k^t \left((1+\lambda_k)C_k S_k C_k^t + V_k\right)^{-1} C_k S_k A_k^t$$

Now, in terms of $L_k = P_k - S_k$ and

$$\Psi_k = S_k C_k^t \left((1+\lambda_k)C_k S_k C_k^t + V_k\right)^{-1} C_k S_k \; ; \; \Phi_k = A_k \Psi_k A_k^t$$
$$\mathcal{R}_k = A_k^{-t}(P_k - S_k)^{-1} \; ; \; \hat{A}_k = (1+\lambda_k)\mathcal{R}_k\Phi_k$$

we manipulate (8.92)-(8.93) to reach

$$
\begin{aligned}
L_{k+1} &= (1+\lambda_k)\left(A_k L_k A_k^t + \Lambda_k\right) \quad ; \quad L_{k-\tau} = 0 \quad \forall k \in [0,\tau] \\
\Lambda_k &= \Phi_k \\
&\quad - (1+\lambda_k)\left\{ A_k(P_k - L_k)\Phi_k^t \mathcal{R}_k^t + \mathcal{R}_k \Phi_k (P_k - L_k) A_k^t \right\} \\
&\quad + \left\{ (1+\lambda_k)^2 \mathcal{R}_k \Phi_k (2P_k - L_k)\Phi_k^t \mathcal{R}_k^t \right\}
\end{aligned}
$$

(8.108)

By iterating on (8.108) and (8.92), it follows that $L_k = P_k - S_k > 0 \ \forall k > 0$.

It can be shown in the general case that manipulation of (8.92)-(8.101) yields:

$$
L_{k+1} = (1+\lambda_k)[A_k(I + \mu_k L_k Y_k)L_k A_k^t + \Pi_k]; \quad L_{k-\tau} = 0 \quad \forall k \in [0,\tau]
$$

In this case, Π_k depends on $A_k, H_{1,k}, H_{2,k}, D_k, C_k, P_k$. The derivation of Π_k requires tedious mathematical manipulations and it is therefore omitted. Note that P_k does not depend on the filter matrices and the structure of X_k is identical to that of the joint covariance matrix of the state of a certain system and its standard H_2-optimal estimator. By similarity to the standard H_2−optimal filter, an estimate of z_k in (8.68) will be given by $\hat{z}_k = C_{1,k}\hat{x}_k$.

Remark 8.8: In the delayfree case ($A_{dk} = 0$), we supress the parameter λ_k and observe that (8.102) reduces to the recursive Kalman filter for the system

$$
\begin{aligned}
x_{k+1} &= \hat{A}_k\, x_k + \hat{w}_k & \text{(8.109)} \\
y_k &= C_k\, x_k + \hat{v}_k & \text{(8.110)}
\end{aligned}
$$

where $\hat{w}_k$ and $\hat{v}_k$ are zero-mean white noise sequences with covariance matrices $\hat{W}_k$ and $\hat{\Gamma}_k$, respectively, and having cross-covariance matrix $\hat{M}_k$. Hence, our approach to robust filtering in **Theorem 8.2** corresponds to designing a standard Kalman filter for a related discrete-time system which captures all admissible uncertainties and time-delay, but does not involve parameter uncertainties. In this regard, the matrix δA_k reflects the effect of uncertainties ($\Delta A_k, \Delta C_k$) and time delay factor D_k on the structure of the filter.

8.3.4　Steady-State Filter

In this section, we investigate the asymptotic properties of the recursive Kalman filter of Section 4. We consider the uncertain time-delay system

$$
\begin{aligned}
x_{k+1} &= [A + H\Delta_k E]\, x_k + A_d\, x_{k-\tau} + w_k \\
&= A_\Delta\, x_k + D\, x_{k-\tau} + w_k \quad\quad (8.111) \\
y_k &= [C + H_c\Delta_k E]\, x_k + v_k \\
&= C_\Delta\, x_k + v_k \quad\quad (8.112)
\end{aligned}
$$

where Δ_k satisfies (8.68). In the sequel, we assume that A is a Schur matrix; that is $|\lambda(A)| < 1$. The matrices $A \in \Re^{n \times n}$, $C \in \Re^{m \times n}$ are constant matrices representing the nominal plant. The uncertain parameter matrix Δ_k is, however time-varying. In this regard, the objective is to design a shift-invariant *a priori* estimator of the form

$$
\hat{x}_{k+1} = \hat{A}\hat{x}_k + K_o\, y_k \quad\quad (8.113)
$$

that achieves the following asymptotic performance bound

$$
\lim_{k \to \infty} \mathcal{E}\left\{ (\hat{x}_k - x_k)(\hat{x}_k - x_k)^t \right\} \leq S \quad\quad (8.114)
$$

Theorem 8.7: *Consider the uncertain time-delay system (8.111)-(8.112). If for some scalars $\mu > 0, \lambda > 0$, there exist stabilizing solutions $P \geq 0, S \geq 0$ for the AREs*

$$
\begin{aligned}
P &= (1+\lambda)\{A\,(I + \mu PY)\,PA^t\} + (1+\lambda^{-1})DPD^t + \hat{W}; \quad (8.115) \\
S &= (1+\lambda)\hat{A}\,(I + \mu SY)\,S\hat{A}^t + (1+\lambda)\delta A\,(I + \mu PY)\,P\delta A^t \\
&\quad + (1+\lambda)\mu\hat{A}PYS\hat{A}^t + (1+\lambda)\hat{A}\mu_k SYP\delta A^t \\
&\quad - \hat{M}^t\left(\hat{\Gamma} + \hat{V}\right)^{-1}\hat{M} \quad\quad (8.116)
\end{aligned}
$$

$$
\begin{aligned}
Y &= E^t\left(I - \mu EPE^t\right)^{-1} E \\
\hat{W} &= W + (1+\lambda)\mu^{-1}H_1 H_1^t \quad\quad (8.117) \\
\hat{V} &= V + (1+\lambda)\mu^{-1}H_2 H_2^t \;;\; \hat{\Gamma} = (1+\lambda)CSC^t \quad\quad (8.118) \\
\hat{M} &= (1+\lambda)[CSA^t + \mu SYP\delta A^t + \mu H_2 H_1^t] \quad\quad (8.119)
\end{aligned}
$$

Then the estimator (8.113) is a stable quadratic (SQ) estimator and achieves (8.114) with

$$
\hat{A} = (1+\lambda)^{-1}\{\mathcal{T} - \mathcal{X}\mathcal{Z}\}\,\mathcal{R}^{-1}S^{-1} \quad\quad (8.120)
$$

$$K_o = \hat{M}\left\{\hat{\Gamma} + \hat{V}\right\}^{-1} \tag{8.121}$$

$$\mathcal{T} = (1+\lambda)A[I + \mu PY]S \; ; \; \mathcal{Z} = (1+\lambda)\mu C(P-S)YS \tag{8.122}$$

$$\mathcal{X} = (1+\lambda)A[I + \mu PY](P-S)C^t + (1+\lambda)\mu^{-1}HH_c^t \tag{8.123}$$

Proof: To examine the stability of the closed-loop system, we augment (8.111)-(8.113) with ($w_k = 0, v_k = 0$) to obtain:

$$
\begin{aligned}
\xi_{k+1} &= \widehat{A}_\Delta\, \xi_k + \widehat{D}\, \xi_{k-\tau} \\
&= \begin{bmatrix} A_\Delta & 0 \\ A_\Delta - \hat{A} - K_o(C_\Delta - C) & \hat{A} - K_o C \end{bmatrix} \xi_k \\
&\quad + \begin{bmatrix} A_d & 0 \\ A_d & 0 \end{bmatrix} \xi_{k-\tau}
\end{aligned}
\tag{8.124}
$$

Introduce a discrete Lyapunov-Krasovskii functional

$$V_k = \xi_k^t X \xi_k + \sum_{j=k-\tau}^{k-1} \xi_j^t (1+\lambda^{-1}) \widehat{D}^t X \widehat{D} \xi_j \tag{8.125}$$

for some $\lambda > 0$. By evaluating the first-order difference $\Delta V_k = V_{k+1} - V_k$ along the trajectories of (8.125) and arranging terms, we get:

$$
\begin{aligned}
\Delta V_k &= \xi_k^t [\widehat{A}_\Delta^t X \widehat{A}_\Delta - X] \xi_k + \xi_{k-\tau}^t \widehat{D}^t X \widehat{A}_\Delta \xi_k \\
&\quad + \xi_k^t \widehat{A}_\Delta^t X \widehat{D} \xi_{k-\tau} \\
&\quad + \xi_{k-\tau}^t \widehat{D}^t X \widehat{D} \xi_{k-\tau} + (1+\lambda^{-1}) \xi_k^t \widehat{D}^t X \widehat{D} \xi_k \\
&\quad - (1+\lambda^{-1}) \xi_{k-\tau}^t \widehat{D}^t X \widehat{D} \xi_{k-\tau} \\
&\leq \xi_k^t [\widehat{A}_\Delta^t X \widehat{A}_\Delta - X + (1+\lambda^{-1}) \widehat{D}^t X \widehat{D}] \xi_k \\
&\quad + \lambda^{-1} \xi_{k-\tau}^t \widehat{D}^t X \widehat{D} \xi_{k-\tau} + \lambda \xi_k^t \widehat{A}_\Delta^t X \widehat{A}_\Delta \xi_k \\
&\quad + \xi_{k-\tau}^t \widehat{D}^t X \widehat{D} \xi_{k-\tau} - \xi_{k-\tau}^t (1+\lambda^{-1}) \widehat{D}^t X \widehat{D} \xi_{k-\tau} \\
&= \xi_k^t [(1+\lambda) \widehat{A}_\Delta^t X \widehat{A}_\Delta - X + (1+\lambda^{-1}) \widehat{D}^t X \widehat{D}] \xi_k \tag{8.126}
\end{aligned}
$$

Sufficient condition of asymptotic stability $\Delta V_k < 0$, $\xi_k \neq 0$ is implied by

$$(1+\lambda) \widehat{A}_\Delta^t X \widehat{A}_\Delta - X + (1+\lambda^{-1}) \widehat{D}^t X \widehat{D} < 0 \tag{8.127}$$

Using **A.1**, it follows that inequality (8.127) is equivalent to:

$$
\begin{bmatrix}
-X & \sqrt{1+\lambda}\,\widehat{A}_\Delta^{\,t} & \sqrt{1+\lambda^{-1}}\,\widehat{D}^{\,t} \\
\sqrt{1+\lambda}\,\widehat{A}_\Delta & -X^{-1} & 0 \\
\sqrt{1+\lambda^{-1}}\,\widehat{D} & 0 & -X^{-1}
\end{bmatrix} < 0 \iff
$$

$$
\begin{bmatrix}
-X & \sqrt{1+\lambda}\,\widehat{A}_\Delta & \sqrt{1+\lambda^{-1}}\,\widehat{D} \\
\sqrt{1+\lambda}\,\widehat{A}_\Delta^{\,t} & -X^{-1} & 0 \\
\sqrt{1+\lambda^{-1}}\,\widehat{D}^{\,t} & 0 & -X^{-1}
\end{bmatrix} < 0 \tag{8.128}
$$

Application of **A.1** once again to (8.128) yields

$$
(1+\lambda)\,\widehat{A}_\Delta\,X\,\widehat{A}_\Delta^{\,t} -X + (1+\lambda^{-1})\,\widehat{D}\,X\,\widehat{D}^{\,t} < 0 \tag{8.129}
$$

Now by selecting

$$
X = \begin{bmatrix} P & S \\ S & S \end{bmatrix} \tag{8.130}
$$

with P and S being the stabilizing solutions of (8.115) and (8.116), respectively, it follows from **Definition 8.2** and **Theorem 8.5** in the steady-state as $k \to \infty$ that the augmented system (8.124) is asymptotically stable. The guaranteed performance $\mathcal{E}[e_k e_k^t] \le S$ follows from similar lines of argument as in the proof of **Theorem 8.6**.

Remark 8.9: Note that the invertibility of A is needed for the existence of T and δA. In the delayless case $(D \equiv 0$, it follows from (8.115) and (8.116) with $\hat{W} = \bar{B}\bar{B}^t$ that

$$
P = (1+\lambda)\{APA^t + AP[(\mu^{-1}I + EPE^t)^{-1} PA^t\} + \hat{W} \tag{8.131}
$$

which is a bounded real lemma equation (see **A.3**) for the system

$$
\Sigma = \left(A\sqrt{1+\lambda}, \bar{B}, E, 0\right)
$$

Suppose that for $\mu = \mu^+$, the ARE (8.132) admits a solution $P = P^+$. This implies that the $\mathcal{H}_\infty$-norm of Σ is less than $(\mu^+)^{-1/2}$. It then follows, given a λ, that system (8.111)-(8.112) is quadratically stable for some $\mu \le \mu^+$.

8.3.5 Example 8.2

Consider the following discrete-time delay system

$$x_{k+1} = \left(\begin{bmatrix} 0.2 & -0.1 & 0 \\ 0.004 & 0.4 & 0.1 \\ 0 & 0.1 & 0.6 \end{bmatrix} + \begin{bmatrix} 0.1 \\ 0.1 \\ 0.1 \end{bmatrix} \Delta_k [0.5 \ \ 0.4 \ \ 0.2] \right) x_k$$

$$+ \begin{bmatrix} -0.1 & 0 & 0 \\ 0.05 & -0.2 & 0.1 \\ 0 & 0 & -0.1 \end{bmatrix} x_{k-\tau} + w_k$$

$$y_k = \left(\begin{bmatrix} 1.0 & 0 \\ 0 & 1.0 \\ 0 & 0 \end{bmatrix} + \begin{bmatrix} 0.2 \\ 0.3 \end{bmatrix} \Delta_k [0.5 \ \ 0.4 \ \ 0.2] \right) x_k + v_k$$

which is of the type (8.111)-(8.112). We further assume that $W = I$, $V = 0.2I$. To determine the Kalman gains, we solve (8.115)-(8.116) with the aid of (8.117)-(8.122) for selected values of λ, μ. The numerical computation is basically of the form of iterative schemes and the results for a typical case of $\mu = 0.7, \lambda = 0.7$ are given by:

$$P = \begin{bmatrix} 0.141 & 0.005 & 0.003 \\ 0.005 & 0.255 & 0.175 \\ 0.003 & 0.175 & 0.501 \end{bmatrix}, \ S = 10^{-5} \begin{bmatrix} 0.284 & -1.17 & -2.966 \\ -1.17 & 4.813 & 12.208 \\ -2.966 & 12.208 & 30.962 \end{bmatrix}$$

$$K_o = 10^{-6} \begin{bmatrix} -0.841 & -1.309 \\ 3.463 & 5.388 \\ 8.782 & 13.665 \end{bmatrix}, \ \hat{A} = \begin{bmatrix} 0.331 & -0.019 & -0.034 \\ -0.277 & 0.254 & 0.175 \\ -1.641 & -0.897 & 0.961 \end{bmatrix}$$

The developed estimator is indeed asymptotically stable since

$$\lambda(\hat{A}) = \{0.302, 0.48, 0.765\} \in (0, 1)$$

8.4 Notes and References

The results presented in this chapter were mainly based on [21,25] and essentially provided some extensions of the delayless results of [12,44,45] to UTDS. In principle, there are ample other possibilities to follow including the approaches of [10, 17].

Robust Kalman filtering for interconnected (continuous-time or discrete-time) systems, UTDS with uncertain state-delayed matrix, nonlinear UTDS,

UTDS with unknown delay and robust Kalman filtering with unknown covariance matrices are only representative examples of research topics that
indeed deserve further investigation.

λ	μ	P		L		δA	
0.1	0.6	0.574	0.175	0.566	0.176	0.382	0.784
		0.175	0.457	0.176	0.464	0.459	0.919
0.1	0.8	0.535	0.156	0.500	0.136	0.242	0.483
		0.156	0.431	0.136	0.421	0.306	0.612
0.1	0.9	0.525	0.151	0.480	0.124	0.202	0.404
		0.151	0.425	0.124	0.409	0.261	0.523
0.2	0.6	0.564	0.206	0.554	0.210	0.406	0.811
		0.206	0.374	0.210	0.386	0.409	0.818
0.2	0.8	0.525	0.183	0.488	0.167	0.257	0.514
		0.183	0.350	0.167	0.346	0.269	0.537
0.2	0.9	0.515	0.177	0.468	0.153	0.215	0.430
		0.177	0.344	0.153	0.334	0.228	0.456
0.4	0.6	0.603	0.242	0.576	0.238	0.438	0.877
		0.242	0.357	0.238	0.367	0.405	0.810
0.4	0.8	0.564	0.217	0.508	0.192	0.278	0.557
		0.217	0.335	0.192	0.328	0.265	0.530
0.4	0.9	0.555	0.210	0.486	0.177	0.233	0.466
		0.210	0.330	0.177	0.316	0.225	0.449
0.6	0.6	0.665	0.276	0.612	0.259	0.471	0.942
		0.276	0.369	0.259	0.372	0.418	0.836
0.6	0.8	0.629	0.252	0.540	0.210	0.300	0.600
		0.252	0.348	0.210	0.333	0.274	0.548
0.6	0.9	0.622	0.247	0.517	0.195	0.252	0.503
		0.247	0.344	0.195	0.321	0.232	0.464
0.8	0.6	0.751	0.319	0.656	0.281	0.507	1.015
		0.319	0.392	0.281	0.384	0.437	0.874
0.8	0.8	0.723	0.298	0.580	0.229	0.324	0.649
		0.298	0.374	0.229	0.344	0.287	0.574
0.8	0.9	0.726	0.298	0.556	0.213	0.273	0.545
		0.298	0.372	0.213	0.332	0.244	0.487
0.9	0.6	0.806	0.346	0.681	0.292	0.527	1.055
		0.346	0.407	0.292	0.392	0.448	0.897
0.9	0.8	0.790	0.330	0.603	0.239	0.338	0.676
		0.330	0.392	0.239	0.352	0.294	0.589
0.9	0.9	0.805	0.336	0.579	0.222	0.284	0.569
		0.336	0.393	0.222	0.339	0.250	0.501

Table 8.1: Summary of Some of the Computational Results-I

λ	μ	K^t		$\lambda(A)$	
0.1	0.6	0.314	0.301	-3.227	-0.472
0.1	0.8	0.314	0.301	-3.249	-0.897
0.1	0.9	0.315	0.302	-3.258	-1.017
0.2	0.6	0.317	0.295	-3.279	-0.498
0.2	0.8	0.318	0.301	-3.293	-0.913
0.2	0.9	0.318	0.294	-3.298	-1.031
0.4	0.6	0.324	0.296	-3.302	-0.450
0.4	0.8	0.325	0.301	-3.311	-0.880
0.4	0.9	0.326	0.295	-3.415	-1.002
0.6	0.6	0.331	0.300	-3.311	-0.382
0.6	0.8	0.333	0.299	-3.318	-0.834
0.6	0.9	0.335	0.299	-3.321	-0.962
0.8	0.6	0.340	0.304	-3.318	-0.300
0.8	0.8	0.343	0.304	-3.324	-0.779
0.8	0.9	0.345	0.304	-3.326	-0.914
0.9	0.6	0.344	0.307	-3.321	-0.255
0.9	0.8	0.349	0.307	-3.326	-0.747
0.9	0.9	0.351	0.307	-3.328	-0.887

Table 8.2: Summary of Some of the Computational Results-II

Chapter 9

Robust $\mathcal{H}_\infty$ Filtering

9.1 Introduction

In control engineering research, the robust filtering (state estimation) problem arises out of the desire to determine estimates of unmeasurable state variables for dynamical systems with uncertain parameters. Along this way, the robust filtering problem can be viewed as an extension of the celebrated Kalman filter [3,4] to uncertain dynamical systems. The past decade has witnessed major developments in robust and $\mathcal{H}_\infty-$ control theory [1,2,5-8] with some focus on the robust filtering problem using different approaches. In [10-18], a linear $\mathcal{H}_\infty-$filter is designed such that the $\mathcal{H}_\infty-$norm of the system, which reflects the worst case gain of the transfer function from the disturbance inputs to the estimation error output, is minimized. On the other hand, by constructing a state estimator which bounds the mean square estimation error [12], one can develop a robust Kalman filter. Indeed, the $\mathcal{H}_\infty-$ filtering is superior to the standard $\mathcal{H}_2-$ filtering since no statistical assumption on the input is needed. It considers essentially the exogenous input signal to be energy bounded rather than Gaussian.

Despite the significant role of time-delays in continuous-time modeling of physical systems [19,20], little attention has been paid to the filtering (state estimation) problem of time-delay systems. Only recently, some efforts towards bridging this gap have been pursued in [21-27] where a version of the robust Kalman filter has been developed for both continuous- and discrete-time systems (see Chapter 8). This chapter contributes to the further development of the filtering problem for a class of uncertain time-delay systems with bounded energy noise sources. In particular, we investigate the problem of robust $\mathcal{H}_\infty-$ filtering when the uncertainties are real time-varying and norm-bounded and the state-delay is unknown. It pays equal attention to continuous-time and discrete-time systems. For both system representations, we design a linear filter which provides both robust stability and a guaranteed $\mathcal{H}_\infty-$performance for the filtering error irrespective of the parameteric uncertainties and unknown delays.

9.2   Linear Uncertain Systems

9.2.1   Problem Description and Preliminaries

We consider a class of uncertain time-delay systems represented by:

$$(\Sigma_\Delta): \quad \dot{x} = [A + \Delta A(t)]x(t) + [A_d + \Delta A_d(t)]x(t - \tau) + Dw(t)$$
$$= A_\Delta(t)x(t) + A_{d\Delta}(t)x(t - \tau) + Dw(t) \quad (9.1)$$
$$y(t) = [C + \Delta C(t)]x(t) + Nw(t)$$
$$= C_\Delta(t)x(t) + Nw(t) \quad (9.2)$$
$$z(t) = L\,x(t) \quad (9.3)$$

where $x(t) \in \Re^n$ is the state, $w(t) \in \Re^m$ is the input noise which belongs to $\mathcal{L}_2\,[0,\infty)$, $y(t) \in \Re^p$ is the measured output, $z(t) \in \Re^r$, is a linear combination of the state variables to be estimated and the matrices $A \in \Re^{n\times n}$, $B \in \Re^{n\times m}$, $C \in \Re^{p\times n}$, $H \in \Re^{p\times m}$, $A_d \in \Re^{n\times n}$ and $L \in \Re^{r\times n}$ are real constant matrices representing the nominal plant. Here, τ is an unknown constant scalar representing the amount of delay in the state. For all practical purposes, we let $\tau \leq \tau^*$ where τ^* is known. The matrices $\Delta A(t)$, $\Delta C(t)$ and $\Delta A_d(t)$ represent parameteric uncertainties which are given by:

$$\begin{bmatrix} \Delta A(t) \\ \Delta C(t) \end{bmatrix} = \begin{bmatrix} H \\ H_c \end{bmatrix} \Delta_1(t)\, E$$
$$\Delta A_d(t) = H_d\, \Delta_2(t)\, E_d \quad (9.4)$$

where $H \in \Re^{n\times\alpha}$, $H_c \in \Re^{p\times\alpha}$, $H_d \in \Re^{n\times\omega}$, $E \in \Re^{\beta\times n}$ and $E_d \in \Re^{\varphi\times n}$ are known constant matrices and $\Delta_1(t) \in \Re^{\alpha\times\beta}, \Delta_2(t) \in \Re^{\omega\times\varphi}$ are unknown matrices satisfying

$$\Delta_1^t(t)\,\Delta_1(t) \leq I \quad , \quad \Delta_2^t(t)\,\Delta_2(t) \leq I \quad \forall\, t \quad (9.5)$$

The initial condition is specified as $\langle x(0), x(s) \rangle = \langle x_o, \phi(s) \rangle$, where $\phi(.) \in \mathcal{L}_2[-\tau, 0]$.

For system (Σ_Δ), we wish to design an estimator of $z(t)$ of the form:

$$(\Sigma_e): \quad \dot{\hat{x}}(t) = F_o\hat{x}(t) + K_o y(t) \tag{9.6}$$
$$\hat{z}(t) = L\hat{x}(t) \tag{9.7}$$

where $\hat{x}(t) \in \Re^n$ is the estimator state and the matrices $F_o \in \Re^{n \times n}$, $K_o \in \Re^{n \times m}$ are to be determined. From (9.3) and (9.7), we define the estimation error as:

$$e(t) := z(t) - \hat{z}(t) \tag{9.8}$$

In the subsequent development, we adopt the notion of robust stability independent of delay as examined in Chapter 2. With reference to **Lemma 2.1**, we know that system (Σ_Δ) is robustly stable independent of delay if there exist matrices $0 < P = P^t \in \Re^{n \times n}$ and $0 < W = W^t \in \Re^{n \times n}$ satisfying the ARI:

$$PA_\Delta(t) + A_\Delta^t(t)P + W + PA_{d\Delta}(t)W^{-1}A_{d\Delta}^t(t)P < 0 \tag{9.9}$$
$$\forall \; \|\Delta_1\| \le 1, \quad \|\Delta_2\| \le 1$$

The following preliminary result extends **Lemma 2.1** a bit further for the case of constant delay.

Lemma 9.1: *System (Σ_Δ) is robustly stable independent of delay if one of the following equivalent statements hold:*

(1) *There exist scalars $\mu > 0$, $\sigma > 0$ and matrices $0 \le \bar{P} = \bar{P}^t \in \Re^{n \times n}$ and $0 < W = W^t \in \Re^{n \times n}$ satisfying the ARE:*

$$\bar{P}A + A^t\bar{P} + \bar{P}\left\{\mu H H^t + \sigma H_d H_d^t + A_d[W - \sigma^{-1}E_d^t E_d]^{-1}A_d^t\right\}\bar{P}$$
$$+ \mu^{-1}E^t E + W = 0 \tag{9.10}$$

(2) *A is stable and the following $\mathcal{H}_\infty$ norm bound is satisfied*

$$\left\| \begin{bmatrix} \frac{1}{\sqrt{\mu}}E \\ W^{1/2} \end{bmatrix} [sI - A]^{-1}[\sqrt{\mu}H \quad \sqrt{\sigma}H_d \quad A_d(W - \sigma^{-1}E_d^t E_d)^{-1/2}] \right\|_\infty < 1 \tag{9.11}$$

Proof: (1) For some $\mu > 0$, $\sigma > 0$ it follows that by applying **B.1.2** and **A.2** to inequality (9.9) it reduces to:

$$PA + A^tP + P\left\{\mu H H^t + \sigma H_d H_d^t + A_d[W - \sigma^{-1}E_d^t E_d]^{-1}A_d^t\right\}P +$$
$$\mu^{-1}E^t E + W < 0 \tag{9.12}$$

In the light of **A.3.1**, it follows that the existence of a matrix $0 < P = P^t \in \Re^{n \times n}$ satisfying inequality (9.12) is equivalent to the existence of a stabilizing solution $0 \le \bar{P} = \bar{P}^t \in \Re^{n \times n}$ to the ARE (9.10).

(2) Follows from **A.1** applied to inequality (9.12).

Remark 9.1: It is significant to observe that **Lemma 9.1** provides an alternative testable measure of robust stability for the class of uncertain time-delay systems under consideration.

9.2.2   Robust $\mathcal{H}_\infty$ Filtering

In this section, we study the following robust $\mathcal{H}_\infty$ filtering problem:

Given a prescribed level of noise attenuation $\gamma > 0$, find an estimator of the form (9.6)-(9.7) for the system (9.1)-(9.3) such that the following conditions are satisfied:
(a) The augmented system of (Σ_Δ) and (Σ_e) is robustly stable independent of delay;
(b) With zero initial conditions for $(x(t), \hat{x}(t))$,

$$\|e(t)\|_2 \; < \; \gamma \; \|w(t)\|_2 \tag{9.13}$$

for all admissible uncertainties satisfying (9.4)-(9.5).

It is well-known [7] that an alternative condition to **(a)** and **(b)** above is that

$$H(x, w, t) \; = \; \dot{V}(x_t) \; + \; \{e^t(t)e(t) - \gamma^2 w^t(t)w(t)\} \quad < 0 \tag{9.14}$$

for all admissible $\Delta_1(t), \Delta_2(t)$ satisfying (5), where $H(x, w, t)$ and $V(x_t)$ are, respectively, the Hamiltonian and the Lyapunov functional associated with the system under consideration.

Now, by considering system (Σ_Δ) and system (Σ_e), it is easy to obtain the augmented system:

$$
\begin{aligned}
(\Sigma_{\Delta e}): \quad \dot{\xi}(t) \; &:= \; \begin{bmatrix} \dot{x}(t) \\ \dot{x}(t) - \dot{\hat{x}}(t) \end{bmatrix} \; \in \Re^{2n} \\
&= \; A_{\Delta a}\,\xi(t) + E_{\Delta a}\xi(t - \tau) + B_a w(t) \\
e(t) \; &= \; L_a\,\xi(t) \\
&= \; [0 \quad L]\xi(t) \tag{9.15}
\end{aligned}
$$

where

$$A_a = \begin{bmatrix} A & 0 \\ A - F_o - K_o C & F_o \end{bmatrix}, \ H_a = \begin{bmatrix} H \\ H - K_o H_c \end{bmatrix}, \ H_b = \begin{bmatrix} H_d \\ H_d \end{bmatrix}$$

$$E_{da} = \begin{bmatrix} A_d & 0 \\ A_d & 0 \end{bmatrix}, \ E_a = [E \ 0], \ N_a = [E_d \ 0], \ B_a = \begin{bmatrix} D \\ D - K_o N \end{bmatrix}$$

$$(9.16)$$

Theorem 9.1: *Given a prescribed level of noise attenuation* $\gamma > 0$. *If for some scalars* $\mu > 0$, $\sigma > 0$ *there exist matrices* $0 < \hat{X} = \hat{X}^t \in \Re^{2n \times 2n}$ *and* $0 < W = W^t \in \Re^{2n \times 2n}$ *satisfying the ARE*

$$\hat{X}A_a + A_a^t\hat{X} + H_a^t H_a + \hat{X}B(\mu,\sigma,\gamma)B^t(\mu,\sigma,\gamma)\hat{X} + W = 0 \qquad (9.17)$$
$$B(\mu,\sigma,\gamma)B^t(\mu,\sigma,\gamma) = \mu H_a H_a^t + \sigma H_b H_b^t + E_{da}[W - \sigma^{-1}N_a^t N_a]^{-1}E_{da}^t$$
$$+\gamma^{-2}B_a B_a^t \qquad (9.18)$$

then the robust H_∞–*estimation problem for the system* $(\Sigma_{\Delta e})$ *is solvable with estimator (9.6)-(9.7) and yields.*

$$\|e(t)\|_2 < \gamma \|w(t)\|_2 \qquad (9.19)$$

Proof: Introduce the Lyapunov-Krasovskii functional

$$V(\xi_t) = \xi^t(t)X\xi(t) + \int_{\theta=t-\tau}^{t} \xi^t(\theta)W\xi(\theta)d\theta \qquad (9.20)$$

By evaluating the derivative $\dot{V}(\xi_t)$ along the solutions of (9.15)-(9.16) and grouping similar terms, we express the Hamiltonian $H(\xi, w, k)$ of (9.14) into the form:

$$H(\xi, w, t) =$$
$$\zeta^t(k) \begin{bmatrix} A_{\Delta a}^t X + X A_{\Delta a} + W + H_a^t H_a & X B_a & X E_{\Delta a} \\ B_a^t X & -\gamma^2 I & 0 \\ E_{\Delta a}^t X & 0 & -W \end{bmatrix} \zeta(t)$$
$$= \zeta^t(t)\Omega(X)\zeta(t) \qquad (9.21)$$

where

$$\zeta(t) = [\xi^t(t) \quad w^t(t) \quad \xi^t(t-\tau)]^t \qquad (9.22)$$

For closed-loop system stability with an $\mathcal{H}_\infty$-norm bound constraint, it is required that $H(\xi, w, t) < 0 \;\forall\, \zeta(t) \neq 0$. This is implied by $\Omega(X) < 0$. By A.1, it follows that:

$$X A_{\Delta a} + A_{\Delta a}^t X + H_a^t H_a + X E_{\Delta a} Q^{-1} E_{\Delta a}^t X + \gamma^{-2} B_a B_a^t + W \; < \; 0 \quad (9.23)$$

By applying B.1.2 and B.1.3, inequality (9.23) takes the form:

$$X A_a + A_a^t X + H_a^t H_a + W +$$
$$X \left\{ \mu H_a H_a^t + \sigma H_b H_b^t + E_{da}[W - \sigma^{-1} N_a^t N_a]^{-1} E_{da}^t + \gamma^{-2} B_a B_a^t \right\} X \; < \; 0$$
$$(9.24)$$

In view of (9.18), inequality (9.24) reduces to

$$X A_a + A_a^t X + H_a^t H_a + X B(\mu, \sigma, \gamma) B^t(\mu, \sigma, \gamma) X + W \; < \; 0 \qquad (9.25)$$

Finally, by A.3.1 it follows that the existence of a matrix $0 < X = X^t \in \Re^{n \times n}$ satisfying inequality (9.25) is equivalent to the existence of a stabilizing solution $0 \le \hat{X} = \hat{X}^t \in \Re^{n \times n}$ to the ARE (9.17).

Remark 9.2: It should be observed that **Theorem 9.1** establishes an ARE-based feasibility condition for the robust $\mathcal{H}_\infty$−estimation problem associated with system (Σ_Δ) which requires knowledge about the nominal matrices of the system as well as the gain matrices. In this way, it provides a partial solution to the $\mathcal{H}_\infty$−estimation under consideration.

The next theorem provides explicit formulae for the robust filter and the associated gains.

Theorem 9.2: *Consider the augmented system* ($\Sigma_{\Delta e}$) *for some* $\gamma > 0$. *If for some scalars* $\mu > 0$, $\sigma > 0$ *there exist matrices* $0 < \hat{P} = \hat{P}^t \in \Re^{n \times n}$, $0 < \hat{S} = \hat{S}^t \in \Re^{n \times n}$, $0 < W_p = W_p^t \in \Re^{n \times n}$ *and* $0 < W_s = W_s^t \in \Re^{n \times n}$ *satisfying the AREs*

$$\hat{P} A + A^t \hat{P} + \mu^{-1} E^t E + \hat{P} \hat{B}(\mu, \sigma, \gamma) \hat{B}^t(\mu, \sigma, \gamma) \hat{P} + W_p \; = \; 0 \,(9.26)$$
$$\hat{A} \hat{S} + \hat{A}^t \hat{S} + \hat{S}\{L^t L - \mathcal{G}^t(\mu, \gamma)\mathcal{G}(\mu, \gamma)\}\hat{S} +$$
$$\mathcal{R}(\mu, \sigma, \gamma)\mathcal{R}^t(\mu, \sigma, \gamma) + W_s = 0 \qquad (9.27)$$

then the estimator

$$\dot{\hat{x}}(t) \; = \; \hat{A}\hat{x}(t) + K_o[y(t) - \hat{C}\hat{x}(t)] \qquad (9.28)$$

is a robust $\mathcal{H}_\infty$ *estimator where*

$$
\begin{aligned}
\hat{B}(\mu,\sigma,\gamma)\hat{B}^t(\mu,\sigma,\gamma) &= \mu HH^t + \sigma H_d H_d^t + \gamma^{-2}DD^t \\
&\quad + A_d(W_p - \sigma^{-1}E_d^t E_d)^{-1}A_d^t \qquad (9.29) \\
T(\mu,\gamma) &= \mu H_c H^t + \gamma^{-2}ND^t \qquad (9.30) \\
\mathcal{V}(\mu,\gamma) &= \mu H_c H_c^t + \gamma^{-2}NN^t \qquad (9.31) \\
\hat{A} &= A + \hat{B}(\mu,\sigma,\gamma)\hat{B}^t(\mu,\sigma,\gamma)\hat{P} \qquad (9.32) \\
\hat{C} &= C + T(\mu,\gamma)\hat{P} \qquad (9.33) \\
K_o &= \hat{S}^{-1}T^t(\mu,\gamma)\mathcal{V}^{-1}(\mu,\gamma) \qquad (9.34) \\
\mathcal{G}^t(\mu,\gamma)\mathcal{G}(\mu,\gamma) &= T^t(\mu,\gamma)\mathcal{V}^{-1}(\mu,\gamma)\hat{C} + \hat{C}^t\mathcal{V}^{-1}(\mu,\gamma) \\
&\quad + T^t(\mu,\gamma)\mathcal{V}^{-1}(\mu,\gamma)T(\mu,\gamma) \qquad (9.35)
\end{aligned}
$$

Proof: From **Theorem 9.1**, we know that system $(\Sigma_{\Delta e})$ is robustly stable with disturbance attentuation γ if for some scalars $\mu > 0$, $\sigma > 0$ there exist matrices $0 < X = X^t \in \Re^{2n \times 2n}$ and $0 < W = W^t \in \Re^{2n \times 2n}$ satisfying the ARI

$$
XA_a + A_a^t X + H_a^t H_a + XB(\mu,\sigma,\gamma)B^t(\mu,\sigma,\gamma)X + W < 0 \qquad (9.36)
$$

where $B(\mu,\sigma,\gamma)B^t(\mu,\sigma,\gamma)$ is given by (9.18). Let us define

$$
X = \begin{bmatrix} P & 0 \\ 0 & Z \end{bmatrix}, \quad W = \begin{bmatrix} W_p & 0 \\ 0 & W_q \end{bmatrix} \qquad (9.37)
$$

By substituting (9.16) into inequality (9.36) and using (9.29) and (9.37), we get:

$$
\Gamma = \begin{bmatrix} \Gamma_1 & \Gamma_2 \\ \Gamma_2^t & \Gamma_3 \end{bmatrix} < 0 \qquad (9.38)
$$

$$
\begin{aligned}
\Gamma_1 &= PA + A^t P + \mu^{-1}E^t E + P\hat{B}(\mu,\sigma,\gamma)\hat{B}^t(\mu,\sigma,\gamma)P + W_p \qquad (9.39) \\
\Gamma_2 &= (A^t - F_o^t - C^t K_o^t)Z + P[\sigma H_d H_d^t + + A_d(W_p - \sigma^{-1}E_d^t E_d)^{-1}A_d^t]Z \\
&\quad + P[\mu H(H^t - H_c^t K_o^t) + \gamma^{-2}D(D^t - N^t K_o^t)]Z \qquad (9.40) \\
\Gamma_3 &= ZF_o + F_o^t Z + W_s + L^t L + Z[\sigma H_d H_d^t + A_d(W_p - \sigma^{-1}E_d^t E_d)^{-1}A_d^t]Z \\
&\quad + Z[\mu(H - K_o H_c)(H^t - H_c^t K_o^t) + \gamma^{-2}(D - K_o N)(D^t - N^t K_o^t)]Z \qquad (9.41)
\end{aligned}
$$

Next, we complete the squares in (9.41) with the help of (9.30)-(9.31) to produce K_o in (9.34). Subsequently we use (9.29)-(9.31) and (9.33) with

$S = Z^{-1}, W_s = Z^{-1} W_q Z^{-1}$ to reach:

$$\Gamma_3 = \hat{A}S + \hat{A}^t S + + S\{L^t L - \mathcal{G}^t(\mu,\gamma)\mathcal{G}(\mu,\gamma)\}S + \hat{B}(\mu,\sigma,\gamma)\hat{B}^t(\mu,\sigma,\gamma) + W_s \tag{9.42}$$

It is easy to see from (9.6), (9.28) and (9.32) that $F_o = \hat{A} - K_o\hat{C}$. Then, we manipulate (9.40) using (9.32)-(9.35) to yield $\Gamma_2 = 0$. This yields Γ in (9.38) a block-diagonal matrix. Therefore, we conclude that a necessary and sufficient condition for $\Gamma < 0$ is:

$$PA + A^t P + \mu^{-1} E^t E + P\hat{B}(\mu,\sigma,\gamma)\hat{B}^t(\mu,\sigma,\gamma)P + W_p < 0 \tag{9.43}$$

$$\hat{A}S + \hat{A}^t S + + S\{L^t L - \mathcal{G}^t(\mu,\gamma)\mathcal{G}(\mu,\gamma)\}S + \\ \hat{B}(\mu,\sigma,\gamma)\hat{B}^t(\mu,\sigma,\gamma) + W_s < 0 \tag{9.44}$$

Finally, by **A.3.1** it follows that the existence of matrices $0 < P = P^t \in \Re^{n \times n}$ and $0 < S = S^t \in \Re^{n \times n}$ satisfying inequalities (9.43) and (9.44), respectively, is equivalent to the existence of stabilizing solutions $0 \le \hat{P} = \hat{P}^t \in \Re^{n \times n}$ and $0 \le \hat{S} = \hat{S}^t \in \Re^{n \times n}$ to the AREs (9.26) and (9.27).

Remark 9.3: It is significant to observe from **Theorems 9.1 and 9.2** that the effect of the unknown state-delay has been conveniently absorbed into the quadratic quantities ($\hat{B}\hat{B}^t$ or BB^t) through the $\sigma-, A_d-$ terms.

Now, we provide a result on the $\mathcal{H}_\infty-$filter design for the following nominal system:

$$(\Sigma_D): \quad \begin{aligned} \dot{x} &= Ax(t) + A_d x(t - \tau) + Bw(t) \\ y(t) &= Cx(t) + Nw(t) \\ z(t) &= L\,x(t) \end{aligned} \tag{9.45}$$

using a linear filter of the form

$$(\Sigma_f): \quad \begin{aligned} \dot{\hat{x}}(t) &= F_o\hat{x}(t) + K_o[y(t) - C_o\hat{x}(t)] \\ \hat{z}(t) &= L\hat{x}(t) \end{aligned} \tag{9.46}$$

where $\hat{x}(t) \in \Re^n$ is the filter state and the matrices $F_o \in \Re^{n \times n}$, $K_o \in \Re^{n \times m}$ are the filter gains to be determined.

Theorem 9.3: *Given a scalar* $\gamma > 0$, *if there exist matrices* $0 < \check{P} = \check{P}^t \in \Re^{n \times n}$, $0 < \check{S} = \check{S}^t \in \Re^{n \times n}$, $0 < W_p = W_p^t \in \Re^{n \times n}$ *and*

$0 < W_s = W_s^t \in \Re^{n \times n}$ *satisfying the AREs*

$$\check{P}A + A^t\check{P} + \check{P}\check{B}(\gamma)\check{B}^t(\gamma)\check{P} + W_p = 0 \tag{9.47}$$

$$\hat{A}\check{S} + \hat{A}^t\check{S} + \check{S}\{L^tL - \mathcal{G}^t(\gamma)\mathcal{G}(\gamma)\}\check{S}$$
$$+ \mathcal{R}(\gamma)\mathcal{R}^t(\gamma) + W_s = 0 \tag{9.48}$$

then the estimator (9.46) is a robust $\mathcal{H}_\infty$ *estimator where*

$$\check{B}(\gamma)\check{B}^t(\gamma) = \gamma^{-2}BB^t + A_dW_p^{-1}A_d^t \tag{9.49}$$

$$T(\gamma) = \gamma^{-2}NB^t, \quad \mathcal{V}(\gamma) = \gamma^{-2}NN^t \tag{9.50}$$

$$\hat{A} = A + \check{B}(\gamma)\check{B}^t(\gamma)\check{P}, \quad \hat{C} = C + T(\gamma)\check{P} \tag{9.51}$$

$$\mathcal{G}^t(\gamma)\mathcal{G}(\gamma) = T^t(\gamma)\mathcal{V}^{-1}(\gamma)\hat{C} + \hat{C}^t\mathcal{V}^{-1}(\gamma)T(\gamma)$$
$$+ T^t(\gamma)\mathcal{V}^{-1}(\gamma)T(\gamma) \tag{9.52}$$

$$K_o = \hat{S}^{-1}T(\gamma)\mathcal{V}(\gamma) \tag{9.53}$$

$$F_o = A + \check{B}(\gamma)\check{B}^t(\gamma)\check{P} - \check{S}^{-1}T^t(\gamma)\mathcal{V}^{-1}(\gamma)T(\gamma)\check{P} \tag{9.54}$$

Proof: Follows from **Theorem 9.2** by setting $H = 0, E = 0, H_d = 0, E_d = 0$.

Remark 9.4: The matrices $\delta A = \hat{A} - A$, $\delta C = \hat{C} - C$ given by (9.32)-(9.33) reflect the effect of the parametric uncertainties $\Delta A(t), \Delta C(t)$, and $\Delta E(t)$ on the structure of the filter. In the absence of the parametric uncertainties, we obtain the nominal time-delay system and hence **Theorem 9.2** reduces to **Theorem 9.3**.

9.2.3 Worst-Case Filter Design

In this section, we extend the results of Section 9.2.2 to the case of worst-case filter design. We will treat a general-version of the problem in which the system matrices are time-varying. By similarity to [14], we consider the initial state $\langle x_o, \phi(s) \rangle$ of system $(\Sigma_{\Delta t})$ is unknown and no *a priori* estimate of its value is assumed, where:

$$(\Sigma_{\Delta t}): \quad \dot{x} = [A(t) + \Delta A(t)]x(t) + [A_d(t) + \Delta A_d(t)]x(t - \tau) + D(t)w(t)$$
$$= A_\Delta(t)x(t) + A_{d\Delta}(t)x(t - \tau) + D(t)w(t) \tag{9.55}$$

$$y(t) = [C(t) + \Delta C(t)]x(t) + N(t)w(t)$$
$$= C_\Delta(t)x(t) + N(t)w(t) \tag{9.56}$$

$$z(t) = L(t)\,x(t) \tag{9.57}$$

and $\Delta_1(t)$, $\Delta_2(t)$ satisfy (9.5) and the matrices $A(t), B(t), C(t), D(t), A_d(t),$ $L(t)$ are time-varying piecewise continuous functions. In this regard, the worst-case $\mathcal{H}_\infty$–filtering problem can be phrased as follows:

Given a weighting matrix $0 < R = R^t$ for the initial state $\alpha_o \equiv \langle x_o, \phi(s) \rangle$ and a scalar $\gamma > 0$, find a linear causal filter for $z(t)$ such that the filtering error dynamics is globally uniformly asymptotically stable and

$$||e||_2 < \gamma \{||w||_2^2 + \alpha_o^t R \alpha_o\}^{1/2} \quad \forall \, 0 \neq (\alpha_o, w) \in \Re^n \oplus \mathcal{L}_2[0, \infty)$$

and for all admissible uncertainties.

Note that the matrix R is a measure of the uncertainty in the initial state α_o of $(\Sigma_{\Delta t})$ relative to the uncertainty in w.

In connection with system $(\Sigma_{\Delta t})$, we introduce for some $0 < W = W^t$, $\mu > 0$, $\sigma > 0$ the following parameterized system (Σ_{Dt}):

$$(\Sigma_{Dt}): \quad \dot{\zeta}(t) = A(t)\zeta(t) + [D(t) \quad \gamma \tilde{B}(t)]\tilde{\eta}(t), \quad \zeta(0) = \zeta_o \tag{9.58}$$

$$\tilde{z}(t) = \begin{bmatrix} L(t) \\ \mu^{-1/2}E(t) \\ W1/2 \end{bmatrix} \zeta(t) \tag{9.59}$$

where ζ_o is an unknown initial state and

$$\tilde{B}(t)\tilde{B}^t(t) = \mu H(t)H^t(t) + A_d(t)[W - \sigma^{-1}E_d^t(t)E_d(t)]^{-1}A_d^t(t)$$
$$+ \sigma H_d(t)H_d^t(t) \tag{9.60}$$

where the matrices $L(t), H_d(t), E_d(t), A_d(t), E(t)$ are the same as in (9.4)-(9.5) such that $[W - \sigma^{-1}E_d^t(t)E_d(t)] > 0 \,\forall t$. For system (Σ_{Dt}), we adopt the following $\mathcal{H}_\infty$–like performance measure:

$$\mathcal{J}(\tilde{z}, \tilde{\eta}, \zeta_o, W, R_\zeta) = \sup_{0 \neq (\zeta_o, \tilde{\eta}) \in \Re^n \oplus \mathcal{L}_2[0, \infty)} \left\{ \frac{||\tilde{z}||_2^2}{||\tilde{\eta}||_2^2 + \zeta_o^t R_\zeta \zeta_o} \right\}^{1/2} \tag{9.61}$$

where $0 < R_\zeta = R_\zeta^t$ is a weighting matrix for ζ_o.

Theorem 9.4: Given a scalar $\gamma > 0$ and a matrix $0 < R = R^t$, system $(\Sigma_{\Delta t})$ satisfying (9.5) is globally, uniformly, asymptotically stable about the origin and $||z||_2 < \gamma \{||w||_2^2 + \alpha_o^t R \alpha_o\}^{1/2}$ for all nonzero $(\alpha_o, w) \in$

$\mathfrak{R}^n \oplus L_2[0, \infty)$ *and for all admissible uncertainties and unknown state-delay if system* (Σ_{Dt}) *is exponentially stable and there exist scalars* $\mu > 0$, $\sigma > 0$ *and a matrix* $0 < W = W^t$ *such that* $[W - \sigma^{-1}E_d^t(t)E_d(t)] > 0 \; \forall \, t$, *and* $J(\tilde{z}, \tilde{\eta}, \varsigma_o, W, R_\varsigma) < \gamma$.

Proof: It can be easily established using the same arguments of [26] and taking into consideration **Remark 9.3**.

A solution to the robust $\mathcal{H}_\infty$–filtering problem can now be stated in terms of a scaled $\mathcal{H}_\infty$–like control problem incorporating unknown initial state and without uncertainties and unknown state-delay. For this purpose, consider the following system:

$$(\Sigma_{Dc}): \quad \dot{x}_c(t) = Ax_c(t) + [D \quad \gamma \bar{B}(\mu, \sigma)]w_c(t) \, , \quad x_c(0) = x_{co} \quad (9.62)$$
$$y_c(t) = Cx_c(t) + [N \quad \gamma \mathcal{D}(\mu)]w_c(t) \quad (9.63)$$
$$z_c(t) = \begin{bmatrix} L \\ \mu^{-1/2}E \\ W^{1/2} \end{bmatrix} x_c(t) + \begin{bmatrix} -I \\ 0 \end{bmatrix} u_c(t) \quad (9.64)$$

where $x_c(t) \in \mathfrak{R}^n$ is the state with x_{co} being unknown, $w_c(t) \in \mathfrak{R}^{m_w}$ is the input disturbance, $y_c(t) \in \mathfrak{R}^p$ is the measured output, $z_c(t) \in \mathfrak{R}^{r_z}$ is the controlled output, $u_c(t) \in \mathfrak{R}^p$ is the control input and $\gamma > 0$ is the *desired* $\mathcal{H}_\infty$–*performance* for the robust filter. The matrices A, C, D, A_d, L are the same as in system (Σ_{Dc}) and

$$\bar{B}(\mu, \sigma) = [\mu^{1/2}H \quad A_d(W - \sigma^{-1}E_d^t E_d)^{-1/2} \quad \sigma^{1/2}H_d] \quad (9.65)$$
$$\mathcal{D}(\mu) = [\mu^{1/2}H_c \quad 0 \quad 0] \quad (9.66)$$

The main result is summarized by the following theorem.

Theorem 9.5: *Consider system* (Σ_Δ) *satisfying (5) and let* $\gamma > 0$ *be a prescribed level of noise attenuation. Let* $\mathcal{F}$ *be a linear time-varying strictly proper filter with zero initial condition. Then the estimate* $z = \mathcal{F}y$ *for some* $0 < R = R^t$ *solves the robust* $\mathcal{H}_\infty-$ *filtering problem for system* (Σ_Δ) *if there exists scalars* $\mu > 0$, $\sigma > 0$ *and a matrix* $0 < W = W^t$ *such that: (1)* $(W - \sigma^{-1}E_d^t E_d) > 0$, *(2) System* (Σ_{Dc}) *under the action of the control law* $u_c = \mathcal{F}y_c$ *is stable and the measure* $J(z_c, w_c, x_{co}, R) < \gamma$.

Remark 9.5: Again, we note from **Theorem 9.5** that the effect of uncertainties and unknown state-delay has been accomodated by the parameterized model (Σ_{Dc}) and more importantly, the robust $\mathcal{H}_\infty$−filtering has now been converted into a scaled output feedback $\mathcal{H}_\infty$−like control problem without uncertainties and unknown delays. The latter problem can be solved by existing results on $\mathcal{H}_\infty$-theory, for example [2,8].

Remark 9.6: We remark that $\mathcal{F}$ is taken time-varying to reflect the fact that α_o is unknown. Also, it is possible to generalize the result of **Theorem 9.5** to the case where all the matrices are piecewise continuous bounded matrix functions.

Remark 9.7: In the special case where the initial state $\alpha_o = 0$, we obtain the following result.

Given a scalar $\gamma > 0$ and let $\mathrm{F}(\mathrm{s})$ be a linear time-invariant strictly proper filter with zero initial condition. System (Σ_Δ) with $\alpha_o = 0$ satisfying (9.4) is globally asymptotically stable and $\|e\|_2 < \gamma \|w\|_2$ for any nonzero $w \in L_2[0,\infty)$ and for all admissible uncertainties and unknown state-delay if there exist scalars $\mu > 0$, $\sigma > 0$ and a matrix $0 < Q = Q^t$ such that $(Q - \sigma^{-1}N^t N) > 0$ and the closed-loop system (Σ_{Dc}) with the control law $u_c = \mathrm{F}(\mathrm{s})y_c$ is stable and with zero initial condition for (Σ_{Dc}), $\|z_c\|_2 < \gamma \|w_c\|_2$ for any nonzero $w_c \in L_2[0,\infty)$.

9.3 Nonlinear Uncertain Systems

In this section, we consider a class of nonlinear continuous-time systems with norm-bounded uncertainty and unknown state-delay. Moreover, we consider the initial state to be unknown. The nonlinearity appears in the form of a known cone-bounded state-dependent and additive term. We investigate the problem of $\mathcal{H}_\infty$−filtering for this class of systems and design a nonlinear filter that guarantees both robust stability and a prescribed $\mathcal{H}_\infty$−performance of the filtering error dynamics for the whole set of admissible systems.

9.3.1 Problem Description and Assumptions

We consider a class of uncertain time-delay systems represented by:

$$(\Sigma_\Delta): \quad \dot{x}(t) \;=\; [A + \Delta A(t)]x(t) + [A_d + \Delta A_d(t)]x(t - \tau) + H_h h[x(t)]$$

$$
\begin{aligned}
&+ Dw(t) \\
&= A_\Delta(t)x(t) + E_{d\Delta}(t)x(t-\tau) + H_h h[x(t)] + Dw(t) \quad (9.67) \\
y(t) &= [C + \Delta C(t)]x(t) + Nw(t) \\
&= C_\Delta(t)x(t) + Nw(t) \quad\quad (9.68) \\
z(t) &= L\,x(t) \quad\quad (9.69)
\end{aligned}
$$

where $x(t) \in \Re^n$ is the state, $w(t) \in \Re^m$ is the input noise which belongs to $\mathcal{L}_2\,[0,\infty)$, $y(t) \in \Re^p$ is the measured output, $z(t) \in \Re^r$, is a linear combination of the state variables to be estimated, $h[.] : \Re^n \longrightarrow \Re^s$ is a known nonlinear vector function and the matrices $A \in \Re^{n \times n}$, $C \in \Re^{p \times n}$, $D \in \Re^{p \times m}$, $A_d \in \Re^{n \times n}$ and $H_h \in \Re^{r \times n}$ are real constant matrices representing the nominal plant. Here, τ is an unknown constant scalar representing the amount of delay in the state. For all practical purposes, we let $\tau \le \tau^*$ where τ^* is known. The matrices $\Delta A(t)$, $\Delta C(t)$ and $\Delta A_d(t)$ are given by (9.4)-(9.5).

Assumption 9.1:
(1) $h[0] = 0$
(2) There exists some $\rho > 0$ such that for any $x_a,\ x_b \in \Re^n$

$$
\|h[x_a] - h[x_b]\| \le \rho\,\|x_a - x_b\|
$$

Assumption 9.2:
$$
(NN^t + H_c H_c^t) > 0
$$

Remark 9.8: Note that **Assumption 9.1** implies that the function $h[.]$ is cone-bounded. **Assumption 3** ensures the existence of solution and in the absence of uncertainties and time-delays, it eventually reduces to $NN^t > 0$, which is quite standard in $\mathcal{H}_\infty$−filtering of nominal linear systems.

The initial condition can be generally specified as $\alpha_o \equiv \langle x(0), \phi(s)\rangle$, where $\phi(.) \in \mathcal{L}_2[-\tau, 0]$, but it will be considered unknown throughout this section.

Our main concern is to determine an estimate $\hat{z}$ of the vector z using the measurements $\mathcal{Y}_t = \{y(\sigma) :\ 0 \le \sigma \le t\}$ and where no a *priori* estimate of the initial state of system (Σ_Δ) is assumed. In this way, we let $\hat{z} = \mathcal{F}\{\mathcal{Y}_t\}$ where $\mathcal{F}$ stands for a nonlinear filter to be designed and introduce $e(t) = z(t) - \hat{z}(t)$. Specifically, the robust $\mathcal{H}_\infty-$ filtering problem of interest can be phrased as follows:

Given system (Σ_Δ), a weighting matrix $0 < R = R^t$ and a prescribed level of noise attenuation $\gamma > 0$, **find** a filter $\mathcal{F}$ such that the filtering error dynamics is globally, uniformly, asymptotically stable and

$$\|e\|_2 < \gamma\{\|w\|_2^2 + x(0)^t Rx(0) + \int_{-\tau}^0 x(s)^t Rx(s)ds\}^{1/2}$$

for any nonzero $(\alpha_o, w) \in \Re^n \oplus \mathcal{L}_2[0, \infty)$ and for all admissible uncertainties and unknown state-delay.

Consider the uncertain nonlinear time-delay system

$$(\Sigma_{\Delta o}): \quad \dot{x}(t) = A_\Delta(t)x(t) + E_{d\Delta}x(t - \tau) + H_h h[x(t)]$$
$$= [A_\Delta + \Delta A(t)]x(t) + [A_d + \Delta A_d(t)]x(t - \tau)$$
$$+ H_h h[x(t)] \tag{9.70}$$

subject to (9.4) and **Assumptions 9.1**. To establish a sufficient stability criterion for this system, we choose a Lyapunov-Krasovskii functional of the form:

$$V(x_t) = x^t(t)Px(t) + \int_{t-\tau}^t x^t(\alpha)Wx(\alpha)d\alpha$$
$$+ \rho^2 \int_0^t x^t(\alpha)x(\alpha)d\alpha - \int_0^t h^t[x]h[x]d\alpha \tag{9.71}$$

where $\rho > 0$, $0 < P = P^t \in \Re^{n \times n}$; $0 < W = W^t \in \Re^{n \times n}$. Observe in view of **Assumption 9.1** that $V(x_t) > 0$ for $x(t) \neq 0$ and $V(x_t) = 0$ when $x(t) \equiv 0$. Now by evaluating the time derivative $\dot{V}(x_t)$ along the solutions of system $(\Sigma_{\Delta o})$ and arranging terms, we get:

$$\dot{V}(x_t) = X^t(t) \Pi(P) X(t) \tag{9.72}$$

where

$$\Pi(P) = \begin{bmatrix} PA_\Delta(t) + A_\Delta^t(t)P + \rho^2 I + W & PE_\Delta(t) & PH_h \\ E_\Delta^t(t)P & -W & 0 \\ H_h^t P & 0 & -I \end{bmatrix}$$

$$X(t) = [x^t(t) \ x^t(t - \tau) \ h^t[x]]^t \tag{9.73}$$

If $\dot{V}(x_t) < 0$ when $x \neq 0$ then $x \to 0$ as $t \to \infty$ and the asymptotic stability of $(\Sigma_{\Delta o})$ is guaranteed. This is implied by the ARI:

$$PA_\Delta(t) + A_\Delta^t(t)P + P\left\{H_h H_h^t + A_{d\Delta}(t)W^{-1}A_{d\Delta}^t(t)\right\}P$$
$$+ \rho^2 I + W < 0 \quad \forall \ \|\Delta_1\| \leq 1, \ \|\Delta_2\| \leq 1, \ \rho > 0 \tag{9.74}$$

Based on this, we have the following result:

Lemma 9.2: *The uncertain time-delay system* $(\Sigma_{\Delta o})$ *is robustly stable independent of delay (RSID) if one of the following equivalent statements hold:*

(1) *There exist a matrix* $0 < W = W^t \in \Re^{n \times n}$ *and scalars* $\mu > 0$, $\sigma > 0$, $\rho > 0$ *satisfying* $Q - \sigma^{-1} N^t N > 0$ *and such that the ARE*

$$\bar{P}A + A^t \bar{P} + \mu^{-1} E^t E + \rho^2 I + W$$
$$+ \bar{P} \left\{ H_h H_h^t + \mu H H^t + \sigma H_d H_d^t + A_d (W - \sigma^{-1} E_d^t E_d)^{-1} A_d^t \right\} \bar{P} = 0$$
$$(9.75)$$

admits a stabilizing solution $0 \leq \bar{P} = \bar{P}^t \in \Re^{n \times n}$.
(2) A *is stable and the following* $\mathcal{H}_\infty$ *norm bound is satisfied for some* $\mu > 0$, $\sigma > 0$, $\rho > 0$ *and* $0 < W = W^t$ *satisfying* $W - \sigma^{-1} E_d^t E_d > 0$,

$$\left\| \begin{bmatrix} \frac{1}{\sqrt{\mu}} E \\ W^{1/2} \end{bmatrix} [sI - A]^{-1} [H_h \ \sqrt{\mu} L_1 \ \sqrt{\sigma} H_d \ A_d (W - \sigma^{-1} E_d^t E_d)^{-1/2}] \right\|_\infty < 1$$
$$(9.76)$$

Proof: (1) It follows by applying **B.1.2** to the term $(PA_\Delta + A_\Delta^t P)$ and **B.1.3** to the term $(PA_{d\Delta} W^{-1} A_{d\Delta}^t P)$ for some $\mu > 0, \sigma > 0$, that inequality (9.74) is implied by:

$$PA + A^t P + \mu^{-1} E^t E + \rho^2 I + W$$
$$+ P \left\{ H_h H_h^t + \mu H H^t + \sigma H_d H_d^t + A_d (W - \sigma^{-1} E_d^t E_d)^{-1} A_d^t \right\} P < 0$$
$$(9.77)$$

By **A.3.1** it follows that the existence of a matrix $0 < P = P^t \in \Re^{n \times n}$ satisfying inequality (9.77) is equivalent to the existence of a stabilizing solution $0 \leq \bar{P} = \bar{P}^t \in \Re^{n \times n}$ to the ARE (9.75).

(2) Follows from **A.3.1** applied to ARE (9.75).

Remark 9.9: It is significant to observe that **Lemma 9.2** provides a testable measure of robust stability for the class of uncertain nonlinear time-delay systems under consideration. It only requires the nominal matrices as well as the structural matrices of the uncertainties.

Now, consider the following time-varying system

$$(\Sigma_t): \quad \begin{aligned} \dot{x}(t) &= A(t)x(t) + D(t)w(t) + E(t)x(t - \tau), \\ z(t) &= C(t)\,x(t) \end{aligned} \tag{9.78}$$

where $x(t) \in \Re^n$ is the state, α_o is an unknown initial state, $w(t) \in \Re^m$ is the input noise which belongs to $\mathcal{L}_2\,[0,\infty)$, $z(t) \in \Re^r$ is the controlled output and the matrices $A(t), B(t), C(t)$ are real piecewise-continuous and bounded. For system (Σ_t) we define the performance measure:

$$J(z, w, \alpha_o, R) = \sup_{0 \neq (\alpha_o, w) \in \Re^n \oplus \mathcal{L}_2 |0,\infty)} \left\{ \frac{\|z\|_2^2}{\|w\|_2^2 + x(0)^t R x(0) + \int_{-\tau}^0 x^t(s) R x(s) ds} \right\}^{1/2} \tag{9.79}$$

where $0 < R = R^t$ is a weighting matrix for the initial state $\langle x(0), \phi(s)\rangle$.

Lemma 9.3: *Given system (Σ_t) and a scalar $\gamma > 0$, the system is exponentially stable and $J(z, w, \alpha_o, R) < \gamma$ if either of the following conditions holds:*

(1) *There exists a bounded matrix function $0 \leq Q(t) = Q^t(t)$, $\forall t \in [0,\infty)$, such that for some $0 < W = W^t\, < \gamma^2 R$*

$$\dot{Q} + A^t Q + QA + Q(\gamma^{-2}DD^t + EW^{-1}E^t)Q + C^t C + W = 0,$$
$$Q(0) < \gamma^2\,R \tag{9.80}$$

and the system $\dot{x}(t) = [A + (\gamma^{-2}DD^t + EW^{-1}E^t)Q]x(t)$ is exponentially stable.

(2) *There exists a bounded matrix function $0 < S(t) = S^t(t)$, $\forall\, t \in [0,\infty)$, satisfying the differential inequality*

$$\dot{S} + A^t S + SA + S(\gamma^{-2}DD^t + EW^{-1}E^t)S + C^t C + W < 0,$$
$$S(0) < \gamma^2\,R\,, \tag{9.81}$$

for some $0 < W = W^t < \gamma^2\,R.$

Proof: Introduce a Lyapunov-Krasovskii functional for system (Σ_t) with $w \equiv 0$:

$$V_+(x_t) = x^t(t)S(t)x(t) + \int_{t-\tau}^t x^t(\alpha)Wx(\alpha)d\alpha \qquad (9.82)$$

$$0 < S(t) = S(t)^t \in \Re^{n \times n} \ \forall \ t; \quad 0 < W = W^t \in \Re^{n \times n}$$

Differentiating (9.82) along the trajectories of system (Σ_t) with $w \equiv 0$, we get:

$$\frac{d}{dt}V_+(x,t) = \begin{bmatrix} x(t) \\ x(t-\tau) \end{bmatrix}^t \begin{bmatrix} \dot{S} + SA + A^tS + W & SE \\ E^tS & -W \end{bmatrix} \begin{bmatrix} x(t) \\ x(t-\tau) \end{bmatrix} \qquad (9.83)$$

By A.1, inequality (9.81) implies that $\frac{d}{dt}V_+(x,t) < 0$ whenever $[x(t) \ x(t-\tau)] \neq 0$. That is, the system is uniformly asymptotically stable.

To show that $||z||_2 < \gamma \{||w||_2^2 + x^t(0)Rx(0) + \int_{-\tau}^0 x^t(s)Rx(s)ds\}^{1/2}$, we introduce

$$J = \int_0^\infty \{z^tz - \gamma^2 w^tw\}dt - \gamma^2\{x^t(0)Rx(0) + \int_{-\tau}^0 x^t(s)Rx(s)ds\} \qquad (9.84)$$

By using (9.78) and completing the squares in (9.84), it follows that

$$
\begin{aligned}
J &= \int_0^\infty \{x^tC^tCx + \frac{d}{dt}x^tSx - \gamma^2 w^tw\}dt + x^t(0)S(0)x^t(0) \\
&\quad - \gamma^2 x^t(0)Rx(0) - \gamma^2 \int_{-\tau}^0 x^t(s)Rx(s)ds \\
&= \int_0^\infty x^t\{\dot{S} + SA + A^tS\}xdt + \int_0^\infty \{w^tD^tSx + x^tSDw\}dt \\
&\quad + \int_0^\infty \{x^tSEx(t-\tau) + x^t(t-\tau)E^tSx - \gamma^2 w^tw\} dt \\
&= \int_0^\infty x^t\{\dot{S} + SA + A^tS + S(\gamma^{-2}DD^t + EW^{-1}E^t)S + C^tC + W\}xdt \\
&\quad - \int_0^\infty \gamma^2[w - \gamma^{-2}D^tSx]^t[w - \gamma^{-2}D^tSx]dt - \int_{-\tau}^0 x^t(s)[\gamma^2 R - W]x(s)ds \\
&\quad - \int_0^\infty [Wx(t-\tau) - E^tSx]^tW^{-1}[Wx(t-\tau) - E^tSx]dt \\
&\quad - x^t(0)[\gamma^2 R - S(0)]x(0) \qquad (9.85)
\end{aligned}
$$

The condition $J < 0$ is implied by inequality (9.81) $\forall t \ [0, \infty)$. Therefore, we conclude that $||z||_2 < \gamma \{||w||_2^2 + x(0)^tRx(0) + \int_{-\tau}^0 x^t(s)Rx(s)ds\}^{1/2}$ for any nonzero $(\alpha_o, w) \in \Re^n \oplus \mathcal{L}_2[0, \infty)$.

Finally, by **A.3.1** it follows that the existence of a matrix $0 < S = S^t \in \Re^{n \times n}$ satisfying inequality (9.81) is equivalent to the existence of a stabilizing solution $0 \le Q = Q^t \in \Re^{n \times n}$ to the ARE (9.80).

Remark 9.10: It is significant to observe that **Lemma 9.3** establishes a version of the bounded real lemma for a class of nonlinear time-varying systems with state-delay.

9.3.2   Robust $\mathcal{H}_\infty$ Filtering Results

We now proceed to study the robust $\mathcal{H}_\infty$ filtering problem for system (Σ_Δ). For this purpose, we introduce the following nonlinear filter:

$$(\Sigma_e): \quad \dot{\hat{x}}(t) = \hat{A}\hat{x} + H_h h[\hat{x}] + \hat{K}[y(t) - \hat{C}\hat{x}]$$
$$= [A + \delta A]\hat{x} + H_h h[\hat{x}] + \hat{K}[y(t) - (C + \delta C)\hat{x}]$$
$$\hat{z} = L\hat{x}, \quad \hat{x}(0) = 0 \tag{9.86}$$

where the matrices $\delta A \in \Re^{n \times n}$, $\delta C \in \Re^{p \times n}$, $\hat{K} \in \Re^{n \times p}$ are to be determined. Now, by defining $\tilde{x} = x - \hat{x}$, we get from (9.67)-(9.69) and (9.86) the dynamics of the state-error:

$$\dot{\tilde{x}}(t) = \{(A + \delta A) - \hat{K}(C + \delta C)\}\tilde{x}(t) + \{\Delta A - \delta A - \hat{K}(\Delta C - \delta C)\}x(t)$$
$$+ \{B - \hat{K}D\}w(t) + H_h\{h[x] - h[\hat{x}]\} \tag{9.87}$$

Then from system (Σ_Δ) and (9.87), we obtain the dynamics of the filtering error $e(t)$:

$$(\Sigma_{\Delta e}) \quad \dot{\xi}(t) := \begin{bmatrix} \dot{x}(t) \\ \dot{\tilde{x}}(t) \end{bmatrix}$$
$$= \{A_a + H_a \Delta_1(t) E_a\} \xi(t) + \{E_{ad} + H_{ad}\Delta_2(t)N_{ad}\}\xi(t - \tau)$$
$$+ B_u w(t) + H_{ah} h_a[x, \hat{x}], \quad \xi_o \equiv \langle \xi(0), \phi(.)\rangle \tag{9.88}$$
$$e(t) = L_e \, \xi(t) \tag{9.89}$$

where

$$A_a = \begin{bmatrix} A & 0 \\ \delta A + \hat{K}\delta C & A + \Delta A - \hat{K}(C + \delta C) \end{bmatrix},$$

$$H_a = \begin{bmatrix} H \\ H - K_o H_c \end{bmatrix}; \; H_{ah} = \begin{bmatrix} H_h & 0 \\ 0 & H_h \end{bmatrix}, \; B_a = \begin{bmatrix} B \\ B - \hat{K}D \end{bmatrix},$$

$$h_a[x, \hat{x}] = \begin{bmatrix} h[x] \\ h[x] - h[\hat{x}] \end{bmatrix}; \; E_{ad} = \begin{bmatrix} A_d & 0 \\ A_d & 0 \end{bmatrix}, \; H_{ad} = \begin{bmatrix} H_d \\ H_d \end{bmatrix},$$

$$\xi_o = \begin{bmatrix} \alpha_o \\ \alpha_o \end{bmatrix}, \; E_a = [E \; 0], \; N_{ad} = [E_d \; 0], L_e = [0 \; L] \tag{9.90}$$

Observe in view of **Assumption 9.1**, that

$$||h[x] - h[\hat{x}]|| \leq \rho \, ||\tilde{x}||, \; ||h[x]|| \leq \rho \, ||x|| \tag{9.91}$$

and hence

$$||h_a[x, \hat{x}]||^2 = ||h[x]||^2 + ||h[x] - h[\hat{x}]||^2 \leq \rho^2 \, ||\xi||^2 \tag{9.92}$$

Now, we are in a position to present the following result.

Theorem 9.6: *Consider system* (Σ_Δ) *satisfying* **Assumptions 9.1** *and* **9.2** *and the nonlinear filter* (Σ_e). *Given a prescribed level of noise attenuation* $\gamma > 0$ *and matrices* $0 < W_p = W_p^t, \, 0 < R = R^t$, *the robust* $\mathcal{H}_\infty$ *filtering problem is solvable if, for some scalars* $\mu > 0, \sigma > 0, \rho > 0$ *satisfying* $W_p > \sigma^{-1} A_d^t A_d$ *the following conditions hold:*

(1) *There exist a matrix* $0 \leq \bar{P} = \bar{P}^t$ *satisfying the ARE:*

$$\bar{P}A + A^t \bar{P} + \bar{P}B(\mu, \sigma, \gamma)B^t(\mu, \sigma, \gamma)\bar{P} + \mu^{-1} E^t E +$$
$$\rho^2 I + W_p = 0 \tag{9.93}$$
$$B(\mu, \sigma, \gamma)B^t(\mu, \sigma, \gamma) =$$
$$\left\{ \mu H H^t + \sigma H_d H_d^t + A_d (W_p - \sigma^{-1} E_d^t E_d)^{-1} A_d^t + \gamma^{-2} D D^t \right\} \tag{9.94}$$

such that $\bar{P} < \gamma^2 R$ *and matrix* $\{A_o + B(\mu, \sigma, \gamma)B^t(\mu, \sigma, \gamma)\bar{P}\}$ *is stable.*
(2) *There exists a bounded matrix function* $0 \leq S(t) = S^t(t), t \in [0, \infty)$
satisfying

$$\dot{S} = \{\hat{A} - \hat{B}\hat{D}^t \hat{R}^{-1} \hat{C}\} S$$
$$+ S\{\hat{A} - \hat{B}\hat{D}^t \hat{R}^{-1} \hat{C}\}^t + S\{\gamma^{-2} \hat{L}^t \hat{L} - \hat{C}^t \hat{R}^{-1} \hat{C}\} S$$
$$+ \hat{B}\{I - \hat{D}^t \hat{R}^{-1} \hat{D}\}\hat{B}^t, \; S(0) = (R - \gamma^{-2} \bar{P})^{-1} \tag{9.95}$$

such that the unforced linear time-varying system

$$\dot{\eta}(t) = \left\{ \hat{A} - \hat{B}\hat{D}^t\hat{R}^{-1}\hat{C} + S[\gamma^{-2}\hat{L}^t\hat{L} - \hat{C}^t\hat{R}^{-1}\hat{C}] \right\} \eta(t) \qquad (9.96)$$

is exponentially stable where

$$\delta A = \{\gamma^{-2}DD^t + \mu HH^t\}\bar{P}, \quad \delta C = \{\gamma^{-2}ND^t + \mu H_c H^t\}\bar{P} \quad (9.97)$$
$$\hat{B} = [D \quad \gamma\sqrt{\mu}H \quad \gamma H_h \quad A_d(W_p - \sigma^{-1}E_d^t E_d)^{-1/2}],$$
$$\hat{L} = \{L^t L + \rho^2 I\}^{1/2} \qquad (9.98)$$
$$\hat{D} = [N \quad \gamma\sqrt{\mu}H_c \quad 0 \quad 0], \quad \hat{R} = \hat{D}\hat{D}^t \qquad (9.99)$$

and the filter gain is given by

$$\hat{K}(t) = \{S(t)\hat{C}^t + \hat{B}\hat{D}^t\}\hat{R}^{-1} \qquad (9.100)$$

Proof: It follows from the standard results of $\mathcal{H}_\infty$ filtering and in line with **Lemma 9.3** that condition **(2)** is a sufficient condition for the solvability of the infinite-horizon $\mathcal{H}_\infty$ filtering for the linear system

$$\begin{aligned}
\dot{\zeta}(t) &= \hat{A}\zeta(t) + \hat{B}\breve{w}(t) + A_d\zeta(t-\tau) \\
\breve{y}(t) &= \hat{C}\zeta(t) + \hat{D}\breve{w}(t) \\
\breve{z}(t) &= \hat{L}\zeta(t)
\end{aligned} \qquad (9.101)$$

where $\zeta(t) \in \Re^n$ is the state with unknown initial value $\zeta_o \equiv \langle\zeta(0), \phi(.)\rangle$, $\breve{w}(t) \in \Re^{m+s+\alpha}$ is the noise signal which is from $\mathcal{L}_2[0,\infty)$, $\breve{y}(t) \in \Re^p$ is the measurement, $\breve{z}(t) \in \Re^r$ is a linear combination of state variables to be estimated and the performance measure for the filtering problem is given by:

$$J(\breve{z} - \bar{z}, \breve{w}, \zeta_o, R) =$$

$$\sup_{0 \neq (\zeta_o, \breve{w}) \in \Re^n \oplus \mathcal{L}_2[0,\infty)} \left\{ \frac{\|z\|_2^2}{\|w\|_2^2 + \zeta^t(0)R\zeta(0) + \int_{-\tau}^0 \zeta^t(s)R\zeta(s)ds} \right\}^{1/2} < \gamma \qquad (9.102)$$

where $\bar{z}$ is an estimate of $\breve{z}$. An $\mathcal{H}_\infty$-filter for $\bar{z}$ can be cast into the form:

$$\begin{aligned}
\dot{\breve{\zeta}}(t) &= \hat{A}\breve{\zeta}(t) + \hat{K}(t)[\breve{y}(t) - \hat{C}\breve{\zeta}(t)] \\
\bar{z} &= \hat{L}\breve{\zeta}
\end{aligned} \qquad (9.103)$$

where $\hat{K}$ is the filter gain as given by (9.100). In the light of **Lemma 9.3**, it is straightforward to argue that condition **(2)** is equivalent to the existence of a bounded solution $0 \leq \Omega(t) = \Omega^t(t)$, $\forall t \in [0, \infty)$ to the Riccati differential equation:

$$\dot{\Omega} + (\hat{A} - \hat{K}\hat{C})^t\Omega + \Omega(\hat{A} - \hat{K}\hat{C}) + \gamma^{-2}\Omega(\hat{B} - \hat{K}\hat{D})(\hat{B} - \hat{K}\hat{D})^t\Omega$$
$$+\check{L}^t\check{L} = 0 \tag{9.104}$$
$$\Omega(0) < \gamma^2 R - \bar{P}$$

and such that the time-varying system:

$$\dot{\eta}(t) = \left\{\hat{A} - \hat{K}\hat{C} + \gamma^{-2}(\hat{B} - \hat{K}\hat{D})(\hat{B} - \hat{K}\hat{D})^t\Omega\right\} \eta(t) \tag{9.105}$$

is exponentially stable. To proceed further, we introduce

$$0 \leq Y(t) = Y^t(t) := \begin{bmatrix} \bar{P} & 0 \\ 0 & \Omega(t) \end{bmatrix}, \quad \forall t \in [0, \infty) \tag{9.106}$$

Using (9.90), (9.93)-(9.94) and (9.104), it is a simple task to verify for some $W = block - diag[W_p \ W_s]$, $R_* = block - diag[R \ R_s]$ that $Y(t)$, $\forall t \in [0, \infty)$, satisfies the Riccati differential equation:

$$\dot{Y} + A_a^t Y + Y A_a + Y\{\gamma^{-2}B_a B_a^t + \mu H_a H_a^t +$$
$$\sigma H_{ad}H_{ad}^t + H_{ah}H_{ah}^t + E_{ad}(W - \sigma^{-1}N_{ad}^t N_a)^{-1}E_{ad}^t\}Y +$$
$$L_e^t L_e + \mu^{-1}E_a^t E_a + \rho^2 I + W = 0 \tag{9.107}$$
$$W_p < \gamma^2 R, \quad Y(0) < \gamma^2 R_* \tag{9.108}$$

and is such that the system

$$\dot{\eta}(t) = [A_a + [\gamma^{-2}B_a B_a^t + \mu H_a H_a^t + \sigma H_{ad}H_{ad}^t$$
$$+ H_{ah}H_{ah}^t + E_{ad}(W - \sigma^{-1}N_{ad}^t N_{ad})^{-1}E_{ad}^t]Y]\eta(t)$$

is exponentially stable. Note that the matrices W_s, R_s are arbitrary and they will not affect the subsequent analysis. Since $\Omega(0) < \gamma^2 R - \bar{P}$, there exists a sufficiently small scalar $\delta > 0$ such that

$$Y(0) < Y_o = \begin{bmatrix} \bar{P} + \delta I & 0 \\ 0 & \gamma^2 R - \bar{P} - \delta I \end{bmatrix} \tag{9.109}$$

From (9.108)-(9.109) and since $\dot{x}(t) = 0, t \leq 0$, it is easy to see that

$$\xi^t(0) Y_o \xi(0) = x^t(0) (\gamma^2 R) x(0)$$
$$\int_{-\tau}^0 \xi^t(s) Y_o \xi(s)ds = \int_{-\tau}^0 x^t(s) (\gamma^2 R) x(s)ds \tag{9.110}$$

By **Lemma 9.3**, it follows from (9.107) that there exists a matrix $0 < Y_*(t) = Y_*^t(t)$, $\forall t \in [0, \infty)$, satisfying $Y_*(0) < Y_o$ and such that

$$\dot{Y}_* + A_a^t Y_* + Y_* A_a + L_e^t L_e + \mu^{-1} E_a^t E_a + \rho^2 I + W +$$
$$Y_* \{\gamma^{-2} B_a B_a^t + \mu H_a H_a^t + \sigma H_{ad} H_{ad}^t\} Y_* +$$
$$Y^* \{H_{ah} H_{ah}^t + E_{ad}(W - \sigma^{-1} N_{ad}^t N_{ad})^{-1} E_{ad}^t\} Y_* < 0 \quad (9.111)$$

Applying **Lemma 9.2**, it follows from (9.111) that

$$\dot{Y}_* + A_{\Delta a}^t Y_* + Y_* A_{\Delta a} + Y_* \{\gamma^{-2} B_a B_a^t + E_{\Delta a} W^{-1} E_{\Delta a}^t\} Y_* +$$
$$L_e^t L_e + \rho^2 I + W < 0 \quad (9.112)$$

where $A_{\Delta a} = A_a + H_a \Delta_1(t) E_a$, $E_{\Delta a} = E_{ad} + H_{ad} \Delta_2(t) N_{ad}$.

Next, we prove the global, uniform, asymptotic stability of the filtering error dynamics of system $(\Sigma_{\Delta e})$. To do this, we introduce a Lyapunov-Krasovskii functional for system $(\Sigma_{\Delta e})$ with $w \equiv 0$:

$$V_*(\xi_t) = \xi^t(t) Y_* \xi(t) + \int_{t-\tau}^t \xi^t(\alpha) W \xi(\alpha) d\alpha$$
$$+ \rho^2 \int_0^t \xi^t(\alpha) \xi(\alpha) d\alpha - \int_0^t h_a^t[x] h_a[x] d\alpha \quad (9.113)$$
$$\rho > 0, \quad 0 < W = W^t \in \Re^{2n \times 2n}$$

Observe in view of **Assumption 9.2** that $V_*(\xi_t) > 0$ whenever $\xi(t) \neq 0$. The time derivative of $V_*(\xi_t)$ along any state trajectory of system $(\Sigma_{\Delta e})$ with $w \equiv 0$ is given by:

$$\frac{d}{dt} V_*(\xi, t) =$$

$$\begin{bmatrix} \xi(t) \\ \xi(t-\tau) \\ h[x] \end{bmatrix}^t \begin{bmatrix} \dot{Y}_* + Y_* A_{\Delta a} \\ +A_{\Delta a}^t Y_* + W & Y_* E_{\Delta a} & Y_* H_a \\ E_{\Delta a}^t Y_* & -W & 0 \\ H_a^t Y_* & 0 & -I \end{bmatrix} \begin{bmatrix} \xi(t) \\ \xi(t-\tau) \\ h[x] \end{bmatrix}$$
$$(9.114)$$

By taking (9.112) into consideration, it follows from (9.114) that along any state trajectory of system $(\Sigma_{\Delta e})$ with $w \equiv 0$, $\frac{d}{dt} V_*(\xi, t) < 0$ whenever $\xi(t) \neq 0$ and $\xi(t - \tau) \neq 0$. Therefore, the equilibrium state $\xi = 0$ is globally, uniformly, asymptotically stable for all admissible uncertainties and unknown delays. Finally, to show that $\|e\|_2 < \gamma \{\|w\|_2^2 + x^t(0)Rx(0) + \int_{-\tau}^0 x^t(s)Rx(s)ds\}^{1/2}$, we introduce

$$J_* = \int_0^\infty \{e^t e - \gamma^2 w^t w\}dt - \gamma^2[x^t(0)Rx(0) + \int_{-\tau}^0 x^t(s)Rx(s)ds] \quad (9.115)$$

By applying **Lemma 9.3** it is straightforward to show that the condition $J_* < 0$ is implied by inequality (9.116) $\forall\, t\, [0,\infty)$. Therefore, we conclude that $\|e\|_2 < \gamma\{\|w\|_2^2 + x^t(0)Rx(0) + \int_{-\tau}^0 x^t(s)Rx(s)ds\}^{1/2}$ for any nonzero $(\alpha_0, w) \in \Re^n \oplus \mathcal{L}_2[0,\infty)$.

Remark 9.11: By examining (9.93)-(9.95), it is clear that $S(t)$ depends on P. In line with $\mathcal{H}_\infty$−control theory, the Riccati equation for $S(t)$ can be replaced by one which is decoupled from that of P subject to a spectral radius constraint. Using the standard results of [27], it can be easily shown that condition **(2)** of **Theorem 9.6** is equivalent to the following conditions:

(2a) There exists a bounded matrix function $0 \leq \Theta(t) = \Theta^t(t), t \in [0,\infty)$ satisfying

$$\begin{aligned}
\dot{\Theta} &= \{A - \hat{B}\hat{D}^t\hat{R}^{-1}C\}S + S\{A - \hat{B}\hat{D}^t\hat{R}^{-1}C\}^t \\
&+ S\{\gamma^{-2}[\hat{L}^t\hat{L} + \mu^{-1}E^t E + \rho^2 I] - C^t\hat{R}^{-1}C_o\}S \\
&+ \hat{B}\{I - \hat{D}^t\hat{R}^{-1}\hat{D}\}\hat{B}^t , \quad \Theta(0) = R^{-1} \quad (9.116)
\end{aligned}$$

such that the unforced linear time-varying system

$$\begin{aligned}
\dot{\eta}(t) &= [A - \hat{B}\hat{D}^t\hat{R}^{-1}C]\eta(t) \\
&+ S\left(\gamma^{-2}[\hat{L}^t\hat{L} + \mu^{-1}E^t E + \rho^2 I] - C^t\hat{R}^{-1}C\right)\eta(t) \quad (9.117)
\end{aligned}$$

is exponentially stable,

(2b) $I - \gamma^{-2}P\Theta(t) > 0, \ \forall \in [0,\infty)$.

In the case when conditions 1, **(2a)** and **(2b)** hold, a robust $\mathcal{H}_\infty$−filter is given by (9.86) but with the filter gain:

$$\hat{K}(t) = \{I - \gamma^{-2}\Theta(t)\bar{P}\}^{-1}\{\Theta(t)C_o^t + \hat{B}\hat{D}^t\}\hat{R}^{-1} \quad (9.118)$$

Remark 9.12: Considering (9.97) and (9.100), we can rewrite the robust $\mathcal{H}_\infty$ filter (9.86) as :

$$(\Sigma_e): \quad \dot{\hat{x}}(t) = [A + \delta A_w]\hat{x} + H_h h[\hat{x}] + B\hat{v}(t)$$

$$+ \quad \hat{K}[y(t) - (C + \delta C_w)\hat{x} - D\hat{v}(t)]$$
$$\hat{z} \quad = \quad L\hat{x} \tag{9.119}$$

where

$$\delta A_w = \mu H H^t \bar{P}, \quad \delta C_w = \mu H_c H^t \bar{P}, \quad \hat{v}(t) = \gamma^{-2} B^t \bar{P} \hat{x}(t) \tag{9.120}$$

An interpretation of (9.119)-(9.120) is that the matrices $\{\delta A_w, \delta C_w\}$ account for the worst-case parameter uncertainty of $\{\Delta A(t), \Delta C(t)\}$ in the presence of the estimated worst-case noise input $\hat{v}(t)$.

Remark 9.13: Observe from (9.100) and (9.120) that the known structural matrices of the parameteric uncertainty are only needed to determine the filter gain. In the absence of nonlinearities, $h[x] \equiv 0$, and state delay factor $E_o \equiv 0$, the robust filter (9.86) reduces to the one derived in [13]. However with $h[x] \equiv 0$ only, the resulting filter extends the result of [13] to the case of unknown state-delay. The effect of state-delay appears in (9.93)-(9.95) through the quantities H_d, A_d and E_d.

9.4 Linear Discrete-Time Systems

We learned from Chapter 1 that time-delays occur quite naturally in discrete-time systems. Little attention however, has been paid to estimating the state of linear uncertain discrete-time systems with delays. A preliminary result to bridge this gap is reported in [24] by developing a robust Kalman filter for a class of discrete uncertain systems with state-delay.

This section builds upon the results of Chapter 4 and extends them to another dimension by considering the $\mathcal{H}_\infty$ estimation of a class of discrete-time systems with real time-varying norm-bounded parameteric uncertainties and unknown state-delay.

9.4.1 Problem Description

We consider a class of uncertain time-delay systems represented by:

$$(\Sigma_\Delta): \quad x(k+1) \quad = \quad [A + \Delta A(k)]x(k) + Dw(k) + A_d x(k-\tau)$$
$$= \quad A_\Delta(k)x(k) + Dw(k) + A_d x(k-\tau)$$
$$y(k) \quad = \quad [C + \Delta C(k)]x(k) + Nw(k)$$
$$= \quad C_\Delta(k)x(k) + Nw(k)$$
$$z(k) \quad = \quad L\,x(k) \tag{9.121}$$

where $x(k) \in \Re^n$ is the state, $w(k) \in \Re^m$ is the input noise which belongs to $\ell_2 [0, \infty)$, $y(k) \in \Re^p$ is the measured output, $z(k) \in \Re^r$, is a linear combination of the state variables to be estimated and the matrices $A \in \Re^{n \times n}$, $C \in \Re^{p \times n}$, $D \in \Re^{p \times m}$, $A_d \in \Re^{n \times n}$ and $L \in \Re^{r \times n}$ are real constant matrices representing the nominal plant. Here, τ is an unknown constant scalar representing the amount of delay in the state. For all practical purposes, we let $\tau \leq \tau^*$ where τ^* is known. The matrices $\Delta A(k)$ and $\Delta C(k)$ are represented by:

$$\begin{bmatrix} \Delta A(k) \\ \Delta C(k) \end{bmatrix} = \begin{bmatrix} H \\ H_c \end{bmatrix} \Delta(k) E, \ \Delta^t(k)\Delta(k) \leq I, \ \forall \, k \geq 0 \qquad (9.122)$$

where $H \in \Re^{n \times \alpha}$, $H_c \in \Re^{p \times \alpha}$ and $E \in \Re^{\beta \times n}$ are known constant matrices and $\Delta(k) \in \Re^{\alpha \times \beta}$ is unknown matrices. The initial condition is specified as $\alpha_o(.) = \langle x_o, \phi(s) \rangle$, where $\phi(.) \in \ell_2[-\tau, 0]$.

In the sequel, we refer to the following systems:

$$(\Sigma_{\Delta o}): \ x(k+1) \ = \ A_\Delta(k)x(k) + A_d x(k - \tau) \qquad (9.123)$$
$$(\Sigma_{\Delta w}): \ x(k+1) \ = \ A_\Delta(k)x(k) + Dw(k) + A_d x(k - \tau)$$
$$z(k) \ = \ L \, x(k) \qquad (9.124)$$

From Chapter 2, we learned that system $(\Sigma_{\Delta o})$ is robustly stable independent of delay (RSID) if given a matrix $0 < W = W^t \in \Re^{n \times n}$, there exists a matrix $0 < P = P^t \in \Re^{n \times n}$ satisfying ARI:

$$A_\Delta^t(k)PA_\Delta(k) + A_\Delta^t(k)PA_d[W - A_d^t PA_d]^{-1}A_d^t PA_\Delta(k)$$
$$-P + W \ < \ 0$$

for all admissible uncertainties satisfying (9.122). Based on this, we have the following result.

Theorem 9.7: *System $(\Sigma_{\Delta w})$ is robustly stable with disturbance attenuation γ if given a matrix $0 < W = W^t \in \Re^{n \times n}$ there exist a matrix $0 < P = P^t \in \Re^{n \times n}$ satisfying the LMI:*

$$\begin{bmatrix} -P+W & 0 & 0 & L^t & A_\Delta^t \\ 0 & -\gamma^2 I & 0 & 0 & D^t \\ 0 & 0 & -Q & 0 & A_d^t \\ L & 0 & 0 & -I & 0 \\ A_\Delta & D & A_d & 0 & -P^{-1} \end{bmatrix} < 0, \ \forall \Delta : ||\Delta||^2 \leq 1 \quad (9.125)$$

Proof: By evaluating the first-order difference $\Delta V_6(x_k)$ of (2.78) along the trajectories of (9.124) and considering the Hamiltonian

$$H(x, w) = \Delta V(x_k) + \{z^t(k)z(k) - \gamma^2 w^t(k)w(k)\}$$

it yields:

$$H(x, w) = h^t(k)\,\Omega\,h(k)$$

$$\Omega(\sigma, k) = \begin{bmatrix} \Omega_1 & \Omega_2 & \Omega_3 \\ \Omega_2^t & \Omega_4 & \Omega_5 \\ \Omega_3^t & \Omega_5^t & \Omega_6 \end{bmatrix}$$

$$h(k) = [x^t(k) \quad w^t(k) \quad x^t(k-\tau)]^t \qquad (9.126)$$

where

$$\Omega_1 = A_\Delta^t P A_\Delta - P + W + L^t L,\; \Omega_3 = A_\Delta^t P A_d,$$
$$\Omega_5 = D^t P A_d\,;\; \Omega_2 = A_\Delta^t P D\,,$$
$$\Omega_6 = -[W - A_d^t P A_d]\,,\; \Omega_4 = -[\gamma^2 I - D^t P D] \qquad (9.127)$$

The stability condition with ℓ_2-gain constraint $H(x, w) < 0$ is implied by $\Omega(\sigma, k) < 0$. By **A.1** and using (9.127), it follows that $\Omega(\sigma, k) < 0$ is equivalent to

$$\begin{bmatrix} -P + W + L^t L & 0 & 0 \\ 0 & -\gamma^2 I & 0 \\ 0 & 0 & -W \end{bmatrix} + \begin{bmatrix} A_\Delta^t \\ D^t \\ A_d^t \end{bmatrix} P[A_\Delta \quad DB \quad A_d] < 0$$

The above inequality holds if and only if

$$\begin{bmatrix} -P + W & 0 & 0 & L^t \\ 0 & -\gamma^2 I & 0 & 0 \\ 0 & 0 & -W & 0 \\ L & 0 & 0 & -I \end{bmatrix} + \begin{bmatrix} A_\Delta^t \\ D^t \\ A_d^t \\ 0 \end{bmatrix} P[A_\Delta \quad D \quad A_d \quad 0] < 0$$

Simple arrangement of the above inequality with the help of **A.1** yields (9.125).

Remark 9.14: We note that LMI (9.125) holds if and only if the ARI:

$$A_\Delta^t \left\{ P^{-1} - A_d Q^{-1} A_o^t - \gamma^{-2} B_o B_o^t \right\}^{-1} A_\Delta - P + L^t L + Q < 0$$

This further reduces, via **B.1.3**, to

$$A^t \left\{ P^{-1} - A_d Q^{-1} A_d^t - \gamma^{-2} BB^t - \mu HH^t \right\}^{-1} A$$
$$-P + L^t L + \mu^{-1} E^t E + Q < 0 \qquad (9.128)$$

Motivated by **Theorem 9.7** and **Remark 9.14**, we state the following result.

Lemma 9.4: *For the uncertain time-delay system (9.121), the following statements are equivalent:*

(1) *System (Σ_{ow}) is robustly stable with disturbance attenuation γ.*
(2) *There exists a matrix $0 < P = P^t$ satisfying the LMI (9.125).*
(3) *There exists a matrix $0 < P = P^t$ satisfying the ARI (9.128).*
(4) *The following $\mathcal{H}_\infty$ norm bound is satisfied*

$$\left\| \begin{bmatrix} \mu^{-1/2} E \\ W^{1/2} \end{bmatrix} [zI - A_o]^{-1} [\gamma^{-1} B \ \mu^{1/2} H \ A_d W^{-1/2}] \right\|_\infty < 1 \qquad (9.129)$$

(5) *There exists a matrix $0 \leq \bar{P} = \bar{P}^t$ satisfying the ARE*

$$A[\bar{P}^{-1} - A_d^t W^{-1} A_d - \gamma^{-2} DD^t - \mu HH^t]^{-1} A - \bar{P}$$
$$+ \mu^{-1} E^t E + L^t L + W = 0 \qquad (9.130)$$

Remark 9.15: It is significant to observe that **Lemma 9.4** establishes a version of the bounded real Lemma **A.3.2** as applied to uncertain discrete-time systems with state-delay. Additionally, it provides alternative numerical techniques for testing the robust stability of the class of discrete systems under consideration.

9.4.2 $\mathcal{H}_\infty$-Estimation Results

The robust $\mathcal{H}_\infty$ state-estimation problem we are going to examine can be phrased as follows:

For system (Σ_Δ), design a linear estimator of $z(k)$ of the form

$$(\Sigma_e): \quad \hat{x}(k+1) = \hat{A}\hat{x}(k) + \hat{K}[y(k) - \hat{C}\hat{x}(k)]$$
$$= [A + \delta A]\hat{x}(k) + \hat{K}[y(k) - (C + \delta C)\hat{x}(k)]$$
$$\hat{z} = L\hat{x}, \quad \hat{x}(0) = 0 \qquad (9.131)$$

such that the estimation error $e(k) := z(k) - \hat{z}(k)$ *is robustly stable* $\forall\, w(k) \in$ $\ell_2[0, \infty)$ *and* $\|z - \hat{z}\|_2 \leq \gamma \|w\|_2$.

In (9.131), $\delta A \in \Re^{n \times n}$, $\delta C \in \Re^{p \times n}$, $\hat{K} \in \Re^{n \times p}$ are the design matrices to be determined.

We now proceed to solve the robust $\mathcal{H}_\infty$-estimation problem. By defining $\tilde{x}(k) = x(k) - \hat{x}(k)$, we get from (9.121)-(9.122) the dynamics of the state-error:

$$
\begin{aligned}
\tilde{x}(k+1) \;=\;& \{(A + \delta A) - \hat{K}(C + \delta C)\}\tilde{x}(k) + \{D - \hat{K}N\}w(k) \\
+\;& \{\Delta A - \delta A - \hat{K}(\Delta C - \delta C)\}x(k) \quad\quad (9.132)
\end{aligned}
$$

Then from system (Σ_Δ) and (9.131), we obtain the dynamics of the filtering error $e(k)$:

$$
\begin{aligned}
(\Sigma_{\Delta e}): \quad \xi(k+1) \;:=\;& \begin{bmatrix} x(k+1) \\ \tilde{x}(k+1) \end{bmatrix} \\
=\;& \{A_a + L_a \Delta_1(t) M_a\}\, \xi(k) + E_a \xi(k - \tau) \\
+\;& B_a w(k), \quad \xi(0) = \xi_o \quad\quad (9.133) \\
e(t) \;=\;& H_a\, \xi(k) \\
=\;& [0 \quad L]\xi(k) \quad\quad (9.134)
\end{aligned}
$$

where

$$
\begin{aligned}
A_a \;&=\; \begin{bmatrix} A & 0 \\ -\delta A + \hat{K}\delta C & A + \delta A - \hat{K}(C + \delta C) \end{bmatrix} \\
L_a \;&=\; \begin{bmatrix} H \\ H - \hat{K}H_c \end{bmatrix}, \quad E_a = \begin{bmatrix} A_d & 0 \\ A_d & 0 \end{bmatrix}, \quad B_a = \begin{bmatrix} D \\ D - \hat{K}N \end{bmatrix} \\
\xi_o \;&=\; \begin{bmatrix} \alpha_o \\ \alpha_o \end{bmatrix}, \quad M_a = [E \quad 0] \quad\quad (9.135)
\end{aligned}
$$

Theorem 9.8: *Given a prescribed level of noise attenuation* $\gamma > 0$ *and a matrix* $0 < \mathcal{Q} = \mathcal{Q}^t \in \Re^{2n \times 2n}$. *If for some scalar* $\mu > 0$ *there exists* *a matrix* $0 < \mathcal{Y} = \mathcal{Y}^t \in \Re^{2n \times 2n}$ *satisfying the LMI:*

$$\begin{bmatrix} -[\mathcal{Y} + \mu L_a L_a^t]^{-1} + \mu^{-1} M_a^t M_a + \mathcal{Q} & 0 & 0 & H_a^t & A_a^t \\ 0 & -\gamma^2 I & 0 & 0 & B_a^t \\ 0 & 0 & -\mathcal{Q} & 0 & E_a^t \\ H_a & 0 & 0 & -I & 0 \\ A_a & B_a & E_a & 0 & -\mathcal{Y} \end{bmatrix} < 0$$

(9.136)

then the robust H_∞-*estimation problem for the system* $(\Sigma_{\Delta e})$ *is solvable with estimator (9.132) and yields.*

$$\|e(k)\|_2 < \gamma \|w(k)\|_2 \tag{9.137}$$

Proof: By **Theorem 9.7**, system $(\Sigma_{\Delta e})$ is QS with disturbance attenuation γ if given a matrix $0 < \mathcal{Q} = \mathcal{Q}^t \in \Re^{2n \times 2n}$ there exists a matrix $0 < \mathcal{P} = \mathcal{P}^t \in \Re^{2n \times 2n}$ satisfying

$$\begin{bmatrix} -\mathcal{P} + \mathcal{Q} & 0 & 0 & H_a^t & A_{\Delta a}^t \\ 0 & -\gamma^2 I & 0 & 0 & B_a^t \\ 0 & 0 & -\mathcal{Q} & 0 & E_a^t \\ H_a & 0 & 0 & -I & 0 \\ A_{\Delta a} & B_a & E_a & 0 & -\mathcal{P}^{-1} \end{bmatrix} < 0 \tag{9.138}$$

By **A.1**, inequality (9.138) holds if and only if

$$\begin{bmatrix} M_a^t \\ 0 \\ 0 \\ 0 \\ 0 \\ 0 \end{bmatrix} \Delta^t(k)[0 \ 0 \ 0 \ 0 \ L_a^t] + \begin{bmatrix} 0 \\ 0 \\ 0 \\ 0 \\ 0 \\ L_a \end{bmatrix} \Delta(k)[M_a \ 0 \ 0 \ 0 \ 0]$$

$$\begin{bmatrix} -\mathcal{P} + \mathcal{Q} + \mu^{-1} M_a^t M_a & 0 & 0 & H_a^t & A_a^t \\ 0 & -\gamma^2 I & 0 & 0 & B_a^t \\ 0 & 0 & -\mathcal{Q} & 0 & E_a^t \\ H_a & 0 & 0 & -I & 0 \\ A_a & B_a & E_a & 0 & -\mathcal{P}^{-1} \end{bmatrix} < 0$$

(9.139)

By **B.1.4**, inequality (9.139) is equivalent to

$$
\begin{bmatrix} \mu^{-1/2}M_a^t \\ 0 \\ 0 \\ 0 \\ 0 \\ 0 \end{bmatrix} [\mu^{-1/2}M_a \;\; 0 \;\; 0 \;\; 0 \;\; 0] +
$$

$$
\begin{bmatrix} 0 \\ 0 \\ 0 \\ 0 \\ \mu^{1/2}L_a \end{bmatrix} [0 \;\; 0 \;\; 0 \;\; 0 \;\; \mu^{1/2}L_a^t] +
$$

$$
\begin{bmatrix}
-\mathcal{P}+\mathcal{Q}+\mu^{-1}M_a^tM_a & 0 & 0 & H_a^t & A_a^t \\
0 & -\gamma^2 I & 0 & 0 & B_a^t \\
0 & 0 & -\mathcal{Q} & 0 & E_a^t \\
H_a & 0 & 0 & -I & 0 \\
A_a & B_a & E_a & 0 & -\mathcal{P}^{-1}
\end{bmatrix} < 0
$$

$$(9.140)$$

for some $\mu > 0$. Rearranging, we get

$$
\begin{bmatrix}
-\mathcal{P}+\mathcal{Q}+\mu^{-1}M_a^tM_a & 0 & 0 & H_a^t & A_a^t \\
0 & -\gamma^2 I & 0 & 0 & B_a^t \\
0 & 0 & -\mathcal{Q} & 0 & E_a^t \\
H_a & 0 & 0 & -I & 0 \\
A_a & B_a & E_a & 0 & -[\mathcal{P}^{-1}-\mu L_a L_a^t]
\end{bmatrix} < 0
$$

$$(9.141)$$

Letting $Y = [P^{-1}-\mu L_a L_a^t]$ in (9.141) we obtain directly the LMI (9.136).

Remark 9.15: It should be observed that **Theorem 9.8** establishes an LMI-feasibility condition for the robust $\mathcal{H}_\infty$−estimation problem associated with system (Σ_Δ) which requires knowledge about the nominal matrices of the system as well as the structural matrices of the uncertainty. In this way, it provides a partial solution to the $\mathcal{H}_\infty$−estimation under consideration.

To facilitate further development, we introduce

$$
\mathcal{R}_1 = \left\{ S_1^{-1} - \mu^{-1}E^tE - A_d[\mathcal{Q}_1 - A_d^t(S_2^{-1}-LL^t)^{-1}A_d]^{-1}A_d^t \right\}^{-1}
$$

$$\mathcal{R}_3 = \left\{ \mathcal{S}_2^{-1} - LL^t - A_d[\mathcal{Q}_1 - A_d^t(\mathcal{S}_1^{-1} - \mu^{-1}E^tE)^{-1}A_d]^{-1}A_d^t \right\}^{-1}$$

$$\mathcal{R}_2 = -\left\{ \mathcal{S}_1^{-1} - A_d\mathcal{Q}_1A_d^t - \mu^{-1}E^tE \right\}^{-1} E\mathcal{Q}_1^{-1}E^t\mathcal{R}_3$$

$$\mathcal{G}_1 = \mu H^tH + \gamma^{-2}DD^t + A^t\mathcal{R}_2^tA$$

$$\mathcal{G}_2 = \mu H_cH^t + \gamma^{-2}ND^t + C\mathcal{R}_2^tA \qquad (9.142)$$

for some matrices $0 < \mathcal{S}_1 = \mathcal{S}_1^t$, $0 < \mathcal{S}_2 = \mathcal{S}_2^t$, $0 < \mathcal{Q}_1 = \mathcal{S}_1^t$ and $0 < \mathcal{Q}_2 = \mathcal{S}_2^t$. Accordingly we define the matrices:

$$\delta A = \mathcal{G}_1A^{-t}\left\{ \mathcal{R}_1 - \mathcal{R}_2^t \right\}^{-1} , \quad \delta C = \mathcal{G}_2A^{-t}\left\{ \mathcal{R}_1 - \mathcal{R}_2^t \right\}^{-1}$$

$$\hat{A} = A + \delta A , \quad \hat{C} = C + \delta C$$

$$\mathcal{T} = \delta C\left\{ \mathcal{R}_3 - \mathcal{R}_2 \right\}\hat{A}^t + C\mathcal{R}_3\hat{A}^t - C\mathcal{R}_2^t\delta A^t$$

$$+ \delta C\left\{ \mathcal{R}_1 - \mathcal{R}_2^t \right\}\delta A^t$$

$$\mathcal{Z} = \hat{C}\mathcal{R}_3\hat{C}^t - \hat{C}\mathcal{R}_2^t\delta\hat{C}^t - \delta C\mathcal{R}_2\bar{C}^t + \delta C\mathcal{R}_1\delta C^t \qquad (9.143)$$

It is important to note that the indicated inverses in (9.142-9.143) exist in view of **A.2** and the selection of matrices $0 < \mathcal{Q}_1$ and $\mathcal{Q}_2$. Observe in (9.143) using (9.135)-(9.142) that $(\mathcal{R}_1 - \mathcal{R}_2) > 0$.

The next theorem establishes the main result.

Theorem 9.9: *Consider the augmented system $(\Sigma_{\Delta e})$ for some $\gamma > 0$ and given matrices $0 < \mathcal{Q}_1 = \mathcal{Q}_1^t \in \Re^{n \times n}$ and $0 < \mathcal{Q}_2 = \mathcal{Q}_2^t \in \Re^{n \times n}$. If for some scalar $\mu > 0$ there exist matrices $0 < \mathcal{S}_1 = \mathcal{S}_1^t \in \Re^{n \times n}$ and $0 < \mathcal{S}_2 = \mathcal{S}_2^t \in \Re^{n \times n}$ satisfying the LMIs*

$$\begin{bmatrix} -\mathcal{S}_1 + \mathcal{Q}_1 & H^t & D^t & A^t \\ H & -\mu^{-1}I & 0 & 0 \\ D & 0 & -\gamma^2I & 0 \\ A & 0 & 0 & -\mathcal{R}_1^{-1} \end{bmatrix} < 0 \qquad (9.144)$$

$$\begin{bmatrix} -\mathcal{S}_2 + \mathcal{Q}_2 \\ -\bar{A}\mathcal{R}_2^t\delta A^t - \delta A\mathcal{R}_2\hat{A} & \delta A & \hat{A} \\ -\mathcal{T}^t\mathcal{Z}^{-1}\mathcal{T} & & \\ \delta A^t & -\mathcal{R}_1^{-1} & 0 \\ \hat{A}^t & 0 & -\mathcal{R}_3^{-1} \end{bmatrix} < 0 \qquad (9.145)$$

then the robust $\mathcal{H}_\infty$-estimation problem for the system $(\Sigma_{\Delta c})$ is solvable with the estimator

$$\hat{x}(k+1) = \hat{A}\hat{x}(k) + T^t Z^{-1}[y(k) - \hat{C}\hat{x}(k)] \tag{9.146}$$

which yields

$$\|e(k)\|_2 < \gamma \|w(k)\|_2 \tag{9.147}$$

Proof: Given a matrix $0 < Q = Q^t \in \Re^{2n \times 2n}$ and by **Theorem 9.8**, it follows that there exists a matrix $0 < P = P^t \in \Re^{2n \times 2n}$ that satisfies LMI (9.136). Applying **A.1**, this is equivalent to:

$$A_a^t \left\{ P^{-1} - E_a Q^{-1} E_a^t - \gamma^{-2} B_a B_a^t - \mu L_a L_a^t \right\}^{-1} A_a - P + H_a^t H_a$$
$$+ \mu^{-1} M_a^t M_a + Q < 0 \tag{9.148}$$

From the results of [13], it follows that (9.148) holds if and only if there exists a matrix $0 < S = S^t \in \Re^{2n \times 2n}$ satisfying

$$\Xi(S) := A_a \left\{ S^{-1} - E_a Q^{-1} E_a^t - \mu^{-1} M_a^t M_a - H_a^t H_a \right\}^{-1} A_a^t$$
$$- S + \mu L_a L_a^t + \gamma^{-2} B_a B_a^t + Q < 0 \tag{9.149}$$

Define

$$S = \begin{bmatrix} S_1 & 0 \\ 0 & S_2 \end{bmatrix}, \quad Q = \begin{bmatrix} Q_1 & 0 \\ 0 & Q_2 \end{bmatrix} \tag{9.150}$$

Expansion of (9.149) using (9.135)-(9.136) and (9.150) yields:

$$\Xi(S) := \begin{bmatrix} \Xi_1(S) & \Xi_2(S) \\ \Xi_2^t(S) & \Xi_3(S) \end{bmatrix} \tag{9.151}$$

where

$$\Xi_1(S) = A\mathcal{R}_1 A^t - S_1 + Q_1 + \mu H H^t + \gamma^{-2} D D^t \tag{9.152}$$

$$\begin{aligned} \Xi_2(S) = \; & A\mathcal{R}_1(-\delta A + \hat{K}\delta C)^t + \mu H H^t - \mu H H_c^t \hat{K}^t - \gamma^{-2} D N^t \hat{K}^t \\ & + A\mathcal{R}_2[A + \delta A - \hat{K}(C + \delta C)]^t \end{aligned} \tag{9.153}$$

$$\begin{aligned} \Xi_3(S) = \; & [A + \delta A - \hat{K}(C + \delta C)]\mathcal{R}_3[A_o + \delta A - \hat{K}(C + \delta C)]^t \\ & + [A + \delta A - \hat{K}(C + \delta C)]\mathcal{R}_2^t(-\delta A + \hat{K}\delta C)^t \\ & + (-\delta A + \hat{K}\delta C')\mathcal{R}_1(-\delta A + \hat{K}\delta C)^t - S_2 + Q_2 \\ & + \mu(H - \hat{K}H_c)(H - \hat{K}H_c)^t + \gamma^{-2}(D - \hat{K}N)(D - \hat{K}N)^t \end{aligned} \tag{9.154}$$

For internal stability with ℓ_2-bound it required that $\Xi(\mathcal{S}) < 0$. Necessary and sufficient conditions to achieve this are

$$\Xi_1(\mathcal{S}) < 0, \quad \Xi_2(\mathcal{S}) = 0, \quad \Xi_2(\mathcal{S}) < 0 \qquad (9.155)$$

It is readily seen from (9.152) that the condition $\Xi_1(\mathcal{S}) < 0$ is equivalent to the LMI (9.144). Using (9.143) in (9.153) and arranging terms, we conclude that the condition $\Xi_2(\mathcal{S}) = 0$ is satisfied. Finally, from (9.142)-(9.144) and using the "completion of squares" argument with some standard algebraic manipulations we conclude that the Kalman gain is given by $\hat{K} = T^t Z^{-1}$ and the LMI (9.145) corresponds to $\Xi_3(\mathcal{S}) < 0$.

Two important special cases follow.

Lemma 9.5: *Consider the uncertain discrete system without delay*

$$
\begin{aligned}
x(k+1) &= [A + \Delta A(k)]x(k) + Dw(k) + A_d x(k-\tau) \\
&= A_\Delta(k)x(k) + Dw(k) + A_d x(k-\tau) \\
y(k) &= [C + \Delta C(k)]x(k) + Nw(k) \\
&= C_\Delta(k)x(k) + Nw(k) \\
z(k) &= L\, x(k) \qquad\qquad\qquad\qquad\qquad\qquad (9.156)
\end{aligned}
$$

If for some $\gamma > 0$ and a scalar $\mu > 0$ there exist matrices $0 < \mathcal{S}_1 = \mathcal{S}_1^t \in \Re^{n \times n}$ and $0 < \mathcal{S}_2 = \mathcal{S}_2^t \in \Re^{n \times n}$ satisfying the LMIs

$$
\begin{bmatrix}
-\mathcal{S}_1 & H^t & D^t & A^t \\
H & -\mu^{-1}I & 0 & 0 \\
D & 0 & -\gamma^2 I & 0 \\
A & 0 & 0 & -\mathcal{R}_4^{-1}
\end{bmatrix} < 0 \qquad (9.157)
$$

$$
\begin{bmatrix}
-\mathcal{S}_2 - T_1^t Z_1^{-1} T_1 & \delta A & \hat{A} \\
\delta A^t & -\mathcal{R}_4^{-1} & 0 \\
\hat{A}^t & 0 & -\mathcal{R}_5^{-1}
\end{bmatrix} < 0 \qquad (9.158)
$$

then the robust $\mathcal{H}_\infty$-estimation problem is solvable with the estimator

$$\hat{x}(k+1) = \hat{A}\hat{x}(k) + T_1^t Z_1^{-1}[y(k) - \hat{C}\hat{x}(k)] \qquad (9.159)$$

which yields

$$\|e(k)\|_2 < \gamma\, \|w(k)\|_2 \qquad (9.160)$$

and where

$$\begin{array}{rcl}
\mathcal{R}_4 & = & \left\{S_1^{-1} - \mu^{-1}E^t E\right\}^{-1} \quad , \quad \mathcal{R}_5 = \left\{S_2^{-1} - LL^t\right\}^{-1} \\
\mathcal{G}_3 & = & \mu H^t H + \gamma^{-2} D^t D \quad , \quad \mathcal{G}_4 = \mu H_c^t H + \gamma^{-2} N D^t \\
\delta A & = & \mathcal{G}_3 A^{-t} \mathcal{R}_4^{-1} \quad , \quad \delta C = \mathcal{G}_4 A^{-t} \mathcal{R}_4^{-1} \\
\hat{A} & = & A + \delta A \quad , \quad \hat{C} = C + \delta C \\
T_1 & = & \hat{C} \mathcal{R}_5 \hat{A}^t + \delta C \mathcal{R}_4 \hat{A}^t \quad , \quad \mathcal{Z}_1 = \hat{C} \mathcal{R}_3 \hat{C}^t + \delta C \mathcal{R}_4 \delta C^t
\end{array}$$

Proof: Follows from **Theorem 9.9** by setting $A_d \equiv 0$ and $\mathcal{Q}_1 = \mathcal{Q}_2 = 0$ in (9.143)-(9.144) and observing that $\mathcal{R}_2 \equiv 0$.

Lemma 9.6: *Consider the system*

$$\begin{array}{rcl}
x(k+1) & = & Ax(k) + Dw(k) + A_d x(k-\tau) \\
y(k) & = & Cx(k) + Nw(k) \\
z(k) & = & L\,x(k)
\end{array} \qquad (9.161)$$

for some $\gamma > 0$ and given matrices $0 < \mathcal{Q}_1 = \mathcal{Q}_1^t \in \Re^{2n \times 2n}$ and $0 < \mathcal{Q}_2 = \mathcal{Q}_2^t \in \Re^{2n \times 2n}$. If there exist matrices $0 < S_1 = S_1^t \in \Re^{n \times n}$ and $0 < S_2 = S_2^t \in \Re^{n \times n}$ satisfying the LMIs

$$\begin{bmatrix}
-S_1 + \mathcal{Q}_1 & D^t & A^t \\
D & -\gamma^2 I & 0 \\
A & 0 & -\mathcal{R}_6^{-1}
\end{bmatrix} < 0 \qquad (9.162)$$

$$\begin{bmatrix}
-S_2 + \mathcal{Q}_2 \\
-\bar{A}\mathcal{R}_2^t \delta A^t - \delta A \mathcal{R}_2 \hat{A} & \delta A & \hat{A} \\
-T_2^t \mathcal{Z}_2^{-1} T_2 \\
\delta A^t & -\mathcal{R}_6^{-1} & 0 \\
\hat{A}^t & 0 & -\mathcal{R}_8^{-1}
\end{bmatrix} < 0 \qquad (9.163)$$

then the robust $\mathcal{H}_\infty-$estimation problem is solvable with the estimator

$$\hat{x}(k+1) = \hat{A}\hat{x}(k) + T_2^t \mathcal{Z}_2^{-1}[y(k) - \hat{C}\hat{x}(k)] \qquad (9.164)$$

which yields

$$\|e(k)\|_2 < \gamma \, \|w(k)\|_2 \qquad (9.165)$$

and where

$$\mathcal{R}_6 = \left\{S_1^{-1} - A_d[\mathcal{Q}_1 - A_d^t(S_2^{-1} - LL^t)^{-1}A_d]^{-1}E^t\right\}^{-1}$$

$$\mathcal{R}_8 = \left\{S_2^{-1} - LL^t - E[\mathcal{Q}_1 - A_d^t S_1 A_d]^{-1} E^t\right\}^{-1}$$

$$\mathcal{R}_7 = -\left\{S_1^{-1} - A_d \mathcal{Q}_1 A_d^t\right\}^{-1} A_d \mathcal{Q}_1^{-1} A_d^t \mathcal{R}_8$$

$$\mathcal{G}_5 = \gamma^{-2} D^t D + A^t \mathcal{R}_7^t A$$

$$\mathcal{G}_6 = \gamma^{-2} N D^t + C \mathcal{R}_7^t A$$

$$\delta A = \mathcal{G}_5 A^{-t} \left\{\mathcal{R}_6 - \mathcal{R}_7^t\right\}^{-1}, \quad \delta C = \mathcal{G}_6 A^{-t} \left\{\mathcal{R}_6 - \mathcal{R}_7^t\right\}^{-1}$$

$$\hat{A} = A + \delta A, \quad \hat{C} = C + \delta C$$

$$\mathcal{T}_2 = \delta C \left\{\mathcal{R}_8 - \mathcal{R}_7\right\} \hat{A}^t + C \mathcal{R}_8 A^t - C_o \mathcal{R}_7^t \delta A^t + \delta C \left\{\mathcal{R}_6 - \mathcal{R}_7^t\right\} \delta A^t$$

$$\mathcal{Z}_2 = \hat{C} \mathcal{R}_8 \hat{C}^t - \hat{C} \mathcal{R}_7^t \delta \hat{C}^t - \delta C \mathcal{R}_7 \bar{C}^t + \delta C \mathcal{R}_6 \delta C^t$$

Remark 9.16: It is interesting to observe that **Lemma 9.5** gives an LMI-based version of the results in [13]. **Lemma 9.6** presents an $\mathcal{H}_\infty$ filter for a class of discrete-time systems with unknown state-delay. Both **Lemma 9.5, 9.6** are new results for state estimation of time-delay systems.

9.5 Linear Parameter-Varying Systems

Stability analysis and control synthesis problems of linear continuous-time systems where the state-space matrices depend on time-varying parameters have received considerable attention recently [33-38]. When dealing with linear parameter-varying (LPV) systems, there have been two basic approaches. One approach developed in [33-35] where it has been assumed that the trajectory of the parameters is not known *a priori* although its value is known through real-time measurements and therefore the state-space matrices depend continuously on these parameters. Quadratic stability has been the main vehicle in the analysis and control synthesis using a single quadratic Lyapunov function. In [35-38], an alternative approach has been pursued where the real uncertain parameters and their rates have been assumed to vary in some prescribed ranges and hence allowing the state-space matrices to depend affinely on these parameters. Affine quadratic stability has been introduced to facilitate the analysis and control synthesis using parameter-dependent quadratic Lyapunov functions. We examine here the problems of stability and $\mathcal{H}_\infty$ filtering for a class of linear parameter-varying discrete-time (LPVDT) systems in which the state-space matrices depend affinely on time-varying parameters and has unknown constant state-delay. We employ the notion of affine quadratic stability using parameter-dependent Lyapunov functions and establish LMI-based procedures for testing the internal stability. Then, we develop a linear parameter-dependent filter such that the estimation error is affinely quadratically stable with a prescribed performance measure. It is shown that the solvability conditions can be expressed into LMIs which are then evaluated at the vertices of the polytopic range of parameter values. The developed results extend those of [29-31] further and study a class of LPVDT systems with state-delay which is frequently encountered in industrial process control.

9.5.1   Discrete-Time Models

We consider a class of linear parameter-varying discrete-time (LPVDT) systems with state-delay represented by:

$$
\begin{aligned}
x(k+1) &= A(\sigma(k))x(k) + B(\sigma(k))w(k) + E(\sigma(k))x(k-\eta), \\
y(k) &= C(\sigma(k))x(k) + D(\sigma(k))w(k) \\
z(k) &= L(\sigma(k))x(k) , \quad x(0) = x_o
\end{aligned}
\tag{9.166}
$$

where $x(k) \in \Re^n$ is the state, $w(k) \in \Re^m$ is the disturbance input which belongs to $\ell_2[0,\infty)$, $z(k) \in \Re^m$ is the controlled output, $y(k) \in \Re^q$ is the measured output , $\sigma(k) = (\sigma_1, ..., \sigma_r) \in \Im \subset \Re^r$ is a vector of uncertain and possibly time-varying parameters with $\Im$ being compact , η is an unknown constant delay and $A(.), B(.), C(.), D(.), L(.), E(.)$ are known real matrix functions and affinely depending on σ, that is:

$$
\begin{bmatrix}
A(\sigma) & B(\sigma) & E(\sigma) \\
C(\sigma) & D(\sigma) & L(\sigma)
\end{bmatrix}
=
\begin{bmatrix}
A_o & B_o & E_o \\
C_o & D_o & L_o
\end{bmatrix}
+ \sum_{j=1}^{r} \sigma_j
\begin{bmatrix}
A_j & B_j & E_j \\
C_j & D_j & L_j
\end{bmatrix}
\tag{9.167}
$$

where $A_o, ..., A_r$;$B_o, ..., B_r$; $C_o, ..., C_r$; $D_o, ..., D_r$; $E_o, ..., E_r$ and $L_o, ..., L_r$ are known constant matrices of appropriate dimensions. From now onwards, we consider the parameter uncertainty σ to be quantified by the range of parameter values and its incremental variation. In the sequel, an admissible parameter vector $\sigma \in \Re^r$ is a time function that satisfies at each instant:

(A1) Each parameter $\sigma_j(k), (j = 1, ..., r)$ is real and ranges between two extreme values $\check{\sigma}_j$ and $\hat{\sigma}_j$:

$$
\sigma_j(k) \in [\check{\sigma}_j , \hat{\sigma}_j]
\tag{9.168}
$$

(A2) The incremental variation $\delta\sigma_j(k)$ is well-defined at all times and

$$\delta\sigma_j(k) \in [\check{\nu}_j \,, \hat{\nu}_j] \tag{9.169}$$

where the bounds $\check{\sigma}_j, \hat{\sigma}_j, \check{\nu}_j, \hat{\nu}_j$ are known for all $(j = 1, ..., r)$.

It follows from **(A1)** and **(A2)** that each of the parameter vector σ and its incremental variation $\delta\sigma$ is valued in an hyper-rectangle of the parameter space $\mathfrak{R}^r$, the vertices of which are in the sets:

$$\mathcal{W} := \{\omega = (\omega_1, ..., \omega_r) : \omega_j \in [\check{\sigma}_j \,, \hat{\sigma}_j]\} \tag{9.170}$$

$$\mathcal{V} := \{\nu = (\nu_1, ..., \nu_r) : \nu_j \in [\check{\nu}_j \,, \hat{\nu}_j]\} \tag{9.171}$$

Note that $\mathcal{W}$ and $\mathcal{V}$ represent the sets of the 2^r corners of the parameter box and the parameter-rate box, respectively.

For the LPVDT system (9.166), we are mainly concerned with obtaining an estimate, $\hat{z}(k)$, of $z(k)$, via a causal linear parameter-dependent filter using the measurement $y(k)$ and which provides a uniformly small estimation error, $z(k) - \hat{z}(k)$, $\forall\, w(k) \in \ell_2[0, \infty)$. Towards our goal, we provide in the next section relevant stability measures that will be used in developing the main results.

9.5.2 Affine Quadratic Stability

Distinct from (9.166) is the linear parameter-varying free (LPVF) system:

$$x(k+1) = A(\sigma(k))x(k) + E(\sigma(k))x(k - \eta) \,; \; x(0) = x_o \tag{9.172}$$

Definition 9.1: *System $(\Sigma_{\sigma o})$ is said to be affinely quadratically stable (AQS) if given a set of $(r + 1)$ matrices $(Q_o, .., Q_r)$ such that $0 < Q_j = Q_j^t \; j = 0, .., r$ and $\mathcal{Q}(\sigma) := Q_o + \sigma_1 Q_1 + .. + \sigma_r Q_r > 0$ there exists a set of $(r + 1)$ matrices $(P_o, ..., P_r)$ and such that $0 < P_j = P_j^t \; j = 0, .., r$ and*

$$\mathcal{P}(\sigma) := P_o + \sigma_1 P_1 + ... + \sigma_r P_r > 0 \tag{9.173}$$

$$\mathcal{R}(\sigma, \delta\sigma) := A^t(\sigma)\{[\mathcal{P}(\sigma) + \mathcal{P}(\delta\sigma) - P_o]^{-1} - E(\sigma)Q{-}1(\sigma)E^t(\sigma)\}^{-1}A(\sigma)$$
$$- \mathcal{P}(\sigma) + \mathcal{Q}(\sigma) < 0 \tag{9.174}$$

hold for all admissible values and trajectories of the parameter vector σ. The function $V(k, \sigma) := x^t(k)\mathcal{P}(\sigma(k))x(k) + \sum_{j=k-\eta}^{k-1} x^t(j)\mathcal{Q}\sigma(j)x(j)$ is then a quadratic Lyapunov function for system $(\Sigma_{\sigma o})$ in the sense that

$$V(k, \sigma) > 0 \quad \forall\, x \neq 0 \,, \quad \Delta V(k, \sigma) < 0$$

for all initial conditons and parameter trajectories $\sigma(k)$.

The following theorem provides an LMI-based procedure to check the AQS for system (9.172).

Theorem 9.10: *Consider system $(\Sigma_{\sigma o})$ where the matrix $A(.)$ depends affinely on σ in the manner of (9.167). Let W, V denote the sets of corners of the parameter box (9.170) and the incremental variation box (9.171), respectively and let $\bar{\sigma} = (|\check{\sigma}_1 + \hat{\sigma}_1|/2, ..., |\check{\sigma}_r + \hat{\sigma}_r|/2)$ denote the average value of the parameter vector such that $A(\bar{\sigma})$ is stable. This system is affinely quadratically stable if given a set of $(r + 1)$ matrices $(Q_o, .., Q_r)$ such that $0 < Q_j = Q_j^t \; j = 0, .., r$ and $Q(\sigma) := Q_o + \sigma_1 Q_1 + .. + \sigma_r Q_r > 0$ there exists a set of $(r + 1)$ matrices $(P_o, ..., P_r)$ such that $0 < P_j = P_j^t \; \forall \; j = 0, ..., r$, satisfying*

$$A^t(\omega)\{[\mathcal{P}(\omega) + \mathcal{P}(\nu) - P_o]^{-1} - E(\omega)Q(\omega)^{-1}E^t(\omega)\}^{-1}A(\omega)$$
$$-\mathcal{P}(\omega) + Q(\omega) < 0 \qquad \forall \, (\omega, \nu) \in W \times V \qquad (9.175)$$
$$\mathcal{P}(\omega) > 0 \qquad \qquad \forall \, \omega \in W \qquad\qquad\qquad (9.176)$$
$$A_j^t\left\{P_o^{-1} - E_o Q_o^{-1} E_o^t\right\}^{-1} A_j \geq 0 \; j = 1, ..., r \qquad (9.177)$$

where

$$\mathcal{P}(\sigma) := P_o + \sigma_1 P_1 + ... + \sigma_r P_r \qquad\qquad (9.178)$$

When the LMI system (9.175)-(9.177) is feasible, a Lyapunov-Krasovskii function for $(\Sigma_{\sigma o})$ and for all trajectories $\sigma(k)$ satisfying (9.168)-(9.169) is then given by $V(x, \sigma) = x^t(k)\mathcal{P}(\sigma)x(k) + \sum_{\alpha=k-\eta}^{k-1} x^t(\alpha)Q(\sigma)x(\alpha)$.

Proof: Note that under assumptions **(A1)**, **(A2)** and (9.178), we have:

$$\mathcal{P}(\sigma + \delta\sigma) = P_o + (\sigma_1 + \delta\sigma_1) P_1 + ... + (\sigma_r + \delta\sigma_r)P_r$$
$$= \mathcal{P}(\sigma) + \mathcal{P}(\delta\sigma) - P_o \qquad\qquad (9.179)$$

From **Definition 9.1**, the positivity constraint (9.174) is affine in σ and hence it holds for all σ in the parameter box W if and only if it holds at all corners. This yields condition (11). For a given $\nu \in V$ the quantity $\mathcal{P}(\nu) - P_o$ is constant. The fact that $A, E, \mathcal{P}$ are affinely dependent on σ and (9.175)-(9.177) hold for all $\omega \in W$ ensure that

$$\mathcal{P}(\sigma) > 0$$
$$-\mathcal{P}(\sigma) + A^t(\sigma)\{[\mathcal{P}(\sigma) + \mathcal{P}(\delta\sigma) - P_o]^{-1} -$$
$$E(\sigma)Q(\sigma)^{-1}E^t(\sigma)\}^{-1}A(\sigma) + Q(\sigma) < 0 \qquad (9.180)$$

for any value of σ in the parameter box (9.170). Since (9.180) holds for any corner $\nu \in \mathcal{V}$ and $\mathcal{P}(\delta\sigma)$ is affine in $\delta\sigma$, it can be deduced by a standard convexity argument that the inequality

$$-\mathcal{P}(\sigma) + A^t(\sigma)\{\mathcal{P}^{-1}(\sigma + \delta\sigma) -$$
$$E(\sigma)\mathcal{Q}(\sigma)^{-1}E^t(\sigma)\}^{-1}A(\sigma) + \mathcal{Q}(\sigma) < 0 \qquad (9.181)$$

holds over the entire incremental parameter box (9.171) since it is satisfied at all its vertices $\nu \in \mathcal{V}$. Using (9.180), it follows that

$$\begin{aligned}
\Delta V(x, \sigma) &= x^t(k)\{A^t(\sigma)\mathcal{P}(\sigma + \delta\sigma)A(\sigma) - \mathcal{P}(\sigma) + \mathcal{Q}(\sigma)\}x(k) \\
&+ x^t(k)\{A^t(\sigma)\mathcal{P}(\sigma + \delta\sigma)E(\sigma) \\
&\quad [\mathcal{Q}(\sigma) - E^t(\sigma)\mathcal{P}^{-1}(\sigma + \delta\sigma)E(\sigma)]^{-1}E^t(\sigma)\mathcal{P}(\sigma + \delta\sigma)\}x(k) \\
&< 0 \qquad (9.182)
\end{aligned}$$

for any parameter trajectory $\sigma(k)$ satisfying (9.168)-(9.169), which establishes AQS. It remains to clarify the multiconvexity constraint (9.177). Given the affine expressions of $A, \mathcal{P}, E$ and $\mathcal{Q}$, it can be easily shown that for any $0 \neq x \in \Re^n$:

$$\begin{aligned}
g(\sigma) &= x^t \Delta V(x, \sigma)x \\
&= \alpha_o + \sum_j \alpha_j \sigma + \sum_{j<k} \beta_{jk}\sigma_j\sigma_k + \sum_j \varphi_j \sigma_j^2 \qquad (9.183)
\end{aligned}$$

for some scalars $(\alpha_o, \alpha_j, \beta_{jk}, \varphi_j, \delta_{jk}, \zeta_j)$. By [4, Lemma 3.1] and with some standard algebraic manipulations, it follows that:

$$\varphi_j = x^t \left(A_j^t \{P_o^{-1} - E_o Q_o^{-1} E_o^t\}^{-1} A_j \right) x \geq 0, j = 1, ..., r$$

This implies that the multiconvexity requirement corresponds to the constraint (9.177). Observe that (9.175) ensures the negativity of $g(\sigma)$ at all corners of the parameter box and hence

$$x^t \Delta V(x, \sigma)x < \qquad \forall x \neq 0$$

over the entire parameter box and therefore we conclude that $\Delta V(x, \sigma) < 0$ for all admissible σ.

Remark 9.17: The relevance of Theorem 9.10 stems from the fact that it replaces the solution of an infinite number of LMIs to determine

$P_o, ..., P_r$ by a finite number of LMIs at the corners of the hyper-rectangles plus the multiconvexity condition (9.177). In turn, these LMIs can be effectively solved by the LMI Toolbox.

A special case of **Theorem 9.10**, when the σ-parameters are constants, is presented below.

Corollary 9.1: *Consider system $(\Sigma_{\sigma o})$ where the matrix $A(.)$ depend affinely on constant parameters $\sigma \in \Re^r$ satisfying (9.168). Let $\mathcal{W}$ denote the set of corners of the parameter box (9.170). This system is affinely quadratically stable if given a set of $(r + 1)$ matrices $(Q_o, .., Q_r)$ such that $0 < Q_j = Q_j^t \; j = 0, .., r$ and $\mathcal{Q}(\sigma) := Q_o + \sigma_1 Q_1 + .. + \sigma_r Q_r > 0$ there exists a set of $(r + 1)$ matrices $(P_o, ..., P_r)$ satisfying $0 < P_j = P_j^t \; \forall \; j = 0, ..., r$, such that $\mathcal{P}(\sigma) := P_o + \sigma_1 P_1 + ... + \sigma_r P_r$ and*

$$A^t(\omega)\{\mathcal{P}^{-1}(\omega) - E(\omega)\mathcal{Q}^{-1}(\omega)E^t(\omega)\}^{-1}A(\omega)$$
$$-\mathcal{P}(\omega) + \mathcal{Q}(\omega) < 0 \qquad \forall \omega \in \mathcal{W} \qquad (9.184)$$
$$\mathcal{P}(\omega) > 0 \qquad \forall \quad \omega \in \mathcal{W} \qquad (9.185)$$
$$A_j^t \left\{ P_o^{-1} - E_o Q_o^{-1} E_o^t \right\}^{-1} A_j \geq 0, j = 1, ..., r \qquad (9.186)$$

When the LMI system (9.184)-(9.186) is feasible, a Lyapunov function for $(\Sigma_{\sigma o})$ and for all trajectories $\sigma(t)$ satisfying (9.169) is then given by

$$V(k, \sigma) = x^t(k)P(\sigma(k))x(k) + \sum_{j=k-\eta}^{k-1} x^t(j)Q(\sigma(j))x(j)$$

Proof: Set $P(\delta\sigma) = P_o$ in **Theorem 9.10**.

Definition 9.2: *System (Σ_σ) is said to be affinely quadratically stable (AQS) with disturbance attenuation γ if given a set of $(r + 1)$ matrices $(Q_o, ..., Q_r)$ such that $0 < Q_j = Q_j^t \; \forall \; j = 0, ..., r$ and $\mathcal{Q}(\sigma) := Q_o + \sigma_1 Q_1 + ... + \sigma_r Q_r > 0$ there exists a set of $(r + 1)$ matrices $(P_o, ..., P_r)$ and such that $0 < P_j = P_j^t \; \forall \; j = 0, ..., r$ and*

$$\mathcal{P}(\sigma) := P_o + \sigma_1 P_1 + ... + \sigma_r P_r > 0 \qquad (9.187)$$
$$A^t(\sigma)\mathcal{P}(\sigma + \delta\sigma)A(\sigma) - \mathcal{P}(\sigma) + L^t(\sigma)L(\sigma)$$
$$+A^t(\sigma)\mathcal{P}(\sigma + \delta\sigma)\mathcal{B}(\sigma, \delta\sigma)\mathcal{P}(\sigma, \delta\sigma)A(\sigma) + \mathcal{Q}(\sigma) < 0$$

$$(9.188)$$

hold for all admissible values and trajectories of the parameter vector $\sigma = (\sigma_1, ..., \sigma_r)$ *where*

$$\begin{aligned}
\mathcal{B}(\sigma, \delta\sigma) &= B(\sigma)[I - \gamma^{-2}B^t(\sigma)\mathcal{P}(\sigma + \delta\sigma)B(\sigma)]^{-1}B^t(\sigma) \\
&+ E(\sigma)Q^{-1}(\sigma)E^t(\sigma)
\end{aligned} \tag{9.189}$$

it then also follows that the function

$$V(x, \sigma) = x^t(k)\mathcal{P}(\sigma)x(k) + \sum_{\alpha=k-\eta}^{k-1} x^t(\alpha)\mathcal{Q}(\sigma)x(\alpha)$$

is a Lyapunov function for system (9.166).

   Theorem 9.11: *Consider system* $(\Sigma_{\sigma o})$ *where the matrix* $A(.)$ *depend affinely on* σ *in the manner of (9.167). Let* $\mathcal{W}, \mathcal{V}$ *denote the sets of corners of the parameter box (9.170) and the incremental variation box (9.171), respectively and let* $\bar{\sigma} = ([\check{\sigma}_1 + \hat{\sigma}_1]/2, ..., [\check{\sigma}_r + \hat{\sigma}_r]/2)$ *denote the average value of the parameter vector such that* $A(\bar{\sigma})$ *is stable. This system is affinely quadratically stable with disturbance attenuation* γ *if given a set of* $(r + 1)$ *matrices* $(Q_o, ..., Q_r)$ *such that* $0 < Q_j = Q_j^t \; \forall \; j = 0, ..., r$ *and* $\mathcal{Q}(\sigma) := Q_o + \sigma_1 Q_1 + ... + \sigma_r Q_r > 0$ *there exists a set of* $(r + 1)$ *matrices* $(P_o, ..., P_r)$ *such that* $0 < P_j = P_j^t \; \forall \; j = 0, ..., r$, *satisfying*

$$A^t(\omega)[\mathcal{P}(\omega) + \mathcal{P}(\nu) - P_o]A(\omega) - \mathcal{P}(\omega) + L^t(\omega)L(\omega)$$
$$+A^t(\omega)[\mathcal{P}(\omega) + \mathcal{P}(\nu) - P_o]\mathcal{B}(\omega, \nu)$$
$$[\mathcal{P}(\omega) + \mathcal{P}(\nu) - P_o]A(\omega) + \mathcal{Q}(\omega) < 0$$
$$\forall \; (\omega, \nu) \in \mathcal{W} \times \mathcal{V} \tag{9.190}$$
$$\mathcal{P}(\omega) > 0 \qquad \forall \omega \in \mathcal{W} \tag{9.191}$$
$$A_j^t\{P_o^{-1} - E_o Q_o^{-1} E_o^t - B_o$$
$$[I - \gamma^{-2}B_o^t P_o B_o]^{-1}B_o^t\}^{-1}A_j$$
$$\geq 0 \; j = 1, .., r \tag{9.192}$$

where

$$\mathcal{P}(\sigma) := P_o + \sigma_1 P_1 + ... + \sigma_r P_r \tag{9.193}$$

When the LMI system (9.190)-(9.192) is feasible, a Lyapunov function for $(\Sigma_{\sigma o})$ *and for all trajectories* $\sigma(k)$ *satisfying (9.168) and (9.169) is then given by* $V(k, \sigma) = x^t(k)\mathcal{P}(\sigma(k))x(k) + \sum_{\alpha=k-\eta}^{k-1} x^t(\alpha)\mathcal{Q}(\sigma)x(\alpha)$.

Proof: Follows by parallel development to **Theorem 9.10**.

A special case of **Theorem 9.11** when the σ-parameters are constants is presented below.

Corollary 9.2: *Consider system $(\Sigma_{\sigma o})$ where the matrix $A(.)$ depends affinely on constant parameters $\sigma \in \Re^r$ satisfying (9.168). Let $\mathcal{W}$ denotes the set of corners of the parameter box (9.170). This system is affinely quadratically stable with disturbance attenuation γ if given a set of $(r+1)$ matrices $(Q_o, ..., Q_r)$ such that $0 < Q_j = Q_j^t \; \forall \; j = 0, ..., r$ and $Q(\sigma) := Q_o + \sigma_1 Q_1 + ... + \sigma_r Q_r > 0$ there exists a set of $(r+1)$ matrices $(P_o, ..., P_r)$ satisfying $0 < P_j = P_j^t \; \forall \; j = 0, ..., r$, such that $\mathcal{P}(\sigma) := P_o + \sigma_1 P_1 + ... + \sigma_r P_r$ and*

$$A^t(\omega)\mathcal{P}(\omega)A(\omega) - \mathcal{P}(\omega) + L^t(\omega)L(\omega) + A^t(\omega)\mathcal{P}(\omega)\mathcal{B}(\omega, \nu)\mathcal{P}(\omega)A(\omega) +$$
$$\mathcal{Q}(\omega) < 0 \qquad \forall \, \omega \in \mathcal{W} \tag{9.194}$$
$$\mathcal{P}(\omega) > 0 \qquad \forall \, \omega \in \mathcal{W} \tag{9.195}$$
$$A_j^t \{ P_o^{-1} - E_o Q_o^{-1} E_o^t - B_o [I - \gamma^{-2} B_o^t P_o B_o]^{-1} B_o^t \}^{-1} A_j \; \geq 0$$
$$j = 1, .., r \tag{9.196}$$

When the LMI system (9.194)-(9.196) is feasible, a Lyapunov function for $(\Sigma_{\sigma o})$ and for all trajectories $\sigma(k)$ satisfying (9.168) is then given by $V(x, \sigma) = x^t(k)\mathcal{P}(\sigma)x(k) + \sum_{\alpha=k-\eta}^{k-1} x^t(\alpha)\mathcal{Q}(\sigma)x(\alpha)$.

Proof: Set $P(\nu) = P_o$ in **Theorem 9.11**.

Next, we proceed to closely examine the filtering problem for the class of polytopic LPV systems described by (9.166)-(9.72) using an $\mathcal{H}_\infty$-setting.

9.5.3   Robust H_∞ Filtering

The filtering problem we address in this paper is as follows:
Given system (9.166), design a linear parameter-dependent filter that provides an estimate, $\hat{z}(t)$, of $z(t)$ based on $\{y(\tau), 0 \leq \tau \leq t\}$ such that the estimation error system is quadratically stable $\forall w(k) \in \ell_2[0, \infty)$

$$\|z - \hat{z}\|_2 \leq \gamma \|w\|_2$$

where $\gamma > 0$ is a given scalar which specifies the level of noise attenuation in the estimation error.

Attention will be focused on the design of an n-th order filter. In the absence of $w(k)$, it is required that $\|x(k) - \hat{x}(k)\|_2 \to 0$, $k \to \infty$ where $\hat{x}(k)$ is the state of the filter. The linear parameter-dependent filter adopted in this work is given by:

$$\begin{aligned}
\hat{x}(k+1) &= A(\sigma)\hat{x}(k) + K(\sigma)\{y(k) - C(\sigma)\hat{x}(k)\} \\
\hat{z}(k) &= L(\sigma)\hat{x}(k); \quad \hat{x}(0) = 0
\end{aligned} \tag{9.197}$$

where $A(.), C(.), L(.)$ are given by (9.167)-(9.168) and $K(\sigma)$ is the Kalman gain matrix to be determined. By defining $\tilde{x}(k) = x(k) - \hat{x}(k)$ and augmenting systems (9.166) and (9.197), it follows that the estimation error, $e(k) = z(k) - \hat{z}(k)$, can be represented by the state-space model:

$$\begin{aligned}
\xi(k+1) &= [x^t(k+1) \quad \tilde{x}^t(k+1)]^t \in \Re^{2n} \\
&= A_a(\sigma)\xi(k) + B_a(\sigma)w(k) + E_a(\sigma)\xi(k-\eta) \\
e(k) &= L_a(\sigma)\xi(k)
\end{aligned} \tag{9.198}$$

where

$$A_a(\sigma) = \begin{bmatrix} A(\sigma) & 0 \\ 0 & A(\sigma) - K(\sigma)C(\sigma) \end{bmatrix}$$

$$E_a(\sigma) = \begin{bmatrix} E(\sigma) & 0 \\ E(\sigma) & 0 \end{bmatrix}, \quad L_a(\sigma) = [0 \quad L(\sigma)]$$

$$B_a(\sigma) = \begin{bmatrix} B(\sigma) \\ B(\sigma) - K(\sigma)D(\sigma) \end{bmatrix}, \quad \xi(0) = \begin{bmatrix} x_o \\ x_o \end{bmatrix}$$

The main result is then summarized by the following theorem.

Theorem 9.12: *Consider system* (Σ_σ) *where* $\sigma(k)$ *is a time-varying parameter satisfying (9.168)-(9.169), let* $\gamma > 0$ *be a given scalar and given affine matrix* $0 < Q = Q^t \in \Re^{2n \times 2n}$ *with* $Q = diag[Q_1 \quad Q_2]$. *Then there exists a linear parameter-dependent filter*

$$\begin{aligned}
\hat{x}(k+1) &= A(\omega)\hat{x}(k) + T(\omega,\nu)S^{-1}(\omega,\nu)\{y(k) - C(\omega)\hat{x}(k)\} & (9.199) \\
\hat{z}(t) &= L(\omega)\hat{x}(t) \quad\quad\quad \forall\,(\omega,\nu) \in \mathcal{W} \times \mathcal{V} & (9.200) \\
T(\omega,\nu) &= A(\omega)\mathcal{Z}(\omega,\nu)E(\omega)Q_1^{-1}(\omega,\nu)E^t(\omega)\mathcal{X}(\omega,\nu)A^t(\omega) \\
&\quad + B^t(\omega)B(\omega) & (9.201) \\
S(\omega,\nu) &= D(\omega)B^t(\omega) + C(\omega)\mathcal{Z}(\omega,\nu)E(\omega)Q_1^{-1}(\omega,\nu)E^t(\omega)\mathcal{X}(\omega,\nu)A^t(\omega) \\
& & (9.202)
\end{aligned}$$

such that the estimation error is affinely quadratically stable and $\|z - \hat{z}\|_2 < \gamma \|w\|_2 \; \forall w(k) \in \ell_2$ *if there exist affine matrices* $\mathcal{X}(\omega, \nu)$ *and* $\mathcal{Z}(\omega, \nu)$ *satisfying the following LMIs:*

$$
\begin{bmatrix}
-\mathcal{X}(\omega,\nu) + \mathcal{Q}_1(\omega) & & \\
-A(\omega)\mathcal{X}(\omega,\nu)A^t(\omega) & B(\omega) & A(\omega)\mathcal{X}(\omega,\nu)E(\omega) \\
B^t(\omega) & -I & 0 \\
A^t(\omega)\mathcal{X}(\omega,\nu)(\omega)E^t(\omega) & 0 & -\mathcal{Q}_1(\omega)
\end{bmatrix} < 0
$$
$$
\forall \; (\omega,\nu) \in \mathcal{W} \times \mathcal{V} \tag{9.203}
$$

$$
\begin{bmatrix}
-\mathcal{Z}(\omega,\nu) + \mathcal{Q}_2(\omega) & & \\
-\bar{A}(\omega)\mathcal{Z}(\omega,\nu)\bar{A}^t(\omega) & \check{A}(\omega) & \breve{A}(\omega) \\
\bar{B}(\omega)\bar{B}^t(\omega) & & \\
\check{A}^t(\omega,\nu) & -\mathcal{Q}_1(\omega) & 0 \\
\breve{A}^t(\omega,\nu) & 0 & -\gamma^2 I + \bar{L}(\omega,\nu)
\end{bmatrix} < 0
$$
$$
\forall \; (\omega,\nu) \in \mathcal{W} \times \mathcal{V} \tag{9.204}
$$

$$
A_j\{X_o^{-1} - E_o Q_1^{-1} E_o^t - B_o
$$
$$
[I - \gamma^{-2} B_o^t X_o B_o]^{-1} B_o^t\}^{-1} A_j^t
$$
$$
\geq 0 \qquad j = 1,..,r \tag{9.205}
$$

where

$$
\begin{align}
\bar{A}(\omega,\nu) &= A(\omega) - T(\omega,\nu)\mathcal{S}^{-1}(\omega,\nu)C(\omega) \notag \\
\check{A}(\omega,\nu) &= \bar{A}(\omega,\nu)\mathcal{Z}(\omega,\nu)E(\omega) \tag{9.206} \\
\breve{A}(\omega) &= \bar{A}(\omega,\nu)\mathcal{Z}(\omega,\nu)L^t(\omega) \tag{9.207} \\
\bar{B}(\omega,\nu) &= B(\omega) - T(\omega,\nu)\mathcal{S}^{-1}(\omega,\nu)D(\omega) \tag{9.208} \\
\bar{L}(\omega,\nu) &= L(\omega)\mathcal{Z}(\omega,\nu)L^t(\omega) \tag{9.209}
\end{align}
$$

and $\mathcal{X}(\omega,\nu) = \mathcal{X}(\omega) + \mathcal{X}(\nu) - X_o$, $\mathcal{Z}(\omega,\nu) = \mathcal{X}(\omega) + \mathcal{Z}(\nu) - Z_o$.

Proof: By **Definition 9.2**, system (9.198) is AQS with disturbance attenuation γ if given an affine matrix $0 < \mathcal{Q}(.) = \mathcal{Q}^t(.) \in \Re^{2n}$ there exists an affine matrix $0 < \mathcal{P}(.) = \mathcal{P}^t(.) \in \Re^{2n}$ satisfying the matrix inequality:

$$
A_a^t(\sigma)\mathcal{P}(\sigma + \delta\sigma)A_a(\sigma) - \mathcal{P}(\sigma)
$$
$$
+A_a^t(\sigma)\mathcal{P}(\sigma + \delta\sigma)\bar{B}_a(\sigma + \delta\sigma)\mathcal{P}(\sigma + \delta\sigma)A_a(\sigma)
$$
$$
+L_a^t(\sigma)L_a(\sigma) + \mathcal{Q}_a < 0 \tag{9.210}
$$

where

$$\bar{B}_a(\sigma, \delta\sigma) =$$
$$B_a(\sigma)[I - \gamma^{-2}B_a^t(\sigma)\mathcal{P}(\sigma + \delta\sigma)B_a(\sigma)]^{-1}B_a^t(\sigma)$$
$$+E_a(\sigma)\mathcal{Q}_a^{-1}(\sigma)E_a^t(\sigma) \tag{9.211}$$

From the results of [39], inequality (9.210) holds if and only if there exists a matrix $0 < \mathcal{Y} = \mathcal{Y}^t \in \Re^{2n}$ satisfying the matrix inequality:

$$\begin{aligned}
\Xi(\sigma + \delta\sigma) &:= A_a(\sigma)\mathcal{Y}(\sigma + \delta\sigma)A_a^t(\sigma) \\
&+ A_a(\sigma)\mathcal{Y}(\sigma + \delta\sigma)\bar{\mathcal{L}}_a(\sigma, \delta\sigma)\mathcal{Y}(\sigma, \delta\sigma)A_a^t(\sigma) \\
&- \mathcal{Y}(\sigma) + B_a(\sigma)B_a^t(\sigma) + \mathcal{Q}_a \\
&< 0
\end{aligned} \tag{9.212}$$

where

$$\bar{\mathcal{L}}_a(\sigma, \delta\sigma) = L_a(\sigma)[I - \gamma^{-2}L_a(\sigma)\mathcal{Y}(\sigma + \delta\sigma)L_a^t(\sigma)]^{-1}L_a^t(\sigma)$$
$$+E_a(\sigma)\mathcal{Q}_a^{-1}(\sigma)E_a^t(\sigma) \tag{9.213}$$

Define

$$\Xi(\sigma + \delta\sigma) = \begin{bmatrix} \Xi_1(\sigma + \delta\sigma) & \Xi_2(\sigma + \delta\sigma) \\ \Xi_2^t(\sigma + \delta\sigma) & \Xi_3(\sigma + \delta\sigma) \end{bmatrix}$$
$$\mathcal{Y}(\sigma + \delta\sigma) = \begin{bmatrix} \mathcal{X}(\sigma + \delta\sigma) & 0 \\ 0 & \mathcal{Z}(\sigma + \delta\sigma) \end{bmatrix} \tag{9.214}$$

Expanding (9.212) using (9.214), we get:

$$\Xi_1(\sigma + \delta\sigma) = A(\sigma)\mathcal{X}(\sigma + \delta\sigma)A^t(\sigma) - \mathcal{X}(\sigma) + B(\sigma)B^t(\sigma)$$
$$A(\sigma)\mathcal{X}(\sigma + \delta\sigma)E(\sigma)\mathcal{Q}_a^{-1}(\sigma)E^t(\sigma)\mathcal{X}(\sigma + \delta\sigma)A^t(\sigma) + \mathcal{Q}_1 \tag{9.215}$$

$$\Xi_3(\sigma + \delta\sigma) = \{A(\sigma) - K(\sigma + \delta\sigma)C(\sigma)\}\mathcal{Z}(\sigma + \delta\sigma)$$
$$\{A^t(\sigma) - C^t(\sigma)K^t(\sigma + \delta\sigma)\}$$
$$+ \{B(\sigma) - K(\sigma + \delta\sigma)D(\sigma)\}\{B^t(\sigma) - D^t(\sigma)K^t(\sigma + \delta\sigma)\} +$$
$$\{A(\sigma) - K(\sigma + \delta\sigma)C(\sigma)\}\mathcal{Z}(\sigma + \delta\sigma)\bar{\mathcal{E}}(\sigma + \delta\sigma)\mathcal{Z}(\sigma + \delta\sigma)$$
$$\{A^t(\sigma) - C^t(\sigma)K^t(\sigma + \delta\sigma)\} \tag{9.216}$$

$$\Xi_2(\sigma + \delta\sigma) = B(\sigma)\{B^t(\sigma) - D^t(\sigma)K^t(\sigma + \delta\sigma)\} +$$
$$A(\sigma)X(\sigma + \delta\sigma)E(\sigma)\mathcal{Q}_a^{-1}(\sigma)E^t(\sigma)\mathcal{Z}(\sigma + \delta\sigma)$$
$$\{A^t(\sigma) - C^t(\sigma)K^t(\sigma + \delta\sigma)\}$$

$$\tag{9.217}$$

where

$$\bar{\mathcal{E}}(\sigma + \delta\sigma) = L^t|\sigma)(\gamma^{-2}I - L(\sigma)\mathcal{Z}(\sigma + \delta\sigma)L^t(\sigma)]^{-1}L(\sigma)$$
$$+ \quad E(\sigma)Q_a^{-1}(\sigma)E^t(\sigma) \qquad (9.218)$$

It is well known that necessary and sufficient conditions for $\Xi(\sigma + \delta\sigma) < 0$ are $\Xi_1(\sigma + \delta\sigma) < 0$, $\Xi_3(\sigma + \delta\sigma) < 0$ and $\Xi_2(\sigma + \delta\sigma) = 0$. Enforcing $\Xi_2(\sigma + \delta\sigma) = 0$ in (9.218) yields:

$$K(\sigma + \delta\sigma) = T(\sigma + \delta\sigma)S^{-1}(\sigma + \delta\sigma) , \ \forall \ (\sigma, \delta\sigma) \in \Im \times \Im \qquad (9.219)$$

as the desired Kalman gain where $T(.,.)$ and $S(.,.)$ are given by (9.201)-(9.202). Applying **Theorem 2**, it follows from (9.215)-(9.216) that the conditions $\Xi_1(\sigma + \delta\sigma) < 0$ and $\Xi_3(\sigma + \delta\sigma) < 0$ yields the LMIs (9.203)-(9.204) plus the multiconvexity requirement (9.205).

A special case of **Theorem 9.12** when the σ-parameters are constants is presented below.

Corollary 9.3: *Consider system (Σ_σ) where $\sigma(k)$ is a constant parameter satisfying (9.168), let $\gamma > 0$ be a given scalar and given affine matrix $0 < Q = Q^t \in \Re^{2n \times 2n}$ with $Q = diag[Q_1 \quad Q_2]$. Then there exists a linear parameter-dependent filter*

$$\hat{x}(k+1) = A(\omega)\hat{x}(k) + T(\omega)S^{-1}(\omega)\{y(k) - C(\omega)\hat{x}(k)\} \qquad (9.220)$$
$$\hat{z}(t) = L(\omega)\hat{x}(t) , \qquad \forall \ \omega \in \mathcal{W} \qquad (9.221)$$
$$T(\omega) = B^t(\omega)B(\omega) + A(\omega)\mathcal{Z}(\omega)E(\omega)Q_1^{-1}(\omega)E^t(\omega)\mathcal{X}(\omega)A^t(\omega)$$
$$S(\omega) = D(\omega)B^t(\omega) + C(\omega)\mathcal{Z}(\omega)E(\omega)Q_1^{-1}(\omega)E^t(\omega)\mathcal{X}(\omega)A^t(\omega)$$
$$(9.222)$$

such that the estimation error is affinely quadratically stable and $\|z - \hat{z}\|_2 < \gamma\|w\|_2 \ \forall w(k) \in \ell_2$ if there exist affine matrices $\mathcal{X}(\omega)$ and $\mathcal{Z}(\omega)$ satisfying the following LMIs:

$$\begin{bmatrix} -\mathcal{X}(\omega) + Q_1(\omega) \\ -A(\omega)\mathcal{X}(\omega)A^t(\omega) & B(\omega) & A(\omega)\mathcal{X}(\omega)E(\omega) \\ B^t(\omega) & -I & 0 \\ A^t(\omega)\mathcal{X}(\omega)(\omega)E^t(\omega) & 0 & -Q_1(\omega) \end{bmatrix} < 0$$
$$\forall \ \omega \in \mathcal{W} \qquad (9.223)$$

$$\begin{bmatrix} -\mathcal{Z}(\omega) + \mathcal{Q}_2(\omega) + \bar{B}(\omega)\bar{B}^t(\omega) & \check{A}(\omega) & \check{A}(\omega) \\ \quad -\bar{A}(\omega)\mathcal{Z}(\omega)\bar{A}^t(\omega) & & \\ \check{A}^t(\omega) & -\mathcal{Q}_1(\omega) & 0 \\ \check{A}^t(\omega) & 0 & -\gamma^2 I + \bar{L}(\omega) \end{bmatrix}$$
$$< 0, \qquad \forall\ \omega \in \mathcal{W} \qquad\qquad (9.224)$$

$$A_j\{X_o^{-1} - E_o \mathcal{Q}_1^{-1} E_o^t - B_o[I - \gamma^{-2} B_o^t X_o B_o]^{-1} B_o^t\}^{-1} A_j^t$$
$$\geq 0 \qquad j = 1, .., r \qquad\qquad (9.225)$$

where

$$\begin{aligned} \bar{A}(\omega) &= A(\omega) - T(\omega)S^{-1}(\omega)C(\omega) \\ \check{A}(\omega) &= \bar{A}(\omega)\mathcal{Z}(\omega)E(\omega) & (9.226) \\ \check{A}(\omega) &= \bar{A}(\omega)\mathcal{Z}(\omega)L(\omega) & (9.227) \\ \bar{B}(\omega) &= B(\omega) - T(\omega)S^{-1}(\omega)D(\omega) & (9.228) \\ \bar{L}(\omega) &= L(\omega)\mathcal{Z}(\omega)L^t(\omega) & (9.229) \end{aligned}$$

Proof: Set $\mathcal{X}(\nu) = X_o$ and $\mathcal{Z}(\nu) = Z_o$ in **Theorem 9.12.**

9.6 Simulation Example

9.6.1 Example 9.1

Consider a discrete-time delay system of the type (9.121) with

$$\begin{aligned} A &= \begin{bmatrix} 0.67 & 0.087 \\ 0 & 1.105 \end{bmatrix}, \; A_d = \begin{bmatrix} 0.2 & 0 \\ 0 & 0.2 \end{bmatrix}, \; D = \begin{bmatrix} 0.096 \\ 0.316 \end{bmatrix} \\ C &= [0.5 \; 0.5], \; H_o = [1 \; 2], \\ N &= 1, \; H = \begin{bmatrix} 0.1 \\ 0.2 \end{bmatrix}, \; H_c = 0.4, \; E = [0.2 \; 0.3] \end{aligned}$$

Our purpose is to provide a numerical illustration of **Theorem 9.9.** First we select $\mathcal{Q}_1$ and $\mathcal{Q}_2$ as

$$\mathcal{Q}_1 = \begin{bmatrix} 5 & 0 \\ 0 & 5 \end{bmatrix}, \; \mathcal{Q}_2 = \begin{bmatrix} 2 & 0 \\ 0 & 2 \end{bmatrix}$$

Then we expand the LMIs (9.144)-(9.145) into nonstandard algebraic Riccati inequalities and solve them using a sequential computational scheme. This

scheme is initialized by dropping out the additional terms thereby obtaining
standard ARIs. By solving the resulting ARIs, one gets an initial feasible solution. Subsequently, by injecting the solutions continuously into the actual
algebraic inequalities, it has been found that a satisfactory feasible solution
can be obtained after few iterations. In one case with an accuary of 10^{-5},
the result of computations are:

$$
S_1 = \begin{bmatrix} 9.0601 & -0.4380 \\ -0.4380 & 13.0666 \end{bmatrix}, \quad
S_2 = \begin{bmatrix} 140.5205 & 10.6250 \\ 10.6250 & 125.4550 \end{bmatrix}
$$

for $\mu = 0.25$ where the associated matrices are given by:

$$
\mathcal{R}_1 = \begin{bmatrix} 1.1417 & -1.0963 \\ -1.0963 & 1.1733 \end{bmatrix}, \quad
\mathcal{R}_2 = \begin{bmatrix} 1.3145 & -0.4566 \\ -0.4566 & 0.9339 \end{bmatrix},
$$

$$
\mathcal{R}_3 = \begin{bmatrix} 1.2225 & -1.0433 \\ -1.0433 & 0.7652 \end{bmatrix}, \quad
\hat{A} = \begin{bmatrix} 0.3237 & 0.6978 \\ -1.3861 & -0.2980 \end{bmatrix},
$$

$$
T^t = \begin{bmatrix} -9.6089 \\ -8.7517 \end{bmatrix}, \quad
\hat{C}^t = \begin{bmatrix} -0.2041 \\ -0.1840 \end{bmatrix}, \quad
\mathcal{Z} = -25.5851
$$

Hence, from (9.146) the $\mathcal{H}_\infty$ estimator is described by:

$$
\begin{aligned}
\hat{x}(k+1) &= \begin{bmatrix} 0.3237 & 0.6978 \\ -1.3861 & -0.2980 \end{bmatrix} \hat{x}(k) \\
&+ \begin{bmatrix} 0.3756 \\ 0.3421 \end{bmatrix} [y(k) - [-0.2041 \quad -0.1840]\hat{x}(k)]
\end{aligned}
$$

9.7   Notes and References

Admittedly, the available results on robust $\mathcal{H}_\infty$−filtering for time-delay systems are few. The material reported in this chapter was basically delay-
independent. Some potential research areas include developing delay-dependent
filter results, exploring the possibility of reducing the computational load of
the filter, treating linear parameter-varying continuous systems with delay
and investigating nonlinear systems. In addition, some of the ideas that are
worth deep examination are found in [14-18, 45].

Chapter 10

Interconnected Systems

10.1   Introduction

One of the fundamental problems of signal and systems theory is the estimation of state-variables of a dynamic system (filtering) using available (past) noisy measurements. The celebrated Kalman filtering approach [3,4] is, by now, deeply entrenched in the control literature and offers the best filter algorithm based on the minimization of the variance of the estimation error. This type of estimation relies on knowledge of a perfect dynamic model for the signal generation system and the fact that power spectral density of the noise is known. In many cases, however, only an approximate model of the system is available. In such situations, it has been known that the standard Kalman filtering methods fail to provide a guaranteed performance in the sense of the error variance. Considerable interests have been subsequently devoted to the design of estimators that provide an upper bound to the error variance for any allowable modeling uncertainty [10-12,14-17]. These filters are referred to as *robust filters* and can be regarded as an extension of the standard Kalman filter to the case of uncertain systems. An important class of robust filters is the one that employs the $\mathcal{H}_\infty$−norm as a performance measure. In $\mathcal{H}_\infty$−filtering, the noise sources are arbitrary signals with bounded energy which is appropriate when there is significant uncertainty in the power spectral density of the exogenous signals [18].

On another front of systems research, interconnected uncertain systems are receiving growing interest since they reflect numerous practical situations. Most of the time, the problems of decentralized stability, stabilization and control design have been the main concern [13,40,41]. Recently, interests have been shifted to classes of interconnected uncertain systems where state-delay occurs within the subsystem [42]. The problem of decentralized robust filtering seems to have been overlooked despite its importance in control engineering. This chapter attempts to bridge this gap by considering the problem of decentralized robust filtering for a class of interconnected nonlinear uncertain delay systems. The nonlinearities are unknown cone-bounded and state-dependent, the uncertainties are real unknown and time-varying but norm-bounded and the state-delay is unknown. We adopt a worst-case approach to the filter design [4] based on an unknown initial state and subsystem measurements. It is important to emphasize that the obtained results here complement those of Chapter 7 on the control and stabilization.

10.2   Problem Statement

Consider a class of nonlinear systems (Σ_Δ) composed of n_s coupled subsystems (Σ_{Δ_j}) and modeled in state-space form by

$$
\begin{aligned}
\dot{x}_j(t) \; =& \; [A_j + \Delta A_j(t)]x_j(t) + B_j(t)w_j(t) \\
+& \; [A_{dj} + \Delta A_{dj}(t)]x_j(t - \tau_j) \\
+& \; \sum_{k=1}^{n_s}\{[G_{jk}(t) + \Delta G_{jk}(t)]g_{jk}[x_k]\} \quad (10.1) \\
y_j(t) \; =& \; [C_j + \Delta C_j(t)]x_j(t) + D_j(t)w_j(t) \\
+& \; [M_j + \Delta M_j(t)]m_j[x_j] \quad\quad\quad\quad (10.2) \\
z_j(t) \; =& \; L_j x_j(t) \quad\quad\quad\quad\quad\quad\quad\quad\quad (10.3)
\end{aligned}
$$

where $\forall\; j \in \{1,...,n_s\}$; $x_j(t) \in \Re^{n_j}$ is the state ; $w_j(t) \in \Re^{q_j}$ is the input noise which belongs to $\mathcal{L}_2[0,\infty)$; $y_j(t) \in \Re^{p_j}$ is the measured output; $z_j(t) \in \Re^{r_j}$ is a linear combination of state variables to be estimated, and τ_j is a unknown constant time-delay. The initial condition can be generally specified as $\alpha_o \equiv \langle x(0), \phi(s)\rangle$ where $\phi(.) \in \mathcal{L}_2[0,\infty)$, but it will be considered unknown throughout this work. The matrices $A_j, B_j, C_j, D_j, A_{dj}, M_j, G_{jk}$ and L_j are real and constants of appropriate dimensions but the system matrices $\Delta A_j(t), \Delta G_{jk}(t), \Delta A_{dj}(t), \Delta C_j(t), \Delta M_j(t)$ are uncertain (possibly fast time-varying) and are assumed to be given by:

$$
\begin{bmatrix} \Delta A_j(t) \\ \Delta C_j(t) \end{bmatrix} = \begin{bmatrix} H_{1j} \\ H_{2j} \end{bmatrix} \Delta_{1j}(t)E_{1j} \quad\quad (10.4)
$$

$$
\Delta G_{jk}(t) = H_{jk}\Delta_{jk}(t)E_{jk} \quad\quad\quad\quad (10.5)
$$

$$\Delta A_{dj}(t) = H_{3j}\Delta_{2j}(t)E_{2j} \tag{10.6}$$
$$\Delta M_j(t) = H_{4j}\Delta_{3j}(t)E_{3j} \tag{10.7}$$

where $\{H_{1j}, ..., E_{3j}\}$ are known constant matrices and $\{\Delta_{1j}, \Delta_{jk}, \Delta_{2j}, \Delta_{3j}\}$ are unknown real matrix functions satisfying

$$\Delta_{1j}^t\Delta_{1j} \leq I, \quad \Delta_{2j}\Delta_{2j}^t \leq I$$
$$\Delta_{3j}\Delta_{3j}^t \leq I, \quad \Delta_{jk}\Delta_{jk}^t \leq I \tag{10.8}$$

The mappings $g_{jk} : \Re^{n_k} \to \Re^{\lambda_j}$ and $h_j : \Re^{n_j} \to \Re^{\sigma_j}$ are respectively unknown coupling and local nonlinearities satisfying the following assumption:

Assumption 10.1: There exists known real constant matrices W_{jk} and W_j, $\forall j, k \in 1, ..., n_s$ such that $\forall x_j(t) \in \Re^{n_j}, v_j(t) \in \Re^{n_j}$

$$\|g_{jk}[v_j]\| \leq \|W_{jk}v_j\|, \quad \|m_j[x_j]\| \leq \|W_j x_j\|$$

The problem of interest can be phrased as follows:

Given n_s-pairs $(0 < \gamma_j; 0 < R_j = R_j^t), \forall j$, find linear causal filters $P_j :$ $y_j \to \hat{z}_j, \forall j \in \{1, .., n_s\}$ such that the filtering error $e_j = z_j - \hat{z}_j$ is globally asymptotically stable and

$$\sum_{j=1}^{n_s}\|e_j\|_2 < \gamma_j \sum_{j=1}^{n_s}\{\|w_j\|_2^2 + x_j^t(0)R_j x_j(0)$$
$$+ \int_{-\tau}^0 x_j^t(s)R_j x_j(s)ds\}^{1/2} \tag{10.9}$$

for any nonzero $(\alpha_o, w) \in \Re^n \oplus \mathcal{L}_2[0, \infty)$ and for all admissible uncertainties.

Remark 10.1: Note that the function (10.9) can be viewed as a generalization of the usual H_∞-filtering performance measure to account for unknown initial states where R_j provides a measure of the uncertainty in the initial state α_{jo} relative to the uncertainty in w_j.

We close this section by establishing a version of the strict bounded real lemma (see Appendix A) for a class of time-delay systems. Consider the following time-varying system (Σ_t):

$$\dot{\zeta}(t) = A(t)\zeta(t) + B(t)w(t) + A_d(t)x(t - \tau), \tag{10.10}$$
$$\eta(t) = C(t)\zeta(t) \tag{10.11}$$

where $\zeta(t) \in \Re^n$ is the state, $\zeta_o \equiv \langle\zeta(0), \phi(s)\rangle$; $\phi(.) \in \mathcal{L}_2[-\tau, 0]$ is an unknown initial state, $w(t) \in \Re^m$ is the input, $\eta(t) \in \Re^r$ is the output and the matrices $A(t), B(t), C(t)$ are real piecewise-continuous and bounded. Here τ is an unknown constant time-delay. Associated with system (Σ_t) is the worst-case performance measure:

$$J(\eta, w, \zeta_o, R) = \sup \left\{ \frac{\|\eta\|_2^2}{\|w\|_2^2 + \zeta_o^t R \zeta_o} \right\}^{1/2}$$

$$\zeta_o^t R \zeta_o \equiv \zeta^t(0) R \zeta(0) + \int_{-\tau}^0 \zeta^t(s) R \zeta(s) ds$$

$$0 \neq (\zeta_o, w) \in \Re^n \oplus L_2[0, \infty) \tag{10.12}$$

where $0 < R = R^t$ is a weighting matrix for the initial state $\langle\zeta(0), \phi(s)\rangle$. The following result holds:

Lemma 10.1: *Given system (Σ_t) and a scalar $\gamma > 0$, the system is exponentially stable and $J(\eta, w, \zeta_o, R) < \gamma$ if either of the following conditions holds:*

(1) There exists a bounded matrix function $0 \leq Q(t) = Q^t(t)$, $\forall t \in [0, \infty)$, such that for some $0 < \Gamma = \Gamma^t < \gamma^2 R$

$$\dot{Q} + A^t Q + QA + Q(\gamma^{-2} BB^t + E\Gamma^{-1} E^t)Q$$
$$+C^t C + \Gamma = 0; \quad Q(0) < \gamma^2 R \tag{10.13}$$

and the system $\dot{x}(t) = [A + (\gamma^{-2} BB^t + A_d \Gamma^{-1} A_d^t)Q]x(t)$ is exponentially stable.

(2) There exists a bounded matrix function $0 < S(t) = S^t(t)$, $\forall t \in [0, \infty)$, satisfying the differential inequality

$$\dot{S} + A^t S + SA + S(\gamma^{-2} BB^t + A_d \Gamma^{-1} A_d^t)S$$
$$+C^t C + \Gamma < 0, \quad S(0) < \gamma^2 R \tag{10.14}$$

for some $0 < \Gamma = \Gamma^t < \gamma^2 R$.

Proof: Introduce a Lyapunov-Krasovskii functional for system (Σ_t):

$$V_o(x_t) = \zeta^t(t) S(t) \zeta(t) + \int_{t-\tau}^t \zeta^t(\alpha) W \zeta(\alpha) d\alpha \tag{10.15}$$

$$0 < S(t) = S(t)^t \in \Re^{n \times n} \ \forall \ t; \ \ 0 < \Gamma = \Gamma^t \in \Re^{n \times n}$$

Differentiating (10.15) along the trajectories of system (Σ_t) with $w \equiv 0$, we get:

$$\frac{d}{dt} V_o(x_t) = \chi^t(t) \Pi(S) \chi(t)$$

$$\Pi(s) = \begin{bmatrix} \dot{S} + SA + A^t S + \Gamma & SA_d \\ A_d^t S & -\Gamma \end{bmatrix}$$

$$\chi(t) = [\zeta^t(t) \ \zeta^t(t-\tau)]^t \tag{10.16}$$

By **A.1**, inequality (10.14) implies that $\frac{d}{dt} V_o(x_t) < 0$ whenever $[\zeta(t) \ \zeta(t-\tau)] \neq 0$. That is, the system is uniformly asymptotically stable.

To show that $\|\eta\|_2 < \gamma \{\|w\|_2^2 + \zeta^t(0)R\zeta(0) + \int_{-\tau}^{0} \zeta^t(s)R\zeta(s)ds\}^{1/2}$, we introduce

$$J_o = \int_0^\infty \{z^t z - \gamma^2 w^t w\} dt$$
$$- \gamma^2 \{\zeta^t(0)R\zeta(0) + \int_{-\tau}^0 \zeta^t(s)R\zeta(s)ds\} \tag{10.17}$$

By using (10.10)-(10.12) and completing the squares in (10.17), it follows that

$$J_o = \int_0^\infty \{\zeta^t C^t C \zeta + \frac{d}{dt}\zeta^t S\zeta - \gamma^2 w^t w\} dt$$
$$+ \zeta^t(0)S(0)\zeta(0) - \gamma^2 \zeta^t(0)R\zeta(0)$$
$$- \gamma^2 \int_{-\tau}^0 x^t(s)Rx(s)ds$$
$$= \int_0^\infty \left\{\zeta^t[\dot{S} + SA + A^t S]\zeta - \gamma^2 w^t w\right\} dt$$
$$+ \int_0^\infty \{w^t B^t S\zeta + \zeta^t SBw\} dt$$
$$+ \int_0^\infty \left\{\zeta^t SA_d\zeta(t-\tau) + \zeta^t(t-\tau)A_d^t S\zeta\right\} dt$$
$$= \int_0^\infty \zeta^t N(S)\zeta dt - \zeta^t(0)[\gamma^2 R - S(0)]\zeta(0)$$
$$- \int_0^\infty \{\gamma^2 N_2^t N_2 + N_3^t \Gamma^{-1} N_3\} dt$$
$$- \int_{-\tau}^0 \zeta^t(s)[\gamma^2 R - \Gamma]\zeta(s)dt \tag{10.18}$$

where

$$
\begin{aligned}
N(S) &= \dot{S} + A^t S + SA + C^t C + \Gamma \\
&+ S(\gamma^{-2} BB^t + A_d \Gamma^{-1} A_d^t) S \tag{10.19} \\
N_2 &= [w - \gamma^{-2} B^t S \zeta] \tag{10.20} \\
N_3 &= [\Gamma \zeta(t - \tau) - A_d^t S \zeta] \tag{10.21}
\end{aligned}
$$

The condition $J_o < 0$ is implied by inequality (10.14) $\forall\, t\, [0, \infty)$. Therefore, we conclude that $\|\eta\|_2 < \gamma \{\|w\|_2^2 + \zeta^t(0) R \zeta(0) + \int_{-\tau}^0 \zeta^t(s) R \zeta(s) ds\}^{1/2}$ for any nonzero $(\zeta_o, w) \in \Re^n \oplus L_2[0, \infty)$.

Finally, by **A.3.1** it follows that the existence of a matrix $0 < S = S^t \in \Re^{n \times n}$ satisfying inequality (10.14) is equivalent to the existence of a stabilizing solution $0 \leq Q = Q^t \in \Re^{n \times n}$ to the ARE (10.13).

Remark 10.2: In the case of time-invariant systems, the matrix functions $S(t)$ will be replaced by a constant matrix $0 < S = S^t$ [14]. When the initial state ζ_o is known to be zero, the performance measure (10.11) reduces to the usual H_∞-performance measure

$$
J = \sup_{0 \neq w \in L_2[0, \infty)} \left\{ \frac{\|\eta\|_2}{\|w\|_2} \right\}
$$

10.3 H_∞ Performance Analysis

In this section, we will establish an interconnection between the robust performance analysis problem of system (10.1)-(10.3) and the H_∞-performance analysis of the interconnected system

$$
\begin{aligned}
\dot{\xi}_j(t) &= A_j \xi_j(t) + [B_j \ \ \gamma B_j(\mu_j, \sigma_j, \lambda_{jk})] \bar{w}_j(t) \tag{10.22} \\
\bar{z}_j(t) &= \begin{bmatrix} L_j \\ C_j(\mu_j) \end{bmatrix} \xi_j(t) \tag{10.23}
\end{aligned}
$$

where $j \in \{1, .., n_s\}, \xi_j \in \Re^{n_j}$ is the state; ξ_{jo} is an unknown initial state, $\bar{w}_j(t) \in \Re^{q_j + \alpha_j}$ is the input noise which belongs to $L_2[0, \infty)$ and $\bar{z}_j(t) \in \Re^{r_j + \beta_j}$ is the output. The matrices A_j, B_j, L_j are the same as in (10.1)-(10.3) and

$$
B_j(\mu_j, \sigma_j, \lambda_{jk}) B_j^t(\mu_j, \sigma_j, \lambda_{jk}) =
$$

$$\mu_j H_{1j} H_{1j}^t + \sigma_j H_{3j} H_{3j}^t + A_{dj}(Q_j - \sigma_j^{-1} E_{2j}^t E_{2j})^{-1} A_{dj}^t +$$

$$\sum_{k=1}^{n_s} \{G_{jk}(I - \lambda_{jk}^{-1} E_{jk}^t E_{jk})^{-1} G_{jk}^t + \lambda_{jk} H_{jk} H_{jk}^t \} \tag{10.24}$$

$$C_j^t(\mu_j) C_j(\mu_j) = (\sum_{k=1}^{n_s} W_{kj}^t W_{kj}) + \mu_j^{-1} E_{1j}^t E_{1j} \tag{10.25}$$

where $H_{1j}, E_{1j}, E_{2j}, H_{3j}, W_{jk}, E_{jk}$ are as in (10.4)-(10.7) and **Assumption 10.1**. The scalars $\mu_j, \sigma_j, \lambda_{jk}$ are positive scaling parameters such that

$$\lambda_{jk}^{-1} E_{jk}^t E_{jk} < I, \quad \sigma_j^{-1} E_{3j}^t E_{3j} < Q_j$$

The following theorem summarizes the main result.

Theorem 10.1: *Consider system (10.1)-(10.3) satisfying conditions (10.4)-(10.7) and **Assumption 10.1**. Given scalars $\{\gamma_1 > 0, ..., \gamma_{n_s} > 0\}$ and matrices $\{0 < Q_j = Q_j^t; 0 < R_j = R_j^t\}, \forall j \in \{1,..,n_s\}$ such that $Q_j < \gamma_j^2 R_j$, system (10.1)-(10.3) is globally uniformly asymptotically stable about the origin and*

$$\sum_{j=1}^{n_s} \|z_j\|_2 < \gamma_j \sum_{j=1}^{n_s} \{\|w_j\|_2^2 + x_j^t(0) R_j x_j(0)$$

$$+ \int_{-\tau}^0 x_j^t(s) R_j x_j(s) ds\}^{1/2} \tag{10.26}$$

for any nonzero $(\alpha_{jo}, w_j) \in \Re^{n_j} \oplus \mathcal{L}_2[0, \infty)$ and for all admissible uncertainties if system (10.21)-(10.22) is exponentially stable and there exist scaling parameters $\mu_j, \sigma_j, \lambda_{jk}, \forall j, k \in \{1,..,n_s\}$ satisfying $\lambda_{jk}^{-1} E_{jk}^t E_{jk} < I$ and $\sigma_j^{-1} E_{3j}^t E_{3j} < Q_j$ and such that

$$\mathcal{J}(\bar{z}, \bar{w}, \zeta_o, R) < \gamma_j \tag{10.27}$$

Proof: We will carry out the analysis at the subsystem level. Application of **Lemma 10.1** shows $\forall j \in \{1, ..., n_s\}$ that there exists a bounded matrix function $0 < P_j(t) = P_j^t(t), \forall t \in [0, \infty); P_j(0) < \gamma_j^2 R_j$ such that

$$\dot{P}_j + A_j^t P_j + P_j A_j + \gamma_j^{-2} P_j B_j B_j^t P_j + C_j^t(\mu_j) C_j(\mu_j)$$

$$P_j \mathcal{B}_j(\mu_j, \sigma_j, \lambda_{jk}) \mathcal{B}_j^t(\mu_j, \sigma_j, \lambda_{jk}) P_j + L_j^t L_j + Q_j < 0 \tag{10.28}$$

By **B.1.2** and using (10.24)-(10.25), it follows that:

$$N_j(P_j) = \dot{P}_j + A_{\Delta j}^t P_j + P_j A_{\Delta j} + \gamma_j^{-2} P_j B_j B_j^t P_j$$

$$+ \sum_{k=1}^{n_s} P_j G_{\Delta jk} G_{\Delta jk}^t P_j + P_j A_{d\Delta j} Q_j^{-1} A_{d\Delta j}^t P_j$$

$$+ \left(\sum_{k=1}^{n_s} W_{kj}^t W_{kj} \right) + L_j^t L_j + Q_j < 0 \qquad (10.29)$$

where

$$
\begin{aligned}
A_{\Delta j} &= A_j + H_{1j} \Delta_{1j} E_{1j} \\
A_{d\Delta j} &= A_{dj} + H_{3j} \Delta_{2j}(t) E_{2j} \\
G_{\Delta jk} &= G_{jk} + H_{jk} \Delta_{jk} E_{jk}
\end{aligned}
\qquad (10.30)
$$

To examine the stability of system (10.1)-(10.3), we introduce the Lyapunov-Krasovskii functional

$$V_j(x_{tj}) = x_j^t P_j x_j + \sum_{k=1}^{n_s} \int_0^t x_j W_{jk}^t W_{jk} x_j ds$$

$$- \sum_{k=1}^{n_s} \int_0^t g_{jk}^t g_{jk} ds + \int_{t-\tau}^t x_j^t(r) Q_j x_j(r) dr \qquad (10.31)$$

In view of **Assumption 10.1**, it is easy to see that $V_j(x_{tj}) > 0$ whenever $x_j \neq 0$. Differentiating (10.31) along the state trajectories of (10.1)-(10.3) with $w_j(t) \equiv 0$ and using the interconnection constraint

$$\sum_{j=1}^{n_s} \sum_{k=1}^{n_s} x_k^t W_{jk}^t W_{jk} x_k = \sum_{k=1}^{n_s} \sum_{k=1}^{n_s} x_j^t W_{kj}^t W_{kj} x_j$$

we get:

$$\frac{d}{dt} V(x_t) = \sum_{j=1}^{n_s} \frac{d}{dt} V_j(x_{tj}) \leq \sum_{j=1}^{n_s} \chi_j^t(t) \check{A}_j \chi_j(t) \qquad (10.32)$$

where $\chi_j(t) = [x_j^t(t) \ \ x_j^t(t-\tau) \ \ g_{j1}^t \ldots g_{jn_s}^t]^t$ and

$$
\check{A}_j = \begin{bmatrix}
\Omega_j & P_j A_{d\Delta j} & P_j \bar{G}_{j1} & \cdots & P_j \bar{G}_{jn_s} \\
A_{d\Delta j}^t P_j & -Q_j & 0 & \cdots & 0 \\
\bar{G}_{j1}^t P_j & 0 & -I & \cdots & 0 \\
\vdots & \vdots & \vdots & \ddots & \vdots \\
\bar{G}_{jn_s}^t P_j & 0 & \cdots & \cdots & -I
\end{bmatrix}
$$

$$\Omega_j = \dot{P}_j + A_{\Delta j}^t P_j + P_j A_{\Delta j} + \sum_{k=1}^{n_s} W_{kj}^t W_{kj}$$

In view of (10.29), it follows that $\frac{d}{dt} V(x_t) < 0$ whenever $x \neq 0$ which, in turn, means that the equilibruim state $x = 0$ is globally, uniformly, asymptotically stable for all admissible uncertainties. Moreover, since $w_j \in \mathcal{L}_2[0, \infty)$ the boundedness of $\|z_j\|_2$ is guaranteed.

Now to show that system (10.1)-(10.3) has the desired performance (10.26), we introduce:

$$
\begin{aligned}
J = & \sum_{j=1}^{n_s} \int_0^\infty \{z_j^t(t) z_j(t) - \gamma^2 w_j^t(t) w_j(t)\} dt \\
& - \gamma^2 \{x_j^t(0) R_j x_j(0) + \int_{-\tau}^0 x_j^t(s) R_j x_j(s) ds\}
\end{aligned}
\tag{10.33}
$$

By completing the squares in (10.33) and using (10.29), it follows that

$$
\begin{aligned}
J = & \sum_{j=1}^{n_s} \{ \int_0^\infty [x_j^t L_j^t L_j x_j + \frac{d}{dt} V_j(x_{tj}) - \gamma^2 w^t w] dt \\
& + x_j^t(0) P_j(0) x_j(0) - \gamma_j^2 x_j^t(0) R_j x_j(0) \\
& + \gamma_j^2 \int_{-\tau}^0 x_j^t(s) R_j x_j(s) ds - V_j(\infty)\} \\
= & \sum_{j=1}^{n_s} \{x_j^t(0) [P_j(0) - \gamma_j^2 R_j] x_j(0) - V_j(\infty)\} \\
& + \sum_{j=1}^{n_s} \int_0^\infty \{x_j^t N_j(P_j) x_j - N_{4j}^t N_{4j}\} dt \\
& - \sum_{j=1}^{n_s} \int_0^\infty \{\gamma_j^2 N_{2j}^t N_{2j} + N_{3j}^t Q_j^{-1} N_{3j}\} dt \\
& - \sum_{j=1}^{n_s} \int_{-\tau}^0 x_j^t(s) [\gamma_j^2 R_j - Q_j] x_j(s) dt
\end{aligned}
\tag{10.34}
$$

where

$$
\begin{aligned}
N_{2j} &= [w_j - \gamma_j^{-2} B_j^t P_j x_j] \\
N_{3j} &= [Q_j x_j(t - \tau) - A_{d\Delta j}^t P_j x_j] \\
N_{4j} &= \sum_{k=1}^{n_s} [g_{jk}(x_k) - G_{\Delta jk}^t P_j x_j]
\end{aligned}
$$

Note that $V_j(\infty) \geq 0 \; \forall t$ and is bounded. In view of (10.29), it follows that $J < 0$ for all nonzero $(\alpha_{jo}, w_j) \in \Re^{n_j} \oplus \mathcal{L}_2[0, \infty)$ and for all admissible uncertainties.

10.4   Robust H_∞ Filtering

In order to construct a robust H_∞ filter for system (Σ_Δ), we introduce the following n_s scaled systems:

$$\dot{\breve{x}}_j(t) = A_j \breve{x}_j(t) + [B_j \;\; \gamma \mathcal{B}_j(\mu_j, \sigma_j, \lambda_{jk})]\breve{w}_j(t) \tag{10.35}$$

$$\breve{z}_j(t) = \begin{bmatrix} L_j \\ \mu_j E_{1j} \\ \mathcal{W}_j \end{bmatrix} \breve{x}_j(t) + \begin{bmatrix} -I \\ 0 \\ 0 \end{bmatrix} u_j(t) \tag{10.36}$$

$$\breve{y}_j(t) = C_j \breve{x}_j(t) + [D_j \;\; \gamma \mathcal{D}_j(\mu_j, \lambda_{jk})]\breve{w}_j(t) \tag{10.37}$$

where $j \in \{1, .., n_s\}$, $\breve{x}_j \in \Re^{n_j}$ is the state; $\breve{\alpha}_{jo}$ is an unknown initial state, $u_j \in \Re^{p_j}$ is the control input, $\breve{w}_j(t) \in \Re^{q_j + \alpha_j}$ is the disturbance input which belongs to $\mathcal{L}_2[0, \infty)$, $\breve{z}_j(t) \in \Re^{n_j + r_j + \beta_j}$ is the controlled output , $\breve{y}_j \in \Re^{p_j}$ is the measured output, $\gamma_j > 0$ is the desired H_∞ performance of the filter $\mathcal{P}_j$, the matrices A_j, B_j, C_j, D_j, L_j are the same as in (10.1)-(10.3) and $\mathcal{B}_j(\mu_j, \sigma_j, \lambda_{jk})\mathcal{B}_j^t(\mu_j, \sigma_j, \lambda_{jk})$ is given by (10.24) with

$$\mathcal{W}_j^t \mathcal{W}_j = \sum_{k=1}^{n_s} W_{kj}^t W_{kj} + W_j^t W_j \tag{10.38}$$

$$\mathcal{D}_j(\mu_j) \mathcal{D}_j^t(\mu_j) = \lambda_j H_{4j} H_{4j}^t + M_j(I - \lambda_j E_{3j}^t E_{3j})^{-1} M_j^t \tag{10.39}$$

where H_{4j}, E_{3j}, W_{kj} are as in (10.4)-(10.7) and **Assumption 10.1**. The scalars $\mu_j, \sigma_j, \lambda_{jk}, \lambda_j$ are positive scaling parameters such that $\lambda_{jk}^{-1} E_{jk}^t E_{jk} < I$, $\sigma_j^{-1} E_{3j}^t E_{3j} < Q_j$ and $\lambda_j E_{3j}^t E_{3j} < I$. Let the estimate $\breve{z}_j = \mathcal{P}_j(y_j)$ be represented by the following realization:

$$\dot{\nu}_j(t) = A_{\nu j}(t)\nu_j(t) + B_{\nu j}(t)y_j(t), \; \nu_j(0) = 0 \tag{10.40}$$

$$\breve{z}_j(t) = K_{\nu j}(t)\nu_j(t) \tag{10.41}$$

where the dimension, π_j, of the filter $\mathcal{P}_j$ and the time-varying matrices $A_{\nu j}$, $B_{\nu j}(t)$ and $K_{\nu j}$ are to be selected. By [27], it follows that the control law $u_j = \mathcal{P}_j(\breve{y}_j)$ of system (10.35)-(10.37) is given by:

$$\dot{\varphi}_j(t) = A_{\nu j}(t)\varphi_j(t) + B_{\nu j}(t)\breve{y}_j(t), \; \varphi_j(0) = 0 \tag{10.42}$$

$$u_j(t) = K_{\nu j}(t)\varphi_j(t) \tag{10.43}$$

The following theorem summarizes the main result.

Theorem 10.2: *Consider system (10.1)-(10.3) satisfying conditions (10.4)-(10.7) and* **Assumption 10.1.** *Let $\mathcal{P}_j : y_j \rightarrow \check{z}_j, \; j \in \{1,..,n_s\}$ denote a set of linear time-varying strictly proper filters with zero initial conditions. Then, given scalars $\{\gamma_1 > 0, ..., \gamma_{n_s} > 0\}$ and matrices $\{0 < Q_j = Q_j^t; 0 < R_j = R_j^t\}, \forall j \in \{1,..,n_s\}$ such that $Q_j < \gamma_j^t R_j$, the estimate $\check{z}_j = \mathcal{P}(y_j)$ solves the decentralized robust H_∞-filtering problem for system (10.1)-(10.3) if there exist scaling parameters $\mu_j, \sigma_j, \lambda_{jk}, \lambda_j, \; \forall j, k \in \{1,..,n_s\}$ satisfying*

 (1) $\lambda_{jk}^{-1} E_{jk}^t E_{jk} < I, \sigma_j^{-1} E_{3j}^t E_{3j} < Q_j$ *and* $\lambda_j E_{3j}^t E_{3j} < I$

 (2) *the closed-loop system (10.42)-(10.43) is globally, uniformly, asymptotically stable about the origin and $\mathcal{J}(\check{z}_j, \check{w}_j, \check{x}_{jo}, R_j) < \gamma_j$.*

Proof: By augmenting (10.35)-(10.37) and (10.42)-(10.43), we obtain the closed-loop system:

$$\dot{\zeta}_{cj}(t) = \widehat{A}_j(t)\zeta_{cj}(t) + [\breve{B}_j(t)\,\gamma_j \widehat{B}_j(t)]\check{w}_j(t) \tag{10.44}$$

$$\check{z}_j(t) = \begin{bmatrix} \check{C}_{1j} \\ \check{C}_{2j} \end{bmatrix} \zeta_{cj}(t) \tag{10.45}$$

where

$$\zeta_{cj}(t) = \begin{bmatrix} \check{x}_j(t) \\ \varphi_j(t) \end{bmatrix}, \; \zeta_{cj}(0) = \begin{bmatrix} \check{x}_j(0) \\ 0 \end{bmatrix}$$

$$\widehat{A}_j = \begin{bmatrix} A_j & 0 \\ B_{\nu j}C_j & A_{\nu j} \end{bmatrix}, \; \breve{B}_j = \begin{bmatrix} B_j \\ B_{\nu j}D_j \end{bmatrix}$$

$$\widehat{B}_j = \begin{bmatrix} B_j(\mu_j, \sigma_j, \lambda_{jk}) \\ B_{\nu j}D_j(\mu_j) \end{bmatrix}$$

$$\check{C}_{1j} = [C_j \; -K_{\nu j}], \; \check{C}_{2j} = \begin{bmatrix} \frac{1}{\sqrt{\mu_j}}E_{1j} & 0 \\ \mathcal{W}_j & 0 \end{bmatrix} \tag{10.46}$$

On the other hand, by introducing $\zeta_{fj}(t) = [x_j^t(t) \; v_j^t(t)]^t$ then the filtering error , $e_j(t) = z_j(t) - \check{z}_j(t)$, associated with system (10.1)-(10.3) and the filter (10.40)-(10.41) is given by:

$$\dot{\zeta}_{fj}(t) = [\widehat{A}_j + \widehat{H}_j \, \Delta 1j \, \widehat{E}_j]\zeta_{fj}(t)$$

$$+ \sum_{j=1}^{n_s} [\widehat{G}_{jk} + \widehat{H}_{jk}\widehat{\Delta}_{jk}\widehat{E}_{jk}] f_{jk}(\psi_{fj})$$

$$+ \ \breve{B}_j(t)\breve{w}_j(t) + \widehat{A}_{d\Delta j}\ \psi_{fj}(t-\tau)$$

$$e_j(t) \ = \ \breve{C}_{1j}\zeta_{fj}(t), \ \zeta_{fj}(0) = [\alpha_{jo}^t \ 0]^t \tag{10.47}$$

where

$$\widehat{H}_j = \left[\begin{array}{c} \sqrt{\mu_j}H_{1j} \\ \sqrt{\mu_j}B_{\nu j}H_{1j} \end{array}\right], \ \widehat{E}_j = [\frac{1}{\sqrt{\mu_j}}E_{1j} \ 0]$$

$$f_{jk}(\psi_{fj}) = \left\{ \begin{array}{ll} g_{jk}(x_k) & \text{if } k \neq j \\ \left[\begin{array}{c} g_{jj}(x_j) \\ m_j(x_j) \end{array}\right] & \text{if } k = j \end{array} \right.$$

$$\widehat{A}_{d\Delta j} = \left[\begin{array}{cc} E_{\Delta j} & 0 \\ 0 & 0 \end{array}\right]$$

$$\widehat{G}_{jk} = \left\{ \begin{array}{ll} \left[\begin{array}{c} G_{jk} \\ 0 \end{array}\right] & \text{if } k \neq j \\ \left[\begin{array}{cc} G_{jj} & 0 \\ 0 & B_{\nu j}M_j \end{array}\right] & \text{if } k = j \end{array} \right.$$

$$\widehat{H}_{jk} = \left\{ \begin{array}{ll} \left[\begin{array}{c} \sqrt{\lambda_{jk}}H_{jk} \\ 0 \end{array}\right] & \text{if } k \neq j \\ \left[\begin{array}{cc} \sqrt{\lambda_{jj}}H_{jj} & 0 \\ 0 & \sqrt{\lambda_{jj}}B_{\nu j}H_{4j} \end{array}\right] & \text{if } k = j \end{array} \right.$$

$$\widehat{\Delta}_{jk}(t) = \left\{ \begin{array}{ll} \Delta_{jk}(t) & \text{if } k \neq j \\ \left[\begin{array}{cc} \Delta_{jj}(t) & 0 \\ 0 & \Delta_{3j}(t) \end{array}\right] & \text{if } k = j \end{array} \right.$$

$$\widehat{E}_{jk} = \left\{ \begin{array}{ll} \frac{1}{\sqrt{\lambda_{jk}}}E_{jk} & \text{if } k \neq j \\ \left[\begin{array}{cc} \frac{1}{\sqrt{\lambda_{jj}}}E_{jj} & 0 \\ 0 & \frac{1}{\sqrt{\lambda_j}}E_{3j} \end{array}\right] & \text{if } k = j \end{array} \right.$$

We note that $\widehat{\Delta}_{jk}(t)\widehat{\Delta}_{jk}^t(t) \leq I \ \forall j,k \in \{1,...,n_s\}$ and $\forall \zeta_{fj} \in \Re^{n_j + \pi_j}$, we have:

$$||f_{jk}(\zeta_{fj})||^2 \ = \ ||g_{jk}(x_j)||^2 \leq ||W_{jk}x_k||^2$$

$$
\begin{aligned}
&= \quad |||[W_{jk}\ 0]\varphi_k||^2 \quad j \neq k \\
||f_{jk}(\zeta_{fj})||^2 &= \quad ||g_{jj}(x_j)||^2 + ||m_j(x_j)||^2 \\
&\leq \quad |||[\tilde{W}_j\ 0]\varphi_j||^2 \quad j = k
\end{aligned}
$$

with

$$
\tilde{W}_j^t \tilde{W}_j = W_{jj}^t W_{jj} + W_j^t W_j
$$

It is easy to verify using (10.24) that

$$
\begin{aligned}
\mathcal{B}_j \mathcal{B}_j^t &= \widehat{H}_j \widehat{H}_j^t + \widehat{G}_{jk} (I - E^t jk E_{jk})^{-1} \widehat{G}_{jk}^t \\
&+ \widehat{H}_{jk} \widehat{E}_{jk}^t \\
\check{C}_{2j}^t \check{C}_{2j} &= [\frac{1}{\sqrt{\mu_j}} E_{1j}\ 0]^t [\frac{1}{\sqrt{\mu_j}} E_{1j}\ 0] \\
&+ [\mathcal{W}_j\ 0]^t [\mathcal{W}_j\ 0]
\end{aligned}
\tag{10.48}
$$

By carefully examining systems (10.44)-(10.46) and (10.47) in the light of (10.48) and condition (2), the results follow immediately from **Theorem 10.1**.

Corollary 10.1: *Consider system (10.1)-(10.3) with $\alpha_o \equiv 0$ and satisfying (10.4)-(10.7) and **Assumption 10.1**. Let $P_j(s), j \in \{1, .., n_s\}$ be a set of linear time-invariant strictly proper filters with zero initial conditions and let $\check{z}_j = P_j(s)y_j$ be the estimate of $z_j, j \in \{1, .., n_s\}$. Then given scalars $\gamma_1 > 0, ..., \gamma_{n_s} > 0$, the filters $P_j(s)$ solve the decentralized robust H_∞ filtering problem for system (10.1)-(10.3) if there exist matrices $0 < Q_j = Q_j^t, j \in \{1, .., n_s\}$ and scaling parameters $\mu_j > 0, \sigma_j, \lambda_{jk}, \lambda_j, j, k \in \{1, .., n_s\}$ such that*

(1) $\lambda_{jk}^{-1} E_{jk}^t E_{jk} < I$, $\sigma_j^{-1} E_{3j}^t E_{3j} < Q_j$ and $\lambda_j E_{3j}^t E_{3j} < I$
(2) the closed-loop system (10.35)-(10.37) with zero initial state under the control action $u_j = P_j(s)\check{y}_j$ is asymptotically stable and $||\check{z}_j||_2 < \gamma_j ||\check{w}_j||_2$ for any nonzero $\check{w}_j \in \mathcal{L}_2[0, \infty)$.

Remark 10.3: We note that **Corollary 10.1** is a special version of **Theorem 10.2** when the system under consideration is time-invariant with zero initial condition. Observe in this case that the robust H_∞−filter is time-invariant whereas it is time varying in the case of **Theorem 10.2**. The key point here is that the decentralized H_∞−filtering for a wide class of interconnected systems with norm-bounded parameteric uncertainties and unknown cone-bounded nonlinearities as well as unknown state-delays can be solved in terms of parameterized output feedback H_∞–conrol problems for n_s linear decoupled systems which do not involve parametric uncertainties and unknown nonlinearities as well as unknown state-delays. The latter problems can be solved using the results of [2,14].

10.5   Notes and References

Indeed, the model treated in section 10.2 represents one of the many different possible characterizations of interconnected time-delay systems subject to uncertain parameters. Extension of the obtained results to other models is a viable research direction. Examination of delay-dependent stability is another research topic.

Bibliography

[1] Basar, T. and P. Bernard, "$\mathcal{H}_\infty$-Optimal Control and Related Minimax Design Problems: A Dynamic Game Approach," Birkhauser, Boston, 1991.

[2] Doyle, J. C., K. Glover, P. P. Khargonekar and B. A. Francis, "State-Space Solutions to Standard H_2 and H_∞ Control Problems," **IEEE Trans. Automatic Control,** vol. 34, 1989, pp. 831-847.

[3] Anderson, B. D. O. and J. B. Moore, **"Optimal Filtering,"** Prentice Hall, New York, 1979.

[4] Kaliath, T., "A View of Three-Decades of Linear Filtering Theory," **IEEE Trans. Information Theory,** vol. IT-20, 1974, pp. 145-181.

[5] Francis, B. A., "A Course in $\mathcal{H}_\infty$ Control Theory," Springer Verlag, New York, 1987.

[6] Kwakernaak, H., "Robust Control and $\mathcal{H}_\infty$-Optimization: Tutorial Paper," **Automatica,** vol. 29, 1993, pp. 255-273.

[7] Zhou, K., "Essentials of Robust Control", Prentice-Hall, New York, 1998.

[8] Stoorvogel, A., "The $\mathcal{H}_\infty$ Control Problem," Prentice-Hall, New York, 1992.

[9] Gahinet, P. and P. Apkarian, "A Linear Matrix Inequality Approach to $\mathcal{H}_\infty$ Control," **Int. J. Robust and Nonlinear Control,** vol. 4, 1994, pp. 421-448.

[10] Haddad, W. M. and D. S. Berstein, "Robust Reduced-Order, Non-strictly Proper State Estimator via the Optimal Projection Equations

with Guaranteed Cost Bound," IEEE Trans. **Automatic Control,** vol. AC-33, 1988, pp. 591-595.

[11] Fu, M., C. E. deSouza and L. Xie, "H_∞-Estimation for Uncertain Systems," Int. J. **Robust and Nonlinear Control,** vol. 2, 1992, pp. 87-105.

[12] Xie, L. and Y. C. Soh, "Robust Kalman Filtering for Uncertain Systems", **Systems and Control Letters",** vol. 22, 1994, 123-129.

[13] Bahnasawi, A. A., A.S. Al-Fuhaid and M.S. Mahmoud, "Decentralized and Hierarchical Control of Interconnected Uncertain Systems," **Proc. IEE Part D,** vol. 137, 1990, pp. 311-321.

[14] Nagpal, K. M. and P. P. Khargonekar, "Filtering and Smoothing in $\mathcal{H}_\infty$ Setting," IEEE Trans. **Automatic Control,** vol. 36, 1991, pp. 152-166.

[15] Shaked, U., "$\mathcal{H}_\infty$ Minimum Error Estimation of Linear Stationary Processes," IEEE Trans. **Automatic Control,** vol. 35, 1990, pp. 554-558.

[16] Fu, M., C E. de Souza and L. Xie, "$\mathcal{H}_\infty$ Estimation for Discrete-Time Linear Uncertain Systems," Int. J. **Robust and Nonlinear Control,** vol. 1, 1991, pp. 11-23.

[17] Bernstein, D. S. and W. M. Haddad, "Steady-State Kalman Filtering with an H_∞ Error Bound," **Systems and Control Letters,** vol. 16, 1991, pp. 309-317.

[18] Li, X. and M. Fu, "A Linear Matrix Inequality Approach to Robust $\mathcal{H}_\infty$ Filtering," IEEE Trans. **Signal Processing,** vol. 45, 1997 pp. 2338-2350.

[19] Malek-Zavarei, M. and M. Jamshidi, **"Time-Delay Systems: Analysis, Optimization and Applications,"** North-Holland, Amsterdam, 1987.

[20] Kolomanovskii, V. and A. Myshkis, **"Applied Theory of Functional Differential Equations,"** Kluwer Academic Pub., New York, 1992.

[21] Mahmoud, M. S., "Robust Kalman Filtering for Time-Delay Systems," **Kuwait University Technical Report ECC-MSM-98-05, Kuwait.**

[22] Mahmoud, M. S., "Robust $\mathcal{H}_\infty$ Control and Filtering for Time-Delay Systems," **Kuwait University Technical Report ECC-MSM-98-06**, Kuwait.

[23] Mahmoud, M. S., "Robust $\mathcal{H}_\infty$ Filtering for Uncertain Systems with Unknown-Delays," **Kuwait University Technical Report ECC-MSM-98-07**, Kuwait.

[24] Mahmoud, M. S. and L. Xie, "$\mathcal{H}_\infty$-Filtering for a Class of Nonlinear Time-Delay Systems," **Kuwait University Technical Report ECC-MSM-98-08**, Kuwait.

[25] Mahmoud, M. S., L. Xie and Y. C. Soh, "Robust Kalman Filtering for Discrete State-Delay Systems," **Kuwait University Technical Report ECC-MSM-98-09**, Kuwait.

[26] Mahmoud, M. S. "Robust $\mathcal{H}_\infty$ Filtering for Uncertain Systems with Unknown-Delays," **J. of Systems Analysis and Modelling Simulation**, vol. 30, 1999.

[27] Xie, L., C. E. de Souza and M. D. Fragoso, "$\mathcal{H}_\infty$ – Filtering for Linear Periodic Systems with Parameteric Uncertainty," **Systems and Control Letters**, vol. 17, 1991, pp. 343-350.

[28] Limbeer, D. J. N., B. D. O. Anderson, P. P. Khargonekar and M. Green, "A Game Theoretical Approach to H_∞ Control of Time-Varying Systems," **SIAM J. Control and Optimization**, vol. 30, 1992, pp. 262-283.

[29] Mahmoud, M. S., "Robust $\mathcal{H}_\infty$ Filtering of Linear Parameter-Varying Systems," **Kuwait University Technical Report ECE-MSM-98-03**, Kuwait.

[30] Mahmoud, M. S., "Linear Parameter-Varying Discrete Time-Delay Systems: Stability and ℓ_2-Gain Controllers," **Kuwait University Technical Report ECE-MSM-98-04**, Kuwait.

[31] Mahmoud, M. S., "Robust $\mathcal{H}_\infty$ Filtering of Linear Parameter-Varying Discrete-Time Systems with State-Delay," **Proc. the 5th Eurpean Control Conference**, Karlsruhe, Germany, 1999, pp.

[32] Petersen, I. R., and D. C. McFarlane, "Optimal Guaranteed Cost Control and Filtering for Uncertain Linear Systems," **IEEE Trans. Automatic Control**, vol. 39, 1994, 1971-1977.

[33] Apkarian, P. , P. Gahinet and G. Becker, "Self-Scheduled H_∞ Control of Linear Parameter-Varying Systems: A Design Example," **Automatica**, vol. 31, 1995, pp. 1251-1261.

[34] Becker, G. and A. Packard, "Robust Performance of Linear Parametrically-Varying Systems using Parameterically-Dependent Linear Feedback," **Systems and Control Letters**, vol. 23, 1994, pp. 205-215.

[35] Boyd, S., L. El-Ghaoui, E. Feron and V. Balakrishnan, "**Linear Matrix Inequalities in System and Control Theory**," vol. 15, SIAM Studies in Appl. Math., Philadelphia, 1994.

[36] Apkarian, P. and P. Gahinet, "A Convex Characterization of Gain Scheduled H_∞ Controllers," **IEEE Trans. Automatic Control**, vol. 40, 1995, pp. 853-864.

[37] Gahinet, P. , P. Apkarian and M. Chilali, "Affine Parameter-Dependent Lyapunov Functions for Real Parametric Uncertainty," **IEEE Trans. Automatic Control**, vol. 41, 1996, pp. 436-442.

[38] Iwasaki, T. and R. E. Skelton, "All Controllers for the General $\mathcal{H}_\infty$ Control Design Problem: LMI Existence Conditions and State-Space Formulas," **Automatica**, vol. 30, 1994, pp. 1307-1317.

[39] de Souza, C. E., M. Fu and Xie, L., "H_∞ Analysis and Synthesis of Discrete-Time Systems with Time Varying Uncertainty," **IEEE Trans. Automatic Control**, vol. 38, 1993, pp. 459-462.

[40] Bahnasawi,A.A., A.S. Al-Fuhaid and M.S. Mahmoud, " A New Hierarchical Control Structure for a Class of Uncertain Discrete Systems," **Control Theory and Advanced Technology**, vol. 6, 1990, pp.1-21.

[41] Mao, C. and J. H. Yang, "Decentralized Output Tracking for Linear Uncertain Interconnected Systems," **Automatica**, vol. 31, 1995, pp. 151-154.

[42] Mahmoud, M. S., and S. Bingulac, "Robust Design of Stabilizing Controllers for Interconnected Time-Delay Systems," **Automatica**, vol. 34, 1998, pp. 795-800.

[43] Therrien, C. W., "**Discrete Random Signals and Statistical Signal Processing**," Prentice-Hall, New York, 1993.

[44] Xie, L., Y. C. Soh and C. E. de Souza, "Robust Kalman Filtering for Uncertain Discrete-Time Systems," IEEE **Trans. Automatic Control**, vol. 39, 1994, pp. 1310-1314.

[45] Theodor, Y. and U. Shaked, "Robust Discrete-Time Minimum Variance Filtering," IEEE **Trans. Signal Processing**, vol. 44, 1996, pp. 181-189.

Appendix A

Some Facts from Matrix Theory

A.1 Schur Complements

(1) Given constant matrices Ω_1, Ω_2, Ω_3 where $0 < \Omega_1 = \Omega_1^t$ and $0 < \Omega_2 = \Omega_2^t$ then $\Omega_1 + \Omega_3^t \Omega_2^{-1} \Omega_3 < 0$ if and only if

$$\begin{bmatrix} \Omega_1 & \Omega_3^t \\ \Omega_3 & -\Omega_2 \end{bmatrix} < 0 \quad or \quad \begin{bmatrix} -\Omega_2 & \Omega_3 \\ \Omega_3^t & \Omega_1 \end{bmatrix} < 0$$

(2) Let Σ_1 be any square matrix. Then $\Sigma_2 > 0$ and $\Sigma_1^t \Sigma_2 \Sigma_1 - \Sigma_2 < 0$ if and only if

$$\begin{bmatrix} -\Sigma_2^{-1} & -\Sigma_1 \\ -\Sigma_1^t & -\Sigma_2 \end{bmatrix} < 0$$

Remark A.1: Although **(2)** can be derived from **(1)** and vice-versa, we have included them for direct use in the respective chapters.

A.2 Matrix Inversion Lemma

For any real nonsingular matrices Σ_1, Σ_3 and real matrices Σ_2, Σ_4 with appropriate dimensions, it follows that

$$(\Sigma_1 + \Sigma_2\Sigma_3\Sigma_4)^{-1} = \Sigma_1^{-1} - \Sigma_1^{-1}\Sigma_2[\Sigma_3^{-1} + \Sigma_4\Sigma_1^{-1}\Sigma_2]^{-1}\Sigma_4\Sigma_1^{-1}$$

A.3   Bounded Real Lemma

A.3.1   Continuous-Time Systems

For any realization $(\mathcal{A}, \mathcal{B}, \mathcal{C})$, the following statements are equivalent:

(1) $\mathcal{A}$ is stable and $\|\mathcal{C}(sI - \mathcal{A})^{-1}\mathcal{B}\|_\infty < 1$;
(2) There exists a matrix $\hat{P} > 0$ satisfying the algebraic Riccati inequality (ARI):

$$\hat{P}\mathcal{A} + \mathcal{A}^t\hat{P} + \hat{P}\mathcal{B}\mathcal{B}^t\hat{P} + \mathcal{C}^t\mathcal{C} < 0$$

(3) The algebraic Riccati equation (ARE)

$$\bar{P}\mathcal{A} + \mathcal{A}^t\bar{P} + \bar{P}\mathcal{B}\mathcal{B}^t\bar{P} + \mathcal{C}^t\mathcal{C} = 0$$

has a stabilizing solution $0 \le \bar{P} = \bar{P}^t$. Furthermore, if these statements hold, then $\bar{P} < \hat{P}$. See [64] for further details.

A.3.2   Discrete-Time Systems

Let $\mathcal{G}(z) \in \Re^{\wp \times \vartheta}$ be a real rational transfer function matrix with realization $\mathcal{G}(z) = \mathcal{C}(zI - \mathcal{A})^{-1}\mathcal{B} + \mathcal{D}$. Then the following statements are equivalent:

(1) A is Schur-stable and $\|\mathcal{C}(zI - \mathcal{A})^{-1}\mathcal{B} + \mathcal{D}\|_\infty < 1$
(2) There exists a matrix $0 < P = P^t$ satisfying the ARI

$$\mathcal{A}^t P \mathcal{A} \;-\; P + (\mathcal{A}^t P \mathcal{B} + \mathcal{C}^t \mathcal{D})[I - (\mathcal{D}^t \mathcal{D} + \mathcal{B}^t P \mathcal{B})]^{-1}(\mathcal{B}^t P \mathcal{A} + \mathcal{D}^t \mathcal{C})$$
$$+\; \mathcal{C}^t \mathcal{C} < 0 \tag{A.1}$$

and such that $I - (\mathcal{D}^t \mathcal{D} + \mathcal{B}^t P \mathcal{B}) > 0$
(3) There exists a stabilizing matrix $0 \le \breve{P} = \breve{P}^t$ satisfying the ARE:

$$\mathcal{A}^t \breve{P} \mathcal{A} \;-\; \breve{P} + (\mathcal{A}^t \breve{P} \mathcal{A} + \mathcal{C}^t \mathcal{D})[I - (\mathcal{D}^t \mathcal{D} + \mathcal{B}^t P \mathcal{B})]^{-1}(\mathcal{B}^t \breve{P} \mathcal{A} + \mathcal{D}^t \mathcal{C})$$
$$+\; \mathcal{C}^t \mathcal{C} = 0 \tag{A.2}$$

Moreover $\breve{P} \le P$. The stabilizing solution $\breve{P}$ renders the matrix $\mathcal{A} + \mathcal{B}[I - (\mathcal{D}^t \mathcal{D} + \mathcal{B}^t \breve{P} \mathcal{B})]^{-1}(\mathcal{B}^t \breve{P} \mathcal{A} + \mathcal{D}^t \mathcal{C})$ Schur-stable. See [52] for further details.

Appendix B

Some Algebraic Inequalities

B.1 Matrix-Type Inequalities

Let Σ_1, Σ_2, Σ_3 be real constant matrices of compatible dimensions and $H(t)$ be a real matrix function satisfying $H^t(t)H(t) \leq I$. Then the following inequalities hold:

B.1.1 $\quad \Sigma_1^t\Sigma_2 + \Sigma_2^t\Sigma_1 \leq \zeta\Sigma_1^t\Sigma_1 + \zeta^{-1}\Sigma_2^t\Sigma_2, \quad \zeta > 0$

B.1.2 $\quad \Sigma_1 H(t)\Sigma_2 + \Sigma_2^t H^t(t)\Sigma_1^t \leq \rho^{-2}\Sigma_1\Sigma_1^t + \rho^2\Sigma_2^t\Sigma_2, \quad \rho > 0$

B.1.3 $\quad \forall \rho > 0$ such that $\rho^2\Sigma_2^t\Sigma_2 < I$

$$(\Sigma_3 + \Sigma_1 H(t)\Sigma_2)(\Sigma_3^t + \Sigma_2^t H^t(t)\Sigma_1^t) \leq \rho^{-2}\Sigma_1\Sigma_1^t + \Sigma_3[I - \rho^2\Sigma_2^t\Sigma_2]^{-1}\Sigma_3^t$$

Proof:

B.1.1 Since $\Sigma^t(t)\Sigma(t) \geq 0$ for any real matrix $\Sigma(t)$ then

$$[\zeta^{1/2}\Sigma_1 - \zeta^{-1/2}\Sigma_2]^t[\zeta^{1/2}\Sigma_1 - \zeta^{-1/2}\Sigma_2] \geq 0 \tag{B.1}$$

Expansion of (B.1) yields $\zeta\Sigma_1^t\Sigma_1 - \Sigma_1^t\Sigma_2 - \Sigma_2^t\Sigma_1 + \zeta^{-1}\Sigma_2^t\Sigma_2 \geq 0$ which when rearranged gives

$$\zeta\Sigma_1^t\Sigma_1 + \zeta^{-1}\Sigma_2^t\Sigma_2, \quad \zeta > 0$$

B.1.2 Instead of (B.1) consider the matrix function

$$[\rho^{-1}\Sigma_1^t - \rho H(t)\Sigma_2]^t[\rho^{-1}\Sigma_1 - \rho H(t)\Sigma_2] \geq 0 \tag{B.2}$$

On expanding (B.2), we get
$\rho^{-2}\Sigma_1\Sigma_1^t - \Sigma_1 H(t)\Sigma_2 - \Sigma_2^t H^t(t)\Sigma_1^t + \rho^2\Sigma_2^t H^t(t)H(t)\Sigma_2 \geq 0$ and by rearranging the terms using $H^t(t)H(t) \leq I$, we obtain

$$\Sigma_1 H(t)\Sigma_2 + \Sigma_2^t H^t(t)\Sigma_1^t \leq \rho^{-2}\Sigma_1\Sigma_1^t + \rho^2\Sigma_2^t\Sigma_2, \quad \rho > 0$$

B.1.3 Again, instead of (B.1) consider the matrix function

$$\Sigma(t) = [\rho^{-2}I - \Sigma_2\Sigma_2^t]^{-1/2}\Sigma_2\Sigma_3^t - [\rho^{-2}I - \Sigma_2\Sigma_2^t]^{1/2}H^t(t)\Sigma_1^t \qquad (B.3)$$

Expanding $\Sigma^t(t)\Sigma(t) \geq 0$ using the fact that $\rho^2\Sigma_2^t(t)\Sigma_2(t) < I$, we get

$$\Sigma_3\Sigma_2^tH^t(t)\Sigma_1^t + \Sigma_1H(t)\Sigma_2\Sigma_3^t(t) + \Sigma_1H(t)\Sigma_2\Sigma_2^tH^t(t)\Sigma_1^t(t)$$
$$\leq \Sigma_3\Sigma_2^t[\rho^{-2}I - \Sigma_2\Sigma_2^t]^{-1}\Sigma_2\Sigma_3^t + \rho^{-2}\Sigma_1H(t)H^t(t)\Sigma_1^t$$
$$+(\Sigma_3 + \Sigma_1H(t)\Sigma_2)(\Sigma_3 + \Sigma_1H(t)\Sigma_2)^t - \Sigma_3\Sigma_3^t$$
$$\leq \Sigma_3\Sigma_2^t[\rho^{-2}I - \Sigma_2\Sigma_2^t]^{-1}\Sigma_2\Sigma_3^t + \rho^{-2}\Sigma_1H(t)H^t(t)\Sigma_1^t$$
$$+(\Sigma_3 + \Sigma_1H(t)\Sigma_2)(\Sigma_3 + \Sigma_1H(t)\Sigma_2)^t$$
$$\leq \Sigma_3[I + \Sigma_2^t[\rho^{-2}I - \Sigma_2\Sigma_2^t]^{-1}\Sigma_2]\Sigma_3^t + \rho^{-2}\Sigma_1H(t)H^t(t)\Sigma_1^t \qquad (B.4)$$

Since $[I - \rho^2\Sigma_2^t\Sigma_2]^{-1} = [I + \rho^2\Sigma_2^t[I - \rho^2\Sigma_2^t\Sigma_2]^{-1}\Sigma_2]$ and $H^t(t)H(t) \leq I$ implies $H(t)H^t(t) \leq I$ then

$$(\Sigma_3 + \Sigma_1H(t)\Sigma_2)(\Sigma_3^t + \Sigma_2^tH^t(t)\Sigma_1^t) \leq \rho^{-2}\Sigma_1\Sigma_1^t + \Sigma_3[I - \rho^2\Sigma_2^t\Sigma_2]^{-1}\Sigma_3^t$$

Remark B.1: We remark that inequalities (B.1.1)-(B.1.3) are main vehicles in the stability studies throughout this volume in order to derive parameterized upper bounds and/or to remove uncertainties.

B.1.4: Given matrices Σ_1, Σ_2 and Σ_3 where $\Sigma_1 = \Sigma_1^t$ then

$$\Sigma_1 + \Sigma_3\,\Delta(k)\,\Sigma_2 + \Sigma_2^t\,\Delta(k)\,\Sigma_3^t < 0 \qquad (B.5)$$
$$\forall\Delta : \Delta^t(k)\,\Delta(k) \leq I$$

holds if and only if for some $\varepsilon > 0$

$$\Sigma_1 + \varepsilon^{-1}\Sigma_3\,\Sigma_3^t + \varepsilon\Sigma_2^t\,\Sigma_2 < 0 \qquad (B.6)$$

Remark B.2: It should be noted that **B.1.4** can be proved in line with **B.1.2**.

B.1.5: Let Σ_1, Σ_2, Σ_3 and Γ be given constant matrices with appropriate dimensions such that $\Gamma = \Gamma^t$. Then there exists a matrix $\Omega = \Omega^t$ such that

$$[\Sigma_1 + \Sigma_2\,\Delta(k)\,\Sigma_3]^t\,\Omega\,[\Sigma_1 + \Sigma_2\,\Delta(k)\,\Sigma_3] + \Gamma < 0 \qquad (B.7)$$
$$\forall\Delta : \Delta^t(k)\,\Delta(k) \leq I$$

if and only if there exists a scalar $\mu > 0$ such that the following conditions hold:

(a) $\mu \Sigma_2^t \Omega \Sigma_2 < I$

(b) $\Sigma_1^t \Omega \Sigma_1 + \Sigma_1^t \Omega \Sigma_2 [\mu^{-1} I - \Sigma_2^t \Omega \Sigma_2]^{-1} \Sigma_2^t \Omega \Sigma_1 + \mu^{-1} \Sigma_3^t \Sigma_3 + \Gamma < 0$

Proof: By the Schur complements A.1, inequality (B.7) holds if and only if

$$\begin{bmatrix} -\Omega^{-1} & \Sigma_1 + \Sigma_2 \Delta(k) \Sigma_3 \\ \Sigma_1^t + \Sigma_3^t \Delta^t(k) \Sigma_1^t & \Gamma \end{bmatrix} < 0 \qquad \text{(B.8)}$$

Inequality (B.8) is equivalent to:

$$\begin{bmatrix} -\Omega^{-1} & \Sigma_1 \\ \Sigma_1^t & \Gamma \end{bmatrix} + \begin{bmatrix} \Sigma_2 \\ 0 \end{bmatrix} \Delta(k) \begin{bmatrix} 0 & \Sigma_3 \end{bmatrix} + \begin{bmatrix} 0 \\ \Sigma_3^t \end{bmatrix} \Delta^t(k) \begin{bmatrix} \Sigma_2^t & 0 \end{bmatrix} < 0$$

$$\text{(B.9)}$$

By inequality **B.1.4**, it follows that (B.9) for some $\mu > 0$ is equivalent to

$$\begin{bmatrix} -\Omega^{-1} & \Sigma_1 \\ \Sigma_1^t & \Gamma \end{bmatrix} + \mu^{-1} \begin{bmatrix} 0 & 0 \\ 0 & \Sigma_3^t \Sigma_3 \end{bmatrix} + \mu \begin{bmatrix} \Sigma_2 \Sigma_2^t & 0 \\ 0 & 0 \end{bmatrix} < 0$$

$$= \begin{bmatrix} -\Omega^{-1} + \mu \Sigma_2 \Sigma_2^t & \Sigma_1 \\ \Sigma_1^t & \Gamma + \mu^{-1} \Sigma_3^t \Sigma_3 \end{bmatrix} < 0 \qquad \text{(B.10)}$$

Applying **A.1** again, inequality (B.10) holds if and only if

$$\Gamma + \mu^{-1} \Sigma_3^t \Sigma_3 + \Sigma_1^t [\Omega^{-1} - \mu \Sigma_2 \Sigma_2^t]^{-1} \Sigma_1 < 0 \qquad \text{(B.11)}$$

or

$$\Gamma + \mu^{-1} \Sigma_3^t \Sigma_3 + \Sigma_1^t \Omega [I - \mu \Sigma_2 \Sigma_2^t \Omega]^{-1} \Sigma_1 < 0 \qquad \text{(B.12)}$$

By **A.2**, inequality (B.11) reduces

$$\Sigma_1^t \Omega \Sigma_1 + \Sigma_1^t \Omega \Sigma_2 [\mu^{-1} I - \Sigma_2^t \Omega \Sigma_2]^{-1} \Sigma_2^t \Omega \Sigma_1 + \mu^{-1} \Sigma_3^t \Sigma_3 + \Gamma < 0 \quad \text{(B.13)}$$

which corresponds to the matrix expression of condition (2) as desired.

B.1.6: For the linear system

$$\dot{x}(t) = A\, x(t) \qquad \text{(B.14)}$$

where $\lambda(A) \in C^-$, it follows [2] that

$$\|e^{At}\| \leq c\, e^{-\eta t}, \quad t \geq 0, c \geq 1, \eta > 0 \qquad \text{(B.15)}$$

B.1.7 Bellman-Gronwall Lemma: Continuous Systems

Let $\alpha(t), \beta(t), \gamma(t)$ and $\mu(t) \geq 0$ be real continuous functions. If

$$\alpha(t) \leq \beta(t) + \gamma(t) \int_0^t \mu(s)\alpha(s)ds, \quad t \geq 0 \quad \text{(B.16)}$$

then

$$\alpha(t) \leq \beta(t) + \gamma(t) \int_0^t \mu(s)\beta(s)e^{\left\{\int_s^t \mu(r)\gamma(r)dr\right\}}ds, \quad \forall t \geq 0 \quad \text{(B.17)}$$

Special Case: Let $\alpha(t), \mu(t) \geq 0$ be real continuous functions and let σ be a real constant. If

$$\alpha(t) \leq \sigma + \int_0^t \mu(s)\alpha(s)ds, \quad \forall t \geq 0 \quad \text{(B.18)}$$

then

$$\alpha(t) \leq \sigma e^{\left\{\int_0^t \mu(s)ds\right\}}, \quad \forall t \geq 0 \quad \text{(B.19)}$$

B.1.7 Bellman-Gronwall Lemma: Discrete Systems

Let $\{\alpha(k)\}, \{\beta(k)\}$ and $\{\mu(k)\} > 0$ be finitely summable real-valued sequences $\forall k \in \mathcal{Z}_+$. If

$$\alpha(k) \leq \beta(k) + \sum_{j=1}^k \mu(j)\alpha(j) \quad \text{(B.20)}$$

then

$$\alpha(k) \leq \beta(k) + \sum_{j=1}^k \left\{\Pi_{m\in(j,k)}(1+\mu(m))\mu(j)\beta(j)\right\} \quad \text{(B.21)}$$

where $\Pi_{m\in(j,k)}(1+\mu(m))$ is set equal to 1 when $j = k - 1$.

Special Cases:

(a) Let $\{\alpha(k)\}, \{\beta(k)\}$ and $\{\mu(k)\} > 0$ be finitely summable real-valued sequences $\forall k \in \mathcal{Z}_+$. If for some constant μ_M, $\mu(j) \leq \mu_M \forall j$, then

$$\alpha(k) \leq \beta(k) + \mu_M \sum_{j=1}^k (1+\mu(m))^{k-j-1}\beta(j) \quad \text{(B.22)}$$

(b) Let $\{\alpha(k)\}, \{\beta(k)\}$ and $\{\mu(k)\} > 0$ be finitely summable real-valued sequences $\forall k \in \mathcal{Z}_+$. If for some constant β_M, $\beta(j) \leq \beta_M \forall j$, then

$$\alpha(k) \leq \beta_M \sum_{j=1}^{k}(1+\mu(j)) \tag{B.23}$$

The proof of this lemma can be found in [1].

B.2 Vector- or Scalar-Type Inequalities

B.2.1 For any vector quantities u and v of same dimension, it follows that:

$$||u+v||^2 \leq (1+\beta^{-1})||u||^2 + (1+\beta)||v||^2 \tag{B.24}$$

for any scalar $\beta > 0$.
Proof: Since

$$(u+v)^t(u+v) := u^tu + v^tv + 2u^tv \tag{B.25}$$

It follows by taking norm of both sides that:

$$||(u+v)||^2 \leq ||u||^2 + ||v^2|| + 2||u^tv||, \quad \beta > 0 \tag{B.26}$$

But from the traingle inequality $2||u^tv|| \leq \beta^{-1}||u||^2 + \beta||v^2||$, substituting back in $(B.26)$ yields $(B.24)$.

B.2.2 For any scalar quantities z, $a \geq 0$ and $b \neq 0$, it follows that:

$$az - bz^2 \leq a^2/(4b) \tag{B.27}$$

Proof: Since

$$(bz - (1/2)a)(bz - (1/2)a) \geq 0 \tag{B.28}$$

It follows by expansion that

$$b^2z^2 - abz + (1/4)a^2 \geq 0 \Longrightarrow$$
$$bz^2 - az \geq a^2/(4b) \Longrightarrow$$
$$az - bz^2 \leq a^2/(4b) \tag{B.29}$$

Appendix C

Stability Theorems

C.1 Lyapunov-Razumikhin Theorem

Consider the functional differential equation

$$\begin{aligned}
\dot{x}(t) &= f(t, x_t), \quad t \geq t_o \\
x_{t_o}(\theta) &= \phi(t + \theta), \forall \theta \in [-\tau, 0]
\end{aligned} \tag{C.1}$$

where $x_t(t), t \geq t_o$ denotes the restriction of $x(.)$ to the interval $[t - \tau, t]$ translated to $[-\tau, 0]$, that is $x_t(\theta) = \phi(t + \theta), \forall \theta \in [-\tau, 0]$ with $\phi \in C_{n,\tau}$. Let the function $f(t, \phi) : \Re_+ \times C_{n,\tau} \to \Re^n$ be continuous and Lipschitzian in ϕ with $f(t, 0) = 0$. Let $\alpha, \beta, \gamma, \delta : \Re_+ \to \Re_+$ be continuous and nondecreasing functions with

$$\begin{aligned}
\alpha(r), \ \beta(r), \ \gamma(r) &> 0; \quad r > 0 \\
\alpha(0) = 0, \ \beta(0) &= 0 \\
\delta(r) &> r \quad r > 0
\end{aligned}$$

If there exists a continuous function $V : \Re \times \Re^n \to \Re$ such that
(a) $\alpha(\|x\|) \leq V(t, x) \leq \beta(\|x\|), \quad t \in \Re, x \in \Re^n$
(b) $\dot{V}(t, x(t)) \leq -\gamma(\|x\|)$ if
$\quad V(t + \eta, x(t + \eta)) < \delta(V(t, x(t))), \quad \forall \eta \in [-\tau, 0]$
Then the trivial solution of (C.1) is uniformly asymptotically stable.

C.2   Lyapunov-Krasovskii Theorem

Consider the functional differential equation

$$\begin{aligned}
\dot{x}(t) &= f(t, x_t), \ t \geq t_o \\
x_{t_o}(\theta) &= \phi(t + \theta), \forall \theta \in [-\tau, 0]
\end{aligned} \qquad (C.2)$$

where $x_t(t), t \geq t_o$ denotes the restriction of $x(.)$ to the interval $[t - \tau, t]$ translated to $[-\tau, 0]$, that is $x_t(0) = \phi(t + \theta), \forall \theta \in [-\tau, 0]$ with $\phi \in C_{n,\tau}$. Let the function $f : \Re_+ \times C_{n,\tau} \rightarrow \Re^n$ take bounded sets of $C_{n,\tau}$ in bounded sets of $\Re^n$ and $\alpha, \beta, \gamma : \Re_+ \rightarrow \Re_+$ be continuous and nondecreasing functions with

$$\begin{aligned}
\alpha(r), \ \ \beta(r), \ &> 0; \ \ \ r \neq 0 \\
\alpha(0) = 0, \ \beta(0) &= 0
\end{aligned}$$

If there exists a continuous function $V : \Re \times C_{n,\tau} \rightarrow \Re$ such that
(a) $\alpha(||\phi(0)||) \leq V(t, x) \leq \beta(||\phi||_*), \ \ \ t \in \Re, x \in \Re^n$
(b) $\dot{V}(t, \phi) \leq -\gamma(||\phi(0)||)$
then the trivial solution of (C.1) is uniformly stable.
If $\alpha(r) \rightarrow \infty$ as $r \rightarrow \infty$, then the solutions are uniformly bounded.
If $\gamma(r) > 0$ for $r > 0$, then the solution $x = 0$ is uniformly asymptotically stable.
Throughout the book, in applying the Lyapunov-Krasovskii theorem we use a quadratic functional of the form:

$$V(t, x) = x^t(t) P x(t) + \int_{t-\tau}^{t} x^t(\theta) Q x(\theta) \ d\theta \qquad (C.3)$$

where $0 < P = P^t \in \Re^{n \times n}$ and $0 < Q = Q^t \in \Re^{n \times n}$ are weighting matrices. We note that this functional satisfies the conditions of the theorem and in particular we have

$$\lambda_m(P)||x||^2 \leq V(t, x) \leq [\lambda_M(P) + \tau^* \lambda_M(Q)]||x||^2$$

We also note that the first term of the functional takes care of the present state whereas the second term accumulates the effect of the delayed state. While the selection of P is quite standard, the selection of Q is governed by the problem at hand. See Chapter 2 for different forms of Q.

Appendix D

Positive Real Systems

In this section, we give some definitions and technical results on positivity of a class of linear systems without delay which will be used in Chapter 4. The class of systems is given by:

$$(\Sigma_o): \quad \dot{x}(t) \; = \; Ax(t) + Bu(t) \tag{D.1}$$
$$y(t) \; = \; Cx(t) + Du(t) \tag{D.2}$$

Denote the transfer function of (Σ_o) by $T_o(s)$:

$$T_o(s) = C(sI - A)^{-1}B + D \tag{D.3}$$

Based on the results of [3-5], we have the following:

Definition D.1:

(a) The system (Σ_o) is said to be *positive real* (PR) if its transfer function $T_o(s)$ is analytic in $Re(s) > 0$ and satisfies $T_o(s) + T_o^t(s^*) \geq 0$ for $Re(s) > 0$.
(b) The system (Σ_o) is *strictly positive real* (SPR) if its transfer function $T_o(s)$ is analytic in $Re(s) > 0$ and satisfies
$T_o(j\omega) + T_o^t(-j\omega) > 0 \; \forall \, \omega \in [0, \infty)$.
(c) The system (Σ_o) is said to be *extended strictly positive real* (ESPR) if it is SPR and $T_o(j\infty) + T_o^t(-j\infty) > 0$.

Definition D.2: The dynamical system (C.1)-(C.2) is called *passive* if and only if

$$\int_0^\infty u^t(t)y(t) \; > \; \beta \qquad \forall u \in L_2 \tag{D.4}$$

409

where β is some constant depending on the initial condition of the system.

Remark D.1: We recall that the minimal realization of a PR function is stable in the sense of Lyapunov [1,2]. In addition, ESPR implies SPR which further implies PR.

Remark D.2: Bearing in mind that testing the PR conditions of Definition C.1 should be done for all frequencies, an alternative procedure would be desirable to avoid such excessive computational effort. The results of [5] have provided a state-space solution to the positive-real control in terms of algebraic Riccati inequalities.

A version of the positive real lemma to be used in the sequel is now provided.

Lemma D.1 [5]: *Consider system* (Σ_o) *and define the algebraic Riccati inequality (ARI)*

$$PA + A^t P + (C - B^t P)^t (D^t + D)^{-1} (C - B^t P) \; < \; 0 \qquad (D.5)$$

Then the following statements are equivalent:
(1) System (Σ_o) *is ESPR and A is a stable matrix;*
(2) $(D^t + D) > 0$ *and there exists a matrix* $0 < P = P^t \in \Re^{n \times n}$ *solving the ARI (12);*
(3) There exists a matrix $0 < P = P^t \in \Re^{n \times n}$ *solving the linear matrix inequality (LMI)*

$$\begin{bmatrix} PA + A^t P & C^t - PB \\ C - B^t P & -(D + D^t) \end{bmatrix} \; < \; 0 \qquad (D.6)$$

Bibliography

[1] Desoer, C. A. and M. Vidyasagar, "**Feedback Systems: Input-Output Properties**," Academic Press, New York, 1975.

[2] Vidyasagar, M., "**Nonlinear Systems**," Prentice-Hall, New York, 1989.

[3] Anderson, B. D. O. and S. Vongpanitherd, "**Network Analysis and Synthesis: A Modern Systems Theory Approach**," Prentice-Hall, New Jersey, 1973.

[4] Anderson, B. D. O., "A System Theory Criterion for Positive Real Matrices," **SIAM J. Control and Optimization**, vol. 5, 1967, pp. 171-182.

[5] Sun, W., P. P. Khargonekar and D. Shim, "Solution to the Positive Real Control Problem for Linear Time-Invariant Systems," **IEEE Trans. Automatic Control**, vol. 39, 1994, pp. 2034-2046.

[6] de Souza, C. E. and L. Xie, "On the Discrete-Time Bounded Real Lemma with Application in the Characterization of Static State Feedback H_∞ Controllers," **Systems and Control Letters**, vol. 18, 1992, pp. 61-71.

[7] Petersen, I. R., B. D. O. Anderson and E. A. Jonckheere, "A First Principle Solution to the Non-Singular H_∞ Control Problem," **Int. J. Robust and Nonlinear Control**, vol. 1, 1991, pp. 171-185.

Appendix E

LMI Control Software

Linear matrix inequalities (LMIs) have been shown to provide powerful control design tools [1]. The LMI Control Software [2] is so designed to assist control engineers and researchers with a user-friendly interactive environment. Through this environment, one can specify and solve several engineering problems that can formulated as one of the following generic LMI problems:

1. Feasibility Problem: Find a solution $x \in \Re^n$ to the LMI problem

$$A(x) := A_o + \sum_{j+1}^{n} x_j A_j < 0 \qquad (E.1)$$

2. Convex Minimization Problem: Given a convex function $f(x)$, find a solution $x \in \Re^n$ that

$$Minimize \ f(x) \ subject \ to \ (x) < 0 \qquad (E.2)$$

c. Generalized Eigenvalue Problem: Find a solution $x \in \Re^n$ that

$$\begin{array}{cc} & A(x) < \lambda B(x) \\ Minimize \ \lambda \ subject \ to & B(x) > 0 \\ & C(x) < 0 \end{array} \qquad (E.3)$$

LMIs are being solved by efficient convex optimization algorithms [3]. Among several commerically available packages, the **LMI Control Toolbox** offers high-performance software for solving general LMI problems. This is evident in terms of simple specification and manipulation of LMIs (either symbolically with the LMI Editor *lmiedit* or incrementally with the commands *lmivar, lmiterm*), structured-oriented representation (matrix variables) and incorporation of efficient numerical algorithms. The computational engine is formed by three solvers: *feasp, mincx* and *gevp*. In general, the LMI lab can handle any system of LMIs of the form:

$$\mathcal{N}^t \mathcal{L}(X_1, ..., X_K) \mathcal{N} < \mathcal{M}^t \mathcal{R}(X_1, ..., X_K) \mathcal{M} \qquad \text{(E.4)}$$

where $X_1, ..., X_K$ are matrix variables with some prescribed structure, the left and right outer factors N and M are given matrices with identical dimensions and the left and right inner factors L and R are symmetric block matrices with identical block structures, each block being an affine combination of $X_1, ..., X_K$ and their transposes.

The specification of an LMI system involves two-steps:
(1) Declare the dimensions and structure of each matrix variable $X_1, ..., X_K$.
(2) Describe the term content of each LMI.

This computer description is stored as a single vector *LMISYS* and is used by the LMI solvers in all subsequent manipulations. The description of an LMI system starts with *setlmis* and ends with *getlmis*. The function *setlmis* initializes the LMI system description with two possibilities: for a new system, type *setlmis([])* and for adding on to an existing LMI system *LMIMSM*, type *setlmis(LMIMSM)*. In either case, the terminal command is *LMISYS = getlmis* after completing the specification.

Within an LMI file, the matrix variables are declared one at a time with *lmivar* and are specified in terms of their structure. There are two structure types:

Type 1: Symmetric Block Diagonal (SBD) structure. This takes the form

$$X = \begin{bmatrix} D_1 & 0 & ... & 0 \\ 0 & D_2 & ... & \vdots \\ \vdots & ... & ... & 0 \\ 0 & ... & 0 & D_n \end{bmatrix} \qquad \text{(E.5)}$$

where D_j is square and is either zero, a full symmetric matrix, or a scalar matrix $D_j = d \times I; d \in \Re$.

Type 2: Rectangular Structure. This corresponds to arbitrary rectangular matrices without any particular form.

The next step is to specify the term content of each LMI. In this regard, there are three classes:

1. Constant terms: These include fixed matrices like I in $X > I$,
2. Variable terms: These include terms involving a matrix variable like $XA, D^t SD, PXQ$,
3. Outer factors:

In describing the foregoing terms, as a basic rule, we specify *only* the terms in the blocks on or above the diagonal since the inner factors are symmetric.

E.1 Example E.1

Consider the feasibility problem of solving inequality (2.4) using the data

$$A = \begin{bmatrix} -3 & -2 \\ 1 & 0 \end{bmatrix}, \; A_d = \begin{bmatrix} 0 & 0.3 \\ -0.3 & -0.2 \end{bmatrix}, \; Q = \begin{bmatrix} 1 & 0 \\ 0 & 1 \end{bmatrix} \quad \text{(E.6)}$$

for $\tau \in [0.1 - 0.9]$. An LMI-based program is given by

```
setlmis([]);
Qtaw=(1-taw)*eye(2);
p=lmivar(1,[2,1]);
term1=newlmi;
limterm([term1 1 1 p],1,A,'s');
lmiterm([term1 1 1 0],1);
lmiterm([term1 1 2 p],1,b);
lmiterm([term1 2 2 0],-Qtaw);
lmisys=getlmis;
lminbr(lmisys);
(tmin,xfeas)=feasp(lmissys);
pf=dec2mat(lmisys,xfeas,p)
eig(pf);
evlmi=evllmi(lmisys,xfeas);
(lhs1,rhs1)=showlmi(evlmi,2);
eig(lhs1-rhs1)
```

E.2   Example E.2

Consider a discrete system of the type (2.74)-(2.75) with the data

$$A = \begin{bmatrix} 0.1 & 0 & -0.1 \\ 0.05 & 0.3 & 0 \\ 0 & 0.2 & 0.6 \end{bmatrix}, \quad A_d = \begin{bmatrix} -0.2 & 0 & 0 \\ 0 & -0.1 & 0.1 \\ 0 & 0 & -0.2 \end{bmatrix} \quad (E.7)$$

$$Q = \begin{bmatrix} 0.6 & 0 & 0 \\ 0 & 0.6 & 0 \\ 0 & 0 & 0.6 \end{bmatrix}, \quad H = \begin{bmatrix} 0.1 \\ 0 \\ 0.2 \end{bmatrix},$$

$$E = [0.2 \quad 0 \quad 0.3], \quad E_d = [0.4 \quad 0 \quad 0.1] \quad (E.8)$$

An LMI-based program for testing robust stability is given by

```
setlmis([]) ;
p=lmivar(1,[3,1]);
term1=newlmi ;
limterm([term1 1 1 p],-1,1);
lmiterm([term1 1 1 0],Q);
lmiterm([term1 1 1 0],e1*E'*E);
lmiterm([term1 1 2 0],e1*E'*Ed);
lmiterm([term1 1 3 0],A');
lmiterm([term1 2 2 0],-Q);
lmiterm([term1 2 2 0],e1*Ed'*Ed);
lmiterm([term1 2 3 0],Ad');
lmiterm([term1 3 3 inv9p)],-1,1);
lmiterm([term1 3 3 0],e*H*H');
term2=newlmi;
limterm([term2 1 1 0],e*Ed'*Ed);
limterm([term2 1 1 0],-Q);
term3=newlmi;
limterm([term2 1 1 0],e*H*H');
limterm([term2 1 1 inv(p)],-1,1);
lmisys=getlmis;
lminbr(lmisys);
(tmin,xfeas)=feasp(lmissys);
pf=dec2mat(lmisys,xfeas,p);
eig(pf);
evlmi=evllmi(lmisys,xfcas);
```

```
(lhs1,rhs1)=showlmi(evlmi,1);
(lhs2,rhs2)=showlmi(evlmi,2);
(lhs3,rhs3)=showlmi(evlmi,3);
eig(lhs1-rhs1);
eig(lhs2-rhs2);
eig(lhs3-rhs3);
pf
```

Bibliography

[1] Boyd, S., L. El Ghaoui, E. Fern and V. Balakrishnan, "**Linear Matrix Inequalities in Systems and Control Theory**" SIAM Books, Philadelphia, PA, 1994.

[2] Gahinet, P., A. Nemirovski, A. J. Laub and M. Chilai, "**LMI Control Toolbox**" The Math Works, Inc., Boston, MA, 1995.

[3] Nesterov, Y and A. Nemirovski, "**Interior Point Polynomial Methods in Convex Programming: Theory and Applications**" SIAM Books, Philadelphia, PA, 1994.

Author Index

Subject Index